T0318392

# LOCAL ELECTRICITY MARKETS

# LOCAL ELECTRICITY MARKETS

*Edited by*

TIAGO PINTO
*GECAD Research Group, Polytechnic of Porto (ISEP/IPP), Porto, Portugal*

ZITA VALE
*School of Engineering, Polytechnic of Porto (ISEP/IPP), Porto, Portugal*

STEVE WIDERGREN
*Pacific Northwest National Laboratory (PNNL), Richland, WA, United States*

**ACADEMIC PRESS**

An imprint of Elsevier

ELSEVIER

Academic Press is an imprint of Elsevier
125 London Wall, London EC2Y 5AS, United Kingdom
525 B Street, Suite 1650, San Diego, CA 92101, United States
50 Hampshire Street, 5th Floor, Cambridge, MA 02139, United States
The Boulevard, Langford Lane, Kidlington, Oxford OX5 1GB, United Kingdom

MATLAB® is a trademark of The MathWorks, Inc. and is used with permission.
The MathWorks does not warrant the accuracy of the text or exercises in this book.
This book's use or discussion of MATLAB® software or related products does not constitute endorsement or sponsorship by The MathWorks of a particular pedagogical approach or particular use of the MATLAB® software.

**Notices**
Knowledge and best practice in this field are constantly changing. As new research and experience broaden our understanding, changes in research methods, professional practices, or medical treatment may become necessary.

Practitioners and researchers must always rely on their own experience and knowledge in evaluating and using any information, methods, compounds, or experiments described herein. In using such information or methods they should be mindful of their own safety and the safety of others, including parties for whom they have a professional responsibility.

To the fullest extent of the law, neither the Publisher nor the authors, contributors, or editors, assume any liability for any injury and/or damage to persons or property as a matter of products liability, negligence or otherwise, or from any use or operation of any methods, products, instructions, or ideas contained in the material herein.

**British Library Cataloguing-in-Publication Data**
A catalogue record for this book is available from the British Library

**Library of Congress Cataloging-in-Publication Data**
A catalog record for this book is available from the Library of Congress

ISBN: 978-0-12-820074-2

For Information on all Academic Press publications
visit our website at https://www.elsevier.com/books-and-journals

*Publisher:* Brian Romer
*Acquisitions Editor:* Graham Nisbet
*Editorial Project Manager:* Leticia M. Lima
*Production Project Manager:* Kamesh Ramajogi
*Cover Designer:* Mark Rogers

Typeset by MPS Limited, Chennai, India

# Contents

# I

## Distributed energy resources as enablers of local electricity markets

### 1. New electricity markets. The challenges of variable renewable energy

Ana Estanqueiro and António Couto

### 2. Integration of electric vehicles in local energy markets

José Almeida and João Soares

### 3. From wholesale energy markets to local flexibility markets: structure, models and operation

Fernando Lopes

## 4. From the smart grid to the local electricity market

Fernando Lezama, Tiago Pinto, Zita Vale, Gabriel Santos and
Steve Widergren

# II

# Local market models and opportunities

## 5. Local market models

Amin Shokri Gazafroudi, Hosna Khajeh, Miadreza Shafie-khah,
Hannu Laaksonen and Juan Manuel Corchado

## 6. Peer-to-peer energy platforms

Liyang Han, Thomas Morstyn and Malcolm D. McCulloch

## 7. Transmission system operator and distribution system operator interaction

Hugo Gabriel Morais Valente, E. Lambert and J. Cantenot

# III

# Enablers for local electricity markets

# IV

# Regulatory framework: Current trends and future perspectives

## 16. An economic analysis of market design: Local energy markets for energy and grid services

L. Lynne Kiesling

## 17. South American Markets—regulatory framework: current trends and future prospects in South America

Rubipiara C. Fernandes, Edison A.C. Aranha Neto and Fabrício Y.K. Takigawa

## 18. Electricity markets and local electricity markets in Europe

Zita Vale, Débora de São José and Tiago Pinto

## 19. Local electricity markets: regulation, opportunities, and challenges in the United Kingdom

Karim L. Anaya

## 20. Competition and restructuring of the South African electricity market

Komla Agbenyo Folly

## 21. Asia electricity markets

Panhong Cheng and Yan Gao

## 22. Current trends and perspectives in Australia

Alan Moran

# List of contributors

**Komla Agbenyo Folly** Department of Electrical Engineering, University of Cape Town, Cape Town, South Africa

**José Almeida** GECAD Research Center, ISEP–School of Engineering of Polytechnic of Porto, Porto, Portugal

**Elvira Amicarelli** ENEL S.p.A., Italia

**Karim L. Anaya** Energy Policy Research Group (EPRG), Cambridge Judge Business School, University of Cambridge, Cambridge, United Kingdom

**Edison A.C. Aranha Neto** Federal Institute of Santa Catarina, Florianópolis, Brazil

**Nataly Bañol Arias** Department of Energy Systems, School of Electrical and Computer Engineering, University of Campinas, Campinas, Brazil

**J. Cantenot** Électricité de France (EDF), Paris, France

**Emre Çelebi** Center for Energy and Sustainable Development, Kadir Has University, Istanbul, Turkey; Industrial Engineering Department, Yeditepe University, Istanbul, Turkey

**Panhong Cheng** School of Management, University of Shanghai for Science and Technology, Shanghai, P.R. China

**Gianfranco Chicco** Dipartimento Energia "Galileo Ferraris," Politecnico di Torino, Torino, Italy

**Juan Manuel Corchado** BISITE Research Group, University of Salamanca, Salamanca, Spain; Department of Electronics, Information and Communication, Faculty of Engineering, Osaka Institute of Technology, Osaka, Japan

**António Couto** National Energy and Geology Laboratory (LNEG), Renewable Energy and Energy Efficiency Unit, Lisboa, Portugal

**Débora de São José** Polytechnic Institute of Porto, Porto, Portugal

**Murat Elhüseyni** Faculty of Engineering and Natural Sciences, Sabanci University, Istanbul, Turkey

**Danial Esmaeili Aliabadi** Helmholtz Centre for Environmental Research - UFZ, Leipzig, Germany

**Ana Estanqueiro** National Energy and Geology Laboratory (LNEG), Renewable Energy and Energy Efficiency Unit, Lisboa, Portugal

**Ricardo Faia** Research Group on Intelligent Engineering and Computing for Advanced Innovation and Developmen (GECAD), Polytechnic of Porto (ISEP/IPP), Porto, Portugal

**Rubipiara C. Fernandes** Federal Institute of Santa Catarina, Florianópolis, Brazil

**John Fredy Franco** School of Energy Engineering, São Paulo State University, Rosana, Brazil

**Yan Gao** School of Management, University of Shanghai for Science and Technology, Shanghai, P.R. China

**Amin Shokri Gazafroudi** BISITE Research Group, University of Salamanca, Salamanca, Spain; Institute for Automation and Applied Informatics, Karlsruhe Institute of Technology (KIT), Eggenstein-Leopoldshafen, Germany

**Madeleine Gibescu** Department of Mathematics and Computer Science, Eindhoven University of Technology, Eindhoven, The Netherlands; Copernicus Institute of Sustainable Development, Utrecht University, Utrecht, The Netherlands

**Liyang Han** Department of Electrical Engineering, Technical University of Denmark, Kongens Lyngby, Denmark

**Hosna Khajeh** School of Technology and Innovations, University of Vaasa, Vaasa, Finland

**L. Lynne Kiesling** Institute for Regulatory Law & Economics, Carnegie Mellon University, Pittsburgh, PA, United States; Engineering & Public Policy, Carnegie Mellon Universitys, Pittsburgh, PA, United States; Wilton E. Scott Institute for Energy Innovation, Carnegie Mellon University, Pittsburgh, PA, United States

**Hannu Laaksonen** School of Technology and Innovations, University of Vaasa, Vaasa, Finland

**E. Lambert** Électricité de France (EDF), Paris, France

**Fernando Lezama** Research Group on Intelligent Engineering and Computing for Advanced Innovation and Developmen (GECAD), Polytechnic of Porto (ISEP/IPP), Porto, Portugal

**Fernando Lopes** LNEG – National Laboratory of Energy and Geology, Lisbon, Portugal

**Ngoc An Luu** The University of Danang - University of Science and Technology, Danang, Vietnam

**Leonardo H. Macedo** Department of Electrical Engineering, São Paulo State University, Ilha Solteira, Brazil

**Andrea Mazza** Dipartimento Energia "Galileo Ferraris," Politecnico di Torino, Torino, Italy

**Malcolm D. McCulloch** Department of Engineering Science, University of Oxford, Oxford, United Kingdom

**Mihail Mihaylov** i.LECO, Geel, Belgium

**Decebal Constantin Mocanu** Department of Computer Science, EEMCS, University of Twente, Enschede, The Netherlands; Department of Mathematics and Computer Science, Eindhoven University of Technology, Eindhoven, The Netherlands

**Elena Mocanu** Department of Computer Science, EEMCS, University of Twente, Enschede, The Netherlands

**Hugo Gabriel Morais Valente** INESC-ID, Department of Electrical and Computer Engineering, Instituto Superior Técnico-IST, Universidade de Lisboa, 1049-001 Lisbon, Portugal

**Alan Moran** Regulation Economics, Melbourne, Australia

**Thomas Morstyn** School of Engineering, University of Edinburgh, Edinburgh, United Kingdom

**Van Hoa Nguyen** Univ. Grenoble Alpes, Grenoble INP, G2Elab, F-38000, Grenoble, France

**Nikolaos G. Paterakis** Department of Electrical Engineering, Eindhoven University of Technology, Eindhoven, The Netherlands

**Tiago Pinto** GECAD Research Group, Polytechnic of Porto (ISEP/IPP), Porto, Portugal

**Iván S. Razo-Zapata** Instituto Tecnológico Autónomo de México (ITAM), Mexico City, Mexico; COCOA Collaborative Innovation, Delft, The Netherlands; LICORE, Queretaro, Mexico

**Rubén Romero** Department of Electrical Engineering, São Paulo State University, Ilha Solteira, Brazil

**Güvenç Şahin** Faculty of Engineering and Natural Sciences, Sabanci University, Istanbul, Turkey

**Gabriel Santos** GECAD, Polytechnic of Porto (ISEP/IPP), Porto, Portugal

**Jan Segerstam** Empower IM Oy, Helsinki, Finland

**Miadreza Shafie-khah** School of Technology and Innovations, University of Vaasa, Vaasa, Finland

**João Soares** GECAD Research Center, School of Engineering of Polytechnic of Porto, Porto, Portugal

**Alejandra Tabares** Department of Electrical Engineering, São Paulo State University, Ilha Solteira, Brazil

**Fabrício Y.K. Takigawa** Federal Institute of Santa Catarina, Florianópolis, Brazil

**Quoc Tuan Tran** Université Grenoble Alpes, CEA LITEN, Grenoble, France

**Zita Vale** School of Engineering, Polytechnic of Porto (ISEP/IPP), Porto, Portugal

**Steve Widergren** Pacific Northwest National Laboratory (PNNL), Richland, WA, United States

# Introduction

*Tiago Pinto[1], Zita Vale[2] and Steve Widergren[3]*

[1]GECAD Research Group, Polytechnic of Porto (ISEP/IPP), Porto, Portugal [2]School of Engineering, Polytechnic of Porto (ISEP/IPP), Porto, Portugal [3]Pacific Northwest National Laboratory (PNNL), Richland, WA, United States

## 1. Introduction

The traditional centralized top-down approach of electricity markets has proven to be insufficient to take full advantage from the participation of small players—both consumers and distributed generation (DG) [1]. Moreover, the tentative reforms of retail markets are not being able to achieve the envisaged goals as they are being built under the same top−down principles as wholesale markets [2]. Electricity prices for smaller consumers still do not reflect the market prices and the introduction of flexible, innovative tariffs adapted to consumers' needs and behaviors, able to promote and fairly remunerate their contribution toward an increasingly efficient energy system are still distant targets. New approaches that are able to bring a closer connection between small consumers and DG and the wholesale electricity market are required promptly [3]. The worldwide investments already made in smart grids (SG), for example, the rollout of smart meters and the investment in communication infrastructure between local players, provides a solid basis for the developments to come

in the next years [4]. A pioneer solution to overcome the current problems has been implemented in the New York electricity market, in the United States [5]. The creation of local electricity markets (LEMs) as part of the regional electricity market is being put into practice, enabling smaller portions of the power network (microgrids) to participate in the electricity market as aggregators of the resources that are part of the portion of the grid. This way, resources can be managed at a local level, enhancing the potential of smaller sized resources, to participate in coordinated system operation using electricity markets facilitated by the microgrid operators [6]. This provides an important incentive for the development of adequate methods to manage resources at lower levels and make their connection with wholesale electricity markets, more effective. See for example, the case study presented in Ref. [7] for further details.

Several investments have also been made in the EU to develop local markets, namely through the simulation and analysis of the potential ways for implementing them. This is resulting in research project investments related to LEMs simulation and development. EMPOWER is developing a LEM place through innovative business models, that incentivizes the participation of consumers and distributed renewable generation [8]. DOMINOES is also working on the development of a local energy market solution that can be applied EU-wide. The path follows the development of new demand response, aggregation, grid management, and peer-to-peer (P2P) trading services [9]. E-REGIO is analyzing, testing, and validating a new way to implement local energy markets around energy storage units and flexible assets supervised by an entity referred to as local system operator (LSO) [10]. P2PQ is creating a Peer-to-Peer (P2P) energy market for prosumers to allocate locally produced energy and acts as the basis for a local energy community. The whole system for the presently 41 pilot households went live in late 2019 [11]. Blockchain technology is the basis of RegHEE and of Quartierstrom, which evaluates technical feasibility, market mechanisms, and user behavior in blockchain-based LEMs. PANEL 2050 is working to create replicable energy networks at local level, by implementing and incentivizing local energy visions, strategies, and action plans for the transition toward low-carbon communities in 2050 [12]. ADAPT has developed a multiagent decision support system to assist the negotiation process of smart grid and electricity market players in both local and wholesale market transactions [13]. Other relevant projects in this domain are Powerpeers [14], Pebbles, Cornwall Local Energy Market [15], SmarterEMC2 [16], and northern European initiatives Oulu Energy Farm Power [17], Energy Collective [18], The Landau Microgrid Project (LAMP), and Helen Solar Power [19].

The large majority of projects related to local markets are still ongoing, so there remains a relevant gap of technological tools, simulators and field. Importantly, there is a lack of a consolidated vision on the concept of LEMs. In fact, the most relevant works in this domain refer to recent studies on potential alternative local market structures, mostly based on the evolution and development of current retail market structures [20,21], works trying to close the gap between wholesale and retail electricity markets, such as transactive energy approaches [22], blockchain-based solutions [23], and recently also some works on P2P trading; see, for example, Ref. [24], which presents a review on current P2P trading projects, and Ref. [25], which provides a literature review on P2P markets. Some of the most relevant research works in this domain include [26], which proposes a local market structure for energy and reserve. This work studies the importance of providing affordable ancillary services to enable reaching a higher levels of participation from renewable energy generators. It also addresses how players can qualify to participate in the market to provide these ancillary services. The work presented in Ref. [27] proposes a two-stage auction-based local market mechanism to allocate physical storage rights. The implications of recently proposed market designs under the current rules in the context of the German market have been studied in Ref. [28], and a novel market design called Tech4all is introduced. In Ref. [29], a flexible large-scale agent-based simulation tool for SG is presented. This simulator, named GridLAB-D is open source and considers different SG resources such as demand response, storage, electric vehicles, and the retail market. The flexibility of the simulator enables it to be used to study multiple questions in the field, related to energy trading, market design, flexibility management and operation of SG. Another simulation system for different types of consumers and their behavior in a SG setting is presented in Ref. [30]. This simulator enables consumer agents to define their actions at each time, deciding among using local generation, reducing their consumption, trading energy with utilities or their neighbors, and using their batteries. Multiagent-based approaches are, in fact, one of the most widely used approaches for the modeling and simulation of electricity market models, mechanisms, and interaction with the SG; see, for example, Refs. [31] and [32] for detailed reviews on this topic. Some works have tackled the problem of modeling the participation process for small participants in electricity markets. For example, Ottesen et al. [33] proposed a model for an aggregator that can sell and buy electricity on behalf of a group of prosumers in the wholesale market. The approach proposed in this work considers an integrated modeling of the bidding process and market modeling for both local and wholesale markets. The optimal participation of residential aggregators in local energy and flexibility market is the topic of Ref. [34], by introducing an optimization model for home energy management systems from an aggregator's standpoint.

The benefits of local markets have already been studied broadly with the recognition that local markets provide crucial characteristics for managing local grid balance efficiently and contributing to consumers' participation on the energy markets. Among these characteristics are the ability of markets to scale easily, the freedom of choice they provide to the participants, the respect for information privacy, and the encouragement of local operational responsibility. As a drawback the low maturity level of introduced concepts often lack important attributes or features that demonstrate comprehensive, viable solutions for the liberalized energy markets.

- Some of the concepts like blockchain concentrate only on trading mechanisms and enabling transactions between consumers, while neglecting the basic energy system characteristics like balance responsibility and demand/supply balance.
- Many solutions enabling local market-like functionalities can be seen just as a gimmick inside a single retailers' energy balance. Buying excess energy from one prosumer and selling it to another consumer requires only a supporting metering solution and the rest is pure billing. This does not contribute to resolving grid congestions and the end customers are unable to get a benefit in two major components of the energy bill, namely grid costs and taxes.
- Transparency of the local market structures is also an issue with many introduced solutions. Enabling transactions locally and integrating properly incentivized behaviour with the organized wholesale markets requires new regulatory policy for information exchange, trading and metering solutions. To push innovation and the penetration of new market structures, access to these markets should be transparent and not restrict competition or favour certain consumers' position in the markets.
- Energy balance responsibility is one key characteristic and requirement in the energy systems that local markets also need to address. Electricity market participants' electricity use/sales should equal their generation/purchase at every instant. In practice, market participant's demand and supply rarely match as demand cannot be forecasted perfectly and generation plans can also be uncertain, e.g., due to changes in weather or equipment deration. At the system level, the balance between demand and supply

also must be maintained every instant. Market participants' imbalances are settled financially afterwards. However, it is not feasible to settle each market participants' financial imbalances at every instant. Instead, the imbalance settlement periods used around the world vary from a few minutes up to 1 hour. Attention is needed to develop coherent plans to harmonize approaches and use a standard imbalance settlement period.

- The regulatory framework needs to be updated and improved. While the smart meters facilitate accurate and efficient balance settlement, the development of metering and control opportunities needs to be addressed carefully in regulatory policy. For example, the handling of imbalances caused by control actions made by nonbalance responsible parties needs resolution in regulatory policy (see, Ref. [35]). Information exchange structures and models to enable interaction between local, retail, and wholesale markets need to be agreed to and solutions must be proposed on how local markets should be incorporated into the current regulatory framework.

Resolving current and emerging issues to create local market solutions that are compatible with the current organized wholesale markets, taking into account the basic nature of the energy systems, and enabling a smooth transition into a more distributed and local energy market environment represents a major challenge for the future. All developed structures and concepts need to be transparent for energy market stakeholders to utilize and coordinate their integration with them.

Although there exists a significant body of literature addressing the topic of local energy markets, this is still a rather new and emerging topic. There is still a deficit of relevant sources of review material on the subject that can provide an organized view on this domain and offer perspectives on future needs and challenges. Moreover, the policy and regulatory framework is very different from country to country, which makes it extremely difficult to identify common paths for undertaking the problems related to the local energy market topic.

This book provides a thorough, yet concise, review on the current status of LEM development all around the world, including the most promising research models and opportunities that are being proposed; an overview of the regulatory issues on the different Continents a discussion on the current infrastructure (both hardware and software) that is implemented and that is expected to facilitate the implementation and wide spread application of local energy markets (e.g., resulting from the investments already made in SG deployment); a review of current applications and practical implementations of LEMs; and a wrap-up and discussion on the most relevant paths for future research and development in this field of study.

# References

[1] B. Gencer, E.R. Larsen, A. van Ackere, Understanding the coevolution of electricity markets and regulation, Energy Policy 143 (2020) 111585.

[2] R. Esplin, B. Davis, A. Rai, T. Nelson, The impacts of price regulation on price dispersion in Australia's retail electricity markets, Energy Policy 147 (2020) 111829.

[3] I. MacGill, R. Esplin, End-to-end electricity market design - some lessons from the Australian National Electricity Market, Electr. J. 33 (9) (2020) 106831.

[4] S. Kakran, S. Chanana, Smart operations of smart grids integrated with distributed generation: a review, Renew. Sustain. Energy Rev. 81 (2018) 524–535.

[5] R. Walton, ConEd virtual power plant shows how New York's REV is reforming utility practices, Utility Dive, [Online]. Available: <https://www.utilitydive.com/news/coned-virtual-power-plant-shows-how-new-yorks-rev-is-reforming-utility-pra/421053/>, (2016).

[6] G. Bade, "Little less talk: with new revenue models, New York starts to put REV into action," Utility Dive, [Online]. Available: <https://www.utilitydive.com/news/little-less-talk-with-new-revenue-models-new-york-starts-to-put-rev-into/420657/>, (2016).

[7] E. Mengelkamp, J. Gärttner, K. Rock, S. Kessler, L. Orsini, C. Weinhardt, Designing microgrid energy markets: a case study: the Brooklyn Microgrid, Appl. Energy 210 (2018) 870–880.

[8] Empower, EMPOWERING a reduction in use of conventionally fueled vehicles using positive policy measures, [Online]. Available: <https://cordis.europa.eu/project/id/636249/fr>, 2018 (accessed 20.09.18).

[9] DOMINOES, Smart distribution grid: a market driven approach for the next generation of advanced operation models and services, [Online]. Available: <http://dominoesproject.eu/about/>, 2018 (accessed 20.09.18).

[10] E-REGIO project, [Online]. Available: <https://www.eregioproject.com/> (accessed 15.07.20).

[11] P2PQ, [Online]. Available: <https://nachhaltigwirtschaften.at/de/sdz/projekte/peer2peer-im-quartier.php>.

[12] Panel2050, Partnership for new energy leadership 2050, [Online]. Available: <https://ceesen.org/panel2050/>, 2018 (accessed 20.09.18).

[13] ADAPT, Adaptive decision support for agents negotiation in electricity market and smart grid power transactions, [Online]. Available: <https://adapt.usal.es/>, 2018 (accessed 20.09.18).

[14] Powerpeers, Choose yourself who you get energy, [Online]. Available: <https://www.powerpeers.nl/>, 2016 (accessed 20.09.18).

[15] Centrica, Centrica to build pioneering local energy market in Cornwall, [Online]. Available: <https://www.centrica.com/media-centre/news/2016/centrica-to-build-pioneering-local-energy-market-in-cornwall/>, 2016 (accessed 20.09.18).

[16] SmartEMC2, Smarter grid: empowering SG market actors through information and communication technologies, [Online]. Available: <http://www.smarteremc2.eu/>, 2018 (accessed 20.09.18).

[17] Oulu Energy Farm Power, [Online]. Available: <https://www.oulunenergia.fi/>, 2016 (accessed 20.09.18).

[18] C. Weinhardt, et al., How far along are local energy markets in the DACH + region? A comparative market engineering approach, in: e-Energy 2019 - Proceedings of the 10th ACM International Conference on Future Energy Systems, 2019, pp. 544–549.

[19] Helen, RYHDY AURINKOSÄHKÖN TUOTTAJAKSI, 2016.

[20] P. Olivella-Rosell, et al., Optimization problem for meeting distribution system operator requests in local flexibility markets with distributed energy resources, Appl. Energy 210 (2018) 881–895.

[21] R.P. Hämäläinen, J. Mäntysaari, J. Ruusunen, P.-O. Pineau, Cooperative consumers in a deregulated electricity market—dynamic consumption strategies and price coordination, Energy 25 (9) (2000) 857–875.

[22] F.A. Rahimi, A. Ipakchi, Transactive energy techniques: Closing the gap between wholesale and retail markets, Electr. J. 25 (8) (2012) 29–35.

[23] ESMT, Blockchain in the energy transition. A survey among decision-makers in the German energy industry, 2016.

[24] C. Zhang, J. Wu, C. Long, M. Cheng, Review of existing peer-to-peer energy trading projects, Energy Procedia 105 (2017) 2563–2568.

[25] T. Sousa, T. Soares, P. Pinson, F. Moret, T. Baroche, E. Sorin, Peer-to-peer and community-based markets: a comprehensive review, Renew. Sustain. Energy Rev. 104 (2019) 367–378. https://doi.org/10.1016/j.rser.2019.01.036.

[26] C. Rosen, R. Madlener, An auction design for local reserve energy markets, Decis. Support. Syst. 56 (2013) 168–179.

[27] D. Thomas, J. Kazempour, A. Papakonstantinou, P. Pinson, O. Deblecker, C.S. Ioakimidis, A local market mechanism for physical storage rights, IEEE Trans. Power Syst. (2020) 1.

[28] A. Lüth, J. Weibezahn, J.M. Zepter, On distributional effects in local electricity market designs—evidence from a German case study, Energies 13 (8) (2020) 1993.

[29] S. Behboodi, D.P. Chassin, N. Djilali, C. Crawford, Transactive control of fast-acting demand response based on thermostatic loads in real-time retail electricity markets, Appl. Energy 210 (2018) 1310–1320.

[30] S. Kahrobaee, R.A. Rajabzadeh, L.-K. Soh, S. Asgarpoor, Multiagent study of smart grid customers with neighborhood electricity trading, Electr. Power Syst. Res. 111 (2014) 123–132.

[31] P. Ringler, D. Keles, W. Fichtner, Agent-based modeling and simulation of smart electricity grids and markets - a literature review, Renew. Sustain. Energy Rev. 57 (2016) 205–215.

[32] J. Soares, T. Pinto, F. Lezama, H. Morais, Survey on complex optimization and simulation for the new power systems paradigm, Complexity (2018) 32.

[33] S.Ø. Ottesen, A. Tomasgard, S.-E. Fleten, Prosumer bidding and scheduling in electricity markets, Energy 94 (2016) 828–843.

[34] C.A. Correa-Florez, A. Michiorri, G. Kariniotakis, Optimal participation of residential aggregators in energy and local flexibility markets, IEEE Trans. Smart Grid 11 (2) (2020) 1644–1656.

[35] SGTF-EG3, Regulatory recommendations for the deployment of flexibility, 2015.

# PART I

# Distributed energy resources as enablers of local electricity markets

CHAPTER

# 1

# New electricity markets. The challenges of variable renewable energy

*Ana Estanqueiro and António Couto*

National Energy and Geology Laboratory (LNEG), Renewable Energy and Energy
Efficiency Unit, Lisboa, Portugal

## 1.1 Introduction

The development and large-scale dissemination of new renewable technologies (as wind and solar power)—referred to as variable renewable energy (VRE)—took place from 1990 onwards in most of the developed countries, in a process led by Europe. This development was anchored on the offer of financial incentives, be it for investment, or for the payment of renewable energy [1]. Those incentives were commonly called "FIT—feed-in tariffs," "guaranteed tariffs," or "green" tariffs, and consisted of ensuring investors a return profit guaranteeing the recovery of investments made, thus minimizing the financial risks and contributing to reducing the cost of the investments in the renewable energy sector [2]. That was the key basis that essentially supported the remarkable growth of the renewable sector in Europe in the past 30 years.

The majority of the European countries resorted to FIT systems that all presented similar characteristics: (1) the tariff values paid were (well) above the average energy value in the electricity markets; (2) this generation was exempted (in most countries) of costs from "network usage fees" or other production costs normally charged on conventional generation; and (3) in some countries the interconnection to the preexisting network was provided at low (or no) cost. Some variations exist in those incentives for generation and renewable technologies; for example, (1) Denmark followed a different, but very effective path by defining minimum percentages of renewable penetration in the energy sold by electric utilities; (2) Spain favored a mixed system, that is, a (pseudo) market share associated with the payment of a (fixed) premium above the spot market value for new renewables, as wind energy; and, outside Europe, (3) the United States has favored a (complex) system of tax incentives. In all countries, the base legislation of FITs usually defined a limited duration of application: in France, it had a maximum duration of 9 years, in Portugal initially it had a

3

duration of 15 years and, later, it was indexed to the level of production of the installation that can vary between 9 and 15 years (depending on the period of installation and the technology) and other countries had similar rules [3]. The essential thing to keep in mind is that in all countries it was assumed that the existence of FIT was transitory and that FITs assumed an incentive role for the European energy and industrial energy sector only during the (possibly average) period of recovery of investment, guaranteeing predictability of risk that attracted private investors, by lowering financing costs.

Regardless of the natural diversity of energy policies in the various Member States, Europe is currently phasing-out FITs. In Portugal the process has already started and it is expected that in 2023 a significant part of wind generation has to move to a market regime; France and Germany have ended most of those "green" contracts; and in Spain FITs were suspended during the 2010 economic crisis. Actually, new trading mechanisms adapted to the technical characteristics of VREs and reflecting their value for the environment and society and that are fair to all energy sectors need to be designed and put into practice.

Section 1.2 provides an overview of the renewable generation physical characteristics, namely, of the temporal variability of these power sources and how the natural complementarity of their primary resources can be exploited to maximize the VRE share into electricity/market environments, without overlooking the security of the energy system. In Section 1.3, different concepts to increase the VRE generation value are presented. In Section 1.4, the challenges of the current market designs for the foreseen nearly 100% renewable power systems are addressed, while in Section 1.5, some of the specifications for new market designs under such levels of renewable penetration are discussed. Finally, in Section 1.6, some final remarks are provided.

## 1.2 Physical characteristics of renewable generation

### 1.2.1 Time variability: daily, seasonal, and annual cycles

VREs generation has natural cycles that result from the variability of their primary energy resources, such as wind, solar irradiation, rainfall, and/or melting. Those cycles are explained by the synoptic weather conditions and their interaction with the different topographic characteristics. Since the principle of stable operation of an electrical system requires a constant balance between production and demand, a large contribution from VRE and their reduced predictability (strongly) increases the challenge of managing the power system from its security and robustness point of view. Thus given the power system and markets' dependency on weather, is crucial to assess the timescale-associated variability with each technology. The renewable energy sources (RES) variability timescales are presented in Fig. 1.1.

*Solar*: Solar irradiance varies in timescales from seconds to years. Part of the seasonal and daily variability depends on the apparent movement of the sun in the sky and it is predictable with high precision with adequate mathematical models. However, for a given site and time, the existence of clouds can induce large variations in the available irradiance with respect to the (theoretical) clear sky irradiation. Although with less impact, aspects such as atmospheric turbidity due to aerosols can also act as a cloud-inducing variability considering irradiation. For photovoltaic systems, additional factors can affect the

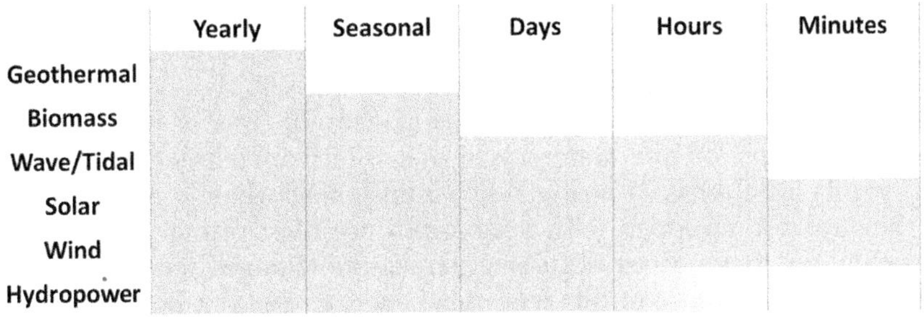

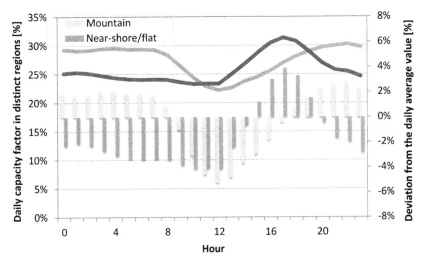

**FIGURE 1.1** Renewable energy variability along different timescales.

**FIGURE 1.2** Daily wind capacity factor in mountain and nearshore/flat regions. The bars depict the hourly deviation with respect to the daily average value.

generation, namely, the temperature of the cell, and obstacles that may shade the system unleashing up/down power ramps [4,5].

*Wind*: The time variability of the wind velocity is originated by the nonhomogeneity of large-scale air masses. Under certain patterns, local effects like sea/land breezes that stem from the sharp contrast between high temperatures over land and lower temperatures over the sea, atmospheric thermal stratification, local terrain, and surface roughness can have a significant impact on the atmospheric airflow [5]. Therefore, in many geographical areas, there is a marked daily and seasonal wind speed profile. Some sites exhibit pronounced diurnal patterns, which depend on location (Fig. 1.2). In mountain regions, the highest wind speed values are typically observed during the nighttime, while for coastal, flat, and nearshore regions, on average, the daily peak wind speed is observed in the mid-afternoon.

Wind power plants convert the kinetic energy of the wind into electrical energy, which they deliver to the grid, and the amount of energy produced is directly proportional to the cube of the wind speed until the nominal conditions are met. Most of the modern wind technology operates with wind speeds normally between 3.0 m/s (cut-in) and 25 m/s (cut-out). During high-speed events, the wind turbine may be turned off for safety reasons. Modern control strategies enable to operate wind turbines above cut-out wind speeds, thus avoiding the curtailment of the available wind power.

*Hydro*: The capacity of hydroelectric power plants (run-of-river or reservoirs) to produce energy depends, mainly, on the water cycle that affects precipitation and ice melting. Hydropower plants have reduced power fluctuations on hourly and even daily timescales, even when they are not equipped with reservoirs since their primary energy source, that is, the water flow on rivers, does not show perceptive changes within small timescales, reflecting the integrating nature of this renewable energy resource, both in time and space. However, high interannual and seasonal scale variability is common, especially in temperate climate regions. The energy production of a run-of-river hydroelectric (RoR) plant misses the regulating and storage capability of the reservoir, thus it depends directly on the streamflow of the river where it operates which is very dependent on the regional (accumulated) weather conditions. In the European winter, the flow of rivers tends to be higher than average, increasing the production of the RoR power plants; conversely, in the summer, the lower availability of the resource leads to less generation by these power plants. In large power plants equipped with reservoirs, the seasonal variability also dictates the resource availability and, consequently, the amount of energy (in the form of potential energy) that can be stored in reservoirs [6].

The generation variability of the dominant renewable technologies is presented in Fig. 1.3. Wind power is the only one without a clearly diurnal cycle, as the hydrocycle is mainly due to plant control, not to the primary resource. For wind, an autocorrelation above 0.5 is expected for time lags below 1 day. This means that wind power generation has some degree of correlation within a 1-day time horizon, and above that, the correlation value tends smoothly to zero indicating a reduced level of predictability for larger time horizons. Due to its primary resource, solar power exhibits the highest variation in the autocorrelation values, and daily cycles with 24 hours are clearly depicted. The hydrogeneration autocorrelation

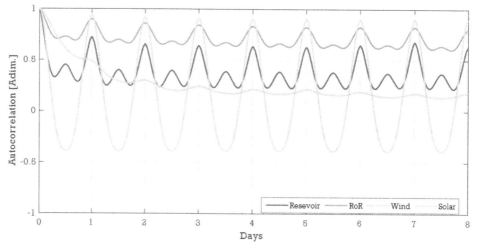

FIGURE 1.3    Autocorrelation of the different RES generation available in the Portuguese power system.

values show two peak points with an identical period—12 and 24 hours. These peaks translate the energy storage/regulation capability of this technology [6]. The lower autocorrelation values from reservoir power plants, when compared with RoR, expose the controllability capabilities of the hydropower.

Regarding the common RES technologies under current deployment, geothermal and biomass are the ones that show the lowest variability, due to their independence from weather cycles. However, the potential of these technologies is still limited to very specific locations, thus preventing economies of scale that could lead to lowering the present installation costs. The power variability of those technologies is globally low and they have the capability of contributing to baseload consumption, as well as providing a wide range of ancillary services, contributing to both short- and long-term flexibility needs. Other sources, such as wave/tidal power, have insignificant variability in the second to the hourly timescale, but their fluctuations become relevant on a daily and seasonal timescale [5]. Variability of wave power tends to be quasi-independent from the time of the day, while the tidal power is associated with the periodic variations in gravitational attraction exerted by celestial bodies in water bodies (e.g., oceans, lakes). Although the wave and tidal generation are more predictable than wind and solar power, their reduced expression on power systems has not yet required their variability to be studied in detail.

Climate change and the existence of extreme events (e.g., successive dry years) is expected to have a great impact on the weather parameters previously presented, and, consequently, on the planning, operation, and reliability of power systems as well as in the electricity markets [7]. In the current energy transition, it is paramount to adopt a long-term and holistic approach that considers and anticipates the variability of these generation sources in a cost-effective way, and with acceptable levels of security and reliability [8].

### 1.2.2 Natural complementarity of renewable energy resources

Understanding the complementarity of the primary resources from VRE power production and their capability to meet the predetermined operational setpoints required by system operators, as well as their capability to participate in electricity markets, is a crucial step toward the decarbonization of electrical power systems without overlooking the security of the energy system and the overall operational costs [8]. This is particularly relevant to enable a smooth energy transition since most of the European power system scenarios to achieve the European Union 2030 and 2050 ambitious targets strongly rely on VRE, such as wind and solar photovoltaic (PV). These technologies are fundamentally different from conventional energy sources due to the stochastic nature of their primary resource. Additionally, cyclic generation patterns exist, for example, the solar power marked daily profile associated to the electricity consumption produces the so-called PV generation "duck curve" [9], depicted in Fig. 1.4.

Those renewable weather-induced patterns raise concerns regarding the capacity of existing power systems to absorb the VRE generated power ramps [9,10], that is, the capability of the conventional power plants to accommodate the ramp rate and range needed to fully exploit variable renewable technologies—a concept normally translated by the "flexibility of the power system"—without (1) increasing the overall costs of the system and (2) reducing the environmental benefits of variable renewables. Associated to the requirements for allocation of extra reserves (i.e., to increase power system flexibility) a strong change in the form in which electricity markets are

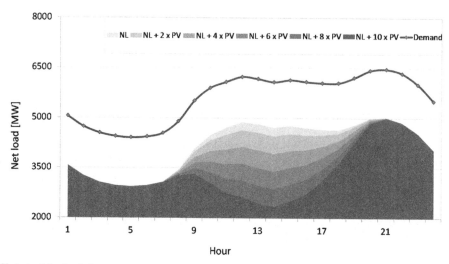

**FIGURE 1.4**  Net load (consumption minus wind and solar generation) daily profile for Portugal and the impact of different levels of solar PV capacity.

designed and operate is also expected. Those steep PV and wind ramps can only be compensated by dispatchable power plant technologies or energy storage units, which expose the power systems to a limited number of players in some periods of the day, thus reducing the number of players and the competitiveness of markets. Additionally, increasing the penetration of zero marginal cost variable renewable technologies as solar and wind may lead to a reduction in wholesale electricity prices due to the so-called "merit–order effect" (see Section 1.4).

Recent studies reveal the existence of VRE generation complementarity on different timescales, from hourly to annual, associated with the diurnal and/or seasonal cycle of solar radiation, which unleashes different weather phenomena across the world, as summarized by Jurasz [11]. For the Iberian Peninsula (IP) regional climate, it was recently concluded [12,13] that the joint operation of wind power with PV would mitigate the wind power variability due to the strong complementarity of the daily profiles of both renewables, promoting the so-called power smoothing effect [14], especially during summer months.

Fig. 1.5 depicts the renewable aggregated generation in Portugal during a 3-year period, including hydropower plants (reservoir and RoR), wind, and solar PV. The daily and monthly correlation values for this RES aggregate (on the country scale) are presented in Table 1.1. It is possible to conclude that, in average terms, on a daily scale, the correlation between hydropower and solar PV is reduced. A similar result is observed for wind power and solar PV. However, on a monthly scale, the complementarity between the solar PV with hydrogeneration is substantial (a cross-correlation factor from −0.42 to −0.48). These results are explained by the hydrogeneration storage capacity, and the planning time of adjustment of hydrogeneration to market conditions. For power systems with very high renewable penetration—typically above 70%—assessing the synergy between hydropower and the remaining renewable technologies (wind and PV) is crucial to maintaining high standards of service quality, namely the long-term security of supply, while minimizing the energy storage requirements, as well as the optimization of its economic performance.

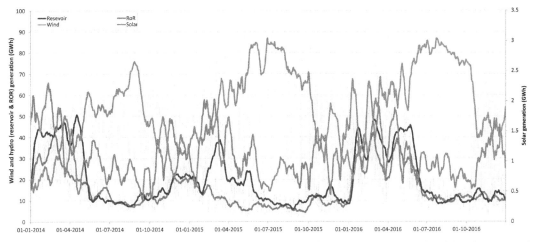

**FIGURE 1.5**  Hydro (reservoir and RoR), wind, and solar daily (thin lines) and weekly (thick lines) aggregated generation in Portugal over 3 years.

**TABLE 1.1**  Correlation values for the national aggregated RES generation using daily and monthly average data.

|  | Daily | | | Monthly | | |
|---|---|---|---|---|---|---|
|  | RoR | Wind | Solar PV | RoR | Wind | Solar PV |
| Reservoir | 0.78 | 0.02 | 0.02 | 0.87 | 0.67 | −0.48 |
| RoR | − | 0.19 | 0.01 | − | 0.64 | −0.42 |
| Wind | − | − | 0.01 | − | − | −0.64 |

## 1.3 Enhance the VRE value: from the large European to small local electricity markets

The existing competitive and liberalized markets were extremely effective in reducing the costs of electricity by adopting designs adapted to the technical and operational characteristics of the 20th century conventional dispatchable power plants. Those designs rely on the predictability (or predefinition) of the power produced by a dispatchable plant for each time interval, using a (conventional) power plant characteristic normally referred to as its "guarantee of power." The challenge for the participation of VRE in markets is they do not offer *guarantee of power*, that is, one cannot ensure with high reliability that for the next day (for a day-ahead designed market) a wind or PV solar plant will deliver a certain quantity of energy during one hour. Both the time and spatial variability of their renewable resources, as well as the difficulties associated to forecasting them, hamper decisively that objective for VRE plants. That has, as consequence, a strong reduction of VRE market value, since markets design strongly penalizes VRE operation up to a point where it compromises their participation in existing markets. Nevertheless, in recent years novel approaches in the planning, designing, and operating of VRE plants have enabled to reduce VRE variability and, consequently, to minimize the generated power deviation with respect to a forecasted or predefined set-point. This section addresses different approaches that enable to increase the VRE value by taking into consideration their variability and intrinsic stochastic generation features.

### 1.3.1 Spatial smoothing effects

One of the most basic features for wind and solar PV power production, which can cancel in many cases the rapid fluctuations of VRE power production and minimize the impact of their storage variability, is the lack of spatiotemporal correlation of their primary resources—for example, between wind speed and solar irradiation [5,15]. Fig. 1.6 shows the Pearson correlation and standard deviation of the combined generation (on an hourly basis) considering the existing wind power plants' locations in Portugal. A high-resolution mesoscale model was used

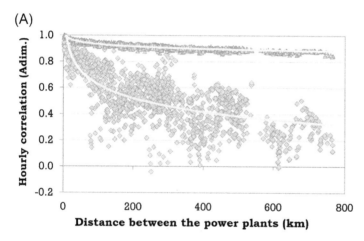

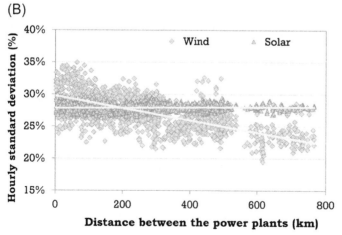

FIGURE 1.6   Hourly (A) correlation and (B) standard deviation between wind power plants in Portugal using 2 years of data (2015—16). The same locations were used to calculate the hourly correlation for the solar PV power production.

to obtain wind power production time series. For comparison, the exact same locations were used to assess the hourly correlation for solar PV power production based on satellite data.

In the solar PV generation case, the smoothing effect for a timescale of 1 hour is small. This behavior can be explained by longitude/latitude dependency from the primary resource rather than the local effects that are strongly present for wind generation, but not for PV. This physical characteristic of renewable primary sources and the use of concepts as virtual (or hybrid) power plants for several VRE power plants widely dispersed, enables the application of an aggregated dispatch strategy and allows VREs to participate in electricity markets as a single entity (with a certain degree of power control capability). This operational strategy is expected to limit the overall VRE generation forecasting errors, thus increasing their value in electricity markets [16,17]. In recent years, several authors have highlighted the benefits of a statistical smoothing effect to improve the VRE power forecast since their errors in a large geographical area are mostly uncorrelated [17,18]. Consequently, the aggregate wind or solar PV power forecast can provide lower errors when compared to the sum of individual power forecasts, reducing the need to balance at high costs the energy on the reserve markets. In principle, and if grid limitations are not a concern, this would favor the enlargement of electricity markets.

## 1.3.2 Aggregation and virtual renewable power plants

As a result of the current energy transition, new energy players are emerging—among them, the aggregating agents of variable renewable generation—who contribute to the provision of new electricity services and auxiliary systems [19,20]. These new agents may—in the near future and assisted by information and communication technologies (ICT)—provide renewable and spatially distributed technologies with "some" control of energy production and/or storage. This technical capability will give VREs some degree of operating flexibility, thus reducing the risk associated with their participation in electricity markets and also contributes to the robustness of the power system as well as its quality of the service.

These days, it is commonly recognized that the aggregate management of several renewable energy sources distributed associated to storage units—in a configuration usually referred to as a *virtual (or hybrid) renewable power plant* (VRPP) [21]—when integrated into active (*smart*) networks, will make an essential contribution to an economically optimized operation of electrical systems in power systems with near 100% renewable-based generation.

Although different definitions can be found in the literature, the VRPP can be described as a flexible representation of a portfolio of different renewable energy technologies, capable of aggregating its capacities and creating a single operation profile, offering a certain degree of guarantee of power, and operating as (close as possible to) a conventional power plant [22]. This concept bases its approach on the development of ICTs systems to aggregate, monitor, and control the status of the distributed energy resources, thus increasing security and granting economic sustainability to its integration in power systems and its participation in an energy market (EM) environment [19,21]. Depending on the objective, the VRPP concept can be divided into two main categories [23]: technical and commercial.

The commercial VRPP aims to empower the participation of small- and medium-scale VREs producers in the electricity and ancillary services markets [24], reducing their risk exposure and optimizing their value within a liberalized market environment. In the VRPP concept, the renewable power plants are combined irrespective of their geographical location and the operational parameters. Thus the VRPP participation in the market environments is usually based on the marginal costs and technical characteristics of the combined renewable power plants considering an optimal schedule (and offer) [22,25].

The benefit of the VRPP solution to enhance the economic performance of renewable producers as well as the power system is well-documented in the current literature [20,22]. Fig. 1.7 shows an optimal VRPP's production profile able to cope with the variability both from local demand and VREs [22]. This particular VRPP is constituted by wind, solar PV, and hydropower plants (both RoR and pumped hydrostorage—PHS). Hydrogeneration, especially PHS units, play a crucial role in regulating VRPP's production profile. As desired, the combined solution is capable of adapting its production to the local demand profile (almost) independently of the primary renewable resource, operating close to a dispatchable power plant (Fig. 1.7B). Moreover, by shifting the exported generation periods from hours of low spot market prices to higher ones an increase in the VRPP's total overall value is observed (Fig. 1.8).

Joint participation in the day-ahead market (DAM) allows increasing the value of the energy produced by the VRE power plants by 29% for wind and 2% for the PV comparing to their stand-alone operation (Fig. 1.8). Compared to market average prices, an 8% increase in revenues is obtained by the VRPP in the DAM. This result indicates a significant increase in the total revenue of renewable power plants, even for the PV power plant

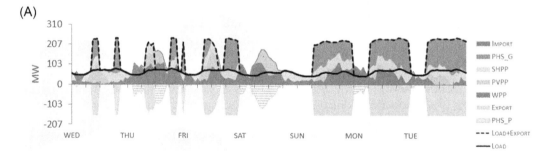

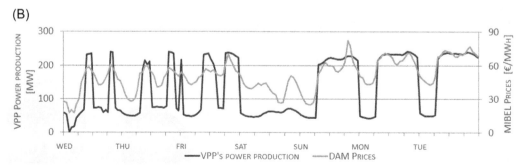

FIGURE 1.7    VRPP (A) production and consumption for 1 week, (B) production profile and spot market prices for 1 week.

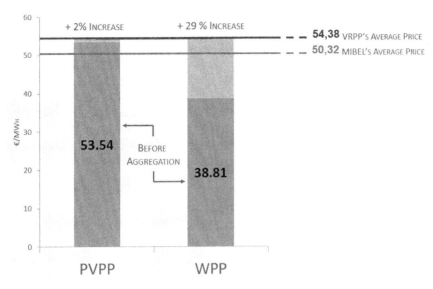

**FIGURE 1.8**  Marginal value of the solar (PVPP), wind (WPP), and virtual renewable (VRPP) power plants.

that, despite producing energy during the day (when DAM prices are higher) can also benefit from the flexibility given by the PHS to shift energy from days of low market prices to higher ones (e.g., from weekends to weekdays).

From the recent Clean Energy for All Europeans Package the RES aggregator emerges as one of the key players in the foreseen energy transition. This new player will act as an intermediary between decentralized/distributed actors and the market by supporting small actors, for example, prosumers, to participate in the electricity market. In general terms, similarly to the VRPP, the aggregator will bring different players within the power system to act as a single entity when participating in the markets or when selling services to power system operators. With aggregation (VRPPs) it will be possible to combine several different units optimizing their operation profiles, costs, and overall value. For consumers and prosumers also participating in VRPPs, the aggregation will have the potential to reduce energy costs, while it reduces the overall power system balancing costs.

VRPPs assume nowadays an extreme relevance, especially when considering the foreseen phase-out of European feed-in tariffs in the near future and the very recent proposals/announcements from the European Commission (EC) within the Clean Energy package, in which the access of the VREs to the grid is proposed as no longer guaranteed.

## 1.3.3 Emerging of local markets

The reduction in VRE technologies' costs, especially solar PV, and the advances in ICT and power electronics solutions are transforming electric grids into a decentralized management system due to the active participation of large numbers of electricity consumers—the so-called prosumers. The proliferation of prosumers and new concepts, such as peer-to-peer (P2P) negotiation of electricity, enable decentralized and competitive markets that are not controlled by the conventional electric utility companies, in which their energy surplus could be traded with neighbors, thus allowing maximization of the VREs value. Further benefits are expected with this type of market since (1) the consumers may negotiate the electricity at a fair (local) price; (2) it can promote the VRE deployment in a sustainable form while reducing dependence on subsidies; and (3) they enable the power systems to become more resilient and efficient as the production is located near to the consumption [26,27]. Moreover, local electricity markets can adopt different trading concepts, such as P2P, auctions, or others more adapted to the natural, environmental, and societal characteristics of each region. These markets were regulated at European level in the recent Clean Energy package presented by the EC [27] and are expected to be deployed to their full potential during the next decade (2021–30).

## 1.4 The challenges of the current market designs for ~100% renewable power systems

The current design of the electric energy markets—defined in a period where conventional power technologies dominated—is based on the marginal cost [the cost necessary to (incrementally) produce a megawatt] in the formation of its hourly price. The market price is defined by the intersection of the supply curve and the energy demand curve. To participate in the daily market, producers take into account their marginal costs and the expected production quantity, creating a stack of offers. Since VRE technologies have very low marginal costs when compared to conventional technologies, there is a tendency for a decrease in the energy costs verified in wholesale markets with a large participation of VREs—the order merit effect [28]. This trend is already very scrutinized in the literature for markets with high penetration of VRE [29,30]. Fig. 1.9A underlines this situation for the MIBEL case by employing the methodology presented by Hirth [31]. In the situation presented, a large share of wind power tends to have low prices, which reduce the market value of wind energy. Fig. 1.10 depicts this impact on a daily scale by showing the average day-ahead prices binned by different levels of wind power production. From this figure, it is possible to observe that low prices in the day-ahead market are associated with a high level of wind generation levels, and vice versa. The quantification of this impact for wind power producers during the first half of 2016 for the MIBEL market was also investigated in detail in [29]. Using data from the market and an agent-based simulation tool MATREM, the authors obtained an average energy price decrease of nearly €17/MWh (−35% regarding the no-wind case for an average wind penetration share of 28%) in the period under analysis.

For certain levels of VRE penetration the effect of the merit order is positive for consumers due to a reduction in electricity market prices. This leads to a reduction in producers profitability ("self-cannibalization effect") [31]. This situation may lessen the incentive to invest in new capacity deployment needed to ensure a stable electricity supply as well as to accomplish the ambitious European renewable target, being an actual concern in power systems with high to very high variable renewable participation.

Fig. 1.9A also reveals the market value of different technologies in the Portuguese power system. Hydroelectric power plants show a behavior similar to wind technology,

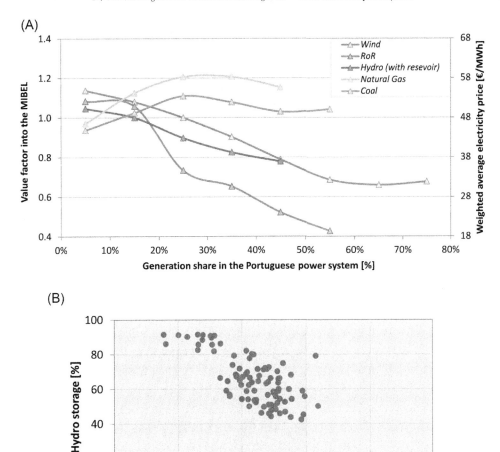

FIGURE 1.9 (A) Value factor and weighted average electricity price of main technologies in the Portuguese power system during 2014–18; (B) monthly day-ahead (DAM) average prices versus hydropower storage capacity.

that is, the increase of hydropower share in generation mix as well as the storage capacity decreases its market value (see Fig. 1.9B). In this case, the value of this technology is usually designated as "water value." This value represents an opportunity cost since the production of reservoir hydropower plants is normally adjusted to resource availability, market conditions, and/or (perception of) consumption needs. The value of this technology into the market decreases less than the RoR technology. As expected, the fossil technologies (coal and natural gas) increase their market value as the share of the energy mix increases.

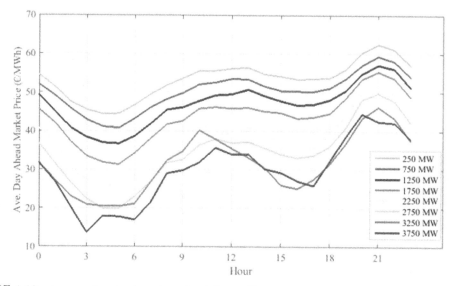

**FIGURE 1.10** Average day-ahead market price daily profile according to different levels of wind power production (bin width of 500 MW).

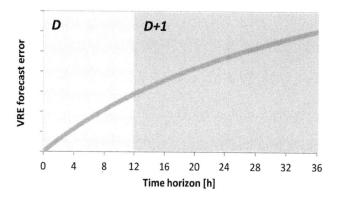

**FIGURE 1.11** Dependence of the VRE forecast error due to the time horizon.

Another key issue that can compromise the profitability of VRE producers and prevent a full and fair integration of VRE into electricity markets is the current market designs. Most of the electricity markets require, at 12 noon on day $D$, the forecast of electricity production for the 24 hours on the next day $(D + 1)$ from all participating producers, whether their production comes from easily adjustable dispatchable power plants (with storage), or from renewable "stochastic" energy sources, such as wind and solar PV. Despite the recent improvements observed in the power forecasts of renewable systems, large errors are still observed, especially for long time horizons (that involve a time lag up to 36 hours in some cases) as currently still required by electricity markets (Fig. 1.11). Those VREs forecast errors produced an imbalance in the power system compensated by other producers, usually coming from conventional power sources that act as a reserve to maintain the supply/demand equilibrium and the stability of power systems. The use of these operational reserves has an associated cost, known as "balancing cost," that needs to be supported by the producers that deviate from the initial scheduled. In the current electricity market designs this could reduce substantially the revenues of VRE producers.

The reality is, although there is a positive impact induced by the renewable technologies in the market price of electricity, and a reduction of energy volumes (and thus, profit margins) by conventional fossil fuel-based power plants in the so-called "energy markets," the fact markets are designed and organized in a "day-ahead" mode, prevents the major players in the near future from taking the fair return of their (large) role in the energy, since they are penalized by the inadequate market design. This market imperfection will enable the fossil fuel-based players to recover their business margins in the energy reserve markets, with negative consequences for both the overall cost of electricity as well as for the $CO_2$ emissions, since most fossil fuel plants have very high emissions when cycling for reserve markets participation, thus compromising the overall objective of renewable technologies development and the decarbonization of the electricity sector.

## 1.5 The need for new market designs and trading models under ~ 100% renewable power systems

The expected high-share of VRE into power systems will lead to significant changes in the EM behavior. Future EMs will be characterized by more variable and decentralized electricity production, while at the same time, cross-board dependence between countries, new technological opportunities for consumers to actively participate in electricity markets through demand response, self-consumption or storage, and new market players (e.g., energy communities) will emerge [32]. In 2016 a new legislative package, the *Clean Energy for all Europeans*, was presented by European Commission aiming to facilitate an energy transition (to clean decarbonized sources) and the operation of the European Internal Market for Electricity (EIME). Some of the most important guidelines include gate closure of EMs close to the real-time operation, balance responsibility for VRE and participation in the balancing markets, aggregated bidding, reduction of the market time unit from 1 hour to 15 minutes (until 2025) and implicit allocation of the cross-border capacity [33].

Generally speaking, the European EMs are still not complying with the new EC proposal and some barriers have to be suppressed to potentially increase the general welfare of all market participants [33]. Thus, additionally to the previous EC proposal, a well-designed market needs to (1) be able to explore the VRE inherent features, synergies, and capabilities; (2) reflect the real value of renewable generation for the planet and the society; (3) allow new business models that create the necessary conditions for continuity of the electricity/energy sectors investment (from the current stakeholders but also citizens); (4) avoid privilege of technology and flexibility options; and (5) reduce the complexity of the current legislation [34].

Current market designs and business models also lack incentives and opportunities for end-user electricity consumers (for instance the residential buildings) to become prosumers and actively participate in the electricity trading. New trading opportunities need to be available for different spatial scales (especially for the small-scale ones, e.g., local markets) to allowing an active participation of prosumers with reduced production quantity, but meaningful flexibility.

## 1.6 Final remarks

The current electricity markets design, which has existed for more than 20 years, is poorly adapted to the operating characteristics of time-variable power sources such as wind and solar photovoltaic with marginal costs close to zero. These characteristics, examined in this chapter, allow to understand that the weather-driven power sources present a low predictability, especially for long time horizons, and a reduced guarantee of power. These characteristics bring significant risks to these electricity market participants, a factor which needs to be considered in the design of future electricity markets, and also involve new challenges for system operators to keep the security of supply and maintain the desirable system adequacy.

Nondisruptive modifications to increase the renewable power sources penetration can be the first step in the pathway toward near 100% renewable power systems. Examples of these modifications are, for instance, to reduce negotiation periods and change the market gate closure time frames. Nevertheless, for high levels of variable renewables integration, the emergence of new players and concepts (such as the spatiotemporal smoothing effect, generation aggregation, and the virtual/hybrid renewable power plants) are crucial to allow VRE technologies to take more responsibility by providing a full range of ancillary services. To achieve it, innovative market structures, as well as novel operating rules, are needed. At a local scale, new paradigms based on decentralized negotiation concepts (such as peer-to-peer trading) are needed to complement the centralized markets and help to unlock a range of flexibility solutions at the distribution system level.

It is essential to rethink the "market design" and the negotiation mechanisms for the near future, to reflect the real value of renewable generation for the planet and the society, and to develop new business models to create the necessary conditions that guarantee the continuity of the investment in the electric power sectors. The beneficiary of these new designs will be, ultimately, the end consumers and the economies of the countries with a large penetration of renewable generation.

## Acknowledgment

This work has received funding from the EU Horizon 2020 research and innovation program under project TradeRES (grant agreement No 864276).

## References

[1] S. Abolhosseini, A. Heshmati, The main support mechanisms to finance renewable energy development, Renew. Sustain. Energy Rev. 40 (2014) 876–885. Available from: https://doi.org/10.1016/j.rser.2014.08.013.

[2] F. Polzin, F. Egli, B. Steffen, T.S. Schmidt, How do policies mobilize private finance for renewable energy?—A systematic review with an investor perspective, Appl. Energy 236 (2019) 1249–1268. Available from: https://doi.org/10.1016/j.apenergy.2018.11.098.

[3] I. Peña, L. Azevedo, I. Marcelino Ferreira LAF, Lessons from wind policy in Portugal, Energy Policy 103 (2017) 193–202. Available from: https://doi.org/10.1016/j.enpol.2016.11.033.

[4] F.J. Rodríguez-Benítez, C. Arbizu-Barrena, F.J. Santos-Alamillos, et al., Analysis of the intra-day solar resource variability in the Iberian Peninsula, Sol. Energy 171 (2018) 374–387. Available from: https://doi.org/10.1016/j.solener.2018.06.060.

[5] J. Widén, N. Carpman, V. Castellucci, et al., Variability assessment and forecasting of renewables: a review for solar, wind, wave and tidal resources, Renew. Sustain. Energy Rev. 44 (2015) 356–375.

[6] K. Engeland, M. Borga, J.-D. Creutin, et al., Space-time variability of climate variables and intermittent renewable electricity production − a review, Renew. Sustain. Energy Rev. 79 (2017) 600–617. Available from: https://doi.org/10.1016/j.rser.2017.05.046.

[7] J. Peter, How does climate change affect electricity system planning and optimal allocation of variable renewable energy? Appl. Energy 252 (2019) 113397. Available from: https://doi.org/10.1016/j.apenergy.2019.113397.

[8] K. Hansen, C. Breyer, H. Lund, Status and perspectives on 100% renewable energy systems, Energy 175 (2019) 471–480. Available from: https://doi.org/10.1016/j.energy.2019.03.092.

[9] M. Obi, R. Bass, Trends and challenges of grid-connected photovoltaic systems - a review, Renew. Sustain. Energy Rev. 58 (2016) 1082–1094. Available from: https://doi.org/10.1016/j.rser.2015.12.289.

[10] M. Lacerda, A. Couto, A. Estanqueiro, Wind power ramps driven by windstorms and cyclones, Energies 10 (2017) 1475–1495. Available from: https://doi.org/10.3390/en10101475.

[11] J. Jurasz, F.A. Canales, A. Kies, et al., A review on the complementarity of renewable energy sources: concept, metrics, application and future research directions, Sol. Energy 195 (2020) 703–724. Available from: https://doi.org/10.1016/j.solener.2019.11.087.

[12] S. Jerez, R.M. Trigo, A. Sarsa, et al., Spatio-temporal complementarity between solar and wind power in the Iberian Peninsula, Energy Procedia 40 (2013) 48–57. Available from: https://doi.org/10.1016/j.egypro.2013.08.007.

[13] R. Castro, J. Crispim, Variability and correlation of renewable energy sources in the Portuguese electrical system, Energy Sustain. Dev. 42 (2018) 64–76. Available from: https://doi.org/10.1016/j.esd.2017.10.005.

[14] A. Estanqueiro Impact of wind generation fluctuations in the design and operation of power systems, in: 7th International Workshop on Large Scale Integration of Wind Power and on Transmission Networks for Offshore Wind Farms. Madrid, Spain, 2008, p. 7.

[15] S. Shivashankar, S. Mekhilef, H. Mokhlis, M. Karimi, Mitigating methods of power fluctuation of photovoltaic (PV) sources - a review, Renew. Sustain. Energy Rev. 59 (2016) 1170–1184. Available from: https://doi.org/10.1016/j.rser.2016.01.059.

[16] Y. Liu, X. Gao, J. Yan, et al., Clustering methods of wind turbines and its application in short-term wind power forecasts, J. Renew. Sustain. Energy 6 (2014) 053119. Available from: https://doi.org/10.1063/1.4898361.

[17] J. Miettinen, H. Holttinen, B. Hodge, Simulating wind power forecast error distributions for spatially aggregated wind power plants, Wind. Energy 23 (2020) 45–62. Available from: https://doi.org/10.1002/we.2410.

[18] D.W. van der Meer, J. Munkhammar, J. Widén, Probabilistic forecasting of solar power, electricity consumption and net load: investigating the effect of seasons, aggregation and penetration on prediction intervals, Sol. Energy 171 (2018) 397–413. Available from: https://doi.org/10.1016/j.solener.2018.06.103.

[19] B. Moreno, G. Díaz, The impact of virtual power plant technology composition on wholesale electricity prices: a comparative study of some European Union electricity markets, Renew. Sustain. Energy Rev. 99 (2019) 100–108. Available from: https://doi.org/10.1016/j.rser.2018.09.028.

[20] M.M. Othman, Y.G. Hegazy, A.Y. Abdelaziz, Electrical energy management in unbalanced distribution networks using virtual power plant concept, Electr. Power Syst. Res. 145 (2017) 157–165. Available from: https://doi.org/10.1016/j.epsr.2017.01.004.

[21] K.O. Adu-Kankam, L.M. Camarinha-Matos, Towards collaborative virtual power plants: trends and convergence, Sustain. Energy Grids Nctw. 16 (2018) 217–230. Available from: https://doi.org/10.1016/j.segan.2018.08.003.

[22] A.R. Machado, A. Couto, J. Duque, A. Estanqueiro Enhancing the value of wind and PV generation through optimal aggregation, in: 16th Wind Integration Workshop, 25–27 October. Berlin, Germany, 2017, p. 6.

[23] H. Saboori, M. Mohammadi, R. Taghe Virtual power plant (VPP), definition, concept, components and types, in: 2011 Asia-Pacific Power and Energy Engineering Conference. IEEE, 2011, p. 4.

[24] M. Kolenc, P. Nemček, C. Gutschi, et al., Performance evaluation of a virtual power plant communication system providing ancillary services, Electr. Power Syst. Res. 149 (2017) 46–54. Available from: https://doi.org/10.1016/j.epsr.2017.04.010.

[25] S. Hadayeghparast, A. SoltaniNejad Farsangi, H. Shayanfar, Day-ahead stochastic multi-objective economic/emission operational scheduling of a large scale virtual power plant, Energy 172 (2019) 630–646. Available from: https://doi.org/10.1016/j.energy.2019.01.143.

I. Distributed energy resources as enablers of local electricity markets

[26] A.M. Alabdullatif, E.H. Gerding, A. Perez-Diaz, Market design and trading strategies for community energy markets with storage and renewable supply, Energies 13 (2020) 31. Available from: https://doi.org/10.3390/en13040972.

[27] T. Sousa, T. Soares, P. Pinson, et al., Peer-to-peer and community-based markets: a comprehensive review, Renew. Sustain. Energy Rev. 104 (2019) 367–378. Available from: https://doi.org/10.1016/j.rser.2019.01.036.

[28] F. Sensfuß, M. Ragwitz, M. Genoese, The merit-order effect: a detailed analysis of the price effect of renewable electricity generation on spot market prices in Germany, Energy Policy 36 (2008) 3086–3094. Available from: https://doi.org/10.1016/j.enpol.2008.03.035.

[29] F. Lopes, J. Sá, J. Santana, Renewable generation, support policies and the merit order effect: a comprehensive overview and the case of wind power in Portugal, in: F. Lopes, H. Coelho (Eds.), Electricity Markets with Increasing Levels of Renewable Generation: Structure, Operation, Agent-based Simulation, and Emerging Design, Springer, Cham, 2018, pp. 227–263.

[30] R.P. Odeh, D. Watts, Impacts of wind and solar spatial diversification on its market value: a case study of the Chilean electricity market, Renew. Sustain. Energy Rev. 111 (2019) 442–461. Available from: https://doi.org/10.1016/j.rser.2019.01.015.

[31] L. Hirth, The benefits of flexibility: the value of wind energy with hydropower, Appl. Energy 181 (2016) 210–223. Available from: https://doi.org/10.1016/j.apenergy.2016.07.039.

[32] EC (2017) Regulation of the European parliament and of the council on the internal market for electricity.

[33] H. Algarvio, F. Lopes, A. Couto, et al., Effects of regulating the European Internal Market on the integration of variable renewable energy, Wiley Interdiscip. Rev. Energy Env. 8 (2019) 1–13. Available from: https://doi.org/10.1002/wene.346.

[34] E.M. Ländner, A. Märtz, M. Schöpf, M. Weibelzahl, From energy legislation to investment determination: shaping future electricity markets with different flexibility options, Energy Policy 129 (2019) 1100–1110. Available from: https://doi.org/10.1016/j.enpol.2019.02.012.

# Integration of electric vehicles in local energy markets

*José Almeida*[1] *and João Soares*[2]

[1]GECAD Research Center, ISEP–School of Engineering of Polytechnic of Porto, Porto, Portugal [2]GECAD Research Center, School of Engineering of Polytechnic of Porto, Porto, Portugal

## 2.1 Introduction

The mobility sector is a significant contributor to the pollution of the planet, with large emissions of $CO_2$. The need to reduce these emissions is becoming more and more relevant, and according to the Paris Agreement Treaty [1], the global temperature is to be limited to 2°C above preindustrial times. The transportation sector is responsible for nearly 25% of greenhouse gas emissions [2], so the electrification of this sector will reduce this share substantially.

From year to year, the popularity of electric vehicles (EVs) has been increasing, because it is seen as a part of the solution to make the planet greener. Indeed, 51% of EV owners say that the main reason they purchased an EV was to contribute to a more sustainable future [3]. In addition to this, multiple countries have been creating incentives to promote EV-related research projects and EV-related industries. All these factors are contributing to the large-scale growth of EVs.

However, this growth is having and will have a significant impact on the electric grid, bringing multiple challenges and opportunities. The major drawback caused using EVs is the high demand when charging occurs at the same time as residential or industrial peak power consumption [4]. Since the electric grid must satisfy the demand at every instant, a way of reducing the impact of the extra energy required needs to be studied. Understanding the charging behavior of the EV user is a crucial part of this context, to deal with uncertainty associated with EV charging.

Despite this significant challenge, the massive integration of EVs in the distribution network (DN) will offer some opportunities, mainly to take advantage of the flexibility that the EVs can provide in demand [5,6]. Because of this flexibility, demand response (DR) programs fit into the smart grid (SG) context, since their demand can be altered or even reduced, to avoid the high power consumption hours. This load shift is accomplished through some incentives or penalizations, mainly price based. This is where the concept of local energy markets (LEMs) can be introduced, wherein the SG context [7,8] the aggregators can go to the local market (LM), to make their proposals, so they can get an incentive or to avoid a penalty.

This chapter discusses multiple themes related to the integration of EVs in the LEMs. A brief overview of the different types of EVs that exist is given first, and some statistics of EVs are presented. Then the concepts of DR and EV flexibility are detailed, discussing some studies already carried out into the subjects. The idea of LEM is described, making the connection with EVs. Finally, the main conclusions drawn from this work are detailed.

## 2.2 Electric vehicles

This section presents the various classes of EVs that currently exist, briefly describing each one. Then some statistics are given regarding the total stock of electric cars in the world, market sales, and market share by the end of 2018. Also, some numbers related to the charging stations in China and the United States are shown. This section is finalized with a study that shows the impact of EV penetration on residential houses.

### 2.2.1 Types of electric vehicles

EVs differ from the typical internal combustion engine vehicles (ICEV), given that the ICEV throughout chemical processes in the engine (combustion and ignition) converts energy to work. This means the engine itself converts chemical energy into mechanical energy, which through a system of gears makes the car move [9]. In EVs this is not the case, since the energy stored in a battery is going to supply the electric motor, which then transforms the electrical energy into mechanical energy.

There are currently four different types of EVs circulating on the roads:

- Battery Electric Vehicle (BEV);
- Plug-in Hybrid Electric Vehicle (PHEV);
- Hybrid Electric Vehicle (HEV);
- Fuel-cell Electric Vehicle (FCEV).

The so-called BEVs [10] are fully EVs with no combustion engine that run on a rechargeable battery. This battery is responsible for powering all the vehicle electronics and electric motor. This type of vehicle is zero emissions, which means that they do not emit any greenhouse gas emissions.

The PHEVs use two different types of motors: the electric motor and the internal combustion engine (ICE). The electric motor operates at lower speeds. Then, at higher speeds, the ICE turns on. As the name suggests, the battery of a PHEV charges via its connection to an external source.

HEVs are similar to PHEVs, except that the charging of the battery through an external power source is not possible. Instead, the battery recharges via "regenerative braking" [11]. In this process, the electric motor works as a generator using the heat from the brakes, to supply energy for the battery.

Finally, the FCEVs create electric energy from hydrogen and oxygen. In other words, this type of vehicle does not need a battery to store energy. They have excellent efficiency and emit only water, and because of these factors some say that this is the best class of EVs. This sort of technology is still in development and faces many challenges, so it is expected that its entry into the market will not occur in the short term [12].

## 2.2.2 Electric car stock in the world by the end of 2019

Due to all the factors previously mentioned, the growth of the EV is inevitable. At the end of 2019, it was reported that 7.2 million EVs were circulating on the roads, an increase of more than 2 million vehicles from the previous year. As Fig. 2.1 shows, China was the major contributor to this number, with more than 3 million EVs [13]. It is also possible to extract from the figure that the BEVs account for most of the EVs number, representing approximately 66% of total EVs.

## 2.2.3 Electric vehicles market sales and market share in the world by the end of 2019

In the meantime, car sales have also experienced a significant increase in 2019, with 2.1 million sales. This is more than twice the number registered in 2017, with the BEV being the type of EV that was the most sold. China was again the country with most of the sales, as Fig. 2.2 presents in Ref. [13]. Norway is the number one country in the world when it comes to market share, with 56% by the end of 2019.

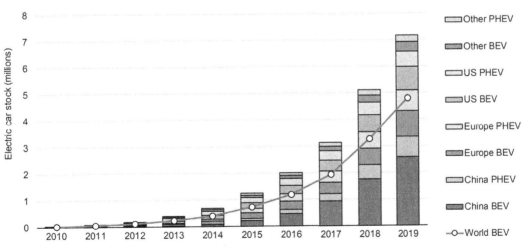

**FIGURE 2.1** Number of electric cars in the major regions, 2010–19. *Source: International Energy Agency (IEA), Global EV Outlook 2020: entering the decade of electric drive? Global EV Outlook 2020, 2020, p. 273.*

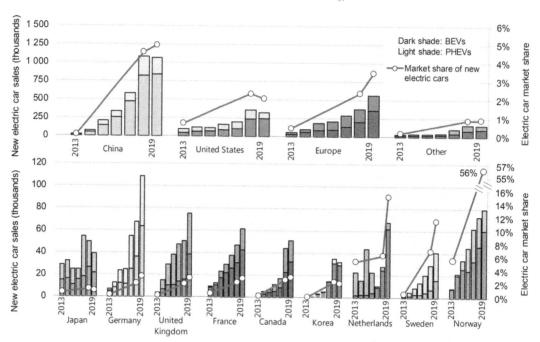

**FIGURE 2.2**  Electric car sales and market share in the major regions, 2013–19. *Source: International Energy Agency (IEA), Global EV Outlook 2020: entering the decade of electric drive? Global EV Outlook 2020, 2020, p. 273.*

## 2.2.4 Electric car stock and market sales by 2030

According to Bunsen et al. [14], two different scenarios are defined to the adoption of EVs until 2030. The *"New Policies Scenario,"* which includes the impact of currently announced policy ambitions, and the *"EV30@30 Scenario,"* representing a scenario in which EVs take up to 30% of all market share (except two-wheelers) by 2030. The policies involved in these scenarios are policies that support the purchase of EVs, through tax incentives and promotion of charging structures. In the same way, companies must increase the number of EVs in their vehicle fleets. The private and public sectors must share information so that there is the rapid integration of this type of vehicle on the streets. This situation is only possible with the encouragement and participation of government entities [15]. As Fig. 2.3 shows, EV stock in the *"New Policies Scenario"* could go up to nearly 150 million by 2030, while the sales exceed the 20 million mark. The *"EV30@30 Scenario"* projects a global EV stock of more than 250 million vehicles by 2030 and more than 40 million EV sales.

## 2.2.5 Electric vehicles charging infrastructures in China and United States

To support this massive growth of EVs, the number of charging infrastructures has also been intensely increasing. In January 2019 there were around 808,000 EV chargers in China, a growth of about 80% in a year [16]. According to Cooper and Schefter [17], the

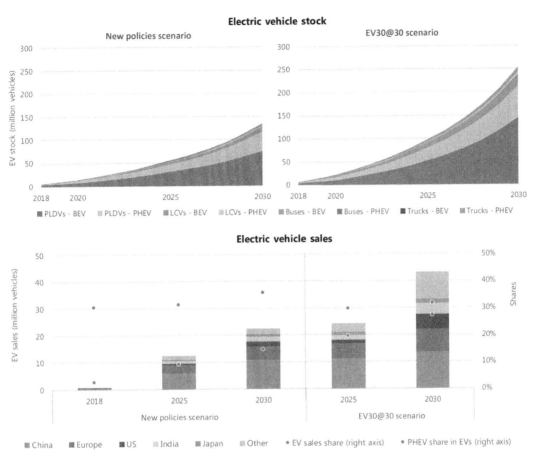

**FIGURE 2.3** Global EV stock and sales by each scenario, 2018–30. *Source: T. Bunsen, et al., Global EV Outlook 2019 to electric mobility, OECD, 2019, p. 232. <iea.org>.*

number of EVs circulating in the United States by 2030 is projected to reach 18.7 million and to support this number, 9.6 million charging ports will be required. Of these 9.6 million, roughly 1% will be fast charging. About 21% will be public, and workplace charging and the remaining 78% will be home charging.

## 2.2.6 Impact of EV penetration on residential homes

Although this mass integration of EVs has already been planned as several studies show, the complexity of the EV still represents a great challenge, mainly to the electric grid, with all the energy required. A study in Ref. [18] using 150 residential homes and 25% of EV was conducted on a typical September day, to show the consumption with and without EV penetration. This study showed that the peak load in the group of houses would increase by 30% with the EV penetration.

## 2.3 Demand response and electric vehicles flexibility

This section gives a brief overview of the DR concept, as well as EV flexibility concept, followed by some studies regarding EV flexibility. Lastly, it presents some specific DR models for EVs in the SG context.

### 2.3.1 Concepts of demand response and electric vehicles flexibility

The energy market works based on offer and demand, which will be reflected in the price of energy. If there is more supply and less demand, naturally, the cost of energy will tend to decrease, as the selling entities will lower the price to see if the demand increases. Consequently, the opposite is also true, that is, when demand is high and supply is low, the cost of energy will tend to rise, as the selling entities will raise the price to try and make the most of this amount of demand.

In the SG context, with the high penetration of renewable energy sources (RES) in the DN, DR programs are essential. This high penetration brings uncertainty because resources are not controllable, that is, the production of electricity is dependent on the climate conditions most of the time (sun, wind, etc.). This factor causes demand not always to be satisfied by supply. It is in these conditions that the DR programs come to play, which allows improving the overall energy efficiency, letting the consumer reduce or shift their load to avoid peak consumption hours so that the power system can be stabilized. The customer, in this case, will accept participating in these DR programs in exchange for some benefit that rewards them for changing their consumption habits.

Deng et al. [19] separated DR programs into three different categories: DR based on a control mechanism, which includes centralized programs and distributed programs; DR based on offered motivations, this one incorporates price-based DR and incentive-based DR; and DR based on decision variables, which contains task scheduling-based DR and energy management-based DR.

The Federal Energy Regulatory Commission (FERC) defines DR as: *"Changes in electric usage by demand-side resources from their normal consumption patterns in response to changes in the price of electricity over time, or to incentive payments designed to induce lower electricity use at times of high wholesale market prices or when system reliability is jeopardized"* [20]. In general, when it comes to SGs, DR can help to incorporate more distributed generation (DG), help to improve the system stability, and, most importantly, brings flexibility to the demand side.

In this context, EVs bring a high elasticity in their load, which is known as EVs flexibility. This flexibility can help to mitigate congestion problems and voltage problems in the grid because it allows the EV user to shift or reduce the vehicle's load consumption and even in some cases inject energy back to the grid.

### 2.3.2 Current challenges

As a result of the possibilities previously mentioned that EVs can bring in terms of demand, several studies have emerged on this topic. Pavic et al. [21] proposed an optimal approach to maximize the benefits of EVs flexibility combining slow home charging with fast charging. This paper proposes multiple charging models [Vehicle-to-Grid (V2G), Grid-to-Vehicle (G2V), Grid-to-Station (G2S), and Station-to-Grid (S2G)], to mitigate the effects of uncontrolled charging, particularly in fast-charging stations. The work also studies the impact of integrating an energy storage system into fast-charging stations. This paper concludes that G2V with reserve services provides high flexibility that can mitigate the problems caused by uncontrolled fast charging in the power system, when charging occurs during an optimal time window. It offers better results when only up to 12% of on-road EVs can fast charge.

Ramirez et al. [22] proposed a model that allows optimizing of both the costs of operation of generation units and EV flexibility in the form of smart charging and discharging, also integrating V2G capabilities in the proposed model. This paper presents a large-scale mixed-integer linear optimization problem, where clustering of generation units and EVs with similar characteristics was conducted to avoid nonlinearities and to reduce decision variables and constraints. It also analyzes the impact of multiple EV flexibility enabling costs, with various percentages of EV penetration and numerous values of wind generation. It concluded that the best situation for the system is when the flexibility allowing cost is 0, which makes every EV in the system flexible. This number decreases as the cost increases. The V2G capabilities implemented also present a significant advantage, mainly in the total system reduction cost.

Papadaskalopoulos et al. [23] portray the applicability of EVs and heat pumps in a proposed market pool. Regarding EV and heat pump flexibility, several scenarios were proposed: both technologies without flexibility; only EV flexibility and then only heat pump flexibility; and finally, both technologies with flexibility in their demand. Through these scenarios, the paper analyzes the benefits of the participation of these two types of load in the electricity market. The joint integration of both technologies with 100% flexibility presents the best results in terms of reducing generation costs and reducing the peak demand of the system. These reductions are even more significant when the penetration of these two loads with 100% flexibility increases. The paper by Gerritsma et al. [24] proposed a method to analyze and simulate the time-dependent EV flexibility, using a Dutch case study. The authors study the maximum flexibility of EV consumption in a residential circuit. In this paper, it is considered that EV users, through smart charging (SC), communicate to their aggregator the estimated time of departure for a trip and the amount of energy required. The authors conclude that 59% of EV demand could be delayed for more than 8 hours and 16% for more than 24 hours, which presents high flexibility that can be used to mitigate several problems on the grid.

Knezović et al. [25] describes the services provided by EV flexibility in the DN. This paper also proposes some recommendations to the market in order to improve the provision of distribution system operator (DSO) grid services by EVs.

Sadeghianpourhamami et al. [26] studied the characteristics of charging flexibility of EVs. A clustering of the departure and arrival times is constructed to better understand the charging behavior of EV users. The authors also say that this paper will help to develop DR algorithms that can significantly explore EV flexibility.

In sum, all these studies allow the conclusion that the flexibility of the EVs is an asset to the electrical system. It helps to reduce network operation costs and the demand during peak hours and allows to take more significant advantage of RES.

## 2.3.3 Specific demand response for electric vehicles

Today, there is an increasing need to implement DR programs for the EVs. Because one of the challenges that the EVs bring to the network is that when their independent charging happens, without restrictions or incentives to change this behavior, this phenomenon is known as dumb charging, in which for the electric grid, the EVs are considered as one more charge, like any other device. As there are no restrictions or incentives, EV users usually charge their vehicles at peak load times, which can cause grid congestion problems as well as substantial decreases in bus voltage.

Multiple strategies can be considered when it comes to DR programs for EVs. As mentioned previously, DR programs can be separated in price-based like time-of-use (TOU), real-time pricing (RTP), and incentive-based programs, which in the case of EVs can be SC, V2G, and many others.

Soares et al. [27] proposed multiple DR programs: incentive-based programs being SC; V2G; and trip shifting, which allows the EV aggregator to change the EV loading period. This change is achieved through a list of departure times for trips, provided by EV users. Trip reducing is a particular DR program where EV users submit a reduction in vehicle loading needs to an aggregator, for example. Users seek an incentive by reducing travel needs and minimum battery level requirements. The authors also propose a fuel shifting DR program and optimal price-based model, like Fig. 2.4 shows. Fuel shifting, as mentioned above, is a special DR program that aims at the EV user leaving the charging location, even if the vehicle is not charged to a minimum quantity. From the network's point of view, this situation acts to decrease energy demand peaks and reduce problems and costs. The customer agrees to participate in this program on a cash incentive basis.

Some of these DR programs, such as SC and V2G, will be explained in more detail, as will the TOU and RTP models.

### 2.3.3.1 Time-of-use

The TOU pricing will apply the fixed pricing (as the name indicates, it is always constant throughout the day) in different periods. Through calculation models, this approach, for periods of higher energy consumption, will present a higher tariff compared to periods of less use, where the tax is lower. This encourages the EV user to charge his vehicle in periods of lower consumption, which causes the load at peak hours to decrease, thus alleviating congestion problems in the grid. Some work incorporating the EVs with this DR program has been carried out. Shi et al. [28] propose an optimized charging model using TOU pricing. In this paper, the influence of uncontrolled charging on peak load, network voltages, and losses are also analyzed, as well as the EV response to TOU pricing. The results obtained in this paper have made it possible to conclude that demand decreases during peak hours when there is an optimum load. These results also showed that the voltage quality in the network bus bars improved with this charging.

Sharma and Jain [29] discusses a model for providing regulation services to the system operator through real TOU pricing and dynamic charge scheduling of EV aggregator. In this paper, the simulations were performed for 1000 EVs, considering the time interval of one hour for a whole day. It concluded that the revenue of the EV aggregator is more significant when using a dynamic charging algorithm, where TOU schemes are incorporated,

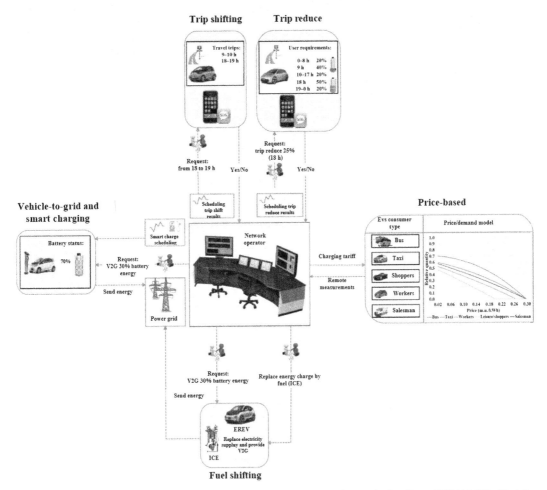

**FIGURE 2.4** Proposed DR programs for EVs. *Source: Proceedings of the First DREAM-GO Workshop Deliverable D7.2.*

versus a static charging algorithm. This is also the case for charging costs, which are lower using the dynamic case.

### 2.3.3.2 Real-time pricing

The RTP strategy reflects the marginal cost of electricity, where the power supplier announces the prices at the beginning of each period of time (e.g., 15 minutes earlier), that is, there is a real-time price change in the electricity market. Considering the literature, some papers involving this strategy and EVs have been published.

Yao et al. [30] discussed an optimal charging schedule in real time for EVs, accommodating DR programs in parking lots. The number of EVs in the charging station is maximized, while minimizing the costs, using dynamic tariffs. The proposed methodology shows good results in terms of cost reduction in the electricity bill when compared to other methods for different values of EVs entering the parking lots. It also shows good results concerning the state of charge of the battery when vehicles leave the car parks.

Canizes et al. [31] proposed an EV user behavior simulator using distributed locational marginal prices (DLMPs), based on the operation and reconfiguration of the grid. The paper also analyzes the impact of dynamic prices to the network and its user, and the effect that these tariffs have on the EV charging, using two different user preference scenarios (price and distance preference). In terms of prices, this paper concludes that dynamic rates are more profitable than fixed rates, especially in the price preference scenario regarding the average price that is charged to EV users.

### 2.3.3.3 Smart charging

As previously mentioned, the uncontrolled charging of EVs (dumb charge) can cause problems for the proper operation of the network. This type of charging has a negative impact on the profit of EV aggregators in their negotiations on the electricity market. It also brings problems for the DSO, as the latter will have to make significant investments to strengthen the network to satisfy the high energy demand that will arise. To mitigate these problems, the EV users need to adhere to a controlled charging strategy, such as SC.

In this charging, the EV and the charging device share connection data and the charging device will then share the data with the charging station operator. The data produced by EVs are collected and processed by their car manufacturers without any guarantee of making this data available to third parties, namely smart charging operators. With the consent of the EV owners, access to this data is an essential prerequisite for implementing SC technology [32]. Then an obstacle arises here that requires the cooperation of the various entities involved in this process (car manufacturers, grid operators, and many others) to design a data access model and communication protocols with cybersecurity and data protection mechanisms [32]. Of course, this procedure must have the full consent of the EV user.

Through smart charging, the charging stations can monitor, manage, and even restrict the charging of vehicles to optimize energy consumption. SC allows users and network operators to schedule charging profiles in a way that benefits the entities involved. The major drawback involving this type of DR model is its high complexity and cost. However, aggregators can convince EV users to switch from uncontrolled charging to SC through economic benefits and convenience of charging, for example, with SC the user could benefit from charging his vehicle outside of peak consumption hours, or discounts on flat rates, to list a few.

### 2.3.3.4 Vehicle-to-grid

The concept of V2G refers to when the EV becomes not only a load but also a source of energy, i.e., it injects energy back into the grid when connected to it. In this case, the EV discharges the energy stored in the battery during peak hours and recharges during off-peak hours, creating a balance in the network.

So V2G is very interesting to the grid in a way because EVs can help with the management of congestion problems in the network (line overloading and voltage problems). Also, EVs have the capability of supplying with regulation services, such as frequency regulation, load shifting, peak shaving, and many others, as shown in Fig. 2.5 [33].

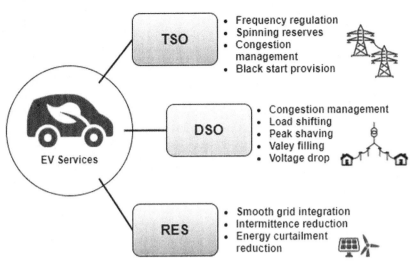

**FIGURE 2.5** EV services provided for different parties in power systems. *Source: N. Banol Arias, S. Hashemi, P.B. Andersen, C. Traeholt, R. Romero, Distribution system services provided by electric vehicles: recent status, challenges, and future prospects, IEEE Trans. Intell. Transp. Syst. 20 (12) (2019) 4277–4296.*

However, EV batteries suffer degradation over time, given the charging and discharging. In V2G mode, their degradation is even worse, due to the constant changes between energy injection into the grid and consumption. Therefore the incentive for EV owners must be higher than in other DR models. This factor makes the implementation of this mode not very feasible in the near future since there have not been yet significant advances to evolve the battery technology necessary for this mode of operation. This mode is seen in the low number of EVs that have V2G functionality.

## 2.4 Electric vehicles in local energy markets

This section discusses the LEM concept and its association with EVs. This section also presents some business models (BMs) integrating EVs that can be applied to the LEM theme.

### 2.4.1 Local market

In the conventional electricity market, the electricity is bought, sold, and traded in the wholesale (WS) and retail markets. In the WS, energy is bought and sold to retailers and final customers, that is, generation companies compete to sell electricity to buyers. In the retail market, there is the purchase and sale of energy to final consumers.

Both markets are suffering from various difficulties due to the massive integration of new sources of electricity, such as RES in DG, which bring considerable uncertainty to the system. This uncertainty increases the burden of day-ahead scheduling, real-time dispatch, and security for the system operator, that is, significant investments will be made to mitigate these situations.

The increase in energy demand, the evolution toward a more customer-oriented market, which is not the case in the two types of the market previously mentioned, and the existence of prosumers (producers and consumers at the same time) in the grid also represent problems for the current energy markets. All these problems can be reduced through joint generation, storage, and DR operation, which can be achieved through the LEMs.

The LM can be considered as a platform where each consumer and prosumer "exchanges" energy, supporting a regional ambit such as a neighborhood. These markets should be established to ensure the high integration of RES in the energy system [34]. This type of market reduces costs as energy can be consumed locally, i.e., there are fewer losses. If there is a surplus of locally generated power, it can be consumed in its vicinity, which leads to a better physical balance locally. The LMs also permit an efficient management of DR programs and allow the development of SGs.

State-of-the-art work has been conducted regarding the LEMs. Lezama et al. [35] analyzed a fully transactive energy system, incorporating LEM. Firstly, they modeled the microgrid resource management considering flexible loads (EVs, energy storage systems (ESSs), DR) and market share. A WS market and a LM were also shaped, as shown in Fig. 2.6, considering bidding options in both domestic and WS markets.

A two-stage stochastic model is used to solve the proposed problem. In the first stage, the WS and LM offers, as well as the dispatchable DG, are considered and in the second-stage, all the uncertainties are considered (EVs, ESSs, DR, etc.). A Monte Carlo simulation was applied to generate several scenarios. Then a scenario reduction technique was applied for each situation to eliminate those with a low probability of occurrence. Finally, this paper analyzes the cost using three different scenarios: no market participation, only WS market participation and the combination of the WS and LM, using flexibility in the load and no flexibility. As expected, the best results are obtained when the two markets come together under flexible loads.

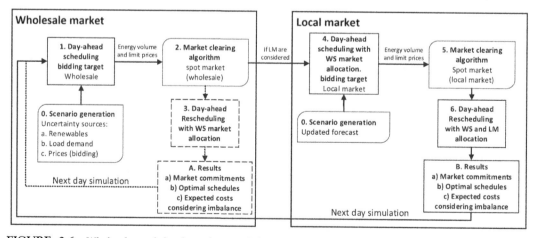

**FIGURE 2.6** Wholesale and local market modulation according to the proposed methodology. *Source: F. Lezama, J. Soares, P. Hernandez-Leal, M. Kaisers, T. Pinto, Z. Vale, Local energy markets: paving the path toward fully transactive energy systems, IEEE Trans. Power Syst. 34 (5) (2019) 4081–4088.*

This paper includes EVs as flexible loads, which proves that their integration into LMs is an added value for cost reduction. That is because the services provided by this type of vehicle, as previously shown, offer excellent reliability, safety, and accessibility of offer in this type of market [36].

## 2.4.2 Business models

LEMs allow for a more active profile of market entities, for example, through peer-to-peer models, using information and communication technologies (ICTs). Lezama et al. [5] proposed several BMs for the flexibility of EVs. These BMs can be integrated into the concept of LEMs because they take into consideration all the aspects mentioned above. This reference presents three BMs, as shown in Fig. 2.7, where the product is EV flexibility. In the first BM, the provider of EV flexibility will be the aggregator, and the customer will be the DSO. The DSO will require this flexibility to avoid any congestion issues. The DSO will have to pay the necessary amount of flexibility, also considering the payment that the aggregator must make to the EV user to get this flexibility. The second BM makes the costumer the EV user and the provider the DSO. The trading of flexibility occurs in a place, through ICTs, where the DSO communicates to the EV user. Finally, the third BM makes the

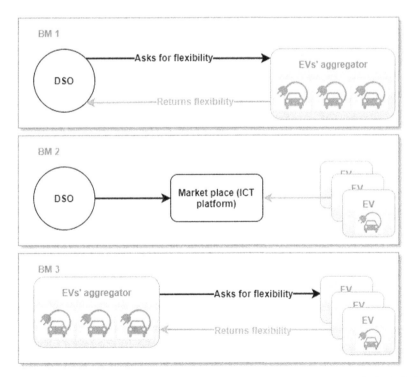

**FIGURE 2.7** Business models using EV flexibility. *Source: F. Lezama, J. Soares, R. Faia, Z. Vale, L.H. Macedo, R. Romero, Business models for flexibility of electric vehicles: evolutionary computation for a successful implementation, in: GECCO 2019 Companion - Proc. 2019 Genet. Evol. Comput. Conf. Companion, 2019, pp. 1873–1878.*

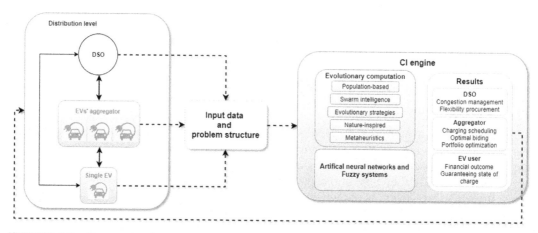

**FIGURE 2.8** Computational intelligence (CI) engine as a decision support system. *Source: F. Lezama, J. Soares, R. Faia, Z. Vale, L.H. Macedo, R. Romero, Business models for flexibility of electric vehicles: evolutionary computation for a successful implementation, in: GECCO 2019 Companion - Proc. 2019 Genet. Evol. Comput. Conf. Companion, 2019, pp. 1873–1878.*

interactions between the EV aggregator and the EV user, where the aggregator will pay the amount of flexibility requested to the EV user, according to the amount sold to the DSO.

The paper also proposes a computational intelligence (CI) engine, which seeks to respond to requests from DSO, EV aggregators, and EV users. Through evolutionary computation (EC), it obtains favorable results in an acceptable time when real problems (nonlinearities, large dimensions, etc.) are being addressed, as shown in Fig. 2.8.

## 2.5 Conclusions

This chapter presented diverse information on multiple concepts such as EVs, DR, and LEM. For each one, several data and developed works were shown that demonstrate the current importance of these themes. Firstly, the concept of EV was introduced, as well as its different types. Then, statistics were presented, which prove the growth of this form of transport from year to year, due to the factors discussed. Also, several modeled DR models for EVs were presented. EVs are a big resource for the DR, due to their high load flexibility, as shown in this chapter. Finally, the concept of LEM was introduced, along with its interaction with EVs through various BMs. It was verified that this is a concept still under study, mainly with the integration of EVs, because the literature is limited on this subject.

## Acknowledgment

This research has received funding from FEDER funds through the Operational Programme for Competitiveness and Internationalization (COMPETE 2020), under Project POCI-01-0145-FEDER-028983; by National Funds through the FCT Portuguese Foundation for Science and Technology, under Projects PTDC/EEI-EEE/28983/2017 (CENERGETIC),CEECIND/02814/2017, and UIDB/000760/2020.

# References

[1] Registration: Status: Text, vol. 21, no. November 2016, 2017, pp. 1−7.

[2] EU, Greenhouse gas emission statistics - emission inventories, Eurostat 63 (3) (2018) 175−180.

[3] Manifesto of Electric Mobility | EVBox. [Online]. Available from: https://info.evbox.com/manifesto-electric-mobility. (accessed 10.02.2020).

[4] N. Daina, A. Sivakumar, J.W. Polak, Electric vehicle charging choices: modelling and implications for smart charging services, Transp. Res. Part C. Emerg. Technol. 81 (2017) 36−56.

[5] F. Lezama, J. Soares, R. Faia, Z. Vale, L.H. Macedo, R. Romero, Business models for flexibility of electric vehicles: evolutionary computation for a successful implementation, in: GECCO 2019 Companion - Proc. 2019 Genet. Evol. Comput. Conf. Companion, 2019, pp. 1873−1878.

[6] A. Schuller, C.M. Flath, S. Gottwalt, Quantifying load flexibility of electric vehicles for renewable energy integration, Appl. Energy 151 (2015) 335−344.

[7] M.E. El-Hawary, The smart grid - state-of-the-art and future trends, Electr. Power Compon. Syst. 42 (3−4) (2014) 239−250.

[8] X. Fang, S. Misra, G. Xue, D. Yang, Smart grid - the new and improved power grid: a survey, IEEE Commun. Surv. Tutor. 14 (4) (2012) 944−980.

[9] Internal Combustion Engine Basics | Department of Energy. [Online]. Available from: https://www.energy.gov/eere/vehicles/articles/internal-combustion-engine-basics. (accessed 10.02.2020).

[10] A. Mahmoudzadeh Andwari, A. Pesiridis, S. Rajoo, R. Martinez-Botas, V. Esfahanian, A review of battery electric vehicle technology and readiness levels, Renew. Sustain. Energy Rev. 78 (2017) 414−430. Available from: https://doi.org/10.1016/j.rser.2017.03.138.

[11] S.E. Lyshevski, C. Yokomoto, Control of hybrid - electric vehicles, in: Proceedings of the American Control Conference, April 4, 1998, pp. 2148−2149. https://doi.org/10.1109/ACC.1998.703007.

[12] C.E. Thomas, Fuel cell and battery electric vehicles compared, Int. J. Hydrog. Energy 34 (15) (2009) 6005−6020. Available from: https://doi.org/10.1016/j.ijhydene.2009.06.003.

[13] International Energy Agency (IEA), Global EV Outlook 2020: entering the decade of electric drive? Global EV Outlook 2020, 2020, p. 273.

[14] T. Bunsen, et al., Global EV Outlook 2019 to electric mobility, OECD, 2019, p. 232. <iea.org>.

[15] EV30@30 campaign | Clean Energy Ministerial | EV30@30 campaign | Advancing Clean Energy Together. <http://www.cleanenergyministerial.org/campaign-clean-energy-ministerial/ev3030-campaign> (accessed 14.08.2020).

[16] A. Hove, D. Sandalow, Electric Vehicle Charging in China and the United States, February 2019.

[17] A. Cooper, K. Schefter, Electric Vehicle Sales Forecast and the Charging Infrastructure Required Through 2030, November 2018, p. 18.

[18] H. Engel, R. Hensley, S. Knupfer, S. Sahdev, The potential impact of electric vehicles on global energy systems, in: McKinsey Cent. Futur. Mobil., no. Exhibit 1, 2018, p. 8.

[19] R. Deng, Z. Yang, M.Y. Chow, J. Chen, A survey on demand response in smart grids: mathematical models and approaches, IEEE Trans. Ind. Inform. 11 (3) (2015) 570−582.

[20] FERC: Industries - Reports on Demand Response & Advanced Metering. [Online]. <https://www.ferc.gov/industries/electric/indus-act/demand-response/dem-res-adv-metering.asp>. (accessed 12.02.2020).

[21] I. Pavic, T. Capuder, I. Kuzle, A comprehensive approach for maximizing flexibility benefits of electric vehicles, IEEE Syst. J. 12 (3) (2018) 2882−2893.

[22] P.J. Ramirez, D. Papadaskalopoulos, G. Strbac, Co-optimization of generation expansion planning and electric vehicles flexibility, IEEE Trans. Smart Grid 7 (3) (2016) 1609−1619.

[23] D. Papadaskalopoulos, G. Strbac, P. Mancarella, M. Aunedi, V. Stanojevic, Decentralized participation of flexible demand in electricity markets - part II: application with electric vehicles and heat pump systems, IEEE Trans. Power Syst. 28 (4) (2013) 3667−3674.

[24] M.K. Gerritsma, T.A. Al Skaif, H.A. Fidder, W.G.J.H.M. van Sark, Flexibility of electric vehicle demand: analysis of measured charging data and simulation for the future, World Electr. Veh. J. 10 (1) (2019) 1−22.

[25] K. Knezović, M. Marinelli, P. Codani, Y. Perez, Distribution grid services and flexibility provision by electric vehicles: a review of options, in: Proc. Univ. Power Eng. Conf., vol. 2015, November 2015.

[26] N. Sadeghianpourhamami, N. Refa, M. Strobbe, C. Develder, Quantitive analysis of electric vehicle flexibility: a data-driven approach, Int. J. Electr. Power Energy Syst. 95 (2018) 451−462.

[27] J. Soares, Z. Vale, N. Borges, Current status and new business models for electric vehicles demand response design in smart grids, pp. 10.

[28] Z. Shi, N. Zhu, J. Yu, The electric vehicle time-of-use price optimization model considering the demand response, in: MATEC Web Conf., vol. 160, 2018, pp. 3−8.

[29] S. Sharma, P. Jain, Integrated TOU price-based demand response and dynamic grid-to-vehicle charge scheduling of electric vehicle aggregator to support grid stability, Int. Trans. Electr. Energy Syst. 30 (1) (2020) 1−6.

[30] L. Yao, W.H. Lim, T.S. Tsai, A real-time charging scheme for demand response in electric vehicle parking station, IEEE Trans. Smart Grid 8 (1) (2017) 52−62.

[31] B. Canizes, J. Soares, Z. Vale, J.M. Corchado, Optimal distribution grid operation using DLMP-based pricing for electric vehicle charging infrastructure in a smart city, Energies 12 (4) (2019).

[32] Access to Electric Vehicles data for smart charging - Florence School of Regulation. <https://fsr.eui.eu/access-to-electric-vehicles-data-for-smart-charging/>. (accessed 10.08.2020).

[33] N. Banol Arias, S. Hashemi, P.B. Andersen, C. Traeholt, R. Romero, Distribution system services provided by electric vehicles: recent status, challenges, and future prospects, IEEE Trans. Intell. Transp. Syst. 20 (12) (2019) 4277−4296.

[34] F. Hvelplund, Renewable energy and the need for local energy markets, Energy 31 (13) (2006) 2293−2302.

[35] F. Lezama, J. Soares, P. Hernandez-Leal, M. Kaisers, T. Pinto, Z. Vale, Local energy markets: paving the path toward fully transactive energy systems, IEEE Trans. Power Syst. 34 (5) (2019) 4081−4088.

[36] F. Teotia, R. Bhakar, Local energy markets: concept, design and operation, in: 2016 Natl. Power Syst. Conf. NPSC 2016, 2017, pp. 1−6.

# From wholesale energy markets to local flexibility markets: structure, models and operation

*Fernando Lopes*

LNEG – National Laboratory of Energy and Geology, Lisbon, Portugal

## 3.1 Introduction

Energy markets (EMs) are built on well-established principles of competition and transparency, representing a good way to guarantee affordable energy prices and secure energy supplies. To a large extent, European markets have adopted a common framework based on general design principles and common rules for the internal market of electricity [1,2]. These rules have subsequently been complemented by legislation concerning electricity trade and grid operation aspects, as well as measures against market abuses [3,4]. These include rules on the separation of energy supply and generation from the operation of transmission networks (unbundling), the independence of national energy regulators, and retail markets. They also guarantee that energy consumers enjoy high standards of consumer protection.

The "common design framework" involves a day-ahead market (DAM) and an intraday-market (IDM), operating together with a bilateral market, and complemented with balancing markets (see, e.g., [5,6]). In short, market participants submit hourly bids to the DAM until a particular hour of day $d$ before the day of operation $(d + 1)$. Market-clearing prices and equilibrium quantities are calculated using EUPHEMIA (acronym for European Union Pan-European Hybrid Electricity Market Integration Algorithm) [7]. The intraday market may involve several sessions similar to the DAM daily auction, and can be cleared several times once the DAM has been cleared.[1] Also, the IDM may operate continuously, 7 days a week, all year around. The bilateral market provides a hedge against

---

[1] In terms of structure, North American markets are similar to European power markets, although they typically include a day-ahead market and a short-term market, referred to as real-time market (RTM), to set prices and schedules for 5-minute intervals [8].

the price volatility of centralized trading. Finally, balancing markets are in place to set prices and schedules to match the imbalances caused by the variability and uncertainty of power systems.

This "common design framework" was set out, however, when the vast majority of generation units were controllable and fuel-based, meaning that production could be shifted in time with limited economic impact. This is no longer true, since a significant part of the traded power comes from renewable energy sources. The unique characteristics of renewable generation—more variable, less predictable, and more decentralized than traditional generation—influence the performance and outcomes of power markets [9]. In particular, large penetrations of renewables reduce market-clearing prices due to their low bid costs. Also, high levels of renewable generation increase market price volatility because of their increased variability.

Noticeably, the increasing levels of renewable generation create unique challenges in the design and operation of power markets [9]. Chief among these is the need to incentivize increasing levels of flexibility in a cost-effective way to manage the rising variability and uncertainty of the net load (i.e., load minus renewable generation) [10]. Also important is the need to ensure revenue sufficiency for achieving long-term reliability [11]. Thus there is a growing need to adapt current market rules to the new market realities.

In 2019, the EU published new rules for the internal market for electricity [12,13]. For wholesale markets, the new rules focus on short-term markets to improve competition and liquidity, dispatch rules adapted to the new market reality, scarcity pricing without price caps, and a remuneration for evolving technologies, such as demand response (DR) and energy storage, more in line with the flexibility provided by such services. For retail markets, the new rules empower consumers and communities to actively participate in the market, by generating electricity, consuming it or selling it back to the market, and mainly by offering DR, energy storage, and specific flexibility products, receiving remuneration directly or through aggregators.

Several improvements and extensions to the "common design framework" have also been proposed during the past few years, such as extended marginal pricing based on the convex-hull approach, explicit products for flexible ramping provision [14], and pay-for-performance regulation (but see [9] for an in-depth discussion). At present, however, it is unclear whether or not such improvements are adequate to incentivize the levels of flexibility required by the rising penetrations of renewables.

In this context, flexibility markets are starting to be recognized as a promising and powerful tool to adequately valorize demand-side flexibility (see, e.g., [15]). There are some ongoing pioneer projects related to flexibility marketplaces with small-scale research demonstrators, such as OSMOSE [16] and WindNODE [17]. Also, a few initiatives are in a piloting phase, notably NODES [18], Enera [19], Piclo Flex [20], and GOPACS [21].

The remainder of the chapter is structured as follows. Section 3.2 describes the centralized and bilateral models underlying most European markets. Section 3.3 analyzes the operation of four central European markets, namely Nord Pool (the Nordic and Baltic market), EPEX Spot (the market for Central Western Europe), MIBEL (the Iberian market), and GME (the Italian market). Following this material, Section 3.4 describes some energy management tools to simulate the operation of power markets. Section 3.5 analyzes the pressing issue of flexibility in system operation and Section 3.6 describes four key flexibility European markets, specifically NODES, Enera, Piclo Fex, and IREMEL. Finally, Section 3.7 presents some concluding remarks.

## 3.2 Wholesale markets: models and operation

European power markets evolved in similar directions, following standard design principles. As noted, the "common design framework" involves four different types of markets: day-ahead and intraday markets, operating together with bilateral markets, and complemented with balancing markets.[2] Fig. 3.1 depicts schematically the various markets.

### 3.2.1 Day-ahead market

This market is the central market for trading energy in advance of the time when the energy is produced and consumed, and is executed every day. For a particular day of operation $d$, the DAM clears typically at 12 noon of day $(d - 1)$.[3] The pricing method is founded on the marginal pricing theory, where the price equals the short-run marginal cost [23]. Pricing plays an important role in the DAM, as it sends monetary signals to all market entities, signals that influence their participation in the market.

Under system marginal pricing (SMP), generators compete to supply demand by submitting bids to the market, which typically involve a price and an energy quantity for every hour of the day of operation. In a more general form, bids to sell energy may include startup costs, as well as operating constraints and the availability of generators. A

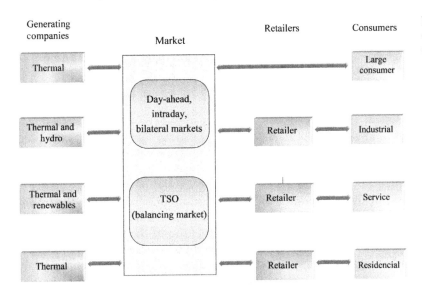

FIGURE 3.1 Typical energy markets and key market players.

---

[2] Notation is somewhat abused here, since day-ahead, intraday, bilateral, and balancing markets are indeed submarkets of power markets [22]. However, submarkets are themselves markets, and thus they will be referred to simply as markets throughout the chapter.

[3] Trading in the day-ahead market is typically done on an hourly basis, although time intervals could be reduced to 15 minutes in the coming years [12].

market operator collects the bids and sorts them according to the price, leading to a supply curve. Buyers submit load purchase offers for every hour of the day under consideration. These offers are also collected by a market operator, who ranks them in order of decreasing price, building a demand curve. The market-clearing price is defined by the intersection of the supply curve with the demand curve. This price is applied to all generators uniformly, regardless of their bids or location. The equilibrium quantity is determined as the sum of all offers that are satisfied at the market-clearing price [6,22].

Under locational marginal pricing (LMP), a more complex variation of SMP, the market operator runs an optimal power flow (OPF) procedure that defines the energy prices at different locations in the system. Thus, LMP involves the pricing of electrical energy according to the location of its injection or withdrawal from the transmission grid. As in SMP, generators submit bids to sell energy and buyers load purchase offers. The main difference now is an optimization process that is subject to various system constraints, such as voltage limits. This optimization may be based on a full alternating current OPF, although a linearized OPF is sometimes considered, to simplify the numerical complexity. Simplified OPF models can result, however, in inefficiencies and cross-subsidies among competing agents.

In the past, European power exchanges (PXs) used different market algorithms, such as COSMOS, SESAM, SIOM, and UPPO, which focused on local features and products, and were typically not able to cover global requirements. This leads to the Price Coupling of Regions project, an initiative of eight European PXs, who developed a single price coupling algorithm, commonly known as EUPHEMIA. Since 2014, this algorithm has been progressively used to calculate energy prices and quantities in different regions across Europe, maximizing the global welfare and increasing the transparency of the computation of prices and flows [7].

### 3.2.2 Intraday market

This market sets prices and schedules some periods ahead to facilitate balancing on advance of real time. Specifically, the intraday market is mainly used for: (1) adjusting energy quantities based on the results of the day-ahead market, (2) managing forecast errors or unforeseen events, (3) adjusting from hourly positions to 30 minutes or even 15 minutes, and (4) offering flexible generation as a substitution for renewables. The traditional trading procedure is based on one or more daily auctions. Recently, a continuous trading procedure has also been adopted in various markets around the world.

In the case of daily auctions, the generic trading procedure is essentially identical to that of the day-ahead market and details are omitted. For day $d$, the intraday market may be cleared one or more times once the DAM has been cleared. Auctions may differ according to the time period under consideration. Specifically, trading may be based on a 60-minute auction, 30-minute auction, or 15-minute auction, thus providing different rebalancing possibilities for market participants. Each auction aims at optimizing the global welfare. The price determined by the market is the price at which all trades will be executed.

Continuous intraday markets run 7 days a week, all year around. Market participants submit bids to sell energy and load purchase offers to the order book of a trading platform (such bids and offers are often referred to as orders). A trading session is a time period during which orders are matched automatically. Unexecuted orders remain in the order book until their expiry or are canceled. The order matching rules ensure that orders are executed at the best price available in the system. Typical contracts for the next day include hourly contracts, 30-minute contracts, and 5-minute contracts. The lead time—that is, the time between the end of the trading session and the start of the delivery period—may range from 60 to 5 minutes.

Generally speaking, the continuous trading matching algorithm, commonly known as SIDC (single intraday coupling) algorithm, involves two key modules: the shared order book module and the capacity management module. The former contains the basic functionality for continuous trading, like order entry, order management, and order matching. The latter provides the functionality for managing and allocating available transmission capacity between all areas in the underlying network [24].

Now, the performance of centralized trading based on auctions for both DAM and IDM is often affected by several important factors, notably price volatility [5]. Also, centralized trading involves basically ex post energy prices, known only after the definition of the dispatch.[4] Accordingly, to hedge against price uncertainty, generators and loads may choose to enter into bilateral trading.

## 3.2.3 Bilateral markets

A bilateral market is a market in which private parties, sellers and buyers, negotiate bilateral agreements for the exchange of electricity under mutually acceptable terms. Such a market provides both standardized financial and physical contracts that span from days to years, notably forwards, futures, options, and swaps [6,22]. Each contract has its own price that depends only on the arrangements between the interested parties. This makes bilateral agreements falling under the category of the pay-as-bid pricing scheme, since the parties are paid according to their offers, rather than on the system marginal price or the marginal nodal price. Also, bilateral trades involve basically ex ante prices, known at the end of the negotiation process.

Market participants enter into bilateral contracts by submitting orders directly to the order book of a trading platform. For each order, several elements should be specified, including the nature (buy or sell), the type of contract, the price, the energy quantity (expressed in a full number of contracts), and the validity period. Trading may take two forms: continuous or auction. In continuous trading, buy and sell orders for each contract likely to interfere with each other generate transactions. This means that the trading platform checks continuously for matching buy and sell orders. When a match is found, the orders for the corresponding contract generate a transaction—that is, a trade executed on the contract—and give rise to a particular market price. Thus, in a given trading session, a number of market prices can be generated.

Auction trading involves a predefined call period during which orders can be introduced in the order book. This period may be of open or closed type. The former involves the display of a provisional equilibrium price corresponding to the situation that results

---

[4] The dispatch consist essentially in a set of market instructions, especially with respect to defining the generators that provide power at any point in time and their output levels [5].

from the submitted orders at any moment. The latter involves no display of information. The call period is followed by the determination of an equilibrium price based on the aggregated values of the buy and sell orders. This means that auction trading involves the generation of a market price only. The orders at a price equal or higher than the equilibrium price generate transactions up to a maximum executable volume. This volume corresponds normally to the total of buy orders at a price equal to or higher than the equilibrium price (or, if lower, to the total of sell orders at a price equal or lower than the equilibrium price). Typically, orders are executed according to specific criteria (e.g., a buy order at the highest price and a sell order at the lowest price benefit from priority for execution).

As noted earlier, a generally accepted strategy for market players to hedge against the price uncertainty of centralized trading is to engage in bilateral contracts that offer more stable rates over time. The drawback, however, is that a fixed bilateral price could be lower than the DAM price, a situation that is disadvantageous to sellers and advantageous to buyers. Alternatively, if the bilateral price is higher than the DAM price, an opposite situation will take place. Thus there is normally a usual linkage between central and bilateral markets [5,22]. Although these two markets are quite different, the centralized market price is typically the leading price indicator for bilateral trades.

### 3.2.4 Balancing markets

In addition to all the aforementioned markets—that is, day-ahead, intraday, and bilateral markets—balancing markets represent the major tool for correcting the imbalances relative to the physical trade of energy, in order to maintain equality between production and consumption, and also to ensure power grid stability. Balancing markets are operated by transmission system operators (TSOs), who are usually neutral and noncommercial organizations. They make use of balancing products—such as primary reserve, secondary reserve or automatic frequency restoration reserve (aFRR), and tertiary reserve or manual frequency restoration reserve (mFRR)—to ensure the maintenance of the system frequency around a nominal value (50 Hz) [25].

Market entities include balancing service providers (BSPs), who are able to provide balancing services to TSOs, as well as balancing responsible parties (BRPs), who are financially responsible for their imbalances. BSPs should qualify for providing bids. BRPs handle balance responsibility to TSOs for production plants, consumption (including grid losses), and physical electricity trading. They should strive to be balanced in real time or help the power system to be balanced [26].

The prices for energy and (some) balancing products are calculated in a similar manner, taking into account the marginal pricing concept (typically, pay-as-cleared). TSOs are responsible for procuring balancing services and thus they constitute the demand side. BRPs may update their schedules until the balancing energy gate closure time, which should be as close as possible to real time. BSPs may submit and update balancing power bids from balancing products until the gate closure time of the procurement process. Thus BSPs with available generation capacity (e.g., producers) may submit upregulation bids to the TSO. Likewise, BSPs able to reduce consumption may submit downregulation bids to the TSO. In the case where the TSO is procuring upregulation, the upregulation orders with lowest prices are activated until the procured quantity is reached. The price of the last upregulated MW sets the upregulation price. A similar procedure is used to find the downregulation price. There is typically a linkage between the balancing market and the day-ahead market (and the IDM as well). Pricing for balancing energy takes normally into account the pricing method in the day-ahead and intraday timeframes.

## 3.3 Key European markets

This section describes four key European markets, namely Nord Pool, EPEX Spot, MIBEL, and GME.

### 3.3.1 The Nordic and Baltic power market (Nord Pool)

This market operates a day-ahead market (Elspot) and an intraday market (Elbas). Elspot accepts several different types of orders according to the trading region, including single hourly orders, block orders, flexible orders, and exclusive groups. Market participants can consider any type of orders or a combination of different types to meet their interests. The Nordic and Baltic areas are divided into bidding areas to handle congestions in the electricity grid. The market closes at 12:00 CET and producers are paid according to their area price. Similarly, all buyers pay the same area price [27,28].

Elbas supplements the day-ahead market and offers 15 minute, 30 minute, hourly, and block products, providing significant flexibility to meet the needs of different market areas. Order types include limit orders, user-defined block orders, predefined orders, and iceberg orders. Trading is essentially continuous, although one or more daily auctions may be conducted by Nord Pool, from time to time. In continuous trading, transactions are matched automatically when concurring orders are registered in the trading platform. Prices are set based on a first-come, first-served principle, where best prices come first—that is, the highest buy price and the lowest sell price. Orders with the same price limit are prioritized by their time stamp oldest first. Trading takes place every day around the clock, 365 days a year, until 1 hour before delivery [27,29].[5]

Elspot and Elbas are complemented by a financial market and local balancing markets. The former is operated by NASDAQ OMX and manages the risks inherent to the two central markets (i.e., Elspot and Elbas). Contracts can be made for up to 6 years and typically consider the Elspot system price as the reference price. The latter are operated by the respective TSOs (e.g., Energinet [30], Svenska kraftnät [31], and Statnett [32]). They are mainly used for making final adjustments and ensuring the correct frequency in the grid and the security of supply. To this end, TSOs buy/"sell" power from/to market participants in the delivery hour on the basis of bids for upward and downward regulation. Specifically, players submit bids for specific prices (€/MWh) and volumes (MW) over specified periods of time. Bids refer to the next day of operation (d) and may or may not

---

[5] Strictly speaking, trading may take place until either one hour before delivery, or 30 minutes or even 5 minutes before delivery, depending on the trading region [26].

cover the entire day. The prices and volumes should be specified—hour by hour—separately for upward and downward regulation. Bids are submitted during day $d-1$ (e.g., until 17 hours) and the entered prices and volumes can be adjusted up to 45 minutes prior to the upcoming delivery hour [30−35].

The market price is determined according to the marginal price principle and calculated on an hourly basis. It is set at the price of the most recently activated bid provided that bottlenecks or other problems do not hinder the free exchange of power between different market areas. Players must be able to fully activate a given bid in a specific period of time from receipt of the activation order (typically, in a maximum of 15 minutes) [30−35].

### 3.3.2 The market for Central Western Europe (EPEX SPOT)

This market operates organized day-ahead and intraday markets for 12 European countries of the Multi-Regional Coupling area (apart from Switzerland).[6] The DAM considers two daily auctions: an hourly auction and an half-hourly auction in Great Britain (GB). The order book for the hourly auction opens 45 days in advance and closes 1 day before delivery at 12:00 CET (11:00 for Switzerland). Tradable products include 24 hourly contracts corresponding to the 24 delivery hours of the following day ($d$). Prices must be included between a minimum and a maximum value for each market area, typically $-500$ and $3000$ €/MWh, respectively. Following the hourly auction, the half-hourly auction in GB provides market participants with the opportunity to balance physical portfolios to the half-hour delivery. Tradable products include 30-minute contracts with delivery on the following day. The order book closes at 15:30 (GMT) of day $d-1$ and market results are published from 15:45 onwards [36−38].

The intraday market offers both continuous and auction trading in 12 European countries. Continuous trading takes place 24 hours a day, all year around. Tradable products for the next day include 1-hour contracts, 30-minute contracts, and 15-minute contracts. The order book opens in day $d-1$, at 00:00 (GMT) in GB, 14:00 (CET) in the Nordics, and 15:00 (CET) in central Europe. Trades are executed as soon as two orders entered into the M7 platform match automatically. The lead time is 30 minutes for France, 15 minutes for GB (for 30-minute contracts), and 5 minutes for Austria, Belgium, Germany, the Netherlands, and Switzerland. The IDM considers various daily auctions, namely a 15-minute auction in Germany, two 30-minute auctions in GB (one at 17:30 of $d-1$ and another at 8:00 of $d$), and two 60-minute auctions in Switzerland (at 16:30 of $d-1$ and 11:15 of $d$, respectively). The 15-minute auction was introduced in 2014 and counts for 13% of the intraday traded volumes. The two GB auctions are coupled with Ireland and the traded volume represent 20% of the total intraday volume in GB. The Swiss auctions are the latest addition (April 2019) and were created to provide new rebalancing opportunities for market participants. They are coupled with Italy and allow players to benefit from the pooling of liquidity built from the existing Italian M12 and M16 auctions [36−38].

---

[6] Strictly speaking, EPEX SPOT operates a day-ahead market, an intraday market, a local flexibility market, and an organized market for French capacity guarantees. Also, EPEX SPOT offers physical fulfilment services in partnership with EEX [36]. Section 3.6 presents a detailed description of the flexibility market.

The DAM and the IDM are complemented by a power derivatives market (EEX) and different balancing markets operated separately by the system operators of the respective countries (e.g., RTE [39], 50 Hertz [40], and Tennet [41]). For instance, RTE (Réseau de Transport d'Électricité), the French TSO, maintains a balancing mechanism to ensure the security of the power system at all times. Balance responsible entities may submit, modify, or withdraw offers for day $d$ at one of the gate closures: 4:00 p.m., 10:00 p.m., and 11:00 p.m. of day $d-1$. Offers involve essentially the following: a balancing direction (upward/downward), a time-period, a price that may vary according to six time slots (e.g., 12:00–6:00 a.m. or 2:00–5:00 p.m.), and usage conditions. RTE selects the offers that correspond to the real-time balancing requirements, based on economic precedence and conditions for use. Offers are remunerated by considering the pay-as-bid pricing scheme. For each half-hour period, the balancing mechanism gives a reference price applicable for settling the imbalances, based on the average weighted prices of upward and downward balancing offers or the EPEX SPOT price [39,42].

### 3.3.3 The Iberian market (MIBEL)

The day-ahead market sets scheduling quantities and energy prices at 12 noon of day $d-1$, for the 24-hours of day $d$. Market players may trade energy regardless of whether they are in Portugal or Spain. Sale bids may be simple or incorporate complex conditions in terms of their content, including indivisibility, load gradients, minimum income, and scheduled stop. Purchase bids are typically simple bids, indicating a price and an amount of power. Sale and purchase bids are accepted according to their merit order, until the interconnection between Portugal and Spain is fully occupied. In case the capacity of the interconnection permits the flow of the energy traded in a certain hour of day $d$, the price for that hour will be the same for both countries. Otherwise, if the interconnection is fully occupied, the price-setting algorithm (EUPHEMIA) will be run separately for the two countries and there will be a different price of electricity for Portugal and Spain. The underlying mechanism is referred to as market splitting and is commonly used in Europe [43,44].

The intraday market runs after the daily market and involves six trading sessions based on auctions, similar to the DAM auction, where hourly prices and energy quantities are determined by the points where the supply and demand meet. The first session opens at 2:00 p.m. and closes at 3:00 p.m. of day $d-1$, involving a schedule horizon of 24 hours. The last session opens at 9:00 a.m. of day $d$ and closes at 9:50 a.m. of the same day, involving a schedule horizon of 12 hours. Thus this market permits to readjust previous commitments (for purchasing and selling energy) up to 4 hours to real time. Sale and purchase bids may be simple or incorporate complex conditions, according to their content. This means that the market operator may run a conditioned matching method considering specific conditions (e.g., indivisibility and load gradient) [43,44].

The intraday market also operates continuously. Market participants have the possibility to better manage their imbalances by gaining access to market liquidity at the national level and also benefiting from the liquidity available in markets in other areas of Europe. The purchase and sale bids introduced by players in one country may be matched by orders submitted in a similar manner by players in other European countries. The opening of the negotiation for all contracts for the next day $d$, in the price areas of Portugal and Spain, is made after the end of the first auction of day $d-1$, provided the system operator has published the definitive daily-ahead schedule for day $d$. Matched orders give rise to firm transactions (i.e., sale orders imply a delivery obligation and purchase orders a purchase obligation) [43,45].

The day-ahead and intraday markets are complemented by a financial market operated by OMIP (The Portuguese Electricity Market Operator) [46]. Also, the DAM and IDM are complemented by balancing markets in Portugal and Spain, operated by REN [47] and REE [48], respectively. For instance, in Portugal, the primary regulation ancillary services are mandatory and the system operator notifies market players every year about the minimum load variation percentage together with the maximum response speed in the event of different frequency deviations. To assign secondary regulation energy band ancillary services, the system operator publishes the hourly requirements for the next day together with the provisional daily viable schedule. This is followed by a period of bid reception and price definition. The submission of offers opens at 7:00 p.m. and closes and 7:45 p.m. of day $d-1$. For the assignation of the tertiary regulation ancillary services, market agents should send their hourly bids for maximum production level changes until 8:00 p.m. of day $d-1$. The offers may be adjusted up to 20 minutes after the publication of the final hourly program (PHF) of the different intraday sessions [49].

### 3.3.4 The Italian electricity market (GME)

The day-ahead Italien market (MGP) hosts most of the electricity trades and the intraday market (MI) allows market participants to modify the schedules defined in the MGP by submitting additional bids in different sessions. More specifically, MGP trades hourly energy blocks for the next day $d$. This market opens at 8:00 a.m. of the ninth day before the day of delivery and closes at 12:00 p.m. of the day before the day of delivery ($d-1$). The intraday market takes place in seven sessions (MI1 to MI7). The sessions are organized in the form of implicit auctions of electricity, in sequence, and with different closing times. For instance, MI1 opens at 12.55 p.m. of day $d-1$ and closes at 3:00 p.m. of the same day. And MI7 opens at 5:30 p.m. of $d-1$ and closes at 3:45 p.m. of day $d$ [50,51].

MGP and MI operate together with a daily products market (MPEG), where players can trade both base and peak energy products with the obligation of delivery. This market operates continuously on weekdays. The sessions take place from 8:00 a.m. to 5:00 p.m. of day $d-2$, and also from 8:00 a.m. to 9.00 a.m. of day $d-1$. The products to be delivered on Saturday, Sunday, and Monday are traded on the corresponding Friday [50].

MGP, MI, and MPEG are complemented by a forward electricity market (MTE) and an ancillary services market (MSD). The latter is operated by the Italian system operator (Terna [52]) and involves a scheduling substage (ex ante MSD) and a balancing market (MB). Both markets include multiple sessions. For instance, MB takes place in six sessions: MB1 to MB6. The first session takes into consideration the valid bids and offers that players submitted in the previous ex ante MSD session. For the other five sessions, the submission process opens at 10.30 p.m. of day $d-1$, and closes 1.5 hours before the first hour which may be negotiated in each session. The MB is mainly used to provide the service of secondary control and to balance energy injections and withdrawals into/from the grid in real time [50,51].

## 3.4 Agent-based tools for energy markets

Traditional market models include optimization and equilibrium models (see [53] for a review). Most optimization models focus on a profit maximization problem in which a centralized decision-maker pursues the maximum profit. These models typically consider a single objective function subject to a set of technical and economic requirements (see, e.g., [54]). Equilibrium models represent the global market behavior and take into consideration competition among market participants. They are often formulated as a simultaneous profit maximization program that considers the different players competing in the market. Also, they are based on the Cournot competition concept or the supply function equilibrium approach (see, e.g., [55,56]).

Both optimization and equilibrium models continue to provide very useful insights into the operation of power markets, but present limitations in their ability to adequately analyze existing market forces and the dynamics that characterize actual markets. Indeed, traditional models were developed under the implicit assumption of a centralized decision-making process and are often considered a poor fit to liberalized EMs, where operation decisions are decentralized and strategically taken by different market operators. Also, equilibrium models consider algebraic and/or differential equations that impose limitations on the representation of competition and are typically very hard to solve. And the fact that power systems are based on the operation of generation units with complex constraints only contributes to complicate the situation [53].

The distributed nature and complexity of power markets calls, therefore, for richer and more flexible modeling techniques [57]. Multiagent systems are essentially systems composed of multiple computing elements, known as software agents, that interact to solve problems that are beyond the individual capabilities or knowledge of each agent. Such elements are also typically able to perform flexible autonomous actions and capable of managing cooperative and competitive interactions with other agents, by making use of methods and techniques from artificial intelligence [58,59].

The multiagent approach presents itself as a promising approach to accurately model and study in detail the behavior of power markets over time. Accordingly, it has attracted considerable attention over the last years and a number of energy management tools have emerged, including SEPIA [60], EMCAS [61], NEMSIM [62], AMES [63], PowerACE [64], MASCEM [65], GAPEX [66], and MATREM [67,68]. Also, the work presented in [69] considers software agents to put together real-time simulation and laboratory emulation of resources, focusing deeply on modeling electric power system components for realistic simulation of microgrids and local EMs.

MATREM (an acronym for Multi-Agent TRading in Electricity Markets) is an agent-based simulation tool developed at LNEG (by the author and his group) to help manage both the complexity of wholesale markets and the unique challenges of bilateral contracting in retail markets. The tool operationalizes a power exchange (comprising a day-ahead market and an IDM), a derivatives exchange (comprising a futures market), a marketplace for negotiating tailored (or customized) long-term bilateral contracts, and a balancing market. Also, MATREM supports various types of market entities, notably generating companies (e.g., wind, solar, hydro, and thermal producers), retailers, large and small consumers, aggregators of consumers, virtual power producers (e.g., aggregators of wind producers), market operators, and system operators. These entities are modeled as software agents able to interact with other agents to meet their design objectives.

Two main types of software agents are being considered: market agents and assistant agents. Market agents represent the entities that take part in the different simulated markets. Assistant agents are further categorized into interface managers and intelligent assistants. The former are responsible for managing the interfaces of the various markets. The latter provide support to the user in making strategic decisions. The agents are being developed using the JAVA programming language and the JADE framework [70]. A classification of MATREM according to various dimensions associated with both electricity markets and intelligent agents can be found in [67]. Also, a detailed description of the various markets and entities supported by the tool is presented in [68]. Fig. 3.2 presents a snapshot of the tool and the remainder of this section gives an overview of the simulated markets.

### 3.4.1 MATREM: overview of the simulated markets

MATREM is an agent-based simulation tool for analyzing the behavior and outcomes of power markets, including markets with increasing levels of renewable generation. The DAM is a central market where generation and demand can be traded on an hourly basis. To this end, a market operator agent collects all bids for a given hour $h$ and sorts them according to the price. In a next step, aggregated supply and demand curves are determined. Supply and

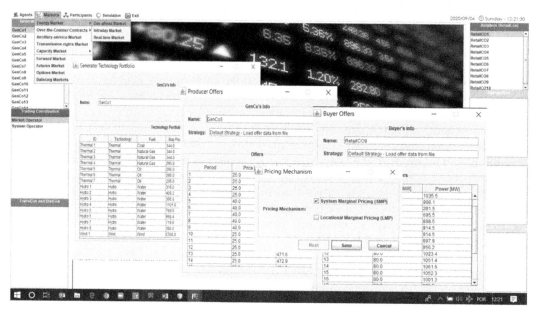

**FIGURE 3.2**  Snapshot of the agent-based system MATREM (for Multi-Agent TRading in Electricity Markets).

demand are then matched by adding up all volumes. The market-clearing price is determined by the last unit necessary to satisfy the demand (see, e.g., [71,72]).

The intraday market is a short-term market and involves several auction sessions. It is used to make adjustments in the positions of participants as delivery time approaches (such as managing forecast errors related to renewable generation and unforeseen events). Both DAM and IDM are based on the marginal pricing theory. Two pricing mechanisms are supported: SMP and LMP.

The futures market is an organized market for both financial and physical products conditioned on delivery at a specific time and place. Such products may span from days to years and typically hedge against the financial risk (i.e., price volatility) inherent to day-ahead and intraday markets. Players enter orders involving either bids to sell or buy energy in an electronic trading platform that supports anonymous operation. The platform automatically and continuously matches the bids likely to interfere with each other.

The balancing market is a market for primary reserve (or frequency control reserve), secondary reserve (or fast active disturbance reserve), and tertiary reserve (or slow active disturbance reserve), for the provision of system services. For the particular case of tertiary reserve, a system operator agent defines the needs of the power system, collects the bids from the market participants, and determines the market prices by considering a simplified version of the SMP algorithm (see, e.g., [73]).

Especially noteworthy is the possibility to negotiate the details of two different types of tailored (or customized) long-term bilateral contracts, namely forward contracts (see, e.g., [74]) and contracts for difference (see, e.g., [75]). Such contracts cover the delivery of large amounts of energy over long periods of time (several months to years). Their terms and conditions are very flexible and can be negotiated privately to meet the objectives of two parties. To this end, market agents are equipped with an interaction model that handles two-party and multiissue negotiation (see, e.g., [76–78]). The negotiation process involves an iterative exchange of proposals and counterproposals. A proposal (or offer) is a set of issue-value pairs, such as "energy price = 50 €/MWh," "contract duration = 12 months," and so on. A counterproposal is a proposal made in response to a previous proposal. Negotiation strategies and tactics are functions that define new values for each issue at stake throughout negotiation. The final result may be either agreement or no agreement. This design feature of the agent-based simulation tool represents, we believe, a new and powerful feature of MATREM.

Overall, the current version of MATREM allows the conducting of different studies on market design and operation (see [79] for a study to investigate the merit order effect of the deployment of wind power in Portugal, and [80] for another study to analyze the impact of DR on the Iberian electricity market price). Also, MATREM was recently updated with new market products [81]. Furthermore, the author (and his group) is conducting a study at LNEG to investigate the use of local flexibility markets and to extend the system with a new local flexibility market.

## 3.5 The energy transition and flexibility markets

The energy sector is experiencing a rapid and profound change largely driven by the rising levels of renewable generation, the increasing deployment of decentralized resources, and the so-called Internet of Things (IoT):

- Renewable generation has grown significantly during the past decade, surpassing all expectations, and this growth is expected to continue during the coming years (see, e.g., [9,79]).
- Traditional (fossil-fueled) resources connected to the transmission grid are increasingly being phased-out and, at the same time, nontraditional resources connected to the distribution grid, such as wind and solar power plants, are increasingly becoming part of the supply mix (see, e.g., [9,79]).
- Microgrids—that is, groups of interconnected loads and decentralized resources within clearly defined electrical boundaries that act as single controllable entities with respect to the grid—are being increasingly implemented in power systems (see, e.g., [82]).
- End users are increasingly transforming from passive consumers into prosumers, who want to actively participate in the EM, either individually or through aggregation services (see, e.g., [83]).
- Energy conversion technologies, such as power-to-gas units and fuel cells, are starting to become market-ready; also, the market introduction of electric vehicles (EVs) is starting to become economically viable (see, e.g., [82,83]).
- The society is generally becoming more digitalized, giving rise to an Internet of Things, and thus making possible the exchange of information between any device and entity in the power system (see, e.g., [83–85]).

Put simply, three megatrends, commonly referred to as the "three-Ds"—decarbonization, decentralization, and digitalization—are shaping the energy landscape.

Clearly, the rapid expansion of renewables is challenging power systems in terms of flexibility for short-term operation [10]. Renewable generation adds variability and uncertainty to power systems at multiple timescales. And besides variability and uncertainty, renewable generation is nonsynchronous to the electrical frequency and location constrained, meaning that it may be located far from load centers.

The rise of renewables is also being accompanied by a growing demand for electrical energy, mainly driven by the electrification of large parts of the residential sector, as well as the electrification of the transport sector (notably, the trend toward EVs). The electrification of these sectors can significantly influence the peak loads on the system. In other words, the combined increase in electrification and intermittent generation makes the need for flexibility in power systems more of a pressing issue.

Furthermore, the increasing deployment of decentralized resources able to offer demand-side flexibility is thought provoking due to their incorporation in a cost-effective manner. Most electric devices and processes offer excellent opportunities to provide flexibility to the system. For instance, the storage provided through EVs can help to support renewable production, by storing the excess of renewable energy to prevent curtailment and discharging it to the grid when additional supply is required. Also, space heating is a relatively slow process in which the thermal buffering of buildings and storage vessels can be used to enable the shifting of energy load over the day. And many other examples could be presented.[7] However, cost-effective ways must be found to manage and harness

---

[7] An interesting piece of work related to demand response and distributed generation is presented in [86], pointing out the complementary of both approaches.

the flexibility potential offered by decentralized resources. The central question is how to ensure that decentralized resources are used when and where they are most needed, in a way that also meets the grid management needs of system operators [87].

Generally speaking, flexibility is the ability of a resource, whether any component or collection of components of the power system, to respond to the known and unknown changes of power system conditions at various operational timescales [10]. It is an ability that is sold in the context of a specific product, rather than as a separate commodity. For instance, flexibility considered in intraday markets takes the form of energy blocks, and flexibility considered in balancing markets transforms into regulating power [15].

Flexibility has been a central issue in EMs since their inception. Accordingly, several traditional mechanisms to incentivize flexibility are in place in most existing markets (which operate mainly according to the "common design framework" described earlier in this chapter). These include efficient centralized scheduling and pricing, day-ahead profit guarantees, make-whole payments, and optimized balancing markets [10].

Also, several EMs around the world have recently proposed new mechanisms to incentivize increased levels of flexibility to better manage the rising variability and uncertainty of the net load. These include extended marginal pricing based on the convex-hull approach, new design elements for balancing markets (e.g., pay-for-performance regulation), explicit products for flexible ramping provision, and the use of emerging technologies to provide flexibility, such as DR and energy storage [10].

At present, however, existing markets have not (yet) converged on specific design elements to incentivize flexibility in system operation. Also, it is important to stress that new design elements such as convex-hull pricing and pay-for-performance ancillary services are in their infancy, and their impact on a changing power system should be analyzed further. Simply put, it is currently unclear whether the aforementioned new market design elements provide sufficient incentives to ensure the adequate levels of flexibility.

Flexibility markets refer generally to markets that are used by distributed system operators (DSOs) and possibly also TSOs to redispatch their grids, as well as to local markets and peer-to-peer trading [88]. Such markets are starting to be recognized as a promising tool to incentivize decentralized resources to trade flexibility products, mitigating the impacts of high levels of renewable generation (see Fig. 3.3). There are a number of ongoing R&D projects related to flexibility markets (see, e.g., OSMOSE [16] and WindNODE [17]). Several flexibility market platforms are emerging, such as NODES [18], Enera [19], Piclo Flex [20], and GOPACS [21]. Heer and Reek introduce three market platforms (NODES, Enera, and GOPACS) in a USEF white paper on flexibility platforms [15]. The authors state that these three pioneering initiatives are essentially in a piloting phase. Also, Schittekatte and Meeus [88] point out that most of the existing literature on flexibility markets focuses on their conceptualization and make a systematic analysis of four market platforms (NODES, Piclo Flex, Enera, and GOPACS).

Overall, despite the existence of some exemplary initiatives, most of them use market mechanisms locally and are not connected or integrated into the existing sequence of markets, such as intraday and balancing markets organized by TSOs.

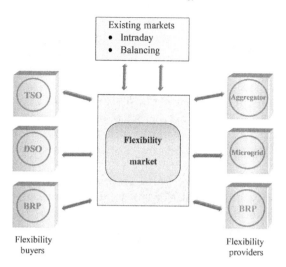

**FIGURE 3.3** Generic flexibility market interacting with other existing markets, notably intraday and balancing markets. The main flexibility buyers include TSOs, DSOs, and BRPs, and the key flexibility providers involve microgrids, aggregators, and BRPs.

## 3.6 Key flexibility market platforms

This section gives an overview of four pioneering flexibility market platforms, namely NODES [18], Enera [19], Piclo Flex [20], and IREMEL [89].

### 3.6.1 The NODES marketplace

NODES is an open, integrated market platform for all flexibility providers, BRPs, and grid operators [90]. The platform was established in 2018 as a joint venture between the European power exchange Nord Pool [27] and the Norwegian utility Agder Energi [91].[8] The platform is currently installed in several locations, including Germany and Norway [92].

The main objective of NODES is to facilitate the trade of local flexibility by taking into consideration the localization of the resources. To this end, two key features of the platform are as follows [18]: the possibility to identify local flexibility through a location tag and the capability to give a value to local flexibility by putting a price tag on it. These features open new opportunities to grid operators. In particular, DSOs can contract local flexibility to solve grid issues. Also, TSOs can access local flexibility that is normally excluded from traditional markets (currently, traditional markets—such as intraday and balancing markets—are essentially targeting large flexibility sources, meaning that most local flexibility is unable to participate).[9]

Market entities include a market operator, flexibility requesters (or flexibility buyers), and flexibility providers. NODES takes the role of an independent market operator, who

---

[8] Agder Energi is a Norwegian energy group involved in power generation, distribution, and trading, as well as in services for customers [87].

[9] The interface between NODES and the intraday market is currently in place. However, the integration of NODES with the balancing market is not yet completed [90].

provides transparent pricing and secure trading. TSOs, distribution system operators, and BRPs constitute the demand side.[10] This means that TSOs and DSOs can procure flexibility on the same platform. Flexibility providers include mainly BRPs, although aggregators, retailers, microgrids and other flexibility owners (e.g., flexible loads) may be considered.

Flexibility buyers may submit offers to the NODES marketplace indicating their willingness to pay for the activation of flexibility at particular locations in the grid. Flexibility providers act on behalf of owners of flexibility assets and may submit bids for sale to NODES.[11] They need to have a model with asset owners and technology that enables the activation of flexibility by those who buy it in the market. Flexibility assets (e.g., consumers) need to be tagged with their location and can be aggregated by flexibility providers.

Flexibility providers can differentiate their bids depending on whether flexibility may be sold locally or centrally [18]. Selling locally, at specific grid locations, may be risky, since flexibility is typically needed locally only a few hundred hours a year. Thus selling centrally may be advantageous, enabling flexibility to be used for rebalancing contractual positions in the intraday market (or for balancing purposes by TSOs). Also, flexibility providers can customize their bids by using a wide range of parameters. These include location, availability, time, profile, and order parameters. Accordingly, a catalogue of flexibility bids can be built up. Flexibility buyers can filter bids from such a catalogue and choose the ones that best fulfill their needs. For instance, DSOs may be more interested in increased consumption than in curtailment. And TSOs may prefer activation time and ramping. Flexibility products in NODES are therefore not standardized. This could be advantageous, allowing the request of very specific flexibility needs, and in some cases to better value flexibility. However, nonstandard flexibility products may not allow for sufficient levels of liquidity, since they increase the difficulty of building up a merit order to organize competition. Nonstandard products do not typically promote price transparency nor transparent competition between flexibility providers [88].

## 3.6.2 Enera and the EPEX SPOT local flexibility market

Enera is part of a funding program of the Federal Ministry for Economic Affairs and Energy, called "Smart Energy Showcases—Digital Agenda for the Energy Transition" (SINTEG). The program involves 32 consortium partners, including business, research, and government. The main goal is to develop and demonstrate scalable standard solutions for an environment-friendly, secure, and affordable power supply (involving large penetrations of renewables). To this end, four different perspectives are being considered: new roles and responsibilities in the energy sector, extension of the regulatory framework, analysis of network charges, and local flexibility markets [19].

In 2018 the European Power Exchange EPEX SPOT and the energy group EWE AG signed a cooperation agreement within the scope of the Enera project. Both companies commit to launching a local market platform for flexibility resources, together with one of

---

[10] The transmission system operator is not active yet [85].

[11] Currently, flexibility trading is based on a continuous procedure [85] and offers are remunerated by considering the pay-as-bid pricing scheme [91].

the German TSOs (TenneT) and two German DSOs (Avacon Netz and EWE NETZ). The cooperation aims at efficiently tackling the widespread issue of grid congestion by developing a scalable pilot in a showcase region (the windy northwest of Germany). The solution is expected to be demonstrated in a fully digitalized energy system with 100% renewables, thus paving the way to the EM of the future [93].

The EPEX SPOT local market platform is an open and voluntary market-based congestion management platform for flexibility providers, efficiently centralizing local flexibility offers that can be used by transmission and distribution system operators to alleviate congestion. EPEX SPOT acts as a neutral intermediary between flexibility demand from system operators and flexibility supply from decentralized resources. Also, EPEX SPOT supervises price formation and guarantees a high level of transparency [15].

The trading platform operates during the intraday time frame as a separate platform from the existing sequence of markets [88]. The access to the platform is standardized— that is, market participants can use the same application programming interface which they use to trade in the intraday market. Certified flexibility providers include (or act on behalf) of aggregators, storage assets, and different types of power plants. On a continuous base, they can submit sell offers to their respective order book. Flexibility buyers include TSOs and both mid-voltage and low-voltage DSOs. Also on a continuous base, they can submit demand orders to the order book that correspond to the market area from which they need flexibility. Order books are, therefore, locational, meaning that a particular market area corresponds to a specific order book on the platform. The trading procedure is similar to the procedure adopted in many markets across Europe—buy and sell offers in the same order book are automatically and continuously matched. Offers are remunerated by considering the pay-as-bid pricing scheme [94].

Flexibility products are standardized. Product definition is determined by EPEX SPOT in cooperation with grid operators. Most products involve blocks of energy (up or down) for a certain duration and a specific location [88]. Overall, the trading platform aims at creating a powerful coordination between system operators at all grid levels, by relying on a high degree of digitalization and automation. The ambition is to create new opportunities for market participants to value their flexible assets, as well as for system operators to avoid or defer costly grid expansion and allow for a higher reliability, security of supply, and coordination [15].

### 3.6.3 The Piclo flex marketplace

Piclo® (the trading name for Open Utility Ltd) is an independent software company that has been active in the energy industry since 2013 [88]. Among other aspects, the company seeks to demonstrate that a marketplace with visibility and transparency at its heart can enable the efficient and fair procurement of flexibility from the rapidly growing number of flexibility providers. Also, it seeks to prove that an open platform could play a fundamental role in supporting the efficient connection and operation of distributed energy resources (DERs), therefore supporting the growth of renewable generation and catalyze the adoption of EVs and other initiatives in the clean energy transition [20,95].

Accordingly, Piclo® was involved in a project funded by the UK Government Department of Business, Energy and Industrial Strategy's (BEIS). The project led to the development of the Piclo Flex marketplace, which was piloted in June 2018 for buyers and sellers of flexibility in the UK, and was subsequently launched as a commercial offering from March 2019 [88]. Piclo Flex aims at helping Britain to deliver a smart, flexible, and clean energy system, by enabling flexibility providers to promote their services online, as well as by encouraging new initiatives and business models to spring up [20].

The piloting phase revealed diversity in the buyers and sellers who can play a role in the flexibility platform. Specifically, the flexibility providers involved essentially aggregators (39.2% of the total registered providers), electricity suppliers (8% of registrations), and industrial and commercial users (4.2% of registrations). However, the trial also revealed some new agents that are poised to join the marketplace, including community and municipality participants (8.5% of registrations), and some large industrial customers that opted to sign up directly. Also, a relatively large number of participants (35.4% of registrations) represented speculative users of the platform, including potential investors in flexibility services [20].

The flexibility buyers included the six DSOs licensed in Britain to distribute electricity in 14 geographically defined areas—that is, UK Power Networks (UKPN), Scottish and Southern Electricity Networks, Electricity North West Limited, Northern Powergrid, SP Networks, and Western Power Distribution. Each of the DSOs used the platform in different ways. Some simply considered the visibility features to enhance existing processes, while others considered the functionality to run auctions, publishing their requirements in a standardized format. The type of requirement for flexibility varied but the greatest single need was due to reinforcement deferral. The second-largest requirement was for unplanned interruptions, both pre- and postfault. Planned maintenance represented a small proportion of flexibility needs during the trial [95].

In 2018−19, during the piloting phase, Piclo Flex signposted demand for more than 456 MW of flexibility from the six DSOs, demonstrating its potential to create a heat-map of areas of network congestion in Britain. UKPN published pricing signals and revenue ranges to accompany their competitions, allocating a total budget of approximately £12 million. The data revealed the significant variance that may exist between the volume of need and revenue range across the different constraint areas [96].

Overall, Piclo Flex was able to demonstrate that an online platform could be used by DSOs to help manage the electrical grid, using existing distributed assets for reinforcement deferral, unplanned interruptions, and planned maintenance. At present, however, the platform seems to be solely used by DSOs, meaning that the cooperation with the TSO is limited. Since establishing a TSO−DSO coordination within a flexibility market takes time, it seems that Piclo Flex decided to move fast by implementing a DSO-only solution. Nevertheless, flexible resources are allowed to engage in revenue stacking (e.g., by also offering services to the TSO) [88].

### 3.6.4 IREMEL and the Iberian electricity market

IREMEL (Integration of distributed Energy Resources through Local Electricity Markets) is a project launched by OMIE, the Spanish electricity market operator, and IDEA, the Institute for the Diversification and Saving of Energy (Ministry for the Ecological Transition) [89].

The main goals of the project are to develop a local market model to facilitate the efficient integration of DERs, such proactive consumers and storage installations, and also to promote the participation of DERs in solving local congestions and the needs of DSOs. In more detail, the project aims at developing various prototypes of local markets, identifying challenges and opportunities for the proactive role of prosumers and consumers in these markets (either directly or through the figure of the aggregator), demonstrating the viability of new technologies that facilitate the management of distributed resources and their participation in local markets, and leading the innovation in electricity markets, in cooperation with all stakeholders, and in accordance with the new European regulations (e.g., Directives 2018/2001 [97] and 2019/944 [13]).

IREMEL will involve the most relevant categories of flexibility providers, including large and small DSOs, individual DERs, aggregation companies, proactive consumers, battery producers, etc. All types of providers will participate in the different prototypes in order to validate the correct functioning of the aforementioned local market model [98]. Also, two main sets of products will be considered: short-term products (traded on demand, only when the need arises), and long-term products for structural problems (DSOs rely on the availability of one or more DERs to react at short-term notice, and contract this commitment for a relatively long period, such as months or years).

IREMEL will involve five pilots on local flexibility markets in different Spanish areas, with different participants and under different conditions [98]. However, as far as the author is aware, IREMEL is currently in the design phase.

## 3.7 Conclusion

This chapter described the centralized and bilateral models underlying most European markets, and analyzed the operation of four central European markets, namely Nord Pool, EPEX Spot, MIBEL, and GME. Following this material, the chapter analyzed the pressing issue of flexibility in system operation and described several flexibility European market platforms.

The chapter pointed out that the rapid expansion of renewables is challenging power systems in terms of flexibility for short-term operation. It also stated that the increasing deployment of decentralized resources able to offer demand-side flexibility is thought-provoking due to their incorporation in a cost-effective manner. In this context, the chapter claimed that local flexibility markets are a promising and powerful tool to adequately valorize demand-side flexibility.

The literature on flexibility markets is interesting, although narrow in scope. The SmartEn position paper [98] lists five exemplary cases, namely NODES, GOPACS, Enera, IREMEL, and Piclo Flex. Also, Heer and Reek [15] introduce three pioneering projects to develop flexibility markets: NODES, Enera, and GOPACS. The authors point out that these three initiatives are conceptually very similar and aim to reach similar benefits for both buyers and sellers of flexibility. The relation to intraday trading seems to be the main differentiator. Furthermore, the initiatives are essentially in a piloting phase—implementation in operational environments is largely dependent on modifications in regulatory frameworks and organizational changes within system operators.

Schittekatte and Meeus [88] analyzed four pioneering initiatives (NODES, Piclo Flex, Enera, and GOPACS) in terms of six dimensions: integration with existing markets, role and independence of the market operator, existence of a reservation payment, standardization of flexibility products, TSO–DSO cooperation, and DSO–DSO cooperation. The authors point out that most of the existing literature on flexibility markets focuses on their conceptualization and state that they intend to go a step further. Accordingly, they analyze the four pioneering platforms with the six-dimension framework, and also discuss some design choices that go beyond flexibility market design—that is, choices related to market access, settlement, and responsibilities. They conclude that all platforms are operated by a third party and engage (or tend to engage) with multiple DSOs to become the standardized platform provider. The differences among them are essentially related to their integration into other existing markets, the use of reservation payments, the use of standardized products, and the way TSO–DSO cooperation is done. Also, the authors conclude that the participation in all flexibility markets is voluntary and involves a prequalification procedure. Flexibility providers can either act as BRPs themselves or can trade on behalf of BRPs, meaning that contractual arrangements between (independent) aggregators, suppliers, and BRPs are a topic of debate and dependent on the regulatory framework. Finally, the authors state that there is no harmonized approach in calculating the baseline and none of the platforms specify penalties for nondelivery.

As noted earlier, this chapter describes several flexibility European market platforms, specifically NODES, Enera, Piclo and IREMEL. To some extent, the analysis allow us to conclude that more research is necessary to evaluate the potential impacts of local flexibility markets in a continually changing power system, particularly with respect to their capability to incentivize increased levels of flexibility when such flexibility is needed due to the rising penetrations of renewables. Put simply, sophisticated flexibility markets are not just around the corner, but serious attempts are underway across Europe, indicating that practice is moving faster than conceptual debate. There is, therefore, some further work to be done, but we can expect important technological developments during the coming years toward modern power systems.

## Acknowledgment

This work has received funding from the EU Horizon 2020 research and innovation program under project TradeRES (grant agreement N. 864276).

## References

[1] EC, Directive 2003/54/EC of the European Parliament and of the Council, Official Journal of the European Union, L 176/37–L 176/55, 2003.
[2] EC, Directive 2009/72/EC of the European Parliament and of the Council, Official Journal of the European Union, L 211/55–L 211/93, 2009.
[3] EC, Regulation 1227/2011 of the European Parliament and of the Council, Official Journal of the European Union, L 326/1–L 326/16, 2011.
[4] EC, Commission Regulation 2015/1222, Official Journal of the European Union, L 197/24–L 197/72, 2015.
[5] S. Stoft, Power System Economics: Designing Markets for Electricity, IEEE Press and Wiley Interscience, 2002.
[6] D. Kirschen, G. Strbac, Fundamentals of Power System Economics, John Wiley & Sons, Chichester, 2018.

[7] NEMO Committee, EUPHEMIA Public Description: Single Price Coupling Algorithm, Nominated Electricity Market Operators Committee, 2019.

[8] E. Ela, M. Milligan, A. Bloom, J. Cochran, A. Botterud, A. Townsend, et al., Overview of wholesale electricity markets, Electricity Markets with Increasing Levels of Renewable Generation: Structure, Operation, Agent-Based Simulation and Emerging Designs, Springer International Publishing, Cham, 2018, pp. 3–21.

[9] F. Lopes, H. Coelho, Electricity Markets With Increasing Levels of Renewable Generation: Structure, Operation, Agent-Based Simulation and Emerging Designs, Springer International Publishing, Cham, 2018.

[10] E. Ela, M. Milligan, A. Bloom, A. Botterud, A. Townsend, T. Levin, Incentivizing flexibility in system operations, Electricity Markets With Increasing Levels of Renewable Generation: Structure, Operation, Agent-Based Simulation, and Emerging Designs, Springer International Publishing, Cham, 2018, pp. 95–127.

[11] E. Ela, M. Milligan, A. Bloom, A. Botterud, A. Townsend, T. Levin, Long-term resource adequacy, long-term flexibility requirements, and revenue sufficiency, Electricity Markets With Increasing Levels of Renewable Generation: Structure, Operation, Agent-Based Simulation, and Emerging Designs, Springer International Publishing, Cham, 2018, pp. 129–164.

[12] EC, Regulation 2019/943, Official Journal of the European Union, L 158/54 – L 158/124, 2019.

[13] EC, Directive 2019/944, Official Journal of the European Union, L 158/125 – L 158/199, 2019.

[14] O. Abrishambaf, P. Faria, Z. Vale, Ramping of demand response event with deploying distinct programs by an aggregator, Energies 13 (2020) 1389.

[15] H. Heer, W. Reek, USEF white paper: flexibility platforms, Universal Smart Energy Framework, 2018, pp. 1–29. <https://www.usef.energy/news-events/publications/> (accessed July 2020).

[16] OSMOSE, Optimal System Mix of Flexibility Solutions for European Electricity. <https://www.osmose-h2020.eu/> (accessed July 2020).

[17] WindNODE, Showcasing Smart Energy Systems from North-Eastern Germany. <https://www.windnode.de/en/about/overview/> (accessed July 2020).

[18] NODES, A Fully Integrated Marketplace for Flexibility, White Paper, 2018. Available at: <https://nodesmarket.com> (accessed July 2020).

[19] Enera, The Next Big Step Towards a Sustainable World. <https://bremen-energy-research.de/projects/enera> (accessed July 2020).

[20] Piclo®, Flexibility & Visibility: Investment and Opportunity in a Flexibility Marketplace, Open Utility Ltd., 2019. Available at: <https://piclo.energy/> (cited on July 2020).

[21] GOPACS, Grid Operators Platform for Congestion Solutions. <https://en.gopacs.eu/> (accessed July 2020).

[22] F. Lopes, Electricity markets and intelligent agents part I: market architecture and structure, Electricity Markets With Increasing Levels of Renewable Generation: Structure, Operation, Agent-Based Simulation and Emerging Designs, Springer, Cham, 2018, pp. 23–48.

[23] F. Schweppe, M. Caramanis, R. Tabors, R. Bohn, Spot Pricing of Electricity, Kluwer Academic Publishers, Boston, MA, 1988.

[24] ACER, Methodology for the Price Coupling Algorithm, the Continuous Trading Matching Algorithm and the Intraday Auction Algorithm, Agency for the Cooperation of Energy Regulators Decision on Algorithm Methodology: Annex I, 2020.

[25] EC, Regulation 2017/1485, Official Journal of the European Union, L 220/1 – L 220/120, 2017.

[26] EC, Regulation 2017/2195, Official Journal of the European Union, L 312/6 – L 312/53, 2017.

[27] Nord Pool, The Nordic Power Exchange. <https://www.nordpoolspot.com/> (accessed July 2020).

[28] Nord Pool, Day-Ahead Market Regulations, Nord Pool AS, 2017.

[29] Nord Pool, Intraday Market Regulations, Nord Pool AS, 2018.

[30] Energinet, The Danish Transmission System Operator. <https://en.energinet.dk/Electricity> (accessed July 2020).

[31] Svenska kraftnät, The Swedish Transmission System Operator. <https://www.svk.se/en/stakeholder-portal/Electricity-market/> (accessed July 2020).

[32] Statnett, The Norwegian Transmission System Operator. <https://www.statnett.no/en/About-Statnett/> (accessed July 2020).

[33] Entsoe, Nordic Balancing Philosophy, European Network of Transmission System Operators, 2016.

[34] Energinet, Regulation C2: The Balancing Market and Balance Settlement, The Danish System Operator, 2017.

[35] Fingrid, Reserve Products and Reserve Marketplaces, Public Presentation of the Finnish System Operator, 2020.

[36] EPEX SPOT, The European Power Exchange. <http://www.epexspot.com/en> (accessed July 2020).

[37] EPEX SPOT, Operational Rules, The European Power Exchange, 2020.

[38] EPEX SPOT, Trading on EPEX SPOT, The European Power Exchange, 2020.

[39] RTE, Réseau de Transport d'Électricité, The French Transmission System Operator. <https://www.rte-france.com/en> (accessed July 2020).

[40] 50Hertz, A Germany Transmission System Operator. <https://www.50hertz.com/en/> (accessed July 2020).

[41] Amplion, A Germany Transmission System Operator. <https://www.amprion.net/technical-service/> (accessed July 2020).

[42] RTE, Balancing Mechanism, Réseau de Transport d'Électricité, 2010.

[43] OMIE, The Spanish Electricity Market Operator. <https://www.omie.es/en> (accessed July 2020).

[44] OMIE, Day-Ahead and Intraday Electricity Market Operating Rules, The Spanish Electricity Market Operator, 2018.

[45] OMIE, Details of the Intraday Market's Operation, The Spanish Electricity Market Operator, 2020.

[46] OMIP, Trading Rule Book, The Portuguese Electricity Market Operator, 2016.

[47] REN, The Portuguese Transmission System Operator. <https://www.ren.pt/en-GB> (accessed July 2020).

[48] REE. The Spanish Transmission System Operator. <https://www.ree.es/en> (accessed July 2020).

[49] ERSE, Manual de Procedimentos da Gestão Global do Sistema do Setor Elétrico, Entidade Reguladora dos Serviços Energéticos, 2018.

[50] GME. The Italian Electricity Market. <http://www.mercatoelettrico.org/En> (accessed July 2020).

[51] GME, Vademecum of the Italian Power Exchange, The Italian Electricity Market, 2020.

[52] Terna, The Italian Transmission System Operator. <https://www.terna.it/en/> (accessed July 2020).

[53] M. Ventosa, A. Baíllo, A. Ramos, M. Rivier, Electricity market modelling trends, Energy Policy 33 (7) (2005) 897−913.

[54] E.P. Kahn, Numerical techniques for analyzing market power in electricity, Electricity J. 11 (6) (1998) 34−43.

[55] B.F. Hobbs, Linear complementarity models of Nash-Cournot competition in bilateral and POOLCO power markets, IEEE Trans. Power Syst. 16 (2) (2001) 194−202.

[56] C.J. Day, B.F. Hobbs, Oligopolistic competition in power networks: a conjectured supply function approach, IEEE Trans. Power Syst. 17 (3) (2002) 597−607.

[57] A. Weidlich, D. Veit, A critical survey of agent-based wholesale electricity market models, Energy Econ. 30 (2008) 1728−1759.

[58] M. Wooldridge, An Introduction to Multi-Agent Systems, John Wiley & Sons, Chichester, 2009.

[59] S. Russell, P. Norvig, Artificial Intelligence: A Modern Approach, Pearson Education, Inc, New Jersey, 2020.

[60] S.A. Harp, S. Brignone, B.F. Wollenberg, T. Samad, SEPIA: a simulator for the electric power industry agents, IEEE Control. Syst. Mag. 20 (4) (2000) 53−69.

[61] V. Koritarov, Real-world market representation with agents: modelling the electricity market as a complex adaptive system with an agent-based approach, IEEE Power Energy Mag. 2 (4) (2004) 39−46.

[62] D. Batten, G.G. Grozev, NEMSIM: finding ways to reduce greenhouse gas emissions using multi-agent electricity modelling, in: P. Perez, D. Batten (Eds.), Complex Science for a Complex World Exploring Human Ecosystems With Agents, Australian National University Press, Canberra, 2006, pp. 227−252.

[63] J. Sun, L. Tesfatsion, Dynamic testing of wholesale power market designs: an open-source agent-based framework, Comput. Econ. 30 (2007) 291−327.

[64] F. Sensfuß, Assessment of the Impact of Renewable Electricity Generation on the German Electricity Sector: An Agent-Based Simulation Approach (Ph.D. dissertation), Karlsruhe University, 2007.

[65] Z. Vale, T. Pinto, I. Praça, H. Morais, MASCEM - electricity markets simulation with strategically acting players, IEEE Intell. Syst. 26 (2) (2011) 9−17.

[66] S. Cincotti, G. Gallo, The Genoa artificial power-exchange, in: J. Filipe, A. Fred (Eds.), Agents and Artificial Intelligence (ICAART 2012), Springer-Verlag, Berlin, 2013, pp. 348−363.

[67] F. Lopes, Electricity markets and intelligent agents: Part II − Agent architectures and capabilities, Electricity Markets With Increasing Levels of Renewable Generation: Structure, Operation, Agent-Based Simulation and Emerging Designs, Springer, Cham, 2018, pp. 49−77.

[68] F. Lopes, MATREM: an agent-based simulation tool for electricity markets, Electricity Markets With Increasing Levels of Renewable Generation: Structure, Operation, Agent-Based Simulation and Emerging Designs, Springer, Cham, 2018, pp. 189−225.

[69] O. Abrishambaf, P. Faria, L. Gomes, J. Spínola, Z. Vale, J.M. Corchado, Implementation of a real-time micro-grid simulation platform based on centralized and distributed management, Energies 10 (2017) 806.

I. Distributed energy resources as enablers of local electricity markets

[70] F. Bellifemine, G. Caire, D. Greenwood, Developing Multi-Agent Systems With JADE, John Wiley & Sons, Chichester, 2007.

[71] D. Vidigal, F. Lopes, A. Pronto, J. Santana, Agent-based simulation of wholesale energy markets: a case study on renewable generation, in: M. Spies, R. Wagner, A. Tjoa (Eds.), 26th Database and Expert Systems Applications (DEXA 2015), IEEE, 2015, pp. 81–85.

[72] H. Algarvio, A. Couto, F. Lopes, A. Estanqueiro, J. Santana, Multi-agent energy markets with high levels of renewable generation: a case-study on forecast uncertainty and market closing time, in: Distributed Computing and Artificial Intelligence, 13th International Conference, AISC, vol. 474, Springer, Cham, 2016, pp. 339–347.

[73] H. Algarvio, F. Lopes, A. Couto, A. Estanqueiro, Participation of wind power producers in day-ahead and balancing markets: an overview and a simulation-based study, WIREs Energy Environ. 8 (5) (2019) e343.

[74] F. Lopes, A.Q. Novais, H. Coelho, Bilateral negotiation in a multi-agent energy market, Emerging Intelligent Computing Technology and Applications, Springer, Berlin, 2009, pp. 655–664.

[75] F. Sousa, F. Lopes, J. Santana, Contracts for difference and risk management in multi-agent energy markets, Advances in Practical Applications of Agents, Multi-Agent Systems, and Sustainability: The PAAMS Collection (PAAMS 2015), Springer International Publishing, 2015, pp. 339–347.

[76] F. Lopes, N. Mamede, A.Q. Novais, H. H. Coelho, Negotiation in a multi-agent supply chain system, Third Int. Workshop of the IFIP WG 5.7 Special Interest Group on Advanced Techniques in Production Planning & Control, Firenze University Press, 2002, pp. 153–168.

[77] F. Lopes, N. Mamede, A.Q. Novais, H. Coelho, A negotiation model for autonomous computational agents: formal description and empirical evaluation, J. Intell. Fuzzy Syst. 12 (2002) 195–212.

[78] F. Lopes, N. Mamede, A.Q. Novais, H. Coelho, Negotiation tactics for autonomous agents, in: 12th International Workshop on Database and Expert Systems Applications (DEXA), IEEE, 2001, pp. 1–5.

[79] F. Lopes, J. Sá, J. Santana, Renewable generation, support policies and the merit order effect: a comprehensive overview and the case of wind power in Portugal, Electricity Markets With Increasing Levels of Renewable Generation: Structure, Operation, Agent-Based Simulation and Emerging Designs, Springer, Cham, 2018, pp. 227–263.

[80] F. Lopes, H. Algarvio, Demand response in electricity markets: an overview and a study of the price-effect on the Iberian daily market, Electricity Markets With Increasing Levels of Renewable Generation: Structure, Operation, Agent-Based Simulation and Emerging Designs, Springer, Cham, 2018, pp. 265–303.

[81] H. Algarvio, F. Lopes, A. Couto, A. Estanqueiro, J. Santana, Variable renewable energy and market design: new market products and a real-world study, Energies 12 (23) (2019) 4576.

[82] D. Ton, M. Smith, The U.S. department of energy's microgrid initiative, Electr. J. 25 (2012) 84–94.

[83] GridBeyond, Energy Trends in the Global Marketplace, GridBeyond Publication. <https://gridbeyond.com/> (accessed July 2020).

[84] USEF, The Framework Explained, Universal Smart Energy Framework, 2015, pp. 1–55. <https://www.usef.energy/download-the-framework/a-flexibility-market-design/> (accessed July 2020).

[85] L. Gomes, J. Spínola, Z. Vale, J.M. Corchado, Agent-based architecture for demand-side management using real-time resources' priorities and a deterministic optimization algorithm, J. Clean. Prod. 241 (2019).

[86] Z. Vale, H. Morais, P. Faria, C. Ramos, Distribution system operation supported by contextual energy resource management based on intelligent SCADA, Renew. Energy 52 (2013) 143–153.

[87] Europex, A Market-Based Approach to Local Flexibility – Design Principles, Position Paper of the Association of European Energy Exchanges, 2020. <https://www.europex.org/publications/> (accessed July 2020).

[88] T. Schittekatte, L. Meeus, Flexibility markets: Q&A with project pioneers, Uti. Policy 63 (2020) 101017.

[89] OMIE, Modelo de Funcionamento de los Mercados Locales de Electricidad, Project IREMEL, Spanish Electricity Market Operator, 2019. <https://www.omie.es/en/proyecto-iremel> (accessed July 2020).

[90] NODES, Flexibility Market Platform. <https://nodesmarket.com> (accessed July 2020).

[91] Agder Energi, Agder Energy Group. <https://www.ae.no/en/> (accessed July 2020).

[92] D. Engelbrecht, A. Schweer, R. Gehrcke, E. Lauen, B. Deuchert, J. Wilczek, et al., Demonstration of a Market-Based Congestion Management Using a Flexibility Market in Distribution Networks, International ETG-Congress, 2019, pp. 306–311.

[93] Press Release, Enera Project: EWE and EPEX SPOT to Create Local Market Platform to Relieve Grid Congestions, Essen, Oldenburg and Paris, 2018. <https://www.ewe.com/en/media/press-releases> (accessed July 2020).

[94] J. Radecke, J. Hefele, L. Hirth, Markets for Local Flexibility in Distribution Networks, Working Paper of ZBW — Leibniz Information Centre for Economics, Kiel, Hamburg, 2019.

[95] Piclo®, Energy On Trial: Piloting a Flexibility Marketplace to Upgrade our Energy System, Open Utility Ltd, 2019. Available at: <https://piclo.energy/> (cited on July 2020).

[96] UKPN, Flexible Zones — Revenue Range per Area, UK Power Networks, 2019.

[97] EC, Directive 2018/2001, Official Journal of the European Union, L 328/82 — L 328/209, 2018.

[98] SmartEn, Design Principles for (Local) Markets for Electricity System Services, Smart Energy Position Paper, 2019. <https://smarten.eu/design-principles-for-local-markets-for-electricity-system-services/> (accessed July 2020).

# From the smart grid to the local electricity market

*Fernando Lezama[1], Tiago Pinto[2], Zita Vale[3], Gabriel Santos[4] and Steve Widergren[5]*

[1]Research Group on Intelligent Engineering and Computing for Advanced Innovation and Developmen (GECAD), Polytechnic of Porto (ISEP/IPP), Porto, Portugal [2]GECAD Research Group, Polytechnic of Porto (ISEP/IPP), Porto, Portugal [3]School of Engineering, Polytechnic of Porto (ISEP/IPP), Porto, Portugal [4]GECAD, Polytechnic of Porto (ISEP/IPP), Porto, Portugal [5]Pacific Northwest National Laboratory (PNNL), Richland, WA, United States

## 4.1 Introduction

The centralized top-down approach of Electricity Markets (EM) has limitations for harnessing the full advantage from the participation of flexible customer with distributed energy resources (DER)—in load, storage, and distributed generation. It is difficult to create and to maintain customer resource models, and they require costly installation and maintenance of behind-the-meter equipment, large and complex optimization software, and solutions to challenging security and information privacy concerns. Moreover, the tentative reforms of retail markets are not able to achieve the policymakers' envisaged goals as they are being built under the same top-to-bottom principles as wholesale markets. The introduction of dynamic and innovative tariffs, adapted to customers' needs and behaviors that fairly remunerate their contribution toward an increasingly efficient energy system, are still a distant target. New approaches that can bring a closer connection between customers' DER and the wholesale EM are required promptly. The worldwide investments already made in smart grids (SG), for example, the rollout of smart meters and the investment in communication infrastructure between local players, provide a basis for the developments to come in the upcoming years [1]. A pioneering solution to overcome the current problems is currently being implemented in the New York EM, in the United States [2]. The creation of Local EM (LEM) as part of the Regional EM is being put into practice, enabling smaller portions of the power network (microgrids) to participate in the

EM as aggregators of the resources that are part of that microgrid. In this way, flexibility of operation can be traded and managed at a local level, enabling the engagement and investment in DER, while microgrid operators facilitate their participation in EM [3]. This provides important incentives for the development of effective methods to manage resources at lower levels and make their coordinated operation represented in the whole-sale EM. See for instance the case study in Ref. [4] for details.

LEMs bring benefits to the system that result from the distributed nature of the customers' flexibility resources [5]. Altogether, local markets enable the efficient management of all DER, increasing the efficiency of the power energy system as a whole and bringing concrete benefits to customers under the form of lower energy prices and increased reliability. The challenge for the coming years is to resolve the emerging challenges by creating interim LEM solutions that are compatible with the current organized markets, and that enable a smooth transition to a more decentralized coordination structure that fully utilizes a LEM approach.

This chapter provides an overview of the advances achieved toward the widespread implementation of SG, including the developments in both technological and infrastructure dimensions. Transactive energy benefits from such technological and infrastructure developments and presents a distributed decision-making coordination approach using automated energy transactions that are enabled by the intelligence and connectivity benefits of SG. The way in which the advances in SG infrastructure and technology, allied to the progresses in the field of transactive energy are contributing to the emergence of the LEM concept is discussed in this chapter.

## 4.2 Emergence and widespread of the smart grid

The current global trends, including global warming, an increase in carbon emissions, and the growing world population and power demand, has led governments, energy utilities, and research institutions to take concrete actions toward the reduction of energy consumption and the use of renewable resources [6]. Due to this, the electric grid has been evolving to greater levels of information and communications technology (ICT)-enabled automation under the banner of SG. A SG is an advanced power network that incorporates two-way communication and computer intelligence for efficient control, reliability, and safety [7]. The related technologies, systems, devices, processes, and methods have come a long way, becoming a revolution in the energy domain [8]. SG research encompasses forecasting methods, wide-area protection and self-healing grids, reliability and power quality enhancement, power flow optimization, storage techniques, large-scale integration of renewables, and cloud computing control, among many other axes that are possible thanks to advanced ICT infrastructure [8–10]. In particular, the National Institute of Standards and Technology (NIST) defines SG as an electric power system that integrates a variety of digital computing and communication technologies and services into the power system infrastructure, going beyond smart meters for home and business as the bidirectional flow of energy using two-way communication and control for new functionalities [11]. This opens the possibility of exploring different research directions with colossal potential and margin of improvements.

One of SG's more exciting features, tightly related to the emergence of LEM and transactive energy concepts, is that SG challenges the traditional electric grid's hierarchical nature of control to consider the possibilities of a decentralized coordination paradigm [12]. This enables the customer and locally managed DERs to become active participants in the energy exchange as participants in system operation. Customers can either consume or produce energy depending on their DERs (sometimes called prosumers) [13,14]. Small self-sustained power networks, known as microgrids (see Fig. 4.1), are receiving special attention as a complementary solution to achieve efficient energy management and high reliability in small areas [15]. Microgrids are intended to balance load, storage, and generation effectively [e.g., through renewable energy sources (RES), energy storage systems (ESS), demand flexibility, and DG] and operate in both nonautonomous (i.e., grid-connected) and autonomous (i.e., stand-alone) modes. Hence, SG at the system level and SG at the microgrid level can provide different benefits to the overall system performance (e.g., the integration of RES without requiring a redesign of the distribution system or energy supply in the case of power shortage [16]).

In the context of Europe, the European Commission (EC) reports that "Europe's transition to a low-carbon society is becoming the new reality on the ground" [17]. The penetration of RES for the distributed generation of electricity has been successively increasing in the EU and worldwide [18,19]. In fact, renewables seem to be the most resilient source of energy to Covid-19 lockdown measures, largely unaffected while demand has fallen for other uses of energy [20]. In concrete numbers, the quantity of renewable-based energy generation within the EU-28 increased overall by 64.0% between 2007 and 2017 [18], with wind and solar power contributing to the renewable energy mix with 13.8% and 6.4% shares, respectively, of EU-28's renewable energy produced in 2017 [18]. Despite these advancements toward a sustainable energy sector, the emergency and widespread deployment of SG enabling technology in Europe is only happening in terms of RES penetration and the installation of smart meters [21]. Smart meters are crucial as enablers of more

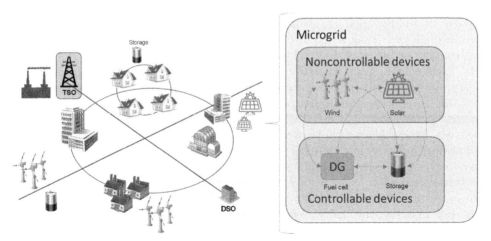

**FIGURE 4.1**    The smart grid and microgrid. *Source: Adapted from F. Lezama et al., Agent-based microgrid scheduling: an ICT perspective. Mob. Netw. Appl. (2017). https://doi.org/10.1007/s11036-017-0894-x.*

flexible and dynamic energy tariffs as well as of customers' active participation in different types of Demand Response (DR) programs (DR can be defined as the customers' ability to change their consumption patterns in the face of specific requests or goals) [22,23]. However, the full potential of SG benefits is yet to be utilized, and it is not clear that the rolled out smart metering equipment is sufficiently open and flexible to be useful for the next generation of SG solutions [21]. Due to the limitation in adopting SG technologies and coordinated operations approaches, a significant part of the renewable-based production capacity is being wasted, compromising the investment return in solar and wind energy. In this context, LEM and transactive energy approaches (coordinating the use of DERs) have the potential to be significant contributors to increased energy efficiency, improved reliability, and to achieving sustainability and environmental targets [24,25]. In addition, efforts to overcome technical and economic barriers around the world for the adoption of ICT infrastructure that supports SGs can achieve a promising future for customers and the electric industry in general [7].

## 4.3 Technological and infrastructure developments

Creating SGs involves the deployment of smart components and the ability to integrate new technology with existing control and information technologies in the power grid to offer greater efficiencies in operation and new capabilities to both utilities and end-users [7]. SGs require modern communication systems to allow the interaction among the smart components and a variety of actors, with different roles, goals, and objectives within the system. Because of this, SGs represent complex, socioeconomic environments that require a great deal of analysis and planning to function correctly. Therefore the deployment of reliable and pervasive ICT infrastructure must be carefully planned to properly manage the structure and operation of SGs [9,14].

Based on the chosen architecture of the communications network infrastructure, different technologies (e.g., wireline or wireless) can be proposed for communication in SGs. The variety of technologies available for SG and microgrids communication goes from wireline technologies [such as power line communications, digital subscriber line (DSL), optical networks or Ethernet] to wireless technologies [such as wireless local area network, WiMAX, Zigbee, Long Term Evolution (LTE)]. The technology that fits better in a one scenario may not be suitable for other scenarios. Studies on different standards and protocols available for SG/microgrid communication can be found in Refs. [7,10,11]. A strategic requirement in the choice of infrastructure to enable SG communication is to ensure a reliable network for establishing robust real-time data through wide area networks (WANs) to the distribution feeder and customer level [7,11].

When deciding the communications networking, some considerations to have in mind are the installation costs, ease of implementation, and suitability in terms of communication range, security, availability, scalability, and upgradability to evolve as technology changes. For instance, the use of existing infrastructure (e.g., DSL or mobile networks) could be considered a practical and cost-effective option. However, such technologies have their own drawbacks, for example, reliability and cybersecurity issues, or high costs for data transmission performance and bandwidth [14].

For the communication between devices in SGs/microgrids, wireless networks have installation advantages over wireline solutions (e.g., optical networks) due to their high flexibility and scalability. These characteristics fit well in situations with a large number of DER sites, or in cases where there is high sensitivity to the integration cost multiplier because of the many small devices that must be connected. This leads to solution approaches in which plug-and-play characteristics are highly valued [26].

Take, for instance, a multiagent system (MAS) put on top of an SG communication infrastructure to enable a LEM. The communication between the agents in such a MAS should be reliable and secure to guarantee transparent transactions between agents representing real customers. Moreover, the realization of a decentralized control should be based on open communication of agents using common protocols and standards that facilitate interoperability [14,27–32]. Three categories of SG/microgrid communication networks include Home Area Networks (HAN), Field Area Networks (FAN), and WAN. The technology used for the communication of DER in SGs depends directly on the application and type of SG network. For instance, to enable agent-based communication between distribution system operations or energy service aggregators to customer facilities for LEM or transactive energy purposes, FANs are preferred due to the distributed communication nature required for interfacing to facility management systems that may manage one or more DER devices or systems. For communicating with responsive devices or systems within a customer facility, HANs are preferred because of the performance and features tailored to a building or site. On the other hand, WANs are not suitable for the operation of local resources but can be very effective in grid-connected mode to retrieve information from external sources (such as Transmission System Operators, balancing responsible parties or Distribution System Operators) [14]. As a summary, communication technologies for SGs/microgrid communication can be found in Table 4.1.

With vast numbers of different devices and communication protocols available, future SG infrastructure must rely on the standardization of communication infrastructure (e.g., smart metering platforms) and techniques to enable smooth functioning and interoperability. For instance, advanced metering infrastructure (AMI) is currently using many Internet standards to ensure interoperability between different manufacturers' products [10]. Extensive activities are carried out all around the world to achieve the goal of standardization of interfaces and communications protocols that shape the SG.

All in all, establishing ICT architectures for the communications infrastructure of SGs is critical when companies and R&D institutions push toward developing and deploying innovative approaches and new concepts such as LEM and transactive energy. Most of today's LEM and transactive energy-inspired deployments fall into the category of pilots or demonstrations. The size of the systems tends to be limited in numbers of participants and scope as the performance of communications systems, DER and market agent designs, and grid–DER service agreements are tested. Beyond the technology and algorithmic concerns, public policy in the electric industry is uncertain of the proper path forward for creating a fair and equitable environment with value propositions that attract operating organizations, customers, and technology solutions–provider businesses to participate. Much of the research focus remains on the development and testing of adequate simulation platforms and tools to investigate the performance results and participant impacts. While ICT infrastructure represents the technical foundation on which LEM and

**TABLE 4.1**   Information communication networking technologies for microgrid control.

| Technology | Data rate | Coverage range | Applications | Limitations |
|---|---|---|---|---|
| ZigBee | 250 Kbps | 30–50 m | AMI, HAN, MAS communication | Low data rate, short range |
| WiMAX | Up to 75 Mbps | 10–50 km (LOS), 1–5 km (NLOS) | AMI, demand response, HAN, MAS communication | Not widespread |
| Wi-Fi | 1 Mbps | 50 m | AMI, HAN, MAS communication | Medium to lower consumption |
| Bluetooth | 1–54 Mbps | 10 m | AMI, HAN, MAS communication | High terminal cost, high linking time |
| GSM | Up to 14.4 Kbps | 1–10 km | AMI, demand response, HAN, MAS communication | Low data rate |
| GPRS | Up to 170 Kbps | 1–10 km | AMI, demand response, HAN, MAS communication | Low data rate |
| 3G | 384 Kbps–2 Mbps | 1–10 km | AMI, demand response, HAN, MAS communication | Costly spectrum |
| LTE | 300 Mbps | up to 5 km | AMI, demand response, HAN, MAS communication | Costly spectrum |
| WiGig | 7 Gbps | 1–10 m | Multigigabit connectivity | Compatibility and short range |
| Giga-IR | 1 Gbps | 1 m | Sensor communication | Short range |
| Li-Fi | >1 Gbps | up to 10 m | Replacement for other technologies | Only works in direct line of sight |
| 5G | 10 Gbps | up to 2 km | SG applications | Under standardization |

*AMI*, Advanced metering infrastructure; *GPRS*, General Packet Radio Service; *GSM*, Global System for Mobile Communications; *LOS*, Line-of-sight; *NLOS*, Non-line-of-sight.

*Adapted from F. Lezama et al., Agent-based microgrid scheduling: an ICT perspective. Mob. Netw. Appl. (2017). https://doi.org/10.1007/s11036-017-0894-x.*

transactive energy approaches are built, greater experience is needed to validate market-based coordination designs and business value propositions so that policy maker anxiety is reduced, stakeholder participation is healthy and sustaining, and the reliability and capabilities of the system mature [13].

## 4.4 Transactive energy

Transactive energy emerged as a term to describe the application of market-based trading mechanisms as a form of distributed decision-making for coordinating the flexibility in responsive (or controllable) demand, generation, and storage in the distribution system with electric system operations. While the flexibility could be owned by any party, the emphasis has been placed on coordinating the operation of customer-owned flexibility with the electric system. Transactive energy gives customers the opportunity to negotiate scheduling energy and other services in a marketplace of participants. While there are variations on the definition of transactive energy in the literature, one of the most accepted ones is the proposed by the GridWise Architecture Council [2], defining transactive energy as "A system of economic and control mechanisms that allows the dynamic balance of supply and demand across the entire electrical infrastructure using value as a key operational parameter." Other definitions, such as the ones found in Refs. [33–36], are variations of this idea depending on differences in the context of the application. The term transactive energy markets is used to define EM in which grid parties, agents, operators, and end-users negotiate the exchange of energy services with

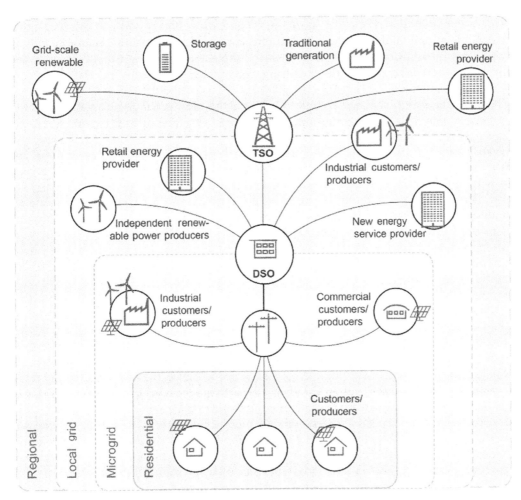

**FIGURE 4.2**   The transactive energy concept. *Source: Adapted from M.R. Knight, Stochastic impacts of metaheuristics from the toaster to the turbine [technology leaders], IEEE Electrif. Mag. 4 (4) (2016) 52–40. https://doi.org/10.1109/MELE.2016.2614218.*

their own perspective of value maximization [37,38]. Value is not merely financial but includes the value of comfort or the value of energy to conduct business. Fig. 4.2 depicts a diagram that can be used to separate the power grid into sectors where transactive energy mechanisms can be applied for coordinating system operation.

Transactive energy concepts emerged at the turn of the century. The SG trends toward lower cost computational resources and ubiquitous connectivity meant that the difficulties faced by centralized optimization and control scaling to manage millions of devices could be addressed in a different manner. In addition, rising concerns about cybersecurity and information privacy made central control approaches even more challenging as they crossed into customer premises. By allowing customer choice in how their resources participate in system operation, many issues begin to be addressed.

The ideas draw upon multiagent system designs and market mechanisms to seek equilibrium of operation based on value trade-offs. The initial pilots tested market mechanisms for balancing supply and demand [40], but also considered constraints on energy delivery through the distribution system. Some early implementations included using market mechanisms to reduce energy flows during constrained conditions on the distribution system [41].

While pricing programs such as time-of-use, critical-peak pricing, and real-time pricing have been a part of DR programs for many years, they are different from transactive energy systems in their one-way communications. That is, they encourage load-shifting behavior based on price changes, but there is no negotiation about the quantity of load shifted. The program operator is put in the situation to estimate the response of the customers without information from the customers on the change in the quantity of energy consumed. The advantage of these programs is their simplicity to set up and use one-way communication from the program operator to the customer. The disadvantage is that balancing supply and demand is left to guess work.

An important characteristic of transactive energy systems is the two-way communication of information between participants in the market. The equilibrium between price and quantity among participants (suppliers and consumers) is a fundamental element of the design.

Pilots of transactive energy systems have experimented with different types of market mechanisms. These include double-sided auctions, consensus mechanisms, and bilateral trading systems. In the double-side auction case, price—quantity curves are bid into the market from buyers and sellers. The market is cleared periodically (e.g., every 5 minutes) by finding the price—quantity point where the supply curve crosses the demand curve. The cleared price is then communicated to all participants who know where to operate based on the associated quantity in their price—quantity curve.

In a consensus mechanism, information about the willingness to supply or consume energy is circulated among the market participants. When the difference between the bids and offers among all participants is negligible, the transactive energy system is said to converge. Each participant knows how to operate based on the converged values.

In a bilateral trading system, sometimes called peer-to-peer trading, participants negotiate to supply or consume energy over a scheduled interval with counter parties. These bilateral exchanges can be facilitated by a variety of approaches, including brokerage houses that facilitate matching buyers with sellers.

In all cases, a transactive energy system design must consider the physics of delivering energy between the market participants. Each of the three types of designs mentioned deal with this in different ways, but unless the limitations of energy flow are taken into account, distribution operations can be disrupted, and more importantly, the flexibility in the system cannot be properly used for the benefit of managing distribution constraints. By managing distribution constraints, investments in distribution circuit upgrades can be deferred or potentially eliminated and the savings passed on to the participants. This is sometimes known as using a transactive energy system for nonwire alternatives.

The early transactive energy systems demonstrations experimented with these different approaches to discover the validity of the approach and assumptions about market performance in its accuracy, stability, and resilience. They experimented with market clearing intervals and researchers discovered quickly that energy use planning and forecasting was important to market performance. This led to pilots that also employed variations of forward markets in addition to real-time markets. In addition, whereas scheduling energy use was the primary service that was exchanged in these systems, experimentation was also applied to other trading services, including voltage management.

Other forms of distributed, intelligent system designs were also developed [13]. Peer-to-peer exchanges exist to trade energy produced from solar systems with consumers who desire green energy. A distinguishing aspect of these types of peer-to-peer systems versus a transactive energy bilateral trading system is they do not take into account the physical flow of energy in the electric infrastructure. They are accounting systems that match buyers and sellers, but do not balance supply and demand within delivery constraints in the distribution system. That job is left to the distribution system operator to handle. That job may be made more difficult as unfulfillable exchanges may need to be curtailed. A transactive energy system is intentionally designed to work with the distribution system operator as a participant and be a tool to manage the balance of supply and demand under delivery constraints.

As noted in Fig. 4.2, transactive energy systems may be designed for different levels or layers of the system structure. Pilots have been done on distribution circuits where customers were exposed to prices that were a function of wholesale locational marginal prices. That allows local flexibility to reduce energy expenses by (1) consuming more when prices are low and consuming less when prices are high; (2) mitigating physical flow constraints in the transmission system, which are reflected in higher or lower locational marginal prices depending on the location of flow constraints in the transmission system; and (3) managing local distribution constraints by reducing consumption when a circuit becomes overloaded.

Demonstrations have also been done with transactive energy systems in energy communities like campuses and microgrids. These systems may have hundreds or thousands of flexible devices (producing and consuming) that coordinate their operation to address local constraints and minimize energy costs with a utility supplier. Finally, transactive energy systems have also been applied within facilities, such as buildings with complex, interrelated systems for heating and cooling, lighting, electrical vehicle charging, etc. Like a microgrid or larger community of buildings, the flexibility can be managed to serve the business needs of the facility while reducing energy costs with their utility.

ICT platform technology provide important enablers for realizing transactive energy systems. Publish and subscribe message bus systems are commonly found in transactive energy system designs and can be a good fit for multiagent systems. Over the past decade, distributed ledger technology has been demonstrated to have attractive features for transactive energy system applications. The irrefutable auditability of the transactions in the distributed ledger is an obvious feature. Cybersecurity is also a strong feature of distributed ledger platforms with keyless cryptography that can address key management issues that can complicate other cybersecurity technology approaches. In addition, features such as smart contracts (also known as chain code) can help in fulfilling the contractual terms specified in a TE system design. Extensive research is being carried out around the deployment of decentralized TE systems and the infrastructure needed to this end (e.g., see Refs. [42–45]).

TE systems concepts play a key role in the evolution of the energy grid and the implementation of local energy markets. As TE market participation grows, market operators will face diverse challenges, including things like privacy of information and trust by participants when energy is exchanged.

## 4.5 The local electricity market concept

Considering the dynamic transformation that the electrical grid is going through, the infrastructural and technological advances that have enabled the creation of SGs provide a basis for the implementation of transactive energy systems. To do this, several requirements, of a technical and technologic nature, are essential, such as two-way communication, a merging of ICT and electricity grid components, intelligent and remote supervision, AMI and smart metering, among others [46].

Consequently, transactive energy systems are demonstrating how a traditionally centralized electricity market environment, focused almost entirely on the large players, can evolve toward the effective and automated accommodation of all the players involved in the sector. Small consumers, in specific, who are required to take on an active role in future energy systems, are able to take advantage of such a comprehensive, interconnected system, through intelligent automated building/home energy management systems and market models suited to realize the full potential from their participation [47].

The widespread adoption of transactive energy systems, and especially the new market models that result from the advances in the field, are leading to the emergence of a new concept: the LEM. LEM is described in Ref. [48] as "the term used to describe initiatives to establish a marketplace to coordinate the generation, supply, storage, transport, and consumption of energy from decentralized energy resources (e.g., renewable energy generators, storage and demand-side response providers) within a confined geographical area." It is clear that transactive energy systems and LEM walk hand in hand in addressing the same fundamental issues. However, some intrinsic differences can be pointed out, for example, while transactive energy systems focus predominantly in automated solutions, LEM tend to incorporate a broader market-driven perspective, namely by defining different types of price-incentive and trading approaches, while comprising the alignment and cooperation with other nonlocal markets, such as wholesale and regional markets [49].

In fact, experiences in LEM models as part of the regional electricity market have already been conducted. A relevant pioneer study is the Brooklyn Microgrid case study, which focuses on the participation of small power networks, or microgrids, in the electricity market, represented by aggregator entities [3]. Complementarily to the validation of the local market mechanisms themselves, such experiments also enable developing the required means for effective connection with larger scale markets, namely wholesale EM [4].

The success of these initial experiences is engaging regulatory and policymakers to incentivize the evolution of EM toward the widespread implementation of LEM. For example, the EC has launched several policy packages that envisage the creation of LEM as a core part of internal European electricity market, by empowering consumers and placing them in the center of future power systems [50,51].

Consequently, the expected impact of LEM adoption and potential effective market models in a LEM context are currently a priority focus of research and experimentation work [52]. These studies reveal relevant conclusions on the benefits, barriers, and opportunities of LEM. The studies' results consider different dimensions on prominent related issues, such as regulatory, infrastructural, technological, and business models. These different aspects are addressed in dedicated chapters of this book. Market and business models are addressed in Part 2—Local market models and opportunities (Chapters 5—8). Technological solutions are discussed in Part 3—Enablers for local electricity markets (Chapters 9—15). The regulatory framework found today and its envisaged evolution all around the globe are discussed in Part 4—Regulatory framework: current trends and future perspectives (Chapters 16—22).

## 4.6 Conclusion

Despite the strain of technological advances that the electrical grid is experiencing in terms of SG investment and the current emergence of LEM, significant work is required to overcome the technical and regulatory barriers before the implementation of such systems becomes routine. This chapter provides an overview of the electrical grid evolution, and the advances achieved toward widespread implementation of SGs in terms of technological and infrastructure requirements. After that, the concept of transactive energy is described, showing its key value in facilitating automated energy transactions at the local level. In fact, transactive energy takes advantage of the automation and technology advancements (e.g., distributed ledger ICT platforms and advanced metering infrastructure) that are emerging, invigorated by the investment in SGs. Advances boosted by SG development and the relevant findings in the transactive energy domain are facilitating widespread interest in the LEM concept. This is leading to greater focus in research and development activities on the topic.

## References

[1] S. Kakran, S. Chanana, Smart operations of smart grids integrated with distributed generation: a review, Renew. Sustain. Energy Rev. 81 (2018) 524—535. Available from: https://doi.org/10.1016/j.rser.2017.07.045.

[2] R. Walton, ConEd virtual power plant shows how New York's REV is reforming utility practices, 2016. Avialble at: https://www.utilitydive.com/news/coned-virtual-power-plant-shows-how-new-yorks-rev-is-reforming-utility-pra/421053/ (Accessed: 18 February 2021)

[3] G. Bade, Little less talk: With new revenue models, New York starts to put REV into action, 2016. Available at: https://www.utilitydive.com/news/little-less-talk-with-new-revenue-models-new-york-starts-to-put-rev-into/420657/ (Accessed: 18 February 2021)

[4] E. Mengelkamp, et al., Designing microgrid energy markets: a case study: the Brooklyn Microgrid, Appl. Energy 210 (2018) 870–880. Available from: https://doi.org/10.1016/j.apenergy.2017.06.054.

[5] E. Mengelkamp, et al., Trading on local energy markets: a comparison of market designs and bidding strategies, in: Proceedings of the 2017 14th International Conference on the European Energy Market (EEM), IEEE, 2017, pp. 1–6. Available from: https://doi.org/10.1109/EEM.2017.7981938.

[6] V.C. Gungor, et al., A survey on smart grid potential applications and communication requirements, IEEE Trans. Ind. Inform. 9 (1) (2013) 28–42. Available from: https://doi.org/10.1109/TII.2012.2218253.

[7] Y. Yan, et al., A survey on smart grid communication infrastructures: motivations, requirements and challenges, IEEE Commun. Surv. Tutor. 15 (1) (2013) 5–20. Available from: https://doi.org/10.1109/SURV.2012.021312.00034.

[8] M.L. Tuballa, M.L. Abundo, A review of the development of smart grid technologies, Renew. Sustain. Energy Rev. 59 (2016) 710–725. Available from: https://doi.org/10.1016/j.rser.2016.01.011.

[9] I. Colak, et al., A survey on the critical issues in smart grid technologies, Renew. Sustain. Energy Rev. 54 (2016) 396–405. Available from: https://doi.org/10.1016/j.rser.2015.10.036.

[10] A. Ghosal, M. Conti, Key management systems for smart grid advanced metering infrastructure: a survey, IEEE Commun. Surv. Tutor. 21 (3) (2019) 2831–2848. Available from: https://doi.org/10.1109/COMST.2019.2907650.

[11] A. Gopstein, C. Nguyen, C.O'Fallon, D. Wollman, N. Hasting, NIST Framework and Roadmap for Smart Grid Interoperability Standards Release 4.0, 2020.

[12] F. Lezama, et al., Local energy markets: paving the path toward fully transactive energy systems, IEEE Trans. Power Syst. 34 (5) (2018) 4081–4088. Available from: https://doi.org/10.1109/TPWRS.2018.2833959.

[13] O. Abrishambaf, et al., Towards transactive energy systems: an analysis on current trends, Energy Strateg. Rev. 26 (2019) 100418. Available from: https://doi.org/10.1016/j.esr.2019.100418.

[14] F. Lezama, et al., Agent-based microgrid scheduling: an ICT perspective, Mob. Netw. Appl. (2017). Available from: https://doi.org/10.1007/s11036-017-0894-x.

[15] I. Abubakar, et al., Application of load monitoring in appliances' energy management—a review, Renew. Sustain. Energy Rev. 67 (2017) 235–245. Available from: https://doi.org/10.1016/j.rser.2016.09.064.

[16] B.N. Silva, et al., Towards sustainable smart cities: a review of trends, architectures, components, and open challenges in smart cities, Sustain. Cities Soc. 38 (2018) 697–713. Available from: https://doi.org/10.1016/j.scs.2018.01.053.

[17] European Commission, Third Report on the State of the Energy Union, Brussels, 2017.

[18] EUROSTAT, Renewable Energy Statistics—Statistics Explained. 2021. Available at: https://ec.europa.eu/eurostat/statistics-explained/index.php/Renewable_energy_statistics (Accessed: 18 February 2021).

[19] International Energy Agency (IEA), Market Report Series: Renewables 2017. Analysis and Forecasts to 2022. 2017.

[20] IEA, IEA: global energy review. < https://www.iea.org/reports/global-energy-review-2020/renewables >, 2020.

[21] European Commission, Smart Metering Deployment in the European Union. 2021. Available at: https://ses.jrc.ec.europa.eu/smart-metering-deployment-european-union (Accessed: 18 February 2021)

[22] P. Siano, Demand response and smart grids—a survey, Renew. Sustain. Energy Rev. 30 (2014) 461–478. Available from: https://doi.org/10.1016/j.rser.2013.10.022.

[23] Z. Wang, R. Paranjape, Optimal residential demand response for multiple heterogeneous homes with real-time price prediction in a multiagent framework, IEEE Trans. Smart Grid 8 (3) (2017) 1173–1184. Available from: https://doi.org/10.1109/TSG.2015.2479557.

[24] European Commission, Incorporing Demand Side Flexibility, in Particular Demand Response, in Electricity Markets, 2013.

[25] P. Faria, et al., Constrained consumption shifting management in the distributed energy resources scheduling considering demand response, Energy Convers. Manag. 93 (2015) 309–320. Available from: https://doi.org/10.1016/j.enconman.2015.01.028.

[26] K. Wang, et al., Wireless big data computing in smart grid, IEEE Wirel. Commun. 24 (2) (2017) 58−64. Available from: https://doi.org/10.1109/MWC.2017.1600256WC.

[27] IEEE 2030.5 Ecosystem Steering Committee (ESC), Interoperability Maturity Roadmap IEEE Std 2030.5. IEEE SA Whitepaper, 2019.

[28] M. Knight, et al., Interoperability maturity model a qualitative and quantitative approach for measuring interoperability, Pacific Northwest Natl. Lab. PNNL-29683, 2020.

[29] K. Kok, S. Widergren, A society of devices: integrating intelligent distributed resources with transactive energy, IEEE Power Energy Mag. 14 (3) (2016) 34−45. Available from: https://doi.org/10.1109/MPE.2016.2524962.

[30] J. Kolln, et al., Reference interoperability procurement language, Pacific Northwest Natl. Lab. PNNL-28666, 2020.

[31] S. Widergren, et al., Interoperability strategic vision, a GMLC white paper, Pacific Northwest Natl. Lab. PNNL-27320, 2018.

[32] S. Widergren, et al., The plug-and-play electricity era: interoperability to integrate anything, anywhere, anytime, IEEE Power Energy Mag. 17 (5) (2019) 47−58. Available from: https://doi.org/10.1109/MPE.2019.2921742.

[33] N. Atamturk, M. Zafar, Transactive energy: a surreal vision or a necessary and feasible solution to grid problems?, California Public Utilities Commission Policy & Planning Division, October 2014. Available at: https://www.cpuc.ca.gov/uploadedFiles/CPUC_Public_Website/Content/About_Us/Organization/Divisions/Policy_and_Planning/PPD_Work/PPDTransactiveEnergy_30Oct14.pdf

[34] C. Hertzog, Transactive Energy − American Perspectives on Grid Transformations, 2013. Available at: https://www.engerati.com/transmission-distribution/transactive-energy-american-perspectives-on-grid-transformations/ (Accessed 18 February 2021).

[35] J. Jeff, A how-to guide for transactive energy, 2013. Available at: https://www.greentechmedia.com/articles/read/a-how-to-guide-for-transactive-energy (Accessed: 18 February 2021).

[36] S. Yin, et al., Decentralized electricity market with transactive energy—a path forward, Electr. J. 32 (4) (2019) 7−13. Available from: https://doi.org/10.1016/j.tej.2019.03.005.

[37] F. Rahimi, et al., The changing electrical landscape: end-to-end power system operation under the transactive energy paradigm, IEEE Power Energy Mag. 14 (3) (2016) 52−62. Available from: https://doi.org/10.1109/MPE.2016.2524966.

[38] S.M. Sajjadi, et al., Transactive energy market in distribution systems: a case study of energy trading between transactive nodes, in: 2016 North American Power Symposium (NAPS), IEEE, 2016, pp. 1−6. Available from: https://doi.org/10.1109/NAPS.2016.7747895.

[39] M.R. Knight, Stochastic Impacts of Metaheuristics from the Toaster to the Turbine [Technology Leaders], IEEE Electrif. Mag. 4(4) (2016) 52–40. Available from: https://doi.org/10.1109/MELE.2016.2614218.

[40] J.K. Kok, et al., PowerMatcher, in: Proceedings of the Fourth International Joint Conference on Autonomous Agents and Multiagent Systems—AAMAS '05, ACM Press, New York, 2005, p. 75. Available from: https://doi.org/10.1145/1082473.1082807.

[41] D.J. Hammerstrom, et al., Pacific Northwest GridWise Testbed Demonstration Projects; Part I. Olympic Peninsula Project, Richland, WA, 2008. Available from: https://doi.org/10.2172/926113.

[42] S.N. Gourisetti, et al., Blockchain Smart Contracts for Transactive Energy Systems, Richland, WA, 2019. Available from: https://doi.org/10.2172/1658380.

[43] J. Guerrero, et al., Towards a transactive energy system for integration of distributed energy resources: home energy management, distributed optimal power flow, and peer-to-peer energy trading, Renew. Sustain. Energy Rev. 132 (2020) 110000. Available from: https://doi.org/10.1016/j.rser.2020.110000.

[44] P. Siano, et al., A survey and evaluation of the potentials of distributed ledger technology for peer-to-peer transactive energy exchanges in local energy markets, IEEE Syst. J. (2019) 1−13. Available from: https://doi.org/10.1109/JSYST.2019.2903172.

[45] M.F. Zia, et al., Microgrid transactive energy: review, architectures, distributed ledger technologies, and market analysis, IEEE Access. 8 (2020) 19410−19432. Available from: https://doi.org/10.1109/ACCESS.2020.2968402.

[46] L.M. Camarinha-Matos, Collaborative smart grids—a survey on trends, Renew. Sustain. Energy Rev. 65 (2016) 283−294. Available from: https://doi.org/10.1016/j.rser.2016.06.093.

[47] F. Rahimi, F. Albuyeh, Applying lessons learned from transmission open access to distribution and grid-edge transactive energy systems, in: 2016 IEEE Power & Energy Society Innovative Smart Grid

Technologies Conference (ISGT), IEEE, 2016, pp. 1–5. Available from: https://doi.org/10.1109/ISGT.2016.7781236.

[48] Catapult's Markets Policy and Regulation Team, The policy and regulatory context for new local energy markets, 2019. Available at: https://es.catapult.org.uk/reports/the-policy-and-regulatory-context-for-new-local-energy-markets/ (Accessed: 18 February 2021).

[49] R. Bray, B. Woodman, Unlocking local energy markets, in: British Institute of Energy Economics (BIEE) Conference, Oxford, 2018.

[50] European Parliament, Council of the EU: Directive (EU) 2019/944 on Common Rules for the Internal Market for Electricity and Amending Directive 2012/27/EU, Off. J. Eur. Union. L 158 (18) (2019). < http://eur-lex.europa.eu/pri/en/oj/dat/2003/l_285/l_28520031101en00330037.pdf > .

[51] The European Commission, Comission Regulation (EU) 2019/943 of the European Parliament and of the Council of 5 June 2019 on the internal market for electricity, Off. J. Eur. Union 2019 714 (2018) 54–124.

[52] T. Sousa, et al., Peer-to-peer and community-based markets: a comprehensive review, Renew. Sustain. Energy Rev. 104 (2019) 367–378. Available from: https://doi.org/10.1016/j.rser.2019.01.036.

# PART II

# Local market models and opportunities

# Local market models

*Amin Shokri Gazafroudi*[1,2], *Hosna Khajeh*[3],
*Miadreza Shafie-khah*[3], *Hannu Laaksonen*[3] *and*
*Juan Manuel Corchado*[1,4]

[1]BISITE Research Group, University of Salamanca, Salamanca, Spain [2]Institute for Automation and Applied Informatics, Karlsruhe Institute of Technology (KIT), Eggenstein-Leopoldshafen, Germany [3]School of Technology and Innovations, University of Vaasa, Vaasa, Finland [4]Department of Electronics, Information and Communication, Faculty of Engineering, Osaka Institute of Technology, Osaka, Japan

## 5.1 Introduction

Conventionally, power was transferred in a top-down manner from large power plants to the customers and end-users through transmission and distribution networks. End-users used to be ratepayers aggregated by retailers. They were usually not subject to the wholesale market prices. Therefore they were not able to follow system variations. These days, however, consumers have changed from inactive submissive ratepayers to proactive consumers called prosumers who can provide energy and flexibility, leading to the bidirectional flow of power [1]. A number of consumers in distribution systems are equipped with renewable energy resources (RES) which may considerably impact the system security and stability [2]. The variability of RES as well as the uncertainty coming from the prediction error of these resources are considered as an important issue that should be resolved when designing the future power system. Not only the variability of power may result in the security problem, but also the bidirectional flow may cause voltage issues in distribution networks. Congestion management is another factor that should be taken into consideration. As a result, the distribution system will require more flexibility resources to maintain the security of the system at a predefined level. On the one hand, prosumers should be motivated enough to produce green energy and support climate change mitigation-related targets. On the other hand, small prosumers are not large enough to participate in the wholesale electricity markets. Hence, in order to promote renewable resources in the local-level system, prosumers should be given enough motivation to follow system variation, and consumers need to have incentives to install renewable resources such as solar panels [3].

In this way, local markets can be defined as an effective solution to manage local resources [4,5]. These markets are able to adopt distributed energy resources (DER) effectively, cope with fluctuations and integrate prosumers aiming to support the implementation of smart grids [6,7]. Local energy and flexibility markets may provide a huge amount of flexibility for distribution and transmission networks [8,9]. They may decrease the burden on the system operator in terms of balancing the system generation and demand [10]. Many small flexibility resources, for example, energy storage systems, shiftable loads and interruptible loads, are capable of responding to the system request fast and are therefore considered as good potential providers of different flexibility services. Furthermore, prosumers' motivation for electricity production will increase if they are given the opportunity to take part in local markets, make profits, and compete with others through selling their surplus power [11,12].

Local trading is suggested to be performed in different frameworks. In the first one, each prosumer will be able to buy and sell energy independently. In other words, there exists no agent as a mediator or broker intervening in the trading process. A prosumer is allowed to choose the agent from whom it will buy its energy [13–16]. It also has an option to select the seller to whom it will sell. This trading can be implemented as a peer-to-peer (P2P) transaction (distributed approach) or be supervised in the local market environment (decentralized approach). The examples of this kind of trading can be seen in Ref. [17–21].

In the second kind of local trading, a centralized entity is responsible for running the local market, matching supply and demand, and determining local market prices. For example, in Ref. [9] a smart energy provider was proposed to be in charge of scheduling of flexible resources in a local market with a high penetration of renewables. The social surplus of the local market was suggested to be maximized in Ref. [22] through the supervision of a local market operator (LMO). The trader's role in Ref. [23] was matching offers of sellers and buyers. Jelenković and Budrović [24] introduced a distributor acting as a mediator who gathers energy from producers and distributes it to those customers who want to buy energy from the local market. While some studies (e.g., [25]) proposed that a regulator would be in charge of regulating local markets, authors in Refs. [26,27] stated that an auctioneer is the one who can coordinate local market players.

In this chapter, we introduce three approaches for electricity trade in the local market. The first approach, called decentralized, empowers agents to buy/sell electricity autonomously from/to the local market. The model is from the perspective of independent agents, aiming to maximize their profits. Moreover, there exists no direct electricity transaction among players in the decentralized approach. In the second approach which is named the distributed approach, agents are given the option to trade electricity with each other through P2P transactions or the local market. The third proposed approach, however, is the centralized approach in which a LMO is responsible for management the electricity transaction among agents. In this model, flexibility can also be traded in the real-time local market, whereas energy is proposed to be exchanged in the day-ahead timescale. Finally, the performance of the proposed decentralized approach as a fundamental model for local electricity trading will be simulated and studied.

## 5.2 Local electricity trading

### 5.2.1 Decentralized approach

In this section, the energy management problem from perspective of autonomous agents in the local electricity market is presented. Fig. 5.1 displays an architecture for an agent for trading energy with other players in the local electricity market. As seen in Fig. 5.1, each agent is able to sell/buy electricity to/from other players in the local electricity market. Eq. (5.1) represents an objective function for agent $i$.

$$\min_{P_i^G,\ P_i^B,\ P_i^S} \lambda_i^B P_i^B - \lambda_i^S P_i^S \tag{5.1}$$

Here, agent $i$ is willing to minimize its objective function which consists of two terms. The first term $\left(\lambda_i^B P_i^B\right)$ represents the cost of energy bought from the local market. Here, $\lambda_i^B$ and $P_i^B$ present the price and quantity of electricity bought from the local market, respectively.

In the second term $\left(\lambda_i^S P_i^S\right)$, the profit of electricity sold to the local market is represented. This way, $\lambda_i^S$ represents the sold electricity price, and $P_i^S$ states the electricity sold to the local market. Also, $\lambda_i^S$ and $\lambda_i^B$ can be presented as offering and bidding curves by

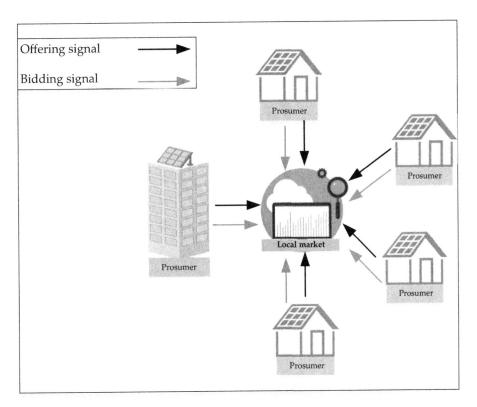

**FIGURE 5.1** An architecture for the decentralized energy trading.

agent $i$. Here, it is assumed that there is a polynomial relation between the offering/bidding electricity price and the sold/bought electricity of agents as represented in Eqs. (5.2) and (5.3). Where $c_i^{0,S}$, $c_i^{1,S}$, $c_i^{0,B}$, and $c_i^{1,B}$ are defined positive appropriate constants.

$$\lambda_i^S = c_i^{0,S} + c_i^{1,S} P_i^S \tag{5.2}$$

$$\lambda_i^B = c_i^{0,B} - c_i^{1,B} P_i^B \tag{5.3}$$

Hence, the objective function of agent $i$ is restated in Eq. (5.4).

$$\min_{P_i^G, P_i^B, P_i^S} c_i^{0,B} P_i^B - c_i^{1,B} \left(P_i^B\right)^2 - c_i^{0,S} P_i^S - c_i^{1,S} \left(P_i^S\right)^2 \tag{5.4}$$

Moreover, there are physical constraints for energy transaction in the system. Agent $i$ is able to produce and consume electricity, so it is called prosumer, that is, proactive consumer.

$$P_i^{max} - P_i^G \geq 0 : \alpha_i \tag{5.5}$$

Eq. (5.5) represents maximum capacity constraint of power generation for agent $i$, representing the power generation and maximum capacity of power generation of agent $i$. Moreover, $\alpha_i$ is a dual variable of Eq. (5.5).

$$P_i^G - L_i - P_i^S + P_i^B \geq 0 : \beta_i \tag{5.6}$$

$$L_i - P_i^G - P_i^B + P_i^S \geq 0 : \gamma_i \tag{5.7}$$

As a prosumer, agent $i$ can transact energy in a two-way direction with other players in the local market. In this way, the main task of agent $i$ is to provide its desired electricity load ($L_i$), then it sells/buys its extra generation/demand to/from other players in the local electricity market, as represented in Eqs. (5.6) and (5.7). $\beta_i$ and $\gamma_i$ are defined as dual variables for Eqs. (5.6) and (5.7), respectively. Besides, $P_i^G$, $P_i^S$, and $P_i^B$ are positive variables which are represented in Eqs. (5.8–5.10), respectively.

$$P_i^G \geq 0 : \delta_i \tag{5.8}$$

$$P_i^S \geq 0 : \mu_i \tag{5.9}$$

$$P_i^B \geq 0 : \nu_i \tag{5.10}$$

In the following, the first-order optimality conditions are obtained:

$$\alpha_i - \beta_i + \gamma_i - \delta_i = 0 \tag{5.11}$$

$$-c_i^{0,S} - 2c_i^{1,S} P_i^S + \beta_i - \gamma_i - \mu_i = 0 \tag{5.12}$$

$$c_i^{0,B} - 2c_i^{1,B} P_i^B - \beta_i + \gamma_i - \nu_i = 0 \tag{5.13}$$

As $\delta_i$, $\mu_i$, and $\nu_i$ are positive variables, Eqs. (5.14–5.16) are obtained based on Eqs. (5.11–5.13), respectively.

$$\alpha_i - \beta_i + \gamma_i \geq 0 \tag{5.14}$$

$$-c_i^{0,S} - 2c_i^{1,S}P_i^S + \beta_i - \gamma_i \geq 0 \tag{5.15}$$

$$c_i^{0,B} - 2c_i^{1,B}P_i^B - \beta_i + \gamma_i \geq 0 \tag{5.16}$$

Thus the energy management problem of agent $i$ could be in the general form of the complementarity programming. Therefore the complementarity conditions are given:

$$P_i^{max} - P_i^G \geq 0 \perp \alpha_i \geq 0, \forall i. \tag{5.17}$$

$$P_i^G - L_i - P_i^S + P_i^B \geq 0 \perp \beta_i \geq 0, \forall i. \tag{5.18}$$

$$L_i - P_i^G - P_i^B + P_i^S \geq 0 \perp \gamma_i \geq 0, \forall i. \tag{5.19}$$

$$\alpha_i - \beta_i + \gamma_i \geq 0 \perp P_i^G \geq 0, \forall i. \tag{5.20}$$

$$-c_i^{0,S} - 2c_i^{1,S}P_i^S + \beta_i - \gamma_i \geq 0 \perp P_i^S \geq 0, \forall i. \tag{5.21}$$

$$c_i^{0,B} - 2c_i^{1,B}P_i^B - \beta_i + \gamma_i \geq 0 \perp P_i^B \geq 0, \forall i. \tag{5.22}$$

## 5.2.2 Distributed (peer-to-peer) approach

Here, we present a distributed approach for electricity trading in the local market considering P2P transactions among agents. Thus each agent is able to trade electricity with other agents as the P2P transactions, or exchange electricity with the local market. Fig. 5.2 illustrates an architecture for P2P trading. Thus Eq. (5.23) represents the objection function for agent $i$ in the distributed approach for local electricity trading.

$$\min_{P_i^G, P_i^B, P_i^S, P_{ij}^B, P_{ij}^S} \lambda_i^B P_i^B - \lambda_i^S P_i^S + \sum_j \left( \lambda_{ij}^{BD} P_{ij}^{BD} - \lambda_{ij}^{SD} P_{ij}^{SD} \right) \tag{5.23}$$

Here, the electricity sold to agent $j$ by agent $i$ is represented by $P_{ij}^{SD}$, and the quantity of electricity bought from agent $i$ by agent $j$ is represented by $P_{ji}^{BD}$.

Moreover, $\lambda_{ij}^{SD}$ and $\lambda_{ij}^{BD}$ represent sold and purchased P2P electricity prices among agents $i$ and $j$. Similar to Eqs. (5.2) and (5.3), the polynomial relation is defined between the quantity of P2P electricity trade and its corresponding price which are presented in Eqs. (5.24) and (5.25).

$$\lambda_{ij}^{SD} = c_i^{0,SD} + c_i^{1,SD}P_{ij}^{SD} \tag{5.24}$$

$$\lambda_{ij}^{BD} = c_i^{0,BD} - c_i^{1,BD}P_{ij}^{BD} \tag{5.25}$$

Hence, the objective function for agent $i$ is redefined in Eq. (5.26).

$$\min_{P_i^G, P_i^B, P_i^S, P_{ij}^B, P_{ij}^S} c_i^{0,B}P_i^B - c_i^{1,B}\left(P_i^B\right)^2 - c_i^{0,S}P_i^S - c_i^{1,S}\left(P_i^S\right)^2$$
$$+ \sum_j \left( c_i^{0,BD}P_{ij}^{BD} - c_i^{1,BD}(P_{ij}^{BD})^2 - c_i^{0,SD}P_{ij}^{SD} - c_i^{1,SD}(P_{ij}^{SD})^2 \right) \tag{5.26}$$

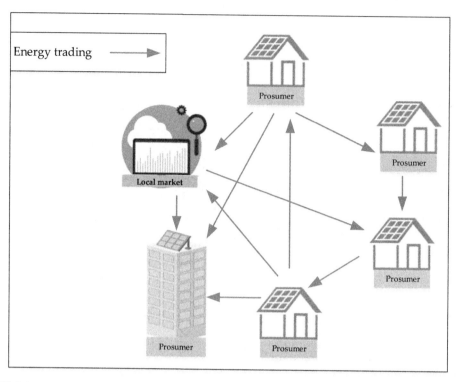

**FIGURE 5.2**  An architecture for the distributed (P2P) trading.

In addition, there are physical constraints for agents in the local electricity network, as represented in Eqs. (5.27–5.29):

$$P_i^G - L_i - P_i^S + P_i^B - \sum_j P_{ij}^{SD} + \sum_j P_{ij}^{BD} \geq 0 : \beta_i \qquad (5.27)$$

$$L_i - P_i^G - P_i^B + P_i^S - \sum_j P_{ij}^{BD} + \sum_j P_{ij}^{SD} \geq 0 : \gamma_i \qquad (5.28)$$

$$P_{ij}^{SD} = P_{ji}^{BD} : \xi_{ij} \qquad (5.29)$$

As seen in Eq. (5.29), the electricity sold to agent $j$ by agent $i$ ($P_{ij}^{SD}$) equals the quantity of electricity bought from agent $i$ by agent $j$ ($P_{ji}^{BD}$). Eqs. (5.30) and (5.31) show that $P_i^G$, $P_i^S$, $P_i^B$, $P_{ij}^{SD}$ and $P_{ij}^{BD}$ are positive variables.

$$P_{ij}^{SD} \geq 0 : \sigma_{ij} \qquad (5.30)$$

$$P_{ij}^{BD} \geq 0 : \varsigma_{ij} \qquad (5.31)$$

Moreover, constraints (5.5) and (5.8–5.10) are considered in the distributed local electricity trading. In this way, the KKT conditions for the distributed electricity problem is given:

$$-c_i^{0,SD} - 2c_i^{1,SD} P_{ij}^{SD} + \beta_i - \gamma_i + \xi_{ij} - \sigma_{ij} = 0 \tag{5.32}$$

$$c_i^{0,BD} - 2c_i^{1,BD} P_{ij}^{BD} - \beta_i + \gamma_i - \xi_{ji} - \varsigma_{ij} = 0 \tag{5.33}$$

Hence, Eqs. (5.32) and (5.33) are obtained because $\sigma_{ij}$ and $\varsigma_{ij}$ are positive variables.

$$-c_i^{0,SD} - 2c_i^{1,SD} P_{ij}^{SD} + \beta_i - \gamma_i + \xi_{ij} \geq 0 \tag{5.32}$$

$$c_i^{0,BD} - 2c_i^{1,BD} P_{ij}^{BD} - \beta_i + \gamma_i - \xi_{ji} \geq 0 \tag{5.33}$$

The distributed local electricity trading is in the general form of a mixed complementarity programming as it consists of an equality constraint [Eq. (5.29)]. Therefore, complementarity constraints for the P2O electricity trading problem are given as:

$$P_i^G - L_i - P_i^S + P_i^B - \sum_j P_{ij}^{SD} + \sum_j P_{ij}^{BD} \geq 0 \perp \beta_i \geq 0, \forall i. \tag{5.34}$$

$$L_i - P_i^G - P_i^B + P_i^S - \sum_j P_{ij}^{BD} + \sum_j P_{ij}^{SD} \geq 0 \perp \gamma_i \geq 0, \forall i. \tag{5.35}$$

$$P_{ij}^{SD} = P_{ji}^{BD} \perp \xi_{ij}, \forall i. \tag{5.36}$$

$$-c_i^{0,SD} - 2c_i^{1,SD} P_{ij}^{SD} + \beta_i - \gamma_i + \xi_{ij} \geq 0 \perp P_{ij}^{SD}, \forall i. \tag{5.37}$$

$$c_i^{0,BD} - 2c_i^{1,BD} P_{ij}^{BD} - \beta_i + \gamma_i - \xi_{ji} \geq 0 \perp P_{ij}^{BD}, \forall i. \tag{5.38}$$

subject to (5.17), (5.20–5.22).

## 5.2.3 Centralized (community-based) approach

This section presents a centralized approach for the electricity transaction in the local market. An architecture for the proposed centralized approach is depicted in Fig. 5.3. This way, the LMO is in charge of managing the electricity trading of agents. Moreover, energy and flexibility are defined as electricity services traded in the local market. Here, two-stage local electricity market model is presented which is cleared in a community-based approach. The first and second stages are called day-ahead and real-time local market. In the day-ahead local market, energy is transacted among agents of the system. However, flexibility, as an electricity commodity, is traded in the real-time local market. Eq. (5.39) represents an objective function of the LMO which is revenue of energy transaction for agents that should be maximized.

$$\max_{P_i^G, P_i^S, P_i^B, F_i^U, F_i^D} \overbrace{\sum_i (\lambda_i^S P_i^S - \lambda_i^B P_i^B)}^{day-ahead\ stage} + \overbrace{\sum_i (\lambda_i^U F_i^U + \lambda_i^D F_i^D)}^{real-time\ stage} \tag{5.39}$$

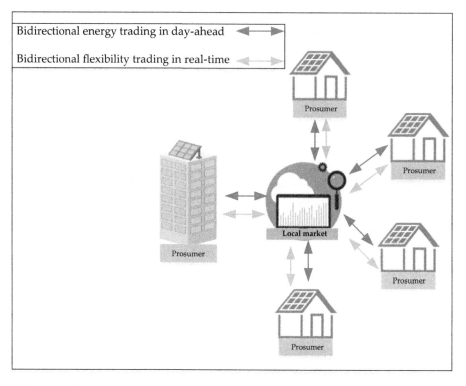

**FIGURE 5.3**    An architecture for the centralized energy and flexibility trading.

Eq. (5.39) represents total revenue for agents in the local market which contains their day-ahead and real-time revenue. Here, $\lambda_i^U$ and $\lambda_i^D$ are upward and downward flexibility price in the real-time local market, and $F_i^U$ and $F_i^D$ are defined as upward and downward flexibility services provided by agent $i$ in the real-time stage. The physical constraints for agents in the day-ahead and real-time stages are presented in the following.

$$P_i^G + P_i^B = L_i + P_i^S, \forall i. \tag{5.40}$$

$$P_i^S \leq u_i P_i^{max}, \forall i. \tag{5.41}$$

$$P_i^B \leq (1 - u_i)L_i, \forall i. \tag{5.42}$$

Eq. (5.40) represents the balancing constraint for energy trade for agent $i$. In Eqs. (5.41) and (5.42), $u_i$ is defined as a binary variable to guarantee that agent $i$ only sells or buys energy from the local market. Moreover, Eqs. (5.5) and (5.8–5.10) are included in the centralized local electricity trading problem. As highlighted before, flexibility, $F_i$, is traded between agents in the real-time local electricity market.

$$F_i = F_i^U - F_i^D, \forall i. \tag{5.43}$$

$$F_i^U = F_i^{U,G} + F_i^{U,C}, \forall i. \tag{5.44}$$

$$F_i^D = F_i^{D,G} + F_i^{D,C}, \forall i. \tag{5.45}$$

$$0 \leq F_i^{U,G} \leq \gamma_i^{U,G} P_i^S, \forall i. \tag{5.46}$$

$$0 \leq F_i^{D,G} \leq \gamma_i^{D,G} P_i^B, \forall i. \tag{5.47}$$

$$0 \leq F_i^{U,C} \leq \gamma_i^{U,C} P_i^S, \forall i. \tag{5.48}$$

$$0 \leq F_i^{D,C} \leq \gamma_i^{D,C} P_i^B, \forall i. \tag{5.49}$$

$$F_i^{U,G} \leq v_i^G P_i^{max}, \forall i. \tag{5.50}$$

$$F_i^{U,C} \leq \left(1 - v_i^G\right) P_i^{max}, \forall i. \tag{5.51}$$

$$F_i^{U,C} \leq v_i^C L_i, \forall i. \tag{5.52}$$

$$F_i^{D,C} \leq \left(1 - v_i^C\right) L_i, \forall i. \tag{5.53}$$

Flexibility splits into upward and downward flexibility as represented in Eq. (5.43). Moreover, both producers and consumers are able to provide flexibility as seen in Eqs. (5.44) and (5.45). Thus $\gamma_i^{U,G}$, $\gamma_i^{D,G}$, $\gamma_i^{U,C}$ and $\gamma_i^{D,C}$ are defined as upward and downward flexibility coefficient provided by producers ($\gamma_i^{U,G}$ and $\gamma_i^{D,G}$) and consumers ($\gamma_i^{U,C}$ and $\gamma_i^{D,C}$) which are between 0 and 1 ($0 \leq \gamma_i^{U,G} + \gamma_i^{U,C} \leq 1$, $0 \leq \gamma_i^{D,G} + \gamma_i^{D,C} \leq 1$) as represented in Eqs. (5.46–5.53). Upward flexibility presents flexibility of agent $i$ as a producer in the local market. However, downward flexibility presents the flexibility of agent $i$ as a consumer in the local network. In other words, flexible producers/consumers have this potential to decrease their desired energy production/consumption. In Eqs. (5.50–5.53), $v_i^G$ and $v_i^C$ are defined as binary variables to guarantee that agent $i$ is able to provide only upward/downward flexibility service from the generation side and demand side, respectively.

$$\sum_i \left(P_i^B - P_i^S - F_i^D + F_i^U\right) = 0, \forall i. \tag{5.54}$$

Eq. (5.52) represents the balancing constraint for local electricity trading. Moreover, the sustainability of the local network is guaranteed by Eq. (5.54). Thus the LMO does not depend on the upstream grid for electricity provision of the agents in the local market. In this way, the centralized local electricity trading is represented in the following:

*max.* (5.54)
s.t. (5.5), (5.8–5.10), (5.40–5.53).

## 5.3 Simulation results

In this section, the results of the local electricity trading problem based on the decentralized approach are represented. In the decentralized study, two agents are considered in the local electricity market. Agent 1 acts as producer and agent 2 acts as a consumer. The system data is presented in Table 5.1. Fig. 5.4 shows the proposed 2-bus test system. According to Eqs. (5.18–5.22) and Table 5.1, the complementarity constraints are obtained.

**TABLE 5.1** Agents data.

| Agent # | Type | $c_i^0$ | $c_i^1$ | $P_i^{max}$ | $L_i$ |
|---|---|---|---|---|---|
| 1 | Producer (S) | 4 | 5.1 | 0.805 | 0 |
| 2 | Consumer (B) | 9.7 | 4.6 | 0 | 1.345 |

| | |
|---|---|
| Complementarity constraints for agent 1: | $0.805 - P_1^G \geq 0 \perp \alpha_1 \geq 0.$ |
| | $P_1^G - 0 - P_1^S + P_1^B \geq 0 \perp \beta_1 \geq 0.$ |
| | $0 - P_1^G - P_1^B + P_1^S \geq 0 \perp \gamma_1 \geq 0.$ |
| | $\alpha_1 - \beta_1 + \gamma_1 \geq 0 \perp P_i^G \geq 0.$ |
| | $-4 - 2 \times 5.1 \times P_1^S + \beta_1 - \gamma_1 \geq 0 \perp P_1^S \geq 0.$ |
| | $0 - 2 \times 0 \times P_1^B - \beta_1 + \gamma_1 \geq 0 \perp P_1^B \geq 0.$ |
| Complementarity constraints for agent 2: | $0 - P_2^G \geq 0 \perp \alpha_2 \geq 0.$ |
| | $P_2^G \quad 1.345 - P_2^S + P_2^B \geq 0 \perp \beta_2 \geq 0.$ |
| | $1.345 - P_2^G - P_2^B + P_2^S \geq 0 \perp \gamma_2 \geq 0.$ |
| | $\alpha_2 - \beta_2 + \gamma_2 \geq 0 \perp P_2^G \geq 0.$ |
| | $-0 - 2 \times 0 \times P_2^S + \beta_2 - \gamma_2 \geq 0 \perp P_2^S \geq 0.$ |
| | $9.7 - 2 \times 4.6 \times P_2^B - \beta_2 + \gamma_2 \geq 0 \perp P_2^B \geq 0.$ |

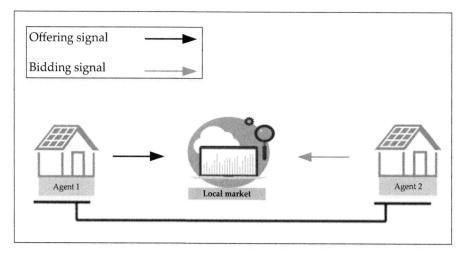

**FIGURE 5.4** The proposed two-bus system trading energy in the decentralized approach.

TABLE 5.2   Outcomes of the decentralized local electricity trading problem.

| Agent # | $P_i^G$ | $P_i^S$ | $P_i^b$ |
|---|---|---|---|
| 1 | 0.805 | 0.805 | 0 |
| 2 | 0 | 0 | 1.345 |

Thus, the results of the decentralized electricity trading in the local market are presented in Table 5.2. As seen in Table 5.2, agent 1 sells its capacity of power generation to the local electricity market, and agent 2 buys its electrical demand from the local market in the decentralized manner.

## Acknowledgments

This work was supported by FLEXIMAR-project (Novel marketplace for energy flexibility), which has received funding from Business Finland Smart Energy Program, 2017—21 (https://www.univaasa.fi/en/research/projects/fleximar/). Moreover, Amin Shokri Gazafroudi acknowledges the support by the CoNDyNet2 project funded by German Federal Ministry of Education and Research under grant number 03EK3055E.

## References

[1] S. Abapour, K. Zare, B. Mohammadi-Ivatloo, Dynamic planning of distributed generation units in active distribution network, IET Gener. Transm. Distrib. 9 (12) (2015) 1455—1463.

[2] M. Khorasany, Y. Mishra, G. Ledwich, Market framework for local energy trading: a review of potential designs and market clearing approaches, IET Gener. Transm. Distrib. 12 (22) (2018) 5899—5908.

[3] A.S. Gazafroudi, F. Prieto-Castrillo, T. Pinto, J.M. Corchado, Organization-based multi-agent system of local electricity market: bottom-up approach, in: F. De la Prieta, et al. (Eds.), Trends in Cyber-Physical Multi-Agent Systems. The PAAMS Collection - 15th International Conference, PAAMS 2017. PAAMS 2017. Advances in Intelligent Systems and Computing, vol. 619, Springer, Cham, 2018.

[4] J. Soares, H. Morais, T. Sousa, Z. Vale, P. Faria, Day-ahead resource scheduling including demand response for electric vehicles, IEEE Trans. Smart Grid 4 (1) (2013) 596—605.

[5] F. Lezama, J. Soares, P. Hernandez-Leal, M. Kaisers, T. Pinto, Z. Vale, Local energy markets: paving the path towards fully transactive energy systems, IEEE Trans. Power Syst. (2018).

[6] W. Saad, Z. Han, H.V. Poor, T. Başar, A noncooperative game for double auction-based energy trading between PHEVs and distribution grids, 2011 IEEE International Conference on Smart Grid Communications (SmartGridComm), IEEE, 2011, pp. 267—272.

[7] M. Rahimiyan, L. Baringo, Strategic bidding for a virtual power plant in the day-ahead and real-time markets: a price-taker robust optimization approach, IEEE Trans. Power Syst. 31 (4) (2015) 2676—2687.

[8] P. Olivella-Rosell, et al., Optimization problem for meeting distribution system operator requests in local flexibility markets with distributed energy resources, Appl. Energy 210 (2018) 881—895.

[9] H. Khajeh, H. Laaksonen, A.S. Gazafroudi, M. Shafie-khah, Towards flexibility trading at TSO-DSO-customer levels: a review, Energies 13 (2020) 165.

[10] J. Villar, R. Bessa, M. Matos, Flexibility products and markets: literature review, Electr. Power Syst. Res. 154 (2018) 329—340.

[11] V. Bertsch, M. Hall, C. Weinhardt, W. Fichtner, Public acceptance and preferences related to renewable energy and grid expansion policy: empirical insights for Germany, Energy 114 (2016) 465—477.

[12] A.S. Gazafroudi, Y. Mezquita, M. Shafie-khah, J. Prieto, J.M. Corchado, Islanded microgrid management based on blockchain communication, Blockchain-Based Smart Grids, Academic Press, 2020, pp. 181—193.

[13] A.S. Gazafroudi, J.M. Corchado, A. Keane, A. Soroudi, Decentralised flexibility management for EVs, IET Renew. Power Gener. 13 (2019) 952—960.

[14] A.S. Gazafroudi, F. Prieto-Castrillo, T. Pinto, J.M. Corchado, Energy flexibility management in power distribution systems: decentralized approach, Proceedings of the 2018 International Conference on Smart Energy Systems and Technologies (SEST), IEEE, 2018, pp. 1–6.

[15] A. Shokri Gazafroudi, J. Prieto, J.M. Corchado, Virtual organization structure for agent-based local electricity trading, Energies 12 (2019) 1521.

[16] A.S. Gazafroudi, M. Shafie-khah, F. Prieto-Castrillo, J.M. Corchado, J.P.S. Catalao, Monopolistic and game-based approaches to transact energy flexibility, IEEE Trans. Power Syst. (2019). 1–1.

[17] C. Long, Y. Zhou, J. Wu, A game theoretic approach for peer to peer energy trading, Energy Procedia 159 (2019) 454–459.

[18] P. Siano, G. De Marco, A. Rolán, V. Loia, A survey and evaluation of the potentials of distributed ledger technology for peer-to-peer transactive energy exchanges in local energy markets, IEEE Syst. J. (2019).

[19] T. Sousa, E. Fallahi, A. Radoszynskil, P. Pinson, Feasibility study on the adoption of peer-to-peer trading integrated on existing retail market and distribution grid, 2019.

[20] Y. Mezquita, A.S. Gazafroudi, J.M. Corchado, M. Shafie-Khah, H. Laaksonen, A. Kamišalić, Multi-agent architecture for peer-to-peer electricity trading based on blockchain technology, 2019 XXVII International Conference on Information, Communication and Automation Technologies (ICAT), IEEE, 2019, pp. 1–6.

[21] T. Sousa, T. Soares, P. Pinson, F. Moret, T. Baroche, E. Sorin, Peer-to-peer and community-based markets: a comprehensive review, Renew. Sustain. Energy Rev. 104 (2019) 367–378.

[22] Y. Xiao, X. Wang, P. Pinson, X. Wang, A local energy market for electricity and hydrogen, IEEE Trans. Power Syst. 33 (4) (2018) 3898–3908. Available from: https://doi.org/10.1109/TPWRS.2017.2779540.

[23] W. Su, A.Q. Huang, A game theoretic framework for a next-generation retail electricity market with high penetration of distributed residential electricity suppliers, Appl. Energy 119 (2014) 341–350.

[24] L. Jelenković, T. Budrović, Simple day-ahead bidding algorithm for a system with microgrids and a distributor, 2015 38th International Convention on Information and Communication Technology, Electronics and Microelectronics (MIPRO), IEEE, 2015, pp. 1103–1108.

[25] N. Rahbari-Asr, U. Ojha, Z. Zhang, M.-Y. Chow, Incremental welfare consensus algorithm for cooperative distributed generation/demand response in smart grid, IEEE Trans. Smart Grid 5 (6) (2014) 2836–2845.

[26] M. Khorasany, Y. Mishra, G. Ledwich, Auction based energy trading in transactive energy market with active participation of prosumers and consumers, 2017 Australasian Universities Power Engineering Conference (AUPEC), IEEE, 2017, pp. 1–6.

[27] W. Tushar, et al., Energy storage sharing in smart grid: a modified auction-based approach, IEEE Trans. Smart Grid 7 (3) (2016) 1462–1475.

# Peer-to-peer energy platforms

*Liyang Han[1], Thomas Morstyn[2] and Malcolm D. McCulloch[3]*

[1]Department of Electrical Engineering, Technical University of Denmark,
Kongens Lyngby, Denmark [2]School of Engineering, University of Edinburgh,
Edinburgh, United Kingdom [3]Department of Engineering Science, University of Oxford,
Oxford, United Kingdom

## 6.1 Introduction

A by-product of the rapid renewable technological development and energy regulations favoring renewable energy is the adoption of energy resources on a distributed level. Small to mid-scale renewable energy generators (photovoltaic panels, solar thermal collectors, micro wind turbines, etc.) have been gaining popularity among electricity end users ever since energy regulations made it possible for end users to use or sell the electricity generated locally, see Ref. [1]. This development served as a catalyst for a chain of changes in the electricity network, which eventually led to the emergence of peer-to-peer (P2P) energy trading. This chapter focuses on P2P platforms from three angles: why, how, and what. First, we review why P2P platforms are being developed and the technological and economic benefits they provide to power networks. Then we discuss the complexity of the P2P platform architectures and different approaches for designing a P2P platform. Finally, we present a few examples of existing P2P platforms, discuss the challenges they face, and provide an outlook into future P2P platforms.

There has been no consensus on the definition of a P2P energy platform. In this chapter, we adopt the definition from Ref. [2] and broadly define a P2P platform as any community-based or consumer-centric electricity platform, where electricity end users cooperate with what they have available for producing, trading, or distributing electricity.

## 6.2 Why do we need P2P?

With the rapidly increasing deployment of distributed energy resources (DERs), the energy distribution network gradually shifted from a unidirectional model to a bidirectional one, see Ref. [3]. At first, it was simply a result of distributed generation shifting from conventional fossil fuel-based generators to renewable sources such as solar and wind generation, see Ref. [4]. Then energy storage became a part of the DERs family, which could also inject energy into the network, see Ref. [5].

This change required the system control to transition from a simple centralized model to a decentralized or distributed one, hence the emergence of microgrid and virtual power plant (VPP) concepts, see Ref. [6]. To keep up with the system operation, energy markets that used to follow a centralized clearing mechanism now needed to be cleared on a distributed level. Smart metering emerged as a technology with capabilities to measure bidirectional energy flow that could facilitate the market clearing on a distributed level, see Ref. [7].

At the same time, smart meters, coupled with smart appliances and smart building systems, offered the possibility for individual end users to actively participate in the energy market. The concept of prosumers started gaining popularity. In this chapter, the definition of prosumers is borrowed from Ref. [8]: proactive consumers with DERs that actively manage their consumption and production of energy.

Initially, because a regular energy end user had little control over their local renewable generation, demand response was the primary way for them to participate in the market, see Ref. [9]. As energy storage technologies matured and electric vehicles gained a larger user base, the idea of energy arbitrage provided another potential market opportunity for prosumers, see Ref. [10].

The increasing flexibility and controllability of energy resources and loads, coupled with the fast advancing information and communication technologies (ICT), inspired the distributed energy networking concept with mass customization and smart communication [11], and subsequently P2P energy networks [12], operated through a P2P market that would empower prosumers by incorporating their energy preferences [2], such as those for financial gains, low greenhouse gas emissions, charitable causes, etc. Fig. 6.1 shows the comparison of a traditional energy network and one with P2P energy trading.

A well-designed P2P market not only satisfies prosumers' monetary or nonmonetary needs but lowers the burden for system operators to manage and maintain the energy

Energy supplier/utility　　　　　　Energy supplier/utility

FIGURE 6.1　An energy network (A) without P2P trading and (B) with P2P trading.

(A)　　　　　　　　　　(B)

balance within the networks. Similar to other sharing economies, such as car sharing, home rental, etc., P2P energy sharing introduces greater efficiencies into the energy market, and financial benefits to the individual prosumers, see Ref. [13].

## 6.3 How do we design a P2P platform?

Based on existing literature on P2P energy trading platforms, we identify three main components that are part of the architecture of a P2P platform.

The first component is the P2P market structure, which sets the general objectives of the P2P platform. The main goal should be increasing the welfare, either financial or social, of the participants. Depending on the overarching grid market structure, this P2P market is built upon the motivation behind P2P trading (e.g., profit-driven, clean energy preference, social equity, or community-driven, etc.), this market structure could be a simple price-based scheme where prosumers are assumed price takers, or a more complex one that considers optimal scheduling of resources and the strategic behavior of prosumers.

The second component is the network control, which lays the technical foundation for the P2P platform. The main goal of this component is to maintain the energy balance. In a P2P network, where flexible resources in the network sometimes rely on real-time market signals to adjust their operation, additional control and real-time monitoring of the power network are often necessary to ensure energy supply security. This component is responsible for power flow modeling with network constraints and real-time power balancing.

The third component is the information and communication system, which carries and processes data and provides a user interface for the P2P platform. The main goal of this component is to engage prosumers. It enables users on the P2P platform with options to decide how they would like to participate in the P2P market, including bidding on the platform and setting individual preferences. It is also in charge of recording the cleared P2P transactions.

The operation of a P2P platform requires these three components to constantly interact with and support each other. The decisions made by the market component need to be

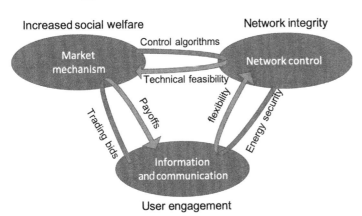

FIGURE 6.2  Three main components of a P2P platform architecture.

monitored and adjusted by the network control component to ensure their technical feasibility. These decisions also inform transactions implemented by the information and communication component. In turn, the information and communication component feeds user bids into the market component, and enables the network control component to adjust the flexible resources within the network for energy balancing. Fig. 6.2 demonstrates the interaction among these three components.

The following subsections will provide more detailed descriptions and examples of these three components.

## 6.3.1 Conceptual component of a P2P platform—P2P market structure

This is the backbone of a P2P platform as it specifies the objectives of the scheme. These objectives are to create value for market participants: the grid and the prosumers. For the grid, some of the objectives are:

- Reduce operation burden induced by the uncertainties of distributed renewable resources.
- Mitigate grid expansion from peak reduction.
- Reduce system losses.
- Enhance system reliability.

For the prosumers, some of the objectives are:

- Reduce energy costs.
- Offer opportunities for social impact.
- Provide financial opportunities for flexible resource operation.
- Automate energy generation and consumption.

P2P market schemes commonly assume that the value of the energy supplied to the prosumers is higher than the prosumer generation, which is often reflected in a higher energy retail price than the feed-in tariff, see Ref. [14]. Various factors contribute to this pricing disparity, including the need to cover the cost of the distribution infrastructure, and distributed generation's high losses and unreliability compared with energy obtained from the wholesale market, see Ref. [15]. This pricing structure is assumed for all the examples in this chapter unless specified otherwise.

We divide these P2P market schemes into two categories: the price-taker model and the game-theoretic model. In a price-taker model, prosumers have zero or little market power and simply accept the P2P market prices as given. In contrast, prosumers are modeled to be strategic in game-theoretic models, where they actively adjust their energy behavior and bids in order to affect the prices of the market for personal economic gains.

Within the price-taker category, there are generally three types of models: community internal price, central market coordinator, and iterative bilateral contract.

Setting an internal price for a prosumer community is the simplest way of financially benefiting participants by increasing the community's energy self-sufficiency. This includes either setting a single internal price for both buying and selling energy within the P2P platform, or setting two different buy and sell prices. All the participants would be exposed to the same internal buy price and the same internal sell price regardless of their energy operation. When the retail price is higher than the feed-in-tariff, as long as the community internal prices are set between the retail price and the feed-in tariff, buyers will be able to buy energy at a cheaper price within the P2P platform, and sellers will be able to sell energy at a higher price within the P2P platform. In the rare cases where dynamic pricing is used and the price of selling electricity back to the grid becomes higher than the buy price, the profit-driven nature of participants will simply prevent them from internally trading electricity temporarily. Long et al. [16], proposed three basic internal pricing schemes: (1) bill sharing, in which the energy import and export prices are averaged and applied to all participants; (2) mid-market rate, in which the internal buy and sell prices are both equal to the average of the retail price and feed-in tariff; and (3) auction-based pricing strategy, in which the internal buy and sell prices are equal to the highest seller offer. Considering the network charges and the P2P platform's operating charges, piecewise functions of the supply demand ratio were used in Ref. [17] to model the internal sell and buy prices. Included in Ref. [18] are prosumers with concave utility functions of economic gain, a common assumption for computational properties that means diminishing marginal returns for energy. The model also includes a local trading center, which determines the internal sell price and buy price by (1) maximizing its own profit if modeled as a for-profit entity, or (2) maximizing the total utility of all the participating energy buyers and energy sellers as a nonprofit entity. A more sophisticated bidding platform was developed in Ref. [19], where each flexible device type has its own bidding strategy, which is adjusted by a home energy management system depending on the user preferences, and sent to the central market platform that sets the internal price to clear the market.

One big disadvantage of the internal price method is that it fails to fairly reward the participating prosumers based on their diverse needs and energy profiles. In other words, there is no price differentiation among the prosumers even though they contribute different amounts of energy sold or bought through the P2P platform. To allow each prosumer to receive prices based on their own operational preferences, the concept of a central market coordinator was explored in several studies. An Energy Broker was proposed in Ref. [20] to model a for-profit coordinator who oversees the bids of the energy sellers and buyers within a local electricity market and sets the final selling and buying prices to maximize its own profit. In Ref. [21], the load aggregator recalculates the bids from energy buyers and sellers with the nodal prices from the local electricity coordinator, which aims to maximize the social surplus through pooling traditional generators, residual loads, and excessive energy from prosumers.

As a result of the misalignment between the participant objectives and the platform objectives (e.g., coordinator profit, social surplus) as described above, the central market coordinator model often leads to prosumers being assigned prices different from their original bids. This may likely result in them deviating from their said operation, leading to inefficiency and instability in the market. To address this issue, a multibilateral economic dispatch optimization problem was formulated in Ref. [22] where bilateral trades are established between all the seller–buyer pairs with trading amounts and prices as variables. The objective is to maximize the social welfare of the P2P community by iteratively solving the optimal trading costs of each bilateral trade. This work provides a distributed framework to analyze the impact of product differentiation reflected on the prosumers' different utility functions. Also focusing on prosumer preferences, another distributed price-directed optimization mechanism using alternating direction method of multipliers (ADMM) was presented in Ref. [23], which enabled prosumers to trade different energy products to satisfy environmental and philanthropic preferences. The results show the significant impact of prosumers' preferences on their final trading prices.

In the game-theoretic category, both cooperative and noncooperative theories have been adopted to facilitate the design of the P2P market.

Using noncooperative game theory, prosumers are often modeled as strategic players in the market, each with the goal to achieve their highest individual utility. An iterative bilateral contract model was applied in Ref. [24], where each agent (a prosumer, a supplier, or a generator) simultaneously bids and accepts contracts in the market to maximize their utility while maintaining their individual energy balance. Through an iterative distributed price-adjustment process, all the contract prices eventually reach a competitive equilibrium. A Nash equilibrium, which guarantees that none of the players has the incentive to change their strategy while the strategies of other players remain fixed, was proved in Ref. [25] to be unique among buyers. A uniqueness of the Nash equilibrium among all sellers, who can be sellers or buyers, is observed empirically in various scenarios simulated in Ref. [26]. To solve the Nash equilibrium of a P2P trading game, the Nikaido−Isoda function was used in Ref. [27], and ADMM was used in Refs. [28,29]. To study the independent behaviors of the sellers and buyers, Lee et al. [25] also proposed to use the framework of Stackelberg games, where players are classified as leaders and followers in a two-level game, where the leaders move first (e.g., set prices, broadcast amount of energy to sell) based on their knowledge of the followers actions (e.g., energy operation, purchasing bids) for a set objective (e.g., own profit, total social welfare). The followers are always the buyers, but the leaders could be either a single central entity (e.g., shared facility controller [30], centralized power system [31], etc.), or all the sellers within the P2P network [32]. The Nash equilibrium and the Stackelberg equilibrium need to be computed in an iterative process, where sellers continuously update the broadcasted energy prices and buyers continuously update their projected operation profiles until convergence [33]. It is shown in Ref. [34] that in P2P noncooperative games analyzed using the Stackelberg game to maximize social welfare, defined as the sum of prosumer utilities, prosumers can achieve 47% higher total utilities than the conventional fixed pricing market scheme.

The noncooperative game-theoretic framework is well established in constructing P2P platforms. In fact, research has started to look into its future indications, for instance the investment in electricity storage, see Ref. [35]. However, the Nash equilibrium is inefficient. In other words, the resulting strategy profiles of the prosumers, none of whom can obtain a higher profit by altering their strategy, are not optimal for maximizing the sum of their profits, see Ref. [36].

Cooperative game-theoretic approaches, on the other hand, encourage the maximization of profits of the grand coalition, which includes all the participants. The key to constructing a stable cooperative game is by allocating the profits to each participant in a way that makes sure that no one can be better off leaving the grand coalition to operate individually or form smaller coalitions, see Ref. [37]. As opposed to being exposed to dynamic pricing in a noncooperative scheme, a known centralized energy price structure is normally assumed where the energy buy price is higher than the sell price from the prosumers' perspective. Through cooperatively scheduling their energy resources, the prosumers as a coalition can achieve energy cost savings, which then are allocated to each participant following a certain algorithm. A cooperative game among microgrids was proposed in Ref. [38] with the objective of minimizing the costs incurred by the losses over the distribution lines. A direct trading cooperative scheme was proposed in Ref. [39] using the Shapley value as the payoff allocation method, which was proved to be in the core [40]—the set of payoff allocations that guarantee that no subgroup of participants has an incentive to abandon the grand coalition. A similar scheme was proposed in Ref. [41], where a more simplified mid-market rate was introduced to facilitate the payoff allocation. To take it one step further, an energy sharing scheme proposed in Ref. [42] included energy storage. With a more sophisticated energy storage model and a time-of-use energy price structure, a cooperative game was introduced in Ref. [43], which proves that the nucleolus allocation scheme provides stabilizing payoffs to incentivize prosumers to participate in the grand coalition.

There are numerous ways of designing a P2P market, and different models can yield significantly different economic impacts on different types of prosumers, see Ref. [44]. The internal price scheme may be easy to implement and simple for customers to understand, but it cannot fairly reflect the values of different types of flexibility, which may cause some participants to defect. A noncooperative game-theoretic scheme may guarantee a stable competitive equilibrium among all participants, but it tends to favor those with more market power (e.g., the Enron Scandal), which may create social equity problems. Cooperative game-theoretic schemes, on the other hand, can maximize social welfare and guarantee benefit for all participants, but they are difficult to scale up because of computational issues related to the value calculation of all the possible coalitions, the number of which is an exponential function of the number of players [40]. When designing a P2P platform, one needs to select the one that fits most with the main objectives of the platform and the types of prosumers that are taking part.

## 6.3.2 Technical component—network controls

While designing a local electricity market is crucial in designing a market platform, one has to work within the laws of physics and make sure robust controls are put in place to accurately translate market transactions to the actual power flow without violating any constraints in the network.

In regards to energy management, previous centralized schemes have focused on working within a range of technical network constraints (voltage limits, thermal limits, operational constraints of DERs), while in designing P2P frameworks, where individuals pursue their own objectives, effort has been mostly focused on setting up an incentive scheme that encourages efficient and robust coordination. However, all the P2P models included in this work are grid connected, meaning that the real-time energy surplus or deficit within the P2P network can be absorbed or supplied by the main grid. However, network constraints will need to be taken seriously if DERs end up providing a significant amount of generation and flexibility.

In order to maintain the network reliability, the distribution system operator as a central authority has been assigned the role in some P2P schemes to check and approve transactions. However, this has the potential to undermine the advantages of a P2P market in terms of autonomy, scalability, and transparency. Research has been done to investigate ways to incorporate network constraints (e.g., voltage limits, congestion), losses, and infrastructure costs into the basic framework of P2P schemes. For example, in the decentralized P2P market proposed in Ref. [45], Voltage Sensitivity Coefficient, Power Transfer Distribution Factors, and Loss Sensitivity Factors are included in the calculation of the external costs, which get passed down to the P2P transactions. This means that it costs more to go through with the transactions that would cause network congestion, significant voltage deviation, and high system losses, thus preventing any violation of the network constraints. Extending the ADMM price negotiation approach, Baroche et al. [46] included the network charges, which significantly increase when the network constraints are reached, into the prosumer cost function to prevent transactions violating the network constraints from being formed. Through updating the network usage charges using an AC OPF model, Kim et al. [47] formulated a cooptimization of the P2P transaction and utility-operated distribution assets to maximize the social welfare while respecting the network constraints.

The P2P network control has to account for the load and DER uncertainties and ensure the network integrity is preserved at all times. Therefore the planned energy operation based on forecasted load and generation require some real-time adjustments. These adjustments are easy to implement if the P2P platform has direct control over flexible resources. Using a grid-connected battery energy storage system, Sossan et al. [48] proposed a real-time model predictive control to compensate for the mismatch between the scheduled operation and the power prosumption realization. In cases where only distributed flexibility is available, Morstyn et al. [24] demonstrated a scalable real-time P2P market in addition to a forward market to ensure the real-time energy balance.

So far, we have discussed the economic backbone of the P2P platform, and how to control the power network following the market will. However, we are still missing one key component—communication. This leads to our discussion on digitizing this P2P platform using modern ICT.

### 6.3.3 Implementation component—information and communication

To align the grid and the market, ICT with data processing capabilities are needed to

- measure and collect distributed bidirectional energy flow data
- control and manage flexible loads and energy resources
- provide an interface to collect prosumer participant inputs
- carry and respond to grid and market signals

These requirements can be met by existing smart grid technologies, such as smart meters and smart appliances.

To align the ICT with the market scheme and network controls, different types of virtual platforms need to be constructed accordingly. For a centralized market platform, such as an energy collective, see Ref. [49], a supervisory third-party needs to be created to facilitate the interface to the prosumers and to different markets. A VPP can create additional value for participating prosumers by aggregating their DER loads and flexibility to participate in the wholesale market, see Ref. [50]. To provide more flexibility to self-organizing prosumers, Morstyn et al. [51] proposed the concept of the federated power plant (FPP), which, as opposed to directly controlling prosumers' energy resources, provides a decentralized market mechanism facilitating energy transactions among them. FPPs can also identify opportunities for upstream grid services and organizes subscribed prosumers to participate in such services.

A local electricity market platform needs to design the communication between peers in a reliable and secure way. The concept packetized energy management has been proposed for aggregating DERs. In this model prosumers send energy packet requests to the central coordinator, be it a VPP [52] or a microgrid community [53], and the central coordinator has the option to accept or reject the request based on grid or market conditions. For a public, decentralized market platform, blockchain with smart contracts has gained attention among researchers and industry experts in recent years. Blockchains are shared and distributed data structures that can securely store digital transactions without the use of a central authority, see Ref. [54]. In the local market context, it offers a technology approach to energy trading in a decentralized fashion. It was shown in Ref. [55] that blockchain lends itself as a vessel to implement the decentralized ADMM market mechanism. From an implementation point of view, Vangulick et al. [56] discussed how the blockchain concept can be adapted to ensure the metered data's accuracy, traceability, privacy, and security.

## 6.4 What do P2P platforms look like—now and the future?

Since the emergence of the concept of P2P trading, trials have been rolled out worldwide. Based on the business model, Morstyn et al. [51] separated P2P platforms into four categories: retail supplier platforms, vendor platforms, microgrid and community platforms, and public blockchain platforms.

### 6.4.1 Past and existing P2P platforms

Retail supplier platforms exist in competitive retail markets, where a retail supplier increases their customer retention by offering opportunities for the customers to obtain more value from their DERs. Besides customer retention, the retail supplier may benefit from an increased level of indirect control over the customers' energy operation through financial incentives. UK-based Piclo[1] and the Netherlands-based Vandebron[2] are pioneers of this model.

Vendor platforms are initiated by energy product vendors with business goals to increase the value of their product. Sonnen,[3] a German home battery system vendor, has developed a P2P trading platform, sonnenCommunity,[4] a community of sonnenBatterie owners who can be matched up through the platform to trade energy without geographic restrictions. Yeloha and Mosaic, both United States-based solar vendors, implemented a

---

[1] https://piclo.energy/.

[2] https://vandebron.nl/.

[3] https://sonnengroup.com/.

[4] https://sonnengroup.com/sonnencommunity/.

scheme where land or property owners can lend their site for PV installation in exchange for a reduction on their energy bill. Unfortunately, both of these projects had to be terminated due to funding issues, see Ref. [57].

Microgrid and community platforms started before the concept of P2P energy trading took shape. Microgrids and community energy initiatives may be based around a shared resource, or shared objective, such as reducing energy costs. Incorporating new P2P concepts into a microgrid design [58], gave a detailed description of the architecture of a P2P microgrid.

Blockchain with smart contracts provides technology options for managing energy trading in a decentralized manner. In recent years, blockchain technology has been incorporated into local electricity market platforms in P2P projects across the world. Run by LO3 Energy, the Brooklyn Microgrid Project pioneered the use of blockchain in the P2P context. Based on TransActive Grid's blockchain architecture, participants can sell their surplus generation to their neighbors and choose from where to buy renewables, see Ref. [59]. It is worth noting that the blockchain Brooklyn Microgrid Project is privately operated by a central entity. Projects that use public blockchain technologies are now starting to be rolled out [54]. In a public blockchain, each user is assigned a secret private key that gives them authority to initiate or accept a trade, and a public key that pseudonymizes this trade that can be then verified by the network, eliminating the need for a trusted third party.

## 6.4.2 Challenges and outlook

Despite the exciting opportunities P2P platforms can provide, numerous risks and challenges need to be addressed before large-scale rollouts of these platforms can become a reality.

Here we summarize these challenges into three categories: technical challenges, regulatory challenges, and social challenges. From a technical perspective, whatever model the P2P platform takes on, it needs to

- protect user data privacy, ensure cyber security;
- have measures in place to account for data latency and recover data losses;
- adopt highly reliable ICT hardware, which in cases of failure can recover in a timely fashion;
- ensure compatibility between the ICT hardware/software and the existing power system control system; and
- be scalable and affordable.

Aside from the above technological challenges, a series of policies and regulations need to be in place to not only open up P2P trading, but also to ensure the fairness and stability of the P2P market. In order to create a healthy environment for developing P2P, the regulatory framework needs to

- allow prosumer-to-prosumer energy trading and sharing;
- enable more flexible tariff designs;

- adapt the currently regulated monopolies that have centralized market control to a more decentralized P2P model;
- prevent and penalize market manipulation;
- set forward rules about user data protection;
- detail the legal and technical responsibilities of different parties;
- encourage technology innovation; and
- ensure social equity so that the system upgrade expenses and network charges do not fall on disadvantaged demographics.

Finally, a wide range of social challenges must be addressed for any successful implementation of a P2P platform. It needs to

- encourage customer adoption not only with financial measures, but with behavioral incentives;
- encourage adoption from the energy suppliers and network operators' point of view;
- ensure at least the same level of energy supply reliability as before;
- ensure social equity and accessibility; and
- reach specific environmental targets such as higher renewables utilization, greenhouse gas mitigation, waste reduction, etc.

Through trial and error, past and existing P2P pilot projects have provided valuable lessons to learn for the next generation of P2P platforms. The following are examples of some new or ongoing academic and industry research projects on P2P. In April 2019, Oxfordshire, UK, Project Local Energy Oxfordshire (LEO) was launched. Out of the £40 million total budget, £13.8 million was funded by the UK government from UK Research and Innovation (UKRI).[5] Led by the distribution network operator SSEN, Project LEO is said to be "one of the most wide-ranging and holistic smart grid trials ever conducted in the UK." Funded by the European Union's Horizon 2020, the 3-year project SHAR-Q started in November 2016 with a focus on sharing distributed storage capacities. The project has since developed a P2P interoperability network that optimizes distributed renewable generation and storage units to create a decentralized collaboration framework.[6] Three pilot projects using the SHAR-Q platform are currently underway, the Solar Lab at ENERC (Portugal), EcoEnergyLand (Austria), and Rafina—Meltemi Campus (Greece).[7] Under the same EU grant, the 4-year project VICINITY started in January 2016 to explore the potential of ICT in creating a decentralized virtual network that helps manage not only energy, but building automation, health, and transportation.[8] The project's plan is to demonstrate the VICINITY platform in four different sites across Europe.

At the same time, there are strong private and industry efforts in developing public P2P platforms to test new market mechanisms and incorporate emerging technologies. Following the iconic Brooklyn microgrid project that tested the use of the TransActive Grid blockchain architecture in P2P trading [59], a few joint initiatives in recent years

---

[5] https://www.oxford.gov.uk/news/article/1050/
oxfordshire_to_trial_40m_industry_first_local_energy_system_project.

[6] http://www.sharqproject.eu/the-project.

[7] http://www.sharqproject.eu/dissemination-material/shar-q-enewsletter-5.

[8] https://vicinity2020.eu/vicinity/content/summary.

emerged out of industry and academic interest to further advance the application of blockchains in P2P platforms. NRGcoin,[9] started off as an industry–academia project, is currently being upscaled in an industrial context through their incentive mechanism NRG-X-Change, see Ref. [60]. Cofounded in early 2017 by Rocky Mountain Institute (RMI)[10] and Grid Singularity,[11] the nonprofit Energy Web Foundation (EWF) launched their blockchain-based platform Energy Web Chain in June 2019, which can be used as the foundational digital infrastructure for blockchain-based decentralized applications.[12] In September 2019, The Energy Origin (TEO),[13] a startup within the French utility Engie became the first application to migrate on to the Energy Web Chain.

At a time when P2P research is advancing at such a high pace worldwide, in order to facilitate knowledge consolidation and transfer, the Global Observatory on P2P, Community Self-Consumption and Transactive Energy Models[14] was launched at University College London on September 2–3, 2019.[15] The Observatory is a 3-year task, Annex of the User-Centred Energy Systems Technology Collaboration Programme (UsersTCP)[16] by the International Energy Agency (IEA). With members from seven countries across Europe, North America, and Australia, the Observatory aims to conduct a systematic study of P2P models, identify challenges and good practices, and eventually generate impact by disseminating findings through publications and forums.

## 6.5  Concluding remarks

In this chapter, we focused on P2P energy trading platforms, a recent development that has drawn much attention from both industry and academia. We followed a systematic approach by looking at P2P platforms from the why, how, and what angles. First, we discussed the technological development in the modern power networks, some of which became drivers for creating a platform for modern prosumers to exchange energy. Then, based on existing research and pilot projects, we provided a description of the architecture of such a P2P platform by dividing it into three components: market mechanism, network controls, and information and communication. Finally, having shown examples of past and existing P2P platforms, we discussed some current technical, regulatory, and social barriers in the way of implementing P2P trading, but we ended the chapter with some recent developments in the field, which could bring us closer to the mass adoption of the sharing economy for energy.

[9] http://nrgcoin.org/about.

[10] https://rmi.org/.

[11] https://gridsingularity.com/.

[12] https://www.energyweb.org/technology/energy-web-chain/.

[13] https://theenergyorigin.com/.

[14] https://userstcp.org/annex/peer-to-peer-energy-trading/.

[15] https://www.ucl.ac.uk/bartlett/energy/events/2019/sep/launch-global-observatory-peer-peer-community-self-consumption-transactive-energy.

[16] https://userstcp.org/.

# References

[1] H. Farhangi, The path of the smart grid, IEEE Power Energy Mag. 8 (2010) 18−28. Available from: https://doi.org/10.1109/MPE.2009.934876.

[2] T. Sousa, T. Soares, P. Pinson, F. Moret, T. Baroche, E. Sorin, Peer-to-peer and community-based markets: a comprehensive review, Renew. Sustain. Energy Rev. 104 (2019) 367−378. Available from: https://doi.org/10.1016/j.rser.2019.01.036.

[3] G. Andersson, M. Geidl, Optimal power flow of multiple energy carriers, IEEE Trans. Power Syst. 22 (2007) 145−155.

[4] J.M. Carrasco, L.G. Franquelo, J.T. Bialasiewicz, E. Galván, R.C. Portillo Guisado, M.Á.M. Prats, et al., Power-electronic systems for the grid integration of renewable energy sources: a survey, IEEE Trans. Ind. Electron. 53 (2006) 1002−1016. Available from: https://doi.org/10.1109/TIE.2006.878356.

[5] J.A.P. Lopes, N. Hatziargyriou, J. Mutale, P. Djapic, N. Jenkins, Integrating distributed generation into electric power systems: a review of drivers, challenges and opportunities, Electr. Power Syst. Res. 77 (2007) 1189−1203. Available from: https://doi.org/10.1016/j.epsr.2006.08.016.

[6] G. Chicco, P. Mancarella, Distributed multi-generation: a comprehensive view, Renew. Sustain. Energy Rev. 13 (2009) 535−551. Available from: https://doi.org/10.1016/j.rser.2007.11.014.

[7] T.J. Lui, W. Stirling, H.O. Marcy, Get smart, IEEE Power Energy Mag. 8 (2010) 66−78. Available from: https://doi.org/10.1109/MPE.2010.936353.

[8] Y. Parag, B.K. Sovacool, Electricity market design for the prosumer era, Nat. Energy 1 (2016) 16032. Available from: https://doi.org/10.1038/nenergy.2016.32.

[9] K. Hamilton, N. Gulhar, Taking demand response to the next level, IEEE Power Energy Mag. 8 (2010) 60−65. Available from: https://doi.org/10.1109/MPE.2010.936352.

[10] K. Worthmann, C.M. Kellett, P. Braun, L. Grüne, S.R. Weller, Distributed and decentralized control of residential energy systems incorporating battery storage, IEEE Trans. Smart Grid 6 (2015) 1914−1923. Available from: https://doi.org/10.1109/TSG.2015.2392081.

[11] P. Mazza, The smart energy network: electricity's third great revolution, Clim. Solut. (2003) 1−22.

[12] H. Beitollahi, G. Deconinck, Peer-to-peer networks applied to power grid. Int. Conf. Risks Secur. Internet Syst. 8 (2007).

[13] L. Einav, C. Farronato, J. Levin, Peer-to-peer markets, Annu. Rev. Econ. 8 (2016) 615−635. Available from: https://doi.org/10.1146/annurev-economics-080315-015334.

[14] W. Tushar, C. Yuen, H. Mohsenian-Rad, T. Saha, H.V. Poor, K.L. Wood, Transforming energy networks via peer-to-peer energy trading: the potential of game-theoretic approaches, IEEE Signal. Process. Mag. 35 (2018) 90−111. Available from: https://doi.org/10.1109/MSP.2018.2818327.

[15] A.J.D. Rathnayaka, V.M. Potdar, T. Dillon, O. Hussain, S. Kuruppu, Goal-oriented prosumer community groups for the smart grid, IEEE Technol. Soc. Mag. 33 (2014) 41−48. Available from: https://doi.org/10.1109/MTS.2014.2301859.

[16] C. Long, J. Wu, C. Zhang, L. Thomas, M. Cheng, N. Jenkins, Peer-to-peer energy trading in a community microgrid, IEEE Power Energy Soc. Gen. Meet. (2018) 1−5. Available from: https://doi.org/10.1109/PESGM.2017.8274546.

[17] N. Liu, X. Yu, C. Wang, C. Li, L. Ma, J. Lei, Energy-sharing model with price-based demand response for microgrids of peer-to-peer prosumers, IEEE Trans. Power Syst. 32 (2017) 3569−3583. Available from: https://doi.org/10.1109/TPWRS.2017.2649558.

[18] Y. Wu, X. Tan, L. Qian, D.H.K. Tsang, W.Z. Song, L. Yu, Optimal pricing and energy scheduling for hybrid energy trading market in future smart grid, IEEE Trans. Ind. Inf. 11 (2015) 1585−1596. Available from: https://doi.org/10.1109/TII.2015.2426052.

[19] W. El-Baz, P. Tzscheutschler, U. Wagner, Integration of energy markets in microgrids: a double-sided auction with device-oriented bidding strategies, Appl. Energy 241 (2019) 625−639. Available from: https://doi.org/10.1016/j.apenergy.2019.02.049.

[20] T. Chen, W. Su, Y.S. Chen, 2017. An innovative localized retail electricity market based on energy broker and search theory. 2017 North Am. Power Symp. NAPS 2017. Available from: https://doi.org/10.1109/NAPS.2017.8107199.

[21] Y. Cai, T. Huang, E. Bompard, Y. Cao, Y. Li, Self-sustainable community of electricity prosumers in the emerging distribution system, IEEE Trans. Smart Grid 8 (2017) 2207−2216. Available from: https://doi.org/10.1109/TSG.2016.2518241.

[22] E. Sorin, L. Bobo, P. Pinson, Consensus-based approach to peer-to-peer electricity markets with product differentiation, IEEE Trans. Power Syst. 34 (2019) 994−1004. Available from: https://doi.org/10.1109/TPWRS.2018.2872880.

[23] T. Morstyn, M. McCulloch, Multi-class energy management for peer-to-peer energy trading driven by prosumer preferences, IEEE Trans. Power Syst. 34 (2018) 4005−4014. Available from: https://doi.org/10.1109/TPWRS.2018.2834472.

[24] T. Morstyn, A. Teytelboym, M.D. McCulloch, Bilateral contract networks for peer-to-peer energy trading, IEEE Trans. Smart Grid 10 (2019) 2026−2035. Available from: https://doi.org/10.1109/TSG.2017.2786668.

[25] J. Lee, J. Guo, J.K. Choi, M. Zukerman, Distributed energy trading in microgrids: a game-theoretic model and its equilibrium analysis, IEEE Trans. Ind. Electron. 62 (2015) 3524−3533. Available from: https://doi.org/10.1109/TIE.2014.2387340.

[26] P. Samadi, V.W.S. Wong, R. Schober, Load scheduling and power trading in systems with high penetration of renewable energy resources, IEEE Trans. Smart Grid 7 (2016) 1802−1812. Available from: https://doi.org/10.1109/TSG.2015.2435708.

[27] N. Zhang, Y. Yan, W. Su, A game-theoretic economic operation of residential distribution system with high participation of distributed electricity prosumers, Appl. Energy 154 (2015) 471−479. Available from: https://doi.org/10.1016/j.apenergy.2015.05.011.

[28] H. Wang, J. Huang, Incentivizing energy trading for interconnected microgrids, IEEE Trans. Smart Grid 9 (2018) 2647−2657. Available from: https://doi.org/10.1109/TSG.2016.2614988.

[29] S. Cui, Y.-W. Wang, J.-W. Xiao, Peer-to-peer energy sharing among smart energy buildings by distributed transaction, IEEE Trans. Smart Grid 10 (2019) 6491−6501. Available from: https://doi.org/10.1109/tsg.2019.2906059.

[30] W. Tushar, B. Chai, C. Yuen, D.B. Smith, K.L. Wood, Z. Yang, et al., Three-party energy management with distributed energy resources in smart grid, IEEE Trans. Ind. Electron. 62 (2015) 2487−2498. Available from: https://doi.org/10.1109/TIE.2014.2341556.

[31] W. Tushar, T.K. Saha, C. Yuen, T. Morstyn, Nahid-Al-Masood, H.V. Poor, et al., Grid influenced peer-to-peer energy trading, IEEE Trans. Smart Grid 1 (2019). Available from: https://doi.org/10.1109/TIE.2016.2554079.

[32] H. Wang, T. Huang, X. Liao, H. Abu-Rub, G. Chen, Reinforcement learning in energy trading game among smart microgrids, IEEE Trans. Ind. Electron. 63 (2016) 5109−5119. Available from: https://doi.org/10.1109/TIE.2016.2554079.

[33] A. Paudel, K. Chaudhari, C. Long, H.B. Gooi, Peer-to-peer energy trading in a prosumer-based community microgrid: a game-theoretic model, IEEE Trans. Ind. Electron. 66 (2019) 6087−6097. Available from: https://doi.org/10.1109/TIE.2018.2874578.

[34] K. Anoh, S. Maharjan, A. Ikpehai, Y. Zhang, B. Adebisi, Energy peer-to-peer trading in virtual microgrids in smart grids: a game-theoretic approach, IEEE Trans. Smart Grid 11 (2019) 1264−1275. Available from: https://doi.org/10.1109/tsg.2019.2934830.

[35] D. Kalathil, C. Wu, K. Poolla, P. Varaiya, The sharing economy for the electricity storage, IEEE Trans. Smart Grid 10 (2019) 556−567. Available from: https://doi.org/10.1109/TSG.2017.2748519.

[36] A. Ghosh, V. Aggarwal, H. Wan, Strategic prosumers: how to set the prices dynamically in a tiered market? IEEE Trans. Ind. Inf. 1 (2018). Available from: https://doi.org/10.1109/TII.2018.2889301.

[37] E. Baeyens, E.Y. Bitar, P.P. Khargonekar, K. Poolla, Coalitional aggregation of wind power, IEEE Trans. Power Syst. 28 (2013) 3774−3784. Available from: https://doi.org/10.1109/TPWRS.2013.2262502.

[38] W. Saad, Z. Han, H.V. Poor, Coalitional game theory for cooperative micro-grid distribution networks, IEEE Int. Conf. Commun. (2011) 6−10. Available from: https://doi.org/10.1109/iccw.2011.5963577.

[39] W. Lee, L. Xiang, R. Schober, V.W.S. Wong, Direct electricity trading in smart grid: a coalitional game analysis, IEEE J. Sel. Areas Commun. 32 (2014) 1398−1411. Available from: https://doi.org/10.1109/JSAC.2014.2332112.

[40] L.S. Shapley, On balanced sets and cores, Nav. Res. Logist. Q. 14 (1967) 453−460. Available from: https://doi.org/10.1002/nav.3800140404.

[41] W. Tushar, T.K. Saha, C. Yuen, P. Liddell, R. Bean, H.V. Poor, Peer-to-peer energy trading with sustainable user participation: a game theoretic approach, IEEE Access. 6 (2018) 62932−62943. Available from: https://doi.org/10.1109/ACCESS.2018.2875405.

[42] P. Chakraborty, E. Baeyens, K. Poolla, P.P. Khargonekar, P. Varaiya, Sharing storage in a smart grid: a coalitional game approach, IEEE Trans. Smart Grid 10 (2019) 4379−4390. Available from: https://doi.org/10.1109/TSG.2018.2858206.

[43] L. Han, T. Morstyn, M. McCulloch, Incentivizing prosumer coalitions with energy management using cooperative game theory, IEEE Trans. Power Syst. 34 (2019) 303−313. Available from: https://doi.org/10.1109/TPWRS.2018.2858540.

[44] E. Mengelkamp, P. Staudt, J. Garttner, C. Weinhardt, Trading on local energy markets: a comparison of market designs and bidding strategies, Int. Conf. Eur. Energy Mark. EEM. (2017). Available from: https://doi.org/10.1109/EEM.2017.7981938.

[45] J. Guerrero, A.C. Chapman, G. Verbic, Decentralized P2P energy trading under network constraints in a low-voltage network, IEEE Trans. Smart Grid 10 (2018) 5163−5173. Available from: https://doi.org/10.1109/TSG.2018.2878445.

[46] T. Baroche, P. Pinson, R.L.G. Latimier, H. Ben Ahmed, Exogenous cost allocation in peer-to-peer electricity markets, IEEE Trans. Power Syst. 34 (2019) 2553−2564. Available from: https://doi.org/10.1109/TPWRS.2019.2896654.

[47] J. Kim, Y. Dvorkin, A P2P-dominant distribution system architecture, IEEE Trans. Power Syst. 35 (2019) 2716−2725. Available from: https://doi.org/10.1109/tpwrs.2019.2961330.

[48] F. Sossan, E. Namor, R. Cherkaoui, M. Paolone, Achieving the dispatchability of distribution feeders through prosumers data driven forecasting and model predictive control of electrochemical storage, IEEE Trans. Sustain. Energy 7 (2016) 1762−1777. Available from: https://doi.org/10.1109/TSTE.2016.2600103.

[49] F. Moret, P. Pinson, S. Member, Energy collectives: a collaborative approach to future consumer-centric electricity markets, IEEE Trans. Power Syst. (2018) 1−2.

[50] F.D. Hatziargyriou, N.D. Kanellos, Control of variable speed wind turbines equipped with synchronous or doubly fed induction generators supplying islanded power systems, IET Renew. Power Gener. 3 (2009) 96−108. Available from: https://doi.org/10.1049/iet-rpg.

[51] T. Morstyn, N. Farrell, S.J. Darby, M.D. McCulloch, Using peer-to-peer energy-trading platforms to incentivize prosumers to form federated power plants, Nat. Energy 3 (2018) 94−101. Available from: https://doi.org/10.1038/s41560-017-0075-y.

[52] L.A.D. Espinosa, M. Almassalkhi, P. Hines, J. Frolik, System properties of packetized energy management for aggregated diverse resources, 20th Power Syst. Comput. Conf. PSCC 2018 (2018). Available from: https://doi.org/10.23919/PSCC.2018.8442954.

[53] P.H.J. Nardelli, H. Alves, A. Pinomaa, S. Wahid, M.D.C. Tome, A. Kosonen, et al., Energy internet via packetized management: enabling technologies and deployment challenges, IEEE Access. 7 (2019) 16909−16924. Available from: https://doi.org/10.1109/ACCESS.2019.2896281.

[54] M. Andoni, V. Robu, D. Flynn, S. Abram, D. Geach, D. Jenkins, et al., Blockchain technology in the energy sector: a systematic review of challenges and opportunities, Renew. Sustain. Energy Rev. 100 (2019) 143−174. Available from: https://doi.org/10.1016/j.rser.2018.10.014.

[55] E. Munsing, J. Mather, S. Moura, Blockchains for decentralized optimization of energy resources in microgrid networks, 1st Annu. IEEE Conf. Control Technol. Appl. CCTA 2017 (2017) 2164−2171. Available from: https://doi.org/10.1109/CCTA.2017.8062773.

[56] D. Vangulick, B. Cornelusse, D. Ernst, Blockchain for peer-to-peer energy exchanges: design and recommendations, 20th Power Syst. Comput. Conf. PSCC 2018 (2018). Available from: https://doi.org/10.23919/PSCC.2018.8443042.

[57] C. Zhang, J. Wu, C. Long, M. Cheng, Review of existing peer-to-peer energy trading projects, Energy Procedia 105 (2017) 2563−2568. Available from: https://doi.org/10.1016/j.egypro.2017.03.737.

[58] C. Zhang, J. Wu, Y. Zhou, M. Cheng, C. Long, Peer-to-peer energy trading in a microgrid, Appl. Energy 220 (2018) 1−12. Available from: https://doi.org/10.1016/j.apenergy.2018.03.010.

[59] E. Mengelkamp, J. Gärttner, K. Rock, S. Kessler, L. Orsini, C. Weinhardt, Designing microgrid energy markets: a case study: the Brooklyn microgrid, Appl. Energy 210 (2018) 870−880. Available from: https://doi.org/10.1016/j.apenergy.2017.06.054.

[60] M. Mihaylov, R. Rădulescu, I. Razo-Zapata, S. Jurado, L. Arco, N. Avellana, et al., Comparing stakeholder incentives across state-of-the-art renewable support mechanisms, Renew. Energy 131 (2019) 689−699. Available from: https://doi.org/10.1016/j.renene.2018.07.069.

# Transmission system operator and distribution system operator interaction

*Hugo Gabriel Morais Valente[1], E. Lambert[2] and J. Cantenot[2]*

[1]INESC-ID, Department of Electrical and Computer Engineering, Instituto Superior Técnico-IST, Universidade de Lisboa, 1049-001 Lisbon, Portugal [2]Électricité de France (EDF), Paris, France

## 7.1 Transmission system operator/distribution system operator cooperation

Electrical energy consumers are becoming actively engaged as market participants through the flexible production and consumption of electrical power. Such consumer participation, individually or organized in energy communities, is expected to increase significantly requiring accelerated transition of distribution network operators (DNOs) to distribution system operators (DSOs). In addition, such transitions will also require major changes to the way DSOs interact with transmission system operators (TSOs) as well as with other power system actors, such as aggregators, fleet operators, balancing service providers, retailers, and significant grid users. Consequently, DSOs and TSOs will be required to reform the way they interact with each other as well as redefining the coordination of operational roles and responsibilities with regard to other emerging industry participants, such as aggregators and DER operators. In this manner, consumers can be provided with secure and scalable solutions that can enable sustainability, affordability, and reliability at a whole system level [1].

To fulfill these objectives, DSOs must enable consumer access to a wide range of market arrangements such as energy provision, ancillary services (AS), system balancing, flexibilities activation, and peer-to-peer negotiation while maintaining the highest standards with regard to security and quality of supply. It is essential that advantage is taken of the opportunity to harness the valuable and increasing amount of energy resources at the distribution level, such as solar panels, wind power, and storage, in order to provide flexible services for

107

**TABLE 7.1**  Overview of expanding DSO tasks [8].

|  | Control of TSO–DSO exchange power | Observe DG | Control DG | Congestion handling | Voltage support with DG |
|---|---|---|---|---|---|
| Passive DSO | TSO | – | – | TSO | – |
| Low DG–no control | TSO | TSO | | TSO | – |
| High DG–no control | TSO | TSO/DSO | | TSO/DSO | – |
| High DG–active power partly controllable | TSO/DSO | TSO/DSO | DSO | TSO/DSO | – |
| High DG–active and reactive power partly controllable | TSO/DSO | TSO/DSO | DSO | TSO/DSO | DSO |

the overall benefit of the power system. Utilizing these resources will facilitate the increasing penetration of renewable energy sources (RES) at a lower cost for consumers by reducing the need to procure services from conventional power generation, reducing the greenhouse gas emissions, and thereby maximizing the rewards for consumers [2,3].

Electrical power system operation is a networked business: TSOs have integrated operational planning processes across their balancing zones and with further market roles, based on standards developed by ENTSO-E [4] and other bodies (spot market coupling, capacity management, nomination etc.) [5]. Also, at a national level, DSOs are continuing to integrate their processes with one another and with further market roles (supplier switching, meter data management, voltage constraints management, congestion management) [6]. Such conventional processes have worked sufficiently in the pre-energy-transition world.

However, RES-based power generation and consequential volatility of electrical power demand, combined with limited ability to precisely forecast supply and demand of power are leading to an increase in congestion both at the distribution and transmission level [7]. Today, predicting congestion locations, their gravity, direction, and duration remain a difficult task under time pressure.

According to [8], the role and responsibilities of DSO are evolving and more control and observability should be in place. The needs are presented in Table 7.1.

In this context of additional challenges and complexity, the need for better cooperation between TSOs and DSOs is being recognized as of paramount importance [9,10]. It is now becoming clear that the interactions among TSOs and DSOs will need to be done on a more regular and harmonized basis. Several cooperation studies have been carried out concerning different aspects of TSO–DSO interaction, like control and automation functions [11,12], performance standards [13,14], cooperation tools [3,15] and requirements [16,17], reactive power management [18–20], AS provision [21,22], and system restoration [8,23].

## 7.2 Cooperative market approaches

Most of the DERs connected to distribution networks have feed-in tariffs and do not participate in electricity markets. This paradigm is changing due to the end of feed-in tariffs as

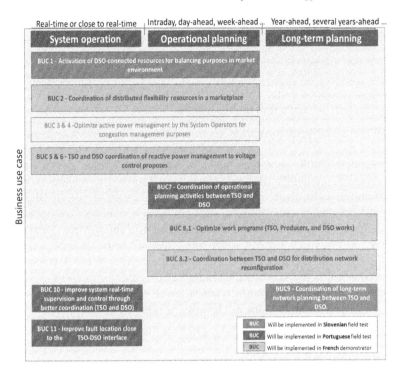

FIGURE 7.1 Business use cases proposed in TDX-ASSIST project.

well as because the new contracts do not assure this kind of mechanism [24]. This means that these energy resources should participate in electricity markets to sell energy or services.

In most of the countries, the electricity markets are managed at the national level by market operators in cooperation with TSOs. In the United States, this role is assured by the Independent System Operators. In a new paradigm, where most of the generation is connected to the distribution system, some coordination mechanism between TSO and DSO should be put in place to assure the quality of supply as well as the same opportunities to all the grid users.

In Ref. [25], five coordination schemes for AS and balancing are proposed: the Centralized AS Market model, the Local AS Market model, the Shared Balancing Responsibility model, the Common TSO–DSO Market model, and the Integrated Flexibility Market model. Each coordinated scheme supposes different management, roles, and information exchange needs. All these markets are focused on the future, considering a high level of cooperation between TSOs and DSOs.

In the TDX-ASSIST project,[1] several services including the coordination between the TSOs and the DSOs have been described. The services were described in different business and system use cases (SUCs) (see Fig. 7.1), each one having detailed information about the exchanged information as well as the role of each player.

---

[1] http://www.tdx-assist.eu/.

Concerning the flexibility markets, it was considered that the flexibility services can be used at the electricity market for the three types of actions:

- portfolio optimization of market players;
- constraints management in transmission and distribution networks (at day-ahead and intraday markets); and
- balancing to ensure power system security (balancing market).

Enabling flexibility services must not lead to market fragmentation or competition distortion. However, the settlement of constraint management and balancing must be clearly separated. Distributed flexibility resource (DFR) providers must be able to sell their services to both TSOs and DSOs. TSO procures flexibility services, from large industrial consumers or aggregators. DSO procures flexibilities also from aggregators and from several types of DERs (distributed generation, demand response, decentralized storage, electric vehicles) connected to the distribution grid with the purpose to maintain quality of service and the security of supply. DSO can procure flexibility services in long and short timescales. Coordination between TSO and DSO is needed since the same flexibility resources may be activated for balancing (TSO services) and congestion constraints management (TSO and DSO services) purposes.

In the project, four scenarios are proposed and detailed in BUC 2—Coordination of DFRs in a marketplace. The proposed scenarios are the following [26]:

- *Scenario 1*: TSO procures the flexibility services and the DSO should validate their activation

DFR are procured by TSO for the purpose of balancing or constraint management in the transmission grid. However, the activation of the flexibilities can jeopardize the security of supply and reduce the quality of services due to, for example, power line congestion, voltage constraints, or overload of the power transformer at the TSO/DSO border. Since the DFR are connected to the distribution network, DSO should be asked to validate the DFRs procured by the TSO, in day-ahead or near real time to avoid constraints in the distribution system. The validation can be done in different ways. The most simple is a validation prior to the clearing process. This analysis considers the existing flexibilities, limiting the ones that can imply violations in technical boundaries. In the second approach, the validation is only performed after the clearing process. This approach has the advantage of focusing only on the procured offers but should have enough computational performance to guarantee the validation in an adequate period of time (Fig. 7.2).

- *Scenario 2*: DSO procures the flexibility services and provides the forecast load/ generation by primary substation

DSO can organize a local procurement of flexibilities to prevent the operational constraints in the distribution network. The constraints can be technical (HV/MV power transformer overload, lines congestion, voltage constraints) but also contractual (agreements between TSO/DSO in the TSO/DSO interconnections). In that case, the DSO is responsible for market-clearing and selects the necessary bids for local use. One important point in this market is that the DFR should inform the DSO about the location of the offers. This aspect will limit the action of the aggregators since it will reduce the possibilities of aggregation.

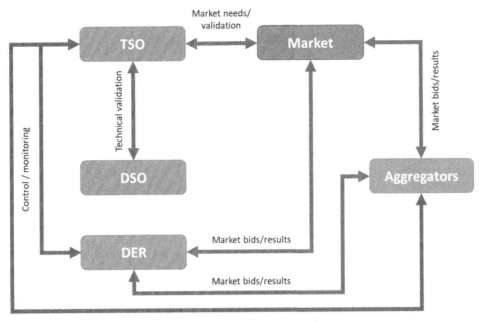

FIGURE 7.2    TDX-ASSIST—BUC2 scenario 1 (National Market vision).

Since the procurement of the DFR will impact the global system operation, the DSO should include the activation of the flexibility offers in the forecasts (load/generation by primary substation) time-series that will be sent to the TSO in day-ahead. This local market can coexist with the national market proposed in scenario 1. Nevertheless, the market operators (TSO and DSO in this case) should found a coordination mechanism to avoid inefficiencies (Fig. 7.3).

- *Scenario 3*: Coordination mechanism between the local and national market

The Flexible Operators connected in the distribution network should place their offer in a local market and the Flexible Operators connected in the transmission network should place their orders in a national market. The clearing mechanism in the local market should provide services for the DSO to solve the distribution network constraints. The offers that are not used in the local market should be technically validated by the DSO and be transferred for the National market. After the clearing process both in local and national markets, the TSO should coordinate the activation of the flexibilities connected in the transmission system and the DSO should coordinate the activation of the flexibilities connected in the distribution network (Fig. 7.4).

- *Scenario 4*: TSO and DSO procure flexibility services in a single flexibility market

To ensure the market liquidity, building a level playing field for different service providers in a single marketplace and the joint procurement of services both for TSOs and DSOs a single procurement system are needed [1]. In this single procurement platform (market) the TSO and the DSO should include their needs for balancing and congestion

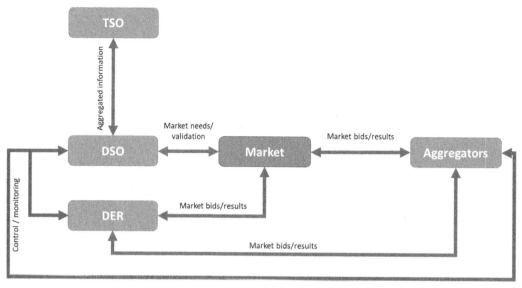

**FIGURE 7.3**    TDX-ASSIST—BUC2 scenario 2 (Local Market vision).

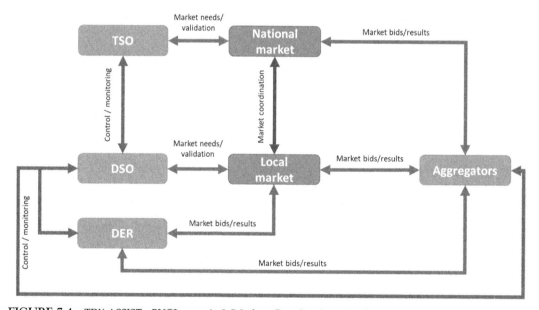

**FIGURE 7.4**    TDX-ASSIST—BUC2 scenario 3 (Markets Coordination vision).

management (TSO) and for congestion and voltage management (DSO). It is important to mention that some voltage constraints can appear in a distribution network even if good reactive management is implemented (Fig. 7.5).

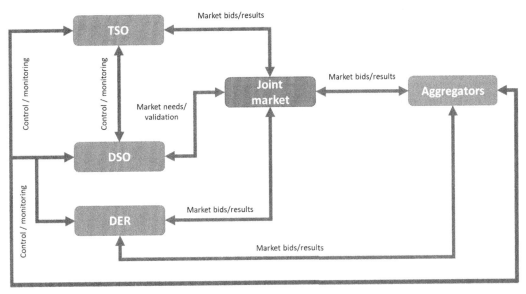

**FIGURE 7.5** TDX-ASSIST—BUC2 scenario 4 (Joint Market vision).

A flexibility platform enables well-structured and organized exchange of data between TSO, DSO, and the flexibility service provider. DSO should have visibility of aggregated resources in the distribution grid and information about individual activations. This information should be available (day-ahead or intraday time frames) to ensure that market schedules are not in conflict with network operation.

The proposed scenarios were modeled according to IEC 62559 and 62913 use case methodology series using Enterprise Architect[2] and the add-on Modeling Smart grid Architecture Unified Solution[3] (MODSARUS[4]). In the elaboration of business use-case (BUC) were defined the objectives, the main assumptions, and the key performance indicators allowing the evaluation of proposed scenarios. Figs. 7.6 and 7.7 present the implementation of BUC 2—Coordination of DFRs in a marketplace proposed in TDX-ASSIST project.

## 7.3 Information exchange needs

The smart grid concept relies on the interaction between the energy sector and ICT (information and communication technology) infrastructure in order to deliver reliable data access for all the stakeholders, assuring adequate cybersecurity levels. The deployment of the innovative services within the electricity network and market raises complex issues and challenges that need the empowerment of more interactions between the stakeholders based on effective and efficient

[2] https://sparxsystems.com/products/ea/.

[3] https://sparxsystems.com/products/3rdparty.html#modsarus.

[4] License free tool—modsarus@edf.fr to get a license.

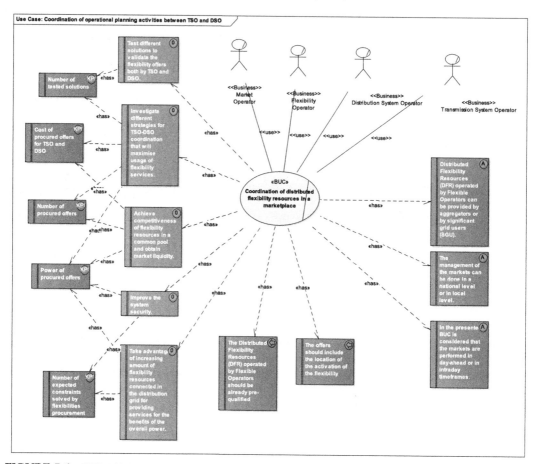

**FIGURE 7.6**    TDX-ASSIST: BUC 2—Coordination of distributed flexibility resources in a marketplace.

ways in handling the data access. However, to gain the full benefits from this collaboration, the stakeholders' needs should be addressed based on the technical and market perspectives [27].

The data sources, access standards, principles, and obligations need to be defined for each time frame and for each stakeholder within an organized and coded manner that addresses the security and protection issue as high priority factors. While every single use case has its own condition, scenarios, and technical criteria, it is important to meet those conditions and scenarios with their specific requirements in order to avoid any conflict between the activities that are handled within the interaction process [27].

Different kinds of information can be exchanged. In Ref. [28] the information is divided into three main categories:

- Structural information: includes all the general and permanent characteristics and attributes of the facility and represents the capabilities of the equipment and is necessary to prepare static and dynamic models of the facilities.

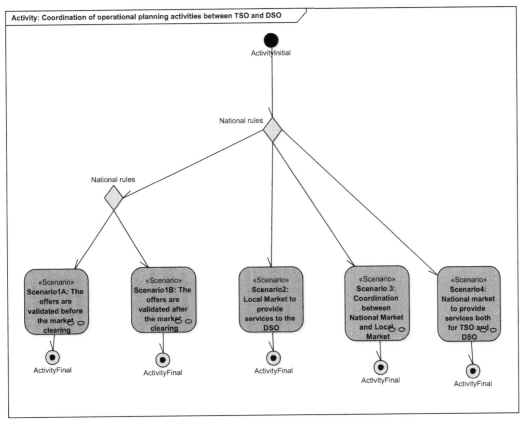

**FIGURE 7.7** BUC 2—Coordination of distributed flexibility resources in a marketplace scenarios.

- Scheduled information: represents the expected behavior of the facilities and network elements in the scheduled time frame and near future, considering the near future up to 1 year according to provisions of the System Operation Guideline. It includes information related to outage planning and generation/ consumption schedules.
- Real-time information: represents the present behavior of the facility.

Moreover, a principle for data access is to create and use harmonized standards. Therefore access to bulk data should be provided in standard machine-readable data formats. Standardization, as well as the interoperability, is a key principle for data access. Interoperability refers to the ability of two or more devices from the same vendor, or different vendors, to exchange information and use that information for correct cooperation. As stated by CEN-CENELEC-ETSI Smart Grid Coordination Group (SG-CG) [29], this definition is extended to "The ability of two or more networks, systems, devices, applications, or components to interwork, to exchange and use information in order to perform required functions."

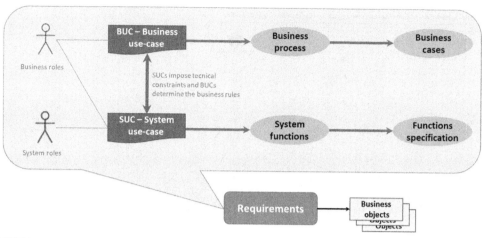

**FIGURE 7.8**    Requirements from use case methodology.

It is possible to distinguish first *semantic*[5] and then *syntactic*[6] levels of interoperability, that is, so as to not only to be able to interchange packets of information[7] but primarily to understand the information contained in those packets. A clear and detailed definition of semantic and syntactic interoperability can be found at Ref. [30]. The smart grid is a means to support a set of applications, and standards are there to enable applications to be deployed. This means that the standardization process should offer a formal path between the application as "requested" by the smart grid and the standards themselves, that is, a "top-down" process. The use of standards allows the replicability and the scalability of the proposed solutions allowing faster and cheaper development of smart grids.

A use case-driven approach is necessary for a top-down development of standards. From a use case perspective, actors, roles, information exchange, and algorithms are identified and requirements are derived. This is the base for a future system approach in standardization.

The use case methodology has been refined by IEC in 62913-1, as described in Fig. 7.8.

For the standardization of the information layer and for any smart grid system, IEC 63097 Smart Grid Roadmap [31] uses the use case methodology. This methodology has been retained by ENTSO-E to support network code requirements. The methodology is interesting since it allows fast identification of gaps between the requirements of a specific use-case and the parameters already defined in different standards. The information exchange requirements should be mapped in the Smart Grid Architecture Model (SGAM) to facilitate the identification of the concerned standards and the replication of use cases, as presented Fig. 7.9.

---

[5] Semantic interoperability—understanding of the concepts contained in the message data structures.

[6] Syntactic interoperability—understanding of data structure in message exchanged between systems.

[7] This relates to the so-called "communication layer" of SGAM.

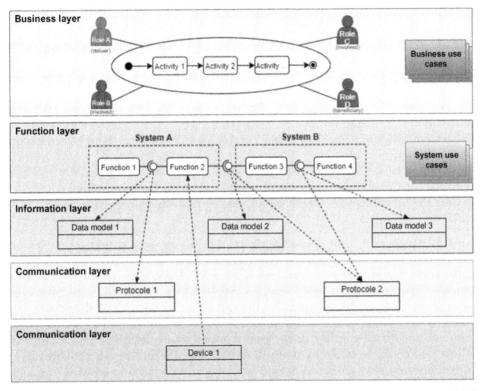

**FIGURE 7.9** Interactions between the use case methodology and the smart grid architecture model. Source: Based on L. Guise, SGCG/M490/G_Smart Grid Set of Standards Version 3.1, CEN-CENELEC-ETSI Smart Grid Coord. Gr., 2014. [32].

**TABLE 7.2** Differences between business and system use cases.

| Type of use case | Description | Actors involved |
|---|---|---|
| Business use cases (BUCs) | Depicts a business process—expected to be system agnostic | Business roles (organizations, organizational entities or physical persons) |
| System use cases (SUCs) | Depicts a function or sub-function supporting one or several business processes | Business roles and system roles (devices, information system) |

For a better comprehension of the methodology, it is important to distinguish the difference between a business use case (BUC) and a SUC. The differences are illustrated in Table 7.2:

The main steps to derive SUCs from BUCs are:

- Identify a needed functionality from BUCs, and this functionality is important for the required implementation.

- Adopt a new point of view: a BUC can be associated with several SUCs, and a SUC can support functionality requested by several BUCs. As an example, a technical validation based in load flow analysis can be used to validate the result of the clearing process of a spot market, to validate the outage planning, or to schedule some capacitor banks.
- Adopt the "implementer"/"developer" attitude: develop smart grid functions which are reusable.
- A SUC is a high-level specification for implementation. This means increasing the detail of information to machine level.
- Adopt new refactoring possibilities (like combining different functions in one system or application).
- Verify if the decisions on activity diagrams are still relevant.
- Identify the core activities independently of the origin of the information.
- Keep only the functionalities which are proposing relevant information, that is, information exchanged.

As already mentioned, four scenarios describing BUCs were proposed in Ref. [26] to improve the coordination between the TSOs and DSOs concerning the flexibility markets. Two of the scenarios are presented in Figs. 7.10 and 7.11. It is important to mention that the scenarios consider generic roles and not entities. This means that, in some systems, the two or more roles can be assured by the same entity. The "flexibility operator" is a role which links, through contractual agreements (flexibility requests, notices, etc.), the role customer and its possibility to provide flexibilities to the roles market and grid; a generic role that could be taken by many stakeholders. This entity can activate all or part of the flexibility resources during a period and regarding a specific location or geographical area.

Concerning the SUCs, the case of national market (scenario 1) was developed. In the proposed architecture, each market actor should do offers, with the identification of the location of the offer. If the offers were provided by Significant Grid User (SGU) connected to the distribution system, the location should be provided at least at HV/MV level. The scenario can be applied for different services requiring a technical validation of the DSO. This scenario is composed of three different processes.

1. The first one describes a generic process to send an offer to the market and to acknowledge the results of the market. The internal process to define the offer is not described in the present scenario.
2. The second scenario describes the process lo lead with the market offers. The market-clearing mechanism is not described in the scenario.
3. Finally, the third scenario describes the process to validate the offers of the market participants with the SGU connected to the distribution system.

Both processes are specific to the French network and may not be applicable in other geographies or in networks with other types of constraints. The interactions between actors and systems are presented in Fig. 7.12.

After the definition of SUCs, it is necessary to analyze the used business objects with the correct level of granularity. Very mild restrictions could increase the complexity of data processing since the presence of several optional fields should be manually tested in

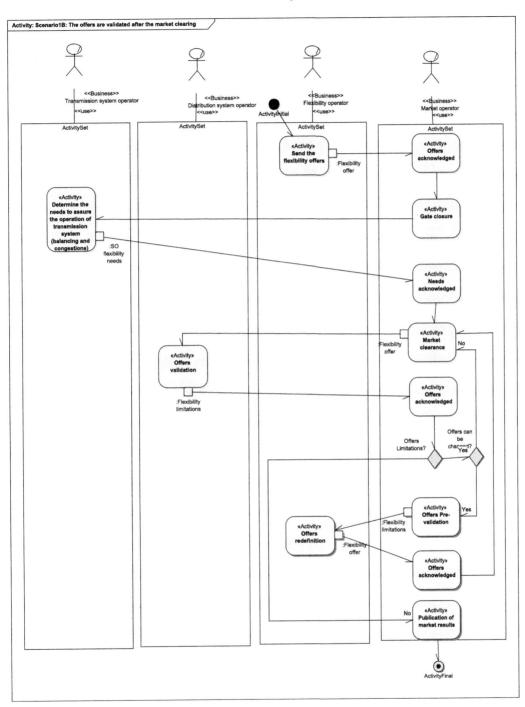

**FIGURE 7.10** Scenario 1—TSO procures the flexibility services and the DSO should validate their activation.

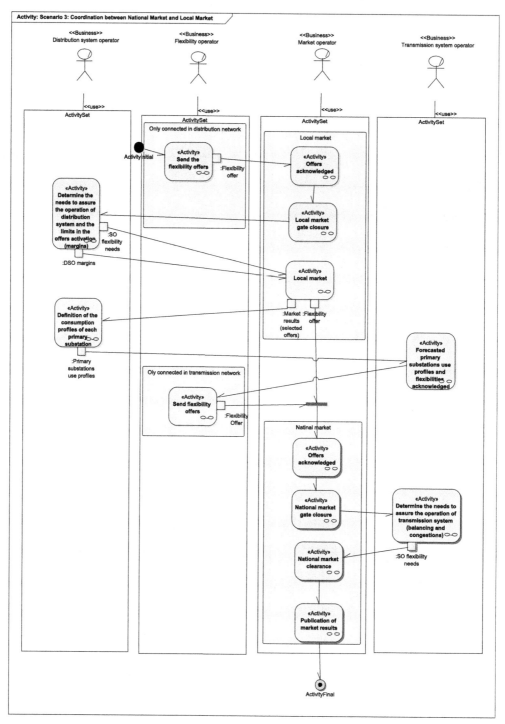

**FIGURE 7.11**    Scenario 3—coordination mechanism between the local and national market.

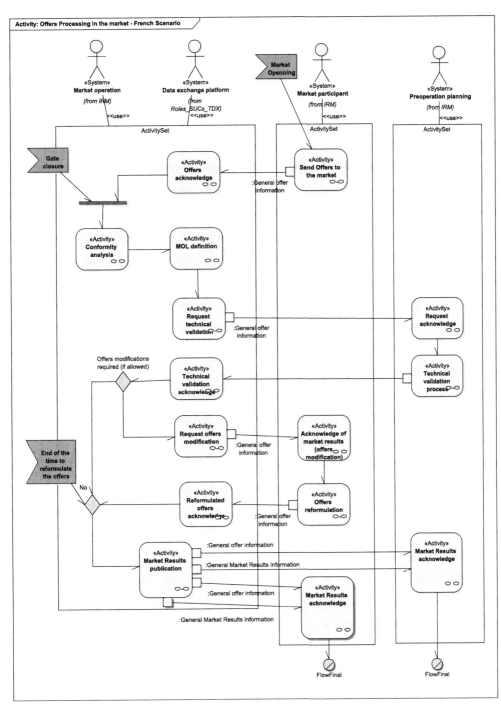

**FIGURE 7.12** SUC—offers processing in the market.

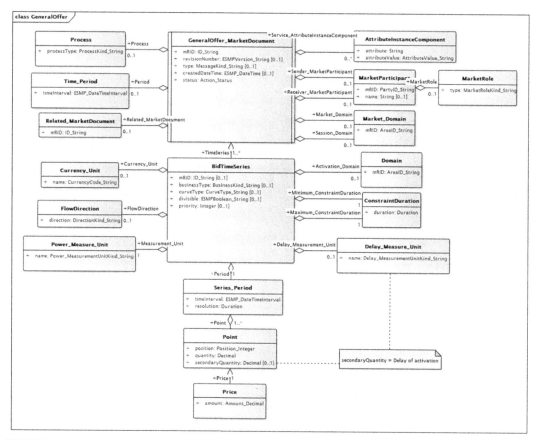

**FIGURE 7.13**   Document contextual model for general offer profile.

each message exchange. On the other hand, a very specialized business object can result in a large number of information models, which can make manipulation of all these objects difficult. Interoperability is therefore improved by choosing an optimum level of granularity of information.

As the defined business objects are not completely defined in the current standards, a specific contextual model specifically created for the TDX-ASSIST project was proposed, although they are still contextualized from IEC CIM (Common Information Model). The "Document Contextual Model for General Offer profile" is presented in Fig. 7.13 and the "Document Contextual Model for Market Results profile" is presented in Fig. 7.14.

Beyond the contextual models, the sending of a message through web services also requires the transmission of information regarding some processing of information, as well as metadata containing the nature of the content, sender and receiver identification, timestamp, etc.

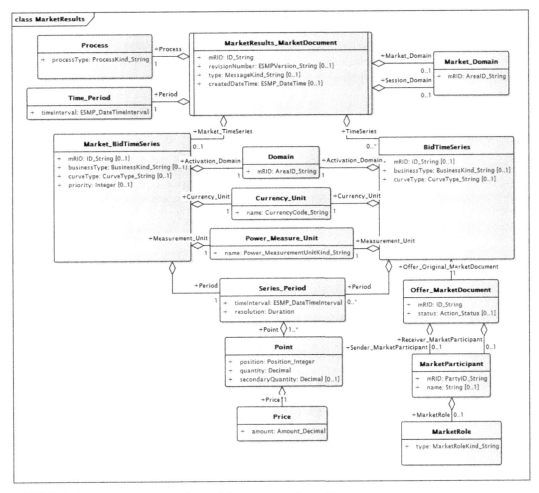

**FIGURE 7.14** Document contextual model for market results profile.

## 7.4 TSO/DSO coordination challenges

The coordination between TSOs and DSOs is "mandatory" in future power systems with high levels of DERs connected to distribution networks. The stockholders are aware of this need and are working on the development of solutions allowing this coordination.

The coordination between TSOs and DSOs can be done at different time horizons from long-term planning to real-time operation. Some of the major needs are network planning, including capacity allocation; outage planning, including the distribution network reconfiguration capabilities; operational planning, including the management of active and reactive limits in TSO/DSO boundaries; and real-time operation, including the situations of an incident. On top of this, the TSOs and DSOs should have common emergency plans allowing the coordination in the restoration process (black-start).

From a business perspective, most of the DERs can participate in different markets and sell services to different players (system operators or aggregators). This new paradigm imposes larger coordination also including market operators, aggregators, and DER owners. Information exchange platforms should be adopted allowing a fast and reliable exchange of information between all players considering the cybersecurity restrictions and the privacy of exchanged data.

## References

[1] E. Lambert, H. Morais, F. Reis, R. Alves, G. Taylor, A. Souvent, et al., Practices and architectures for TSO-DSO data exchange: European landscape, in: Proc. - 2018 IEEE PES Innov. Smart Grid Technol. Conf. Eur. ISGT-Europe 2018, 2018. Available from: https://doi.org/10.1109/ISGTEurope.2018.8571547.

[2] Z. Yuan, M.R. Hesamzadeh, Hierarchical coordination of TSO-DSO economic dispatch considering large-scale integration of distributed energy resources, Appl. Energy (2017). Available from: https://doi.org/10.1016/j.apenergy.2017.03.042.

[3] J. Silva, J. Sumaili, R.J. Bessa, L. Seca, M. Matos, V. Miranda, The challenges of estimating the impact of distributed energy resources flexibility on the TSO/DSO boundary node operating points, Comput. Oper. Res. (2018). Available from: https://doi.org/10.1016/j.cor.2017.06.004.

[4] ENTSO-E, Coordinated balancing areas, 2016. <https://docstore.entsoe.eu/Documents/MCdocuments/balancing_ancillary/2016_04_13_bsg_meeting_coba_v1.pdf>.

[5] ENTSO-E, Enhancing market coupling of SEE Region, 2017. <https://docstore.entsoe.eu/Documents/MCdocuments/170504_ENTSOE_ReportonDAMC_SEE_region.pdf>.

[6] Ecorys, The role of DSOs in a Smart Grid environment, Final Rep. Proj. Ecorys ECN DG ENER, 2014.

[7] G. CEDEC, EDSO, ENTSOE, Eurelectric, TSO—DSO report an integrated approach to active system management with the focus on TSO—DSO coordination in congestion management and balancing, 2019. <https://docstore.entsoe.eu/Documents/Publications/Position papers and reports/TSO-DSO_ASM_2019_190416.pdf>.

[8] C. Roggatz, M. Power, N. Singh, Impact of Distributed Generation on Power System Restoration and Defence, CIGRE, Dublin, 2017.

[9] A. Zegers, H. Brunner, TSO-DSO interaction: an overview of current interaction between transmission and distribution system operators and an assessment of their cooperation in smart grids (Discussion Paper Annex 6, Task 5), Int. Smart Grid Action. Netw. (2014).

[10] EuropeanCommission, Regulation of the European parliament and of the council on the internal market for electricity, Eur. Comm. Winter Packag. (2016).

[11] D. Pudjianto, G. Strbac, D. Boyer, Virtual power plant: managing synergies and conflicts between transmission system operator and distribution system operator control objectives, CIRED - Open. Access. Proc. J. (2017). Available from: https://doi.org/10.1049/oap-cired.2017.0829.

[12] F. Pilo, G. Mauri, B. Bak-Jensen, E. Kämpf, J. Taylor, F. Silvestro, Control and automation functions at the TSO and DSO interface - impact on network planning, CIRED - Open. Access. Proc. J. (2017). Available from: https://doi.org/10.1049/oap-cired.2017.0975.

[13] D. Stanescu, D. Federenciuc, M. Albu, S. Gheorghe, D. Ilisiu, C. Stanescu, Performance standards applied to Romanian TSO and DSO, CIRED - Open. Access. Proc. J. (2017). Available from: https://doi.org/10.1049/oap-cired.2017.0791.

[14] G. CEDEC, EDSO, ENTSOE, Eurelectric, TSO—DSO data management report, Eurelectric (2016). Available from: https://doi.org/10.1021/JP072106N.

[15] M.V. Sebastian, M. Caujolle, B.G. Maraver, J. Pereira, J. Sumaili, P. Barbeiro, et al., LV state estimation and TSO-DSO cooperation tools: results of the French field tests in the evolvDSO project, CIRED - Open. Access. Proc. J. (2017). Available from: https://doi.org/10.1049/oap-cired.2017.0410.

[16] O. Arnaud, M. Chapert, O. Carre, DSO TSO coordination needs induced by smart grids: the ongoing French project between RTE and Enedis, CIRED - Open. Access. Proc. J. (2017). Available from: https://doi.org/10.1049/oap-cired.2017.1182.

[17] C.J. C2/C6.36, System operation emphasizing DSO/TSO interaction and coordination, 2018. <https://e-cigre.org/publication/733-system-operation-emphasizing-dsotso-interaction-and-coordination>.

[18] C2 CRIGRE, Active and Reactive Power Provision Across the TSO − DSO Boundary, CIGRE, Paris, 2018. <https://e-cigre.org/publication/SESSION2018_C2-114>.

[19] J. Morin, F. Colas, X. Guillad, J.Y. Dieulot, S. Grenard, Joint DSO-TSO reactive power management for an HV system considering MV systems support, CIRED - Open. Access. Proc. J. (2017). Available from: https://doi.org/10.1049/oap-cired.2017.0144.

[20] S. Abreu, M. Carvalho, L. Simão, Reactive power management considering stochastic optimization under the portuguese reactive power policy applied to DER in distribution networks, Energies 12 (2019) 4028. Available from: https://doi.org/10.3390/en12214028.

[21] G. Migliavacca, M. Rossi, D. Six, M. Džamarija, S. Horsmanheimo, C. Madina, et al., SmartNet: H2020 project analysing TSO-DSO interaction to enable ancillary services provision from distribution networks, CIRED - Open. Access. Proc. J. (2017). Available from: https://doi.org/10.1049/oap-cired.2017.0104.

[22] H. Morais, B. Goncer, J. Cantenot, E. Lambert, G. Taylor, A. Sovent, et al., Balancing services business use case development for TSO-DSO interoperability demonstration, 2019. Available from: https://doi.org/10.1049/cp.2018.1889.

[23] W.H. Wellssow, D. Raoofsheibani, P. Hinkel, M. Ostermann, S. Loitz, C. Schorn, et al., Operator training for restoration of power systems with high shares of volatile generation, CIGRE Sess. 46 (2016).

[24] C. Hitaj, A. Löschel, The impact of a feed-in tariff on wind power development in Germany, Resour. Energy Econ. (2019). Available from: https://doi.org/10.1016/j.reseneeco.2018.12.001.

[25] SmartNet Project, TSO-DSO Coordination for Acquiring Ancillary Services From Distribution Grids, 2019. <http://smartnet-project.eu/wp-content/uploads/2019/05/SmartNet-Booktlet.pdf>.

[26] TDX-ASSIST, Agreed Models, Use Case List, and Use Case Description in UML, 2018.

[27] Smart Grids Task Force, EG3 First Year Report: Options on Handling Smart Grids Data, 2013.

[28] ENTSO-E, Supporting Document to the All TSOS' Proposal for the Key Organisational Requirements, Roles and Responsibilities Relating to Data Exchange, 2017.

[29] CEN/CENELEC/ETSI Smart Grid Coordination Group, Final Report Standards for Smart Grids, Group, 2011.

[30] J. Frémont, Guidelines for standards implementation, 2012.

[31] IEC, IEC TR 63097, n.d. <https://webstore.iec.ch/publication/27785>.

[32] L. Guise, SGCG/M490/G_Smart Grid Set of Standards Version 3.1, CEN-CENELEC-ETSI Smart Grid Coord. Gr., 2014.

# 8

# Local electricity markets—practical implementations

*Ricardo Faia[1], Fernando Lezama[1] and Juan Manuel Corchado[2,3]*

[1]Research Group on Intelligent Engineering and Computing for Advanced Innovation and Developmen (GECAD), Polytechnic of Porto (ISEP/IPP), Porto, Portugal [2]BISITE Research Group, University of Salamanca, Salamanca, Spain [3]Department of Electronics, Information and Communication, Faculty of Engineering, Osaka Institute of Technology, Osaka, Japan

## 8.1 Introduction

The power system control is becoming challenging at the local level with the rapid growth of distributed energy resources (DERs). In fact, existing control approaches are not ready to face the penetration of millions of distributed generation (DG) units and renewables (e.g., producing variable energy through solar and wind resources) in the electrical network [1]. The role of the traditional consumers is transforming into prosumers that can work as producers and consumers [2]. This transformation was caused by the increasing installation of DGs in the small-scale consumers' facilities, in part as a response to the European Commission (EC) imposed directives [3].

The full liberalization of the European electricity markets allowed end-users to be able to choose their own retailer, enabling competitiveness among retailers. Actually, the retailers have the possibility of buying electricity in the wholesale market or establishing bilateral contracts (BC) with the suppliers. However, prosumers with small electricity production capacity are legally restricted from participating in the wholesale market since a minimum amount of energy is needed for such activity. In this scenario, LEMs are considered as a solution that can circumvent existing restrictions on the participation of small prosumers in wholesale markets [4].

The Clean Energy for all European package [5] from the EC proposes a new electricity market design in order to include more renewables and empower consumers. According to the authors of Refs. [6,7], LEMs empower the household (prosumers and consumers) and small-scale producers, and can provide an increase of DERs installations on the distribution grid. Considering Ref. [6], the emerging LEMs can support the EC to meet the Clean Energy for all European goals. Since there is no consensus on the definition of LEMs in the current literature, in the scope of this work we adopt the definition presented by Mengelkamp [8]: "a local electricity market (LEM) is a market platform for trading locally generated (renewable) electricity among residential agents within a geographically and socially close community. Supply security is ensured through connections to a superimposed energy system." According to the presented definition, LEMs should have mechanisms for complying with the negotiation actions between customers (e.g., households, small-scale producers). These mechanisms make possible the trading of local produced energy, typically provided by DERs, while considering an auxiliary system to provide security in the energy supply (main grid). Therefore the balance between the demand and supply can be obtained at the local level making use of the increase of installed DERs. The electricity transactions resulting from LEM, coordinating demand and supply, can obtain a local energy balance. However, most of the times, the local generation cannot meet with the total demand and there are necessary backup systems (main grid or other sources of demand/generation should be used). The peer-to-peer (P2P) concept has been much explored in the electricity markets to perform the local electricity negotiation. P2P allows a direct negotiation between local market participants [9]. P2P is the ability to trade electricity with one another [10]. Moreover, the emergence of new electricity markets structures at the local level is accepted and considered by the research community of power systems and electricity markets.

With this chapter, we intend to present an overview of the existing application of the LEMs in the literature. To start, we define practical implementation as a realization of a theoretical idea to the case study obtaining results. The case study should be created from real data, the obtained results sometimes are not the best. Based on our literature analyses and knowledge about the field, we formulate the following research question:

> Based on the literature that provides knowledge and information about LEMs, (1) which practical implementations of LEM are currently deployed or ongoing, and (2) what practical implementations are considered in the path toward real-world adoption?

The proposed research question is answered by a detailed literature review performed by the authors of this chapter. The authors do not intend to provide a comprehensive review of LEMs, but rather of their practical implementations. This chapter is structured into five different sections. Section 8.1 presents an initial introduction to the LEMs theme and identifies the focus of the work. Section 8.2 addresses the search methodology used for the selection of relevant literature. Section 8.3 presents the literature review and a simple categorization performed considering the similarities within the literature analyzed. Section 8.4 is a discussion about the results and future direction of research topics are presented. Finally, in Section 8.5 the conclusions are presented.

## 8.2  Search methodology

This literature review, results, and conclusions were obtained following the methodology and guidelines of Ref. [11]. First, we performed a general search about LEMs in which we identified relevant sources within the scope of the topic. Considering these sources, an intensive search was performed around the cited references and a specific time frame. We aimed to select the publications that explicitly included LEMs and electricity trading and sharing at the local level. As a result, the literature search covers the period of January 2010 to December 2019, as it was performed at the beginning of 2020. The research was based on specific keywords related with the LEMs, namely "local electricity markets," "local energy trading," "local energy sharing," and "peer-to-peer electricity market." In our research, we excluded literature related to resources optimization in a specific environment (e.g., smart grid or microgrids) since many of these publications are associated with the term of local and it is even possible to find a kind of market mechanism. However, such a relation does not represent the actual implementation of a LEM following the adopted definition and therefore we consider these publications out of the scope of this work. Three different scientific search engines were used to perform the search: ScienceDirect,[1] IEEE Xplore,[2] and Google Scholar.[3] Also, publication on journals and conferences was added as a search filter on all search engines.

## 8.3  Local electricity market implementation

Considering the search executed attending the keywords identified in Section 8.2, 86 publication in journals and conference proceedings were identified as relevant for the presented review. We narrowed the selection in the function of the title, abstract, keywords, and main contributions, resulting in a final list composed of 50 publications. The selection was performed considering the following steps: only the works that address practical applications were selected. For instance, papers with optimization methods, simulations, and case studies were considered, whereas papers with literature reviews, theoretical concepts, regulatory considerations, or other types were not. Selected publications need to explicitly present a market or trading mechanism that explain how energy is traded/shared. As was stated in the introduction, publications that considered optimization of power flows or microgrids/smart-grids management optimization are excluded from this review. Table 8.1 presents the list of publication and their respective categorization. We selected six features for the categorization of the publications, namely (1) work type, (2) classification, (3) method, (4) organization, (5) trading mechanism, and (6) platform or project. The features consider the following points:

- *Work type*: this feature classifies the publication considering the type of the work developed by the authors. The considered classifications are case studies (CS),

---

[1] https://www.sciencedirect.com/.

[2] https://ieeexplore.ieee.org/.

[3] https://scholar.google.com/.

**TABLE 8.1**    Results of the literature search review, 50 entries.

| References | Work type | Classification | Method | Organization | Trading mechanism | Platform or project |
|---|---|---|---|---|---|---|
| [12] | O | LEM | MOP | C | | |
| [13] | S | LEM | MAS | C | Au | |
| [14] | S | P2P | GT | D | | |
| [15] | S, O | P2P | MOP | C | | |
| [16] | O | P2P | MOP | D | | |
| [17] | S | | MAS | D | | |
| [18] | S, CS | P2P | GT | | Au | |
| [19] | S, O | P2P | MAS, MOP | D | Au | |
| [20] | S, O | LEM | MOP | C | | MASCEM |
| [21] | S, O | P2P | MOP | C | | |
| [22] | O | P2P | MOP | C | | |
| [23] | O, CS | LEM | MOP | C | Au | |
| [24] | O, CS | LEM | MOP | C | | ECO-Trade |
| [25] | S | P2P | MAS | D | BC | |
| [26] | O, CS | P2P | MOP | D | | |
| [27] | S, CS | P2P | Bch | D | Au | LAMP |
| [28] | CS | P2P | Bch | D | Au | Brooklyn Microgrid |
| [29] | S, O | | MOP | C | | SESP |
| [30] | O | P2P | MOP | D | | |
| [31] | S | P2P | MAS | D | | |
| [10] | O, CS | P2P | MOP | C | | |
| [6] | O, CS | LEM | MOP | C | Au | |
| [32] | S | P2P | Bch | D | | |
| [33] | S, O, CS | | MOP | C | | |
| [34] | O | P2P | MOP | C | DT | |
| [35] | S, CS | P2P | GT, MAS | D | | Elecbay |
| [36] | S | P2P | GT | | | |
| [37] | S, O | P2P | MAS, MOP | D | | |
| [38] | O | P2P | MOP | D | | |
| [39] | S, O, CS | | GT | | | |
| [40] | S | LEM | MAS, GT | D | | |

*(Continued)*

**TABLE 8.1** (Continued)

| References | Work type | Classification | Method | Organization | Trading mechanism | Platform or project |
|---|---|---|---|---|---|---|
| [41] | O | P2P | Bch | D | Au | |
| [42] | S | LEM | GT | | | PETCON |
| [43] | O, CS | P2P | MOP | C | DT | |
| [1] | O | LEM | MOP | C | Au | |
| [44] | S, CS | P2P | | | Au | |
| [45] | O | P2P | MOP | C | | |
| [46] | S, CS | P2P | MAS | D | | |
| [47] | S | P2P | MAS | D | Au | |
| [48] | S | P2P | MAS | C | Au | Elecbay |
| [49] | S | LEM | | C | Au | |
| [50] | S | P2P | GT | | DT | |
| [51] | S | LEM | MAS | C | Au | |
| [52] | S | P2P | MAS | D | DT | |
| [53] | O, CS | LEM | MOP | C | Au | |
| [54] | O | P2P | GT | | DT | NOBEL |
| [55] | S, O | LEM | MOP | C | | |
| [56] | S | LEM | MAS | D | Au | |
| [57] | S | LEM | MAS, MOP | D | Au | |
| [58] | S | LEM | MAS | D | Au | |

The publications are sorted by publication year and alphabetic order.

optimization models (O), and simulation approaches (S). By case studies we consider experimental settings based on real data or scenarios, or (near) real implementations. Studies that are tested and validated under more theoretical approaches or based on datasets purely for experimentation purposes are not considered as cases studies.

- *Classification*: this feature separates the publication into two different categories, namely P2P for direct trading (DT) of agents in a local environment, and local energy markets (LEM) addressing the trading at local level without an explicit mechanism for DT between peers.
- *Method*: this feature is used to categorize the publications considering the method used to solve the proposed methodology. We define the following categories: blockchain (Bch), game-theory (GT), multiagent systems (MAS), and mathematical optimization problem (MOP).
- *Organization*: the presented feature has a direct relation with the method used. Two categories are identified, the centralized (C) and decentralized (D) organization of the proposed methodology.

- *Trading mechanism*: in this feature is specified how electricity is distributed. Considering the analysis of publications, we identify the follow mechanisms: auction (Au), BC, and DT. When the trading mechanism is not identified in the publication, even if the market is in place, the column of trading mechanism is not filled in Table 8.1.
- *Platform or project*: this column is created to identify if the publication proposes any platform for negotiation/sharing, or if the publication is linked to a given project.

To appreciate the distribution of works in time, Fig. 8.1 presents the 50 publications from Table 8.1 considering the year of publication and the type of publication (conferences or journals).

As can be seen in Fig. 8.1, the number of publication has been increasing since 2017, with 2018 as the year with more publications in journals, and 2017 with more published articles in conference proceedings. A balance is achieved in 2019, with the higher number of publications considering both journal and conferences. For the "work type," we found 26 papers using O, 32S, and 14 for CS. This classification is based on our appreciation about the contents proposed in each publication.

Now, we discuss some features from Table 8.1 based on the obtained results and findings. For the "classifications" feature, we have 17 references analyzing LEM, 29 references about P2P, and four references without any associated classification. The feature identified in each column is based on authors' proposals in each publication. Considering the P2P classified papers, the concept can be presented as a bilateral agreement between two agents. Considering microgrid environment and the use of game theory approaches, a P2P mechanism is also used in Ref. [35] to simulate the electricity transaction between users. A BC is proposed in Ref. [25] as a market design for P2P energy trading, investigating how this mechanism has an influence on the upstream investments and network efficiency and energy security increase. The energy sharing concept is also explored as a considered P2P market in Ref. [43], in which a demand ratio mechanism is proposed in a microgrid context to avoid

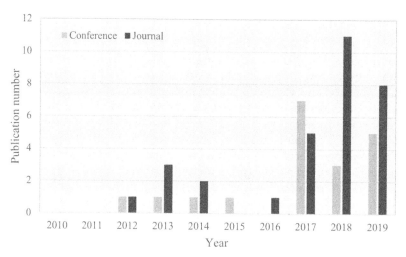

**FIGURE 8.1**  Distribution of selected publication considering the year and the type.

the injection of energy into the main grid. Considering the papers classified as LEM, Ref. [55] considers a LEM to obtain the cheapest price of electricity and maximize the utilization of DERs. In Ref. [13], the authors present another work classified as LEM where congestion management is applied. Local interactions between the distribution system operator and the households were performed to create the LEM. Finally, Ref. [12] considers a LEM to exchange energy within a microgrid and the market to try to maximize the social welfare of the community.

With regard to the second feature, "the method" used in each publication, we found 23 MOP, 14 MAS, 8 GT, and 4 Bch. Considering the publications that present an OP method, we can find different algorithms like linear optimization [10,15,19−21,23,33] and nonlinear optimizations [55,57]. Ref. [15] presents a mixed integer linear problem to address the P2P energy trading, also considering the uncertainty of the impute variables. A multiple mixed integer linear formulation is used in Ref. [23] for energy sharing among microgrids to evaluate the impact of P2P mechanism in the microgrid costs management. In Ref. [55], a nonlinear problem is addressed and formulated as a mixed integer nonlinear problem to maximize the use of DERs considering an islanding microgrid context. Methodologies as bilevel optimization are used in Ref. [12], where microgrid community members can exchange energy and services between themselves in the upper level and the lower level of the problem clears the market. Other different techniques like alternating direction method of multipliers [16,30,37,38], primal-dual gradient descent [24,29], knapsack approximation [45], and consensus + innovation approach [26] are also proposed as approximation methods to optimize the energy exchange and trading in a local context. Considering the publication that uses the MAS method, we found two different types of agents used in MAS, namely the zero intelligent agents (ZIA) and intelligent agents (IA). The work in Ref. [13] proposes ZIA to perform bids to attend a DSO flexibility request. The Refs. [47,51,56] use also ZAI agents in their architectures. The IA are applied in published Refs. [25,46,52,58]. For instance in Ref. [25], the IA are modeled considering mathematical expressions in which the system contains three different models of players: prosumers, suppliers, and generators. In Ref. [19] a different type of agents are considered, the zero intelligence plus traders agents. This type of agent cannot be considered as zero intelligence agents because the agents use an adaptive mechanism that can result in a very similar performance to that of the real players in markets. In publications that use GT, we found a canonical coalition game in Refs. [14,36]. For instance, in Ref. [14] the authors demonstrate a P2P trading scheme considering a framework of motivational psychology. Another publication that uses the coalition game is Ref. [50], in which an asymptotic Shapley value is used to incentivize the abandon of the coalition, to establish a stable operation of DT. Bayesian game theory in Ref. [18] is adopted as the strategy in the energy auction to enable efficient and cost-effective bidding for each buyer, [39,42] propose a Stackelberg game approach that is considered a noncooperative approach to deal with the energy trading in microgrid contexts. The Nash equilibrium is implemented in Ref. [35] as a solution for noncooperative games, in which each player is assumed to know the equilibrium strategies of the other players, and no player has anything to gain by changing only their own strategy. A comparison of collaborative and a noncollaborative approach is presented in Ref. [54]. For the collaborative approach, the authors maximize the social welfare of the power system, whereas in the noncollaborative approach (i.e., considering self-interested customers) each customer determines its interests to maximize its own profit. The Bch method uses blockchain technology. Blockchain is considered a disruptive distributed information and computing paradigm that can create a trustable trading and billing environment without a centralized authority. Studies in Refs. [27,28,32,41] applied Bch technology as a method to emulate the virtual interactions between the prosumers and consumers located at low voltage level. An action mechanism for charging and discharging EV based on Bch is considered in Ref. [41]. The use of Bch technologies secures privacy protection without external parties' need. A P2P trading mechanism using the Bch technology is proposed in Ref. [28].

Considering the "organization" feature, we identified 21 publications with C, 21 for D, and eight with no classification. To make this classification we based it on the method used in each publication. As the centralized (C) organization we classified the classical optimization approaches like linear optimizations and nonlinear. Methods such as MOP (only alternating direction method of multipliers and consensus methodologies), the majority of MAS, and Bch we consider as decentralized (D) organization.

Of the trading mechanisms presented in the literature analysis we identify 18 as Au, 5 as DT, 1 as BC, and 26 didn't specify the trading mechanism used. The continuous double auction is present in Refs. [19,23,41,47,51,53,56,58] as a type of auction mechanism; the asymmetric pool model is implemented in Refs. [13,57]; and the Vickrey Clarke Groves Auction was used in Ref. [18].

## 8.4 Discussion and future directions

In this section, we present a discussion and the future directions based on the analysis of publications presented in Table 8.1. We detected a lack of consensus for a LEM definition in the analyses of publications from Table 8.1. It seems that works are not following a trend toward a single point. In this sense, there is a need for the scientific community to come together and to reach an agreement in the definition of LEM. The differences that distinguish LEM from P2P markets, energy communities, and energy sharing must also be presented. A P2P standard definition is presented in Ref. [59] as the interaction of two agents. Considering the field of electricity markets, P2P models are characterized as an interaction between two agents for the purpose of energy negotiation. On the other hand, these agents may be neighbors residing in the same area or be separated by kilometers of distance (different countries). In view of the above, we consider that P2P energy trading models can be a subset of LEMs if they are applied to agents installed in the same area (e.g., street, neighborhoods, or even cities). Many publications also present the LEM as a mean for trading or sharing electricity/energy in communities. Some applications of LEM have been registered at the community level, for example, sharing energy storage systems between neighborhoods with local generation to achieve better community sustainability. LEM participants assume that a remuneration can be obtained at an economic level. Even if energy sharing may not result in a direct payment or remuneration, Ref. [60] states that an economic remuneration can motivate the sharing of assets. Making the analogy for LEM, the sharing energy models with remuneration and direct focus on trading and economy can be considered as a sub concept of LEM.

One key aspect needed for the implementation of LEMs is related to the development of infrastructure that allows local transactions in a reliable and transparent manner. Currently, some open platforms for local electricity trading are available and can serve as a basis for the study of new business models around the use of flexibility, energy, and electricity, locally. For instance, Piclo [61] is an online platform developed in the UK for peer-to-peer energy local trading between prosumers. It includes basic functions of data analytics and visualization and a matching algorithm to guarantee a balance between generation and consumption locally. On the other hand, Vandebron [62] is another online platform, developed in the Netherlands, for local energy trading where energy consumers can buy electricity directly from independent producers (e.g., farmers with wind turbines or PV generation). In this case, Vandebron plays the role of an energy supplier providing incentives for the exchange of energy between producers and consumers. Considering energy storage systems, the sonnenCommunity [63] is a platform that works as a virtual energy pool of batteries (integrating owners of a battery storage manufacturer in Germany called sonnenBatteries), where PV generation can be stored and shared between the members, taking advantage of the differences in renewable generation from diverse locations. Finally, the Brooklyn Microgrid project [27] is a good example of how a microgrid energy market framework can be put in place considering requirements and components. Seven components are devised in Ref. [27], namely microgrid setup, grid connection, information system, pricing mechanism, market mechanism, EMTS, and legal environment. The Brooklyn Microgrid case study is used to show the importance of compliance with such components, and to demonstrate that blockchains are an eligible technology to operate decentralized microgrid energy markets. However, it is also shown that current regulation can be a barrier in many countries to allow LEM and therefore all the components cannot be satisfied yet. In any case, such initiatives are fundamental toward practical applications of LEM. Real implementations will path the ways toward transactive energy systems and LEM, identifying barriers along the way that can only be studied and mitigated once such mechanisms are put in place.

Some of the papers presented in Table 8.1 show benefits that make the promising LEM to be a good solution to increase the profit of small end-users [12,23,31,34]. Ref. [34] mentioned also an improvement of self-consumption of the PV generated energy and an increase of self-sufficiency. The LEM brings also technical benefits [31]. For instance, Ref. [13] shows results that demonstrate the use of LEM to mitigate problems related to the congestion of distribution networks. In Ref. [10] was demonstrated a very interesting trade-off between the independence of the main grid when the P2P and battery energy systems were used. Considering the electricity cost of the community, the implementation of the two systems mentioned before can be decreased by half. Another result is the decrease of the power exchange between local users and the utility grid [20].

Table 8.2 presents a list of research directions that can be followed in future work. We identified such directions by grouping the future work proposals (found in the 34 articles of Table 8.1) that are similar or refer to a similar work topic.

As can be seen in Table 8.2, we found 11 main groups of potential future works and we developed a small opinion for each one.

**TABLE 8.2**    List of future works proposed by the publications considered in Table 8.1.

| Future works proposals | References |
| --- | --- |
| Consider network constraints in the LEM simulations | [6,24,25,31,42,44,48,51–53] |
| Address regulatory, policy, and social issues regarding LEM implementation | [1,15,19,28,34,39,47] |
| Consider the use of bidding strategies to participate in LEM | [12,13,15,19,28,57] |
| Include in simulations the interactions with different kinds of players | [10,12,13,24,39,51] |
| Integrating LEM with other markets | [10,12,22,28,29] |
| Consider the uncertainties of the variables and risk of the participation in LEM | [25,26,30,31] |
| Use of distributed approaches in LEM simulations | [26,28,32,38] |
| Consider and redesign the actual tariff in LEM | [21,22,25,51] |
| Extend LEM to distribution network level | [42,43,51] |
| Explore the communications issues | [31,52] |
| Study the influence of the forecast methods on LEM results | [56,57] |

- *Considering network constraints in the LEM simulations*—the number of works that propose including the consideration of network constraints in the simulation of LEM is large, which indicates that such consideration is crucial in the future of LEM. Constraints such as bus voltage limits, lines' temperature limits, generators limits, system frequency limits, among others, cannot be neglected in real implementations of LEMs, and thus should be considered in simulations. Also, simulations of LEM should consider the networks characteristics, including power flow analysis or optimal power flow depending on the context.
- *Address regulation, policy, and social issues regarding LEM implementation*—many of the analyzed works mentioned a need for addressing regulation, policy, and social concerns when a LEM is implemented. But before this, a unique definition for LEM should be elaborated, putting all actors on the same page when referring to a LEM implementation. This will help to propose clear and correct regulations and the specification of each entity role. Also, entities regulating the LEM should be properly proposed in the future.
- *Considering the use of bidding strategies to participate in LEM*—with different actors pursuing different goals and with different point of views, a good strategy for each LEM participant should be designed. Therefore the use of bidding methodologies can help the players to find their best option to participate.
- *Including in simulations the interactions with different kind of players*—the analyzed publications only consider some of the players in their simulations, for example, consumers and prosumers in most of the cases. However, many other players will be present in LEM with important implications, such as DSO, aggregators, retailers, or market operators. Thus such players should be modeled and considered in the LEM interactions to analyze with more accuracy the impact of LEM in practice.
- *Integrating LEM with other markets*—the integration of LEM with other markets (namely ancillary services market and wholesale) should be considered. The primary use of

LEM is to solve problems located at the distribution level (i.e., locally), however, LEMs need to be integrated into existing market structures, modifying the interaction of players at different levels, and impacting the activities and business models of upper market levels.

- *Considering the uncertainties of the variables and risk of the participation in LEM*—the consideration of uncertainties and risks in the participation of LEM are important because it can make the player obtain the best market participation avoiding unnecessary risks.
- *Use of distributed approaches in LEM simulations*—this topic of future work can be considered one of the most important steps to realize LEM implementations. Since LEMs have a distributed characteristic (e.g., P2P models), the use of distributed approaches to simulate the LEM interactions should be considered. Blockchain is a promising technology that can be used as a distributed mechanism to simulate the LEM models.
- *Consider and redesign the actual tariff in LEM*—tariff design in electricity markets should be redesigned considering the possibility of participation in LEM. With new considerations, factors such as the use of the networks, losses, costs of electricity, incentives for local players, among others should be considered and included in a proper design of tariffs that distribute the profits in a fair manner.
- *Extend LEM to distribution network level*—although the LEM implementations are considered local, they should cover a large part of the distribution network, encompassing most of the households and small industries. Further studies about the interconnection needed between large areas should be done to assess the implication of LEMs at the large scale.
- *Explore the communications issues*—communications issues are an important topic when the P2P models are implemented due to the high number of access points that can be installed. As a consequence, if the number of access points in the system is large, the system is more susceptible to attacks. Cyber security should be a crucial topic under study before real implementations of LEM are put in place.
- *Study the influence of forecast methods on LEM results*—LEM considers the interaction of small users with renewable generation and storage systems, resulting in a high variability of production and consumption. If forecast methods have a good accuracy, the results of the simulations will be close to reality. Therefore accurate forecast tools are needed to guarantee better participation strategies of users in the market considering the results of the simulations.

## 8.5 Conclusion

We present an organized literature review considering LEM publications in journals and conferences. Considering the time frame of the selected publications, there has been a quick increase in works published since 2017. It is necessary to consider that only publications that included practical applications were considered. A total of 50 publications were analyzed (Table 8.1) and the main characteristics that we consider important for the study of LEM were identified. We identified, in Section 8.4, the trends for future works as a result of the future works proposed by the publications analyzed. Due to this approach, it is likely that other important future works were not mentioned here. LEMs are in an early stage of research, with few real implementations. However, there are already a considerable number of publications in the literature on LEM. We consider as a priority that a definition and a common categorization of existing LEM models should be accepted by the research community.

## Acknowledgment

This work was supported by the Ph.D. Grant of Ricardo Faia from Portuguese National Funds through FCT under Grant SFRH/BD/133086/2017.

## References

[1] D. Holtschulte, et al., Local energy markets in clustering power system approach for smart prosumers, 2017 6th International Conference on Clean Electrical Power (ICCEP), IEEE, 2017, pp. 215–222.

[2] B.P. Koirala, et al., Energetic communities for community energy: a review of key issues and trends shaping integrated community energy systems, Renew. Sustain. Energy Rev. 56 (2016) 722–744.

[3] European Commission, A Roadmap for Moving to a Competitive Low Carbon Economy in 2050.

[4] F. Teotia, R. Bhakar, Local energy markets: concept, design and operation, 2016 National Power Systems Conference (NPSC), IEEE, 2016, pp. 1–6.

[5] European Commission, Clean Energy for All Europeans: Commission Welcomes European Parliament's Adoption of New Electricity Market Design Proposals.

[6] F. Lezama, et al., Local energy markets: paving the path toward fully transactive energy systems, IEEE Trans. Power Syst. 34 (5) (2019) 4081–4088.

[7] G. Mendes, et al., Local energy markets: opportunities, benefits, and barriers, CIRED Work. 0272 (2018) 5.

[8] E.M. Mengelkamp, Engineering Local Electricity Markets for Residential Communities, Karlsruhe Institute of Technology (KIT), 2019.

[9] T. Liu, et al., Energy management of cooperative microgrids with P2P energy sharing in distribution networks, 2015. IEEE International Conference on Smart Grid Communications, Smart Grid Comm 2015, 2016, pp. 410–415. Available from: https://doi.org/10.1109/SmartGridComm.2015.7436335.

[10] A. Lüth, et al., Local electricity market designs for peer-to-peer trading: the role of battery flexibility, Appl. Energy 229 (2018) 1233–1243.

[11] P. Guy, S. Kitsiou, Methods for literature reviews, Handbook of eHealth Evaluation: An Evidence-Based Approach (2017) 157–178.

[12] B. Cornélusse, et al., A community microgrid architecture with an internal local market, Appl. Energy 242 (2019) 547–560.

[13] R. Faia, et al., A local electricity market model for DSO flexibility trading, International Conference on the European Energy Market, EEM, 2019.

[14] W. Tushar, et al., A motivational game-theoretic approach for peer-to-peer energy trading in the smart grid, Appl. Energy 243 (2019) 10–20.

[15] Z. Zhang, et al., A novel peer-to-peer local electricity market for joint trading of energy and uncertainty, IEEE Trans. Smart Grid (2019). 1–1.

[16] C. Orozco, et al., An ADMM approach for day-ahead scheduling of a local energy community, 2019 IEEE Milan PowerTech, PowerTech 2019, 2019.

[17] I. Praca, et al., Analysis and simulation of local energy markets, International Conference on the European Energy Market, EEM, 2019.

[18] C.H. Leong, et al., Auction mechanism for P2P local energy trading considering physical constraints, Energy Procedia (2019) 6613–6618.

[19] J. Guerrero, et al., Decentralized P2P energy trading under network constraints in a low-voltage network, IEEE Trans. Smart Grid (2018).

[20] Z. Zhang, F. Li, Local market design and bidding strategies of prosumers, IEEE Power and Energy Society General Meeting, IEEE, 2019.

[21] M.R. Alam, et al., Peer-to-peer energy trading among smart homes, Appl. Energy 238 (2019) 1434−1443.

[22] J.M. Zepter, et al., Prosumer integration in wholesale electricity markets: synergies of peer-to-peer trade and residential storage, Energy Build. 184 (2019) 163−176.

[23] K. Chen, et al., Trading strategy optimization for a prosumer in continuous double auction-based peer-to-peer market: a prediction-integration model, Appl. Energy 242 (2019) 1121−1133.

[24] M. Khorasany, et al., Two-step market clearing for local energy trading in feeder-based markets, J. Eng. 2019 (18) (2019) 4775−4779.

[25] T. Morstyn, et al., Bilateral contract networks for peer-to-peer energy trading, IEEE Trans. Smart Grid 10 (2) (2019) 2026−2035.

[26] E. Sorin, et al., Consensus-based approach to peer-to-peer electricity markets with product differentiation, IEEE Trans. Power Syst. 34 (2) (2019) 994−1004.

[27] E. Mengelkamp, et al., Decentralizing energy systems through local energy markets: the LAMP-project, MKWI 2018 - Multikonferenz Wirtschaftsinformatik (2018) 924−930.

[28] E. Mengelkamp, et al., Designing microgrid energy markets: a case study: the Brooklyn microgrid, Appl. Energy 210 (2018) 870−880.

[29] M. Khorasany, et al., Distributed market clearing approach for local energy trading in transactive market, IEEE Power and Energy Society General Meeting, IEEE, 2018.

[30] T. Liu, et al., Energy management of cooperative microgrids: a distributed optimization approach, Int. J. Electr. Power Energy Syst. 96 (2018) 335−346.

[31] Y. Zhou, et al., Evaluation of peer-to-peer energy sharing mechanisms based on a multiagent simulation framework, Appl. Energy 222 (2018) 993−1022.

[32] F. Blom, H. Farahmand, On the scalability of blockchain-supported local energy markets, 2018 International Conference on Smart Energy Systems and Technologies, SEST 2018 - Proceedings, IEEE, 2018.

[33] P. Olivella-Rosell, et al., Optimization problem for meeting distribution system operator requests in local flexibility markets with distributed energy resources, Appl. Energy 210 (2018) 881−895.

[34] C. Long, et al., Peer-to-peer energy sharing through a two-stage aggregated battery control in a community microgrid, Appl. Energy 226 (2018) 261−276.

[35] C. Zhang, et al., Peer-to-peer energy trading in a microgrid, Appl. Energy 220 (2018) 1−12.

[36] W. Tushar, et al., Peer-to-peer energy trading with sustainable user participation: a game theoretic approach, IEEE Access. 6 (2018) 62932−62943.

[37] M. Vinyals, et al., A multi-agent system for energy trading between prosumers, Adv. Intell. Syst. Comput. 620 (2018) 79−86.

[38] Y. Zhou, et al., A new framework for peer-to-peer energy sharing and coordination in the energy internet, IEEE Int. Conf. Commun (2017).

[39] C.P. Mediwaththe, et al., Competitive energy trading framework for demand-side management in neighborhood area networks, IEEE Trans. Smart Grid 9 (5) (2018) 4313−4322.

[40] B. Celik, et al., Coordinated neighborhood energy sharing using game theory and multi-agent systems, 2017 IEEE Manchester PowerTech, Powertech 2017, 2017.

[41] J. Kang, et al., Enabling localized peer-to-peer electricity trading among plug-in hybrid electric vehicles using consortium blockchains, IEEE Trans. Ind. Inform. 13 (6) (2017) 3154−3164.

[42] N. Liu, et al., Energy sharing management for microgrids with PV prosumers: a Stackelberg gGame approach, IEEE Trans. Ind. Inform. 13 (3) (2017) 1088−1098.

[43] N. Liu, et al., Energy-sharing model with price-based demand response for microgrids of peer-to-peer prosumers, IEEE Trans. Power Syst. 32 (5) (2017) 3569−3583.

[44] F. Teotia, et al., Modelling local electricity market over distribution network, 2017 7th International Conference on Power Systems, ICPS 2017, 2018, pp. 1−6.

[45] M. Khorasany, et al., Peer-to-peer market clearing framework for DERs using knapsack approximation algorithm, 2017 IEEE PES Innovative Smart Grid Technologies Conference Europe, ISGT-Europe 2017 - Proceedings, IEEE, 2017, pp. 1−6.

[46] Y. Zhou, et al., Performance evaluation of peer-to-peer energy sharing models, Energy Procedia (2017) 817–822.

[47] E. Mengelkamp, et al., Trading on local energy markets: A comparison of market designs and bidding strategies, International Conference on the European Energy Market, EEM, 2017.

[48] C. Zhang, et al., A bidding system for peer-to-peer energy trading in a grid-connected microgrid, Energy Procedia 103 (2016) 147–152.

[49] D. Menniti, et al., Local electricity market involving end-user distributed storage system, 2015 IEEE 15th International Conference on Environment and Electrical Engineering, EEEIC 2015 - Conference Proceedings, 2015, pp. 384–388.

[50] W. Lee, et al., Direct electricity trading in smart grid: a coalitional game analysis, IEEE J. Sel. Areas Commun. 32 (7) (2014) 1398–1411.

[51] M. Ampatzis, et al., Local electricity market design for the coordination of distributed energy resources at district level, IEEE PES Innovative Smart Grid Technologies Conference Europe, 2015.

[52] S. Kahrobaee, et al., Multiagent study of smart grid customers with neighborhood electricity trading, Electr. Power Syst. Res. 111 (2014) 123–132.

[53] Y. Ding, et al., A control loop approach for integrating the future decentralized power markets and grids, 2013 IEEE International Conference on Smart Grid Communications, Smart Grid Comm 2013, 2013, pp. 588–593.

[54] B.G. Kim, et al., Bidirectional energy trading and residential load scheduling with electric vehicles in the smart grid, IEEE J. Sel. Areas Commun. 31 (7) (2013) 1219–1234.

[55] M. Marzband, et al., Experimental validation of a real time energy management system for microgrids in islanded mode using a local day-ahead electricity market and MINLP, Energy Convers. Manag. 76 (2013) 314–322.

[56] P. Goncalves Da Silva, et al., The impact of smart grid prosumer grouping on forecasting accuracy and its benefits for local electricity market trading, IEEE Trans. Smart Grid 5 (1) (2014) 402–410.

[57] C. Rosen, R. Madlener, An auction mechanism for local energy markets: results from theory and simulation, 2012 IEEE Work. Complex. Eng. COMPENG 2012 - Proc (2012) 43–46.

[58] H.S.V.S. Kumar Nunna, S. Doolla, Multiagent-based distributed-energy-resource management for intelligent microgrids, IEEE Trans. Ind. Electron. 60 (4) (2013) 1678–1687.

[59] R. Schollmeier, A definition of peer-to-peer networking for the classification of peer-to-peer architectures and applications, Proc. - 1st Int. Conf. Peer-to-Peer Comput. P2P 2001 (2001) 101–102.

[60] P.A. Albinsson, B.Y. Perera, The Rise of the Sharing Economy: Exploring the Challenges and Opportunities of Collaborative Consumption, 2018.

[61] New York State Energy Planning Board: Shaping the Future of Energy.

[62] Vandebron. <https://vandebron.nl/>. (last accessed 15.04.2020).

[63] Sonnen, sonnenCommunity. <https://sonnengroup.com/sonnencommunity/>. (last accessed 15.04.2020).

# PART III

# Enablers for local electricity markets

# Local energy markets, commercially available tools

## Jan Segerstam
### Empower IM Oy, Helsinki, Finland

## 9.1 Local energy markets, commercially available tools

Local energy markets need tools to enable them to work [1]. These tools are constantly under development and will evolve as the market structures evolve around them. Already now, however, it is evident that a certain level of maturity exists in enabling local energy markets as these have become more prevalent globally in different pilot projects and test implementation.

The notion of local energy markets is actually ages old [2]. It is how it all began, when the first generators were connected to multiple users, for example, a mill and a house to a small hydropower plant at a riverside. The difference today is that we already have a very large and robust energy infrastructure on top that makes the niche for the local market rather different. It needs to justify its existence by adding value to its participants beyond what is being offered by the now traditional grid and the overarching energy markets being operated there.

This chapter is about the tools commercially available to enable local markets to work. In order to be commercially viable, they have to contribute to the basic value of the local market being set up. The chapter deals with the available technologies through a framework that is focused on the participants and the stakeholders in setting up a local market. Thus, it is not a list of definitive tools, but more a toolbox of different capabilities that can then ultimately be adapted to the actual use cases and business models of the specific market being set up.

We start by looking at the market level. What are the tools available to run the actual market? Without them there is nothing, but that can be said of many of the other parts as well. The market will ultimately be a collaborative effort by all stakeholders that crystallizes around the market level functionalities agreed upon by all participants.

We then look at the participant level. What tools enable a participant to run the needed internal processes to be able to participate and what connectivity is available to connect the participant to the market? Participants will compete, and therefore have differing technical solutions, but on a general level all need to implement similar rudimentary functionalities.

Finally, we take a look at the endpoint, the community or the end customer himself. Depending on context, this can mean looking at a single building, piece of equipment, or a vehicle, but it can also mean looking at a block of buildings, a farm of generation assets, a group of end customers, or a fleet of vehicles. All of these are stakeholders that need an interface [3] and connectivity to be able to actually run their energy use and generation processes in accordance with what is being dictated by market transactions.

Bear in mind that these three levels presented earlier are role levels [4] that do not preclude single companies or legal entities performing any multiple or grouping of them. This means that the framework introduced here should not be regarded as restrictive, and should be thought of as a way of giving structure to analyzing the plethora of solutions and implementations already commercially available today. Enjoy the chapter and let your mind explore the possibilities presented and come up with your own analysis of what you can find in your context, or generate your own implementation to extract the value most important to you and your community.

## 9.2 Enabling the local marketplace

The local marketplace is an ecosystem of participants, the marketplace and their enabling technical components. Setting up a local ecosystem requires at least the marketplace and adjoining communication, but other enablers such as datahubs or flexible messaging solutions may bring added value to the ecosystem.

Local markets are today in a period of formation and innovation. This means that the tools used in the different implementations across the world are also different and evolving. A particular challenge is the basic character of local markets. They are by design local, which also means that most development tends to happen locally as the initial business cases concern only local partners.

In order to be scalable, aspects above the single local market must be taken into account. Below we will look at the different commercial tools that can be used to enable a local market while taking into account scaling. These tools presented are a nonexhaustive sampling, chosen to inspire further evaluation and discovery by the reader. A brief overview of tool categories is presented below as a framework for this.

### 9.2.1 The marketplace and messaging

The central element in enabling a local market is the marketplace itself. This marketplace sets the scene for the actual transactions on the market. The trading mechanisms available are those implemented in the chosen platform, and are beyond the scope of this section; however, the samples presented will give an indication of the possibilities available.

From a technical standpoint, it is around the marketplace platform that the market will form. This also means that the technical interfaces for messaging toward other markets and market participants will define the extent of value formation information for the marketplace. In other words, a very simplified market platform will only give limited value information, whereas a more interconnected one will ensure richer value information. Thus a broader and more open platform will arguably enable a more informed indication of the value of the resources being traded.

An important aspect of any central element in a network of participants, such as the local marketplace discussed here, is the management of data privacy. Data privacy legislation in each country provides a starting point and helps to establish the nature of measures each participant, and the marketplace provider as a business entity, needed to set up. From a more practical viewpoint, data privacy in a market is key to its operation in order not to make participants vulnerable to the misuse of participant information for the purpose of manipulating market outcomes. Managing this aspect needs to be proven by the marketplace in order to gain traction and trust among its participants. A secondary, but from a participants point of view, very important aspect is managing the privacy of participant data in order to avoid misuse of information for the purpose of affecting the participant. Such aspirations could manifest themselves in situations where a malevolent entity would like to disrupt the security of supply of a participant or analyze which time-point would be most opportune for physical action against the participant.

The data privacy provisions of any proposed marketplace should be weighed against the scope and purpose of the marketplace to establish a reasonable level of measures that suffices for the participants while not creating a cascading risk to the overarching mechanisms and the overall energy system of which the local marketplace is a part. This means in practice that not only the participants, but the representatives of overarching mechanisms such as relevant network operators and/or marketplaces should be consulted when setting up a mechanism operating within their scope of operation. The subject of data privacy and its wider implementation is beyond the scope of this chapter and information described here is meant to create a placeholder in any implementation project for more in depth analysis.

### 9.2.2 Participant tools

Local means small. Small means less resources, time, and money. This is a line of thought that is sometimes used to justify drastic simplification or the ruling out of any sophisticated mechanisms to establish fair value or optimum use of resources. In modern setups, however, the ubiquitous availability of communications and the low cost and high availability of computing power have made this kind of need for simplification obsolete. That said, great care must be taken and expertise used to extract value out of the myriad of data and web of connections that comes from activating control and information flows to local resources.

The simplification of what local means that what we started with is still defendable, but the challenge it poses is no longer that there would not be technology available or that the cost of operating it would be prohibitive. The challenge is that a great number of initiatives and systems aim to take over and control the ecosystem of local resources. This means that the actual challenge lies in breaking the barriers of the now forming closed systems to ensure access to value and to prevent economic or technical islanding of valuable resources.

Participant tools enable the local market participant to manage the participation on a local market. Crucial components include capabilities to forecast and evaluate outcomes and resource levels of loads and generation and capabilities to divide, settle, and establish individual financial positions of the participating resources.

## 9.2.3 Technical enablement of the market

While participants and marketplaces do form a business ecosystem, they are nothing without the endpoint resources and technical implementation. Key enablers here include messaging between systems of the marketplace, participants, and resources as well as information feeding technologies connected to the actual resources, potential datahubs with shared services, and control systems that ultimately manage the resources at stake based on the market decisions made.

In the next sections we will look at all these in more detail and focus first on the marketplace itself.

### 9.2.3.1 Market tools

A local market needs its heart. Most new local marketplace efforts have built their own technology solution for this. This means that the marketplace tool is usually not sold separately but as an implementation. This will enable market participants to enter into the marketplace and benefit from the capabilities, but limits their capability to control the mechanism itself. From a market enabling standpoint this is actually not a limitation, as by definition, a nonbiased marketplace should be a neutral entity so as not to be biased to cater for the need of any single market participant [5]. Below is a short nonexhaustive description of commercially active marketplaces that are being run or are available for setting up local markets.

deX is a trading platform that allows anyone to trade with energy and flexible loads. The platform is developed for the Australian market by the company Greensync and is available there to both mitigate network control needs and to enable the actual local energy trading between local resource owners. It is an example of an interconnected marketplace. Globally, deX has been contracted for major projects in Europe, Japan, and New Zealand. The company is headquartered in Melbourne, Australia, with offices in London and Amsterdam. To date, over 100 organizations across the world have partnered with deX, including major utilities, leading technology brands, and industry bodies [6].

### 9.2.3.2 Messaging tools

Messaging is a must for making any market work. For a local market, the added challenge is that the number of participants is limited and thus the resources available to develop, standardize, and maintain a messaging structure suited best for that particular market are limited. While larger market scopes warrant use of resources for more elaborate schemes of operation and messaging, the limiting factor in local contexts is ultimately the size and value of a typical transaction. In a small community with 100 participants and typical volumes at kilowatt level, the total value of a transaction could be estimated to float at around 50 Eurocents or less. With one or two transactions per day to cover volume and maybe having structures that allow for retrading, we are still looking at typical volumes of 1–5 Euros per day, creating a revenue of around 500 Euros per day for the market. At 15,000 Euros revenue per month even with nonrealistic transaction fees we are looking at distributing single thousands of Euros per month to cover everything from tools to engagement and messaging. There are of course exceptional opportunities and revenue possibilities, but from an overall viewpoint, messaging must cover the mundane and be feasible even when no exceptional opportunities are present. There simply is no room to invent and create a local market specific modus operandi that is aimed to stay as such. This very simplistic and very rough indication of limitations still outlines why a local market in practice must implement technologies for messaging and indeed other functions by making use of implementations aimed to be scalable and capable of distributed implementation. This also allows for the participation of the community, in whatever limited way possible, in developing the aspects that are key for them, rather than focusing on recreating the mundane mechanics of generic market functions.

Typically messages related to markets include contracting messages as well as messages for measurement data and settlement. These are often implemented with either specific market system API messages or different XML-based messages that are then transferred over known Internet protocols, such as file transfer protocol (ftp) or simple mail transfer protocol (smtp), between participants. These technologies are not necessarily easy to implement in a local environment as they require central technology deployment and operation.

Often this brings about the revelation of blockchain. Being a technology that by design has no central component and that ensures all information is everywhere almost at the same time, it would be logical to think that this is the ultimate marketplace technology [8]. This might even be somewhat defendable from a technology standpoint, but from a market standpoint it actually does not solve all the identified issues. One crucial issue is that all blockchains are different in content and therefore all participants willing to enter must agree to use the content format, that is, it is impossible to be part of a chain if you do not upload and download your data in exactly the form and shape dictated by the chain definition. If one disagrees, one cannot participate at all. This is actually the same thing as agreeing on message content and thus agreeing on an entity that has control over the processes performed. Another crucial issue is that while the technology is distributed, if nobody maintains the blockchain structure and nobody provides the medium for it, there is no usability. Hence, actually you end up with a central component after all, the entity running the blockchain and making decisions on the structure of it. Effectively this is the same as having a central messaging distribution system from a business standpoint. From a technical standpoint, however, when running, the blockchain does not have a central server. This is an advantage in a small environment as there are no implementation or maintenance costs for running any central mechanism. As the community simply uses an existing mechanism for their purpose, they end up only with the marginal cost, which is in some cases even free with the exception of actual communications costs and other technical infrastructure costs. This also allows the blockchain to grow across the Internet in small islands, favoring multiple local market implementations sharing a blockchain definition.

Here we won't go deeper into the definition and technical specifications of local market enabling messaging systems, but we will present a few examples of commercial implementations that can be used to enable local markets.

By using commonly available messaging tools, an Extensible Markup Language (XML) based exchange can be implemented. The list of tools is extensive as these tools have been used for a long time and are in widespread use across multiple business domains. A good example is the well-defined list found at the website of ebXML [9,10]. It is a modular suite of specifications that enables enterprises of any size and in any geographical location to conduct business over the Internet. "Using ebXML, companies now have a standard method to exchange business messages, conduct trading relationships, communicate data in common terms, and define and register business processes" [9,10].

When deciding on blockchain, an initial trust decision has to be made to start on something unless you want to set up an entirely unique chain by yourself. One chain that has gained momentum is Ethereum. It has a well-established set of implementations growing around the world and a catalogue of implementations that will help in choosing a relevant chain implementation to use [11,12].

A simple market implementation could use an Ethereum enabled blockchain market such as Airswap [13] for actual market trading and therefore rely on the Airswap messaging, while handling the energy specific details of the traded commodities separately.

A more elaborate and energy specific implementation could use solutions developed under the Energy Web Foundation that fosters Ethereum blockchain-based energy solutions to be developed for energy sector purposes [14]. Implementations by the Dutch energy distribution company Stedin and many other EWF members can be utilized and handle also the messaging related to the market itself [14].

### 9.2.3.3 Datahub capabilities

Datahubs are central gateways or repositories for data needed to perform business processes [15]. While transactions might be reasonable to distribute everywhere to ensure trust, it would be quite foolish even from a pure resource standpoint to distribute all data about everything to everybody. To illustrate, in a city of 1 million, for energy consumption data alone, that would mean close to 45 billion data entities to be stored at each participating location, assuming 5 years' worth of history. Accommodating these information streams and storage requirements would be challenging to implement, unless either distributed databases, a centralized repository, or a cleaning and archiving mechanism amounting to the same is used. In addition, security becomes a real issue because of the extrapolation opportunities in the data, but these are beyond the scope of this chapter.

Hubs are more prevalent in large environments. From a local market standpoint, hubs meant for larger audiences could be used as enablers and in some cases, they could also be used as checkpoints for the viability of transactions on the local markets [16]. The latter is particularly important when local markets form in network-constrained areas. There a network can provide information on constraints and also approve or deny specific energy transactions that could cause the network to fail for all or part of its users. It is important however to stress that this network function is not a utility function, but something that an unbundled and nonmarket participating neutral network can do when fulfilling its role as an enabler for all markets functioning on top of its hardware.

Using messaging tools or blockchain technologies will enable most functions of a local market and on a small scale might serve even all needs for information exchange. This is also, why there are no publicly available configurable datahubs available to be implemented for small local markets. All are specifically made for running larger markets or are simply generic databases. Of course implementing a database and connecting it via messaging is always a possibility and for that purpose, many possibilities ranging from Oracle products to Microsoft SQL and open source databases such as PostgreSQL are available.

So in essence, when forming a local market, the market enabler entity will face a decision about handling the data involved. By way of simple logic it is inevitable that the market enabler, or a third party they opt to use, will then have to implement a gateway to distributed data or a set up a centralized repository. Having no access to data effectively closes the market. Several ways of implementing gateways exist, as do ways to facilitate centralized data handling. Making use of relevant parts of the overarching mechanisms of the existing markets, such as those implemented in the Danish datahub described in Fig. 9.1, will also be something that should be seriously considered to avoid islanding the nascent market out from the start.

In Fig. 9.1 we see a generalized description of how a nationwide retail energy market can function when enabled by a datahub for information exchange. The Danish datahub shown in Fig. 9.1 handles the critical information needed to enable the single point of contact for the customer in Denmark, the supplier, to provide holistic energy services to the customer. It also creates a level playing field for all suppliers on top of the complex structure of network information, taxes, and required processes that have to be followed in order to be able to supply retail electricity in the country.

## 9.3 Participant architecture

Market participants need ways to make their business processes work. These processes are very similar if not identical to the ones being run by energy suppliers, industrial energy users, producers running participant processes such as forecasting, trading, and control of resources, and networks performing market tasks such as measurement and settlement. Tools for all these processes exist in abundance. The challenge with them is often the scale of implementation needed and the specificity of their implementation toward a selected geographic market area.

An example of a cloud-based energy management, forecasting, and trading platform that is used in the Nordic area is EnerimEMS [18]. This system is a shared platform available through secure access and is used here to illustrate how a market participant could run local market enabled energy trading processes. In Fig. 9.2 you can see how the generic EnerimEMS cloud system can provide all basic functions needed to operate energy management and trading processes, while also allowing customized applications and calculations to be formed. Similar architectures are available in other tools as well.

### 9.3.1 Back office tools

In a pure local market implementation, the participants still need the same functionalities as on larger markets, but most probably use a limited scope of capabilities or rely

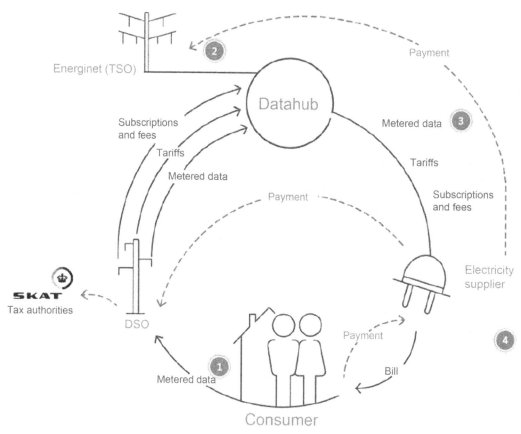

**FIGURE 9.1**    Handling customer transactions in a datahub environment as done by the Danish datahub [17].

heavily on the reporting provided by the marketplace itself. Much of the participant process content will be handled in what could be described as the back office, where the administration of the business takes place.

Back offices for market participants need to plan ahead for markets, feed the front office with information and then integrate the resultant trades and positions from all processes. This is achieved by integrating upward in the value chain.

As an example of commercial implementation of market participant functionality in the back office, we can use the EnerimEMS system. The system can be used to implement the specific portfolios and report calculations of the market participant that correspond to its business processes and market participation strategy. Graphic tools may be used to simplify the development process of implementing the participant's business on this cloud service. Similar tools are available from other vendors as well.

The operations and integration of the participant can be facilitated by an energy management system, as described in Fig. 9.2. The system should be connected to the relevant operations and markets upward in the value chain. For back offices, the capability to extract, tune, and manipulate data with fast and readily available tools like Microsoft Excel is also often a

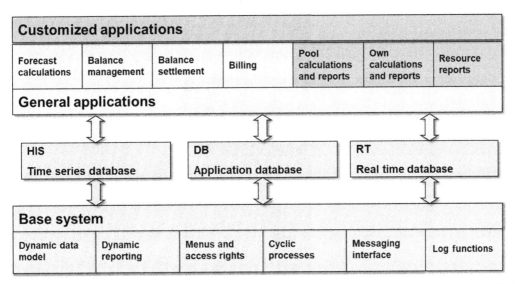

**FIGURE 9.2**   Participant architecture offered by the EnerimEMS cloud-based energy management system [18].

practical solution. It provides the flexibility and the low barrier of implementation needed to be practical enough in a small environment. A sample of such capability is shown in Fig. 9.3 where load data is available through automatically updating linked Excel content that can then be used as is or connected to further Excel functions and analysis such as graphing and own calculations. A balance will eventually be found between implementing back office processes as automated calculations as described in Fig. 9.4 and the editable Excel solutions of Fig. 9.3.

### 9.3.2  Front office tools

Market participants need to engage end customers and energy communities in different ways. This requires both common sales and marketing tools as well as capabilities to handle customer and contract data linked with energy data. Several systems used by energy retailers and networks are available and here we show a sample from EnerimCIS that is a modular platform for enabling customer-related functions in the energy domain. It is most often used by retailers or service providers, but can be used in small communities as well. For microcommunities comprising hundreds of participants or fewer, generic customer relationship management tools could be used as the energy management solutions will be able to handle the underlying energy specific data. In Fig. 9.5 we can see a customer data window from the EnerimCIS system that shows the information of a local market participant and analysis on payment behavior among other content. Use of tools like this will make it easier for the local market to handle growth. Growth in this context should also be understood as the sharing of administrative functions between local markets without actual energy interconnection. This kind of sharing would allow cost efficiencies to develop for the participating local markets.

TimmätFörbr_NÄT.xlsx

EMPOWER — Timmät förbrukning, Snitt 1

| Name | SVK MWh Arvo s | NÄT MWh Arvo s | KALK MWh Arvo s |
|---|---|---|---|
| NÄT 01 | -4,814 | -4,814 | -4,814 |
| NÄT 02 | -0,579 | -0,579 | -0,579 |
| NÄT 03 | -1,882 | -1,882 | -1,882 |
| NÄT 04 | 0,000 9 | 0,000 9 | 0,000 |
| NÄT 05 | -5,150 | -5,150 | -5,150 |
| NÄT 06 | -1,028 | -1,028 | -1,028 |
| NÄT 07 | 0,000 9 | 0,000 9 | 0,000 |
| NÄT 08 | -4,768 | -4,768 | -4,768 |
| NÄT 09 | -4,676 | -4,676 | -4,676 |
| NÄT 10 | -5,400 | -5,400 | -5,400 |
| NÄT 11 | 0,000 9 | 0,000 9 | 0,000 |
| NÄT 12 | 0,000 9 | 0,000 9 | 0,000 |
| NÄT 13 | -6,910 | -6,910 | -6,910 |
| SUM | -35,2 | -35,2 | -35,2 |

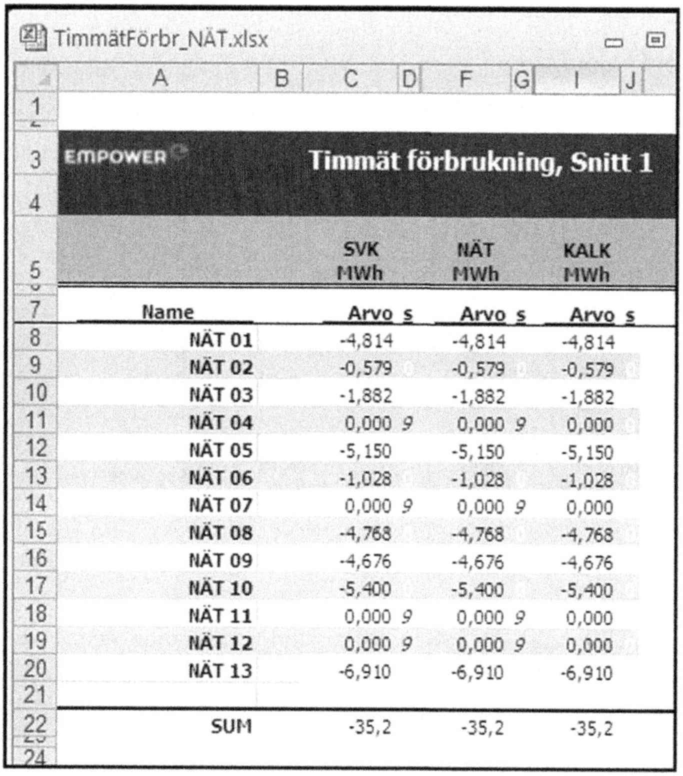

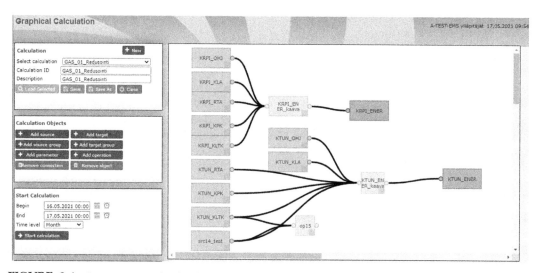

**FIGURE 9.3** Back office interactive reporting of load volumes by Swedish network area (NÄT XX) offered by the EnerimEMS cloud-based energy management system for back office use and Excel integration [18].

**FIGURE 9.4** Interactive graphical calculation tool for automated calculation of trade, load, or generation offered by the EnerimEMS cloud-based energy management system for back office use and Excel integration [18].

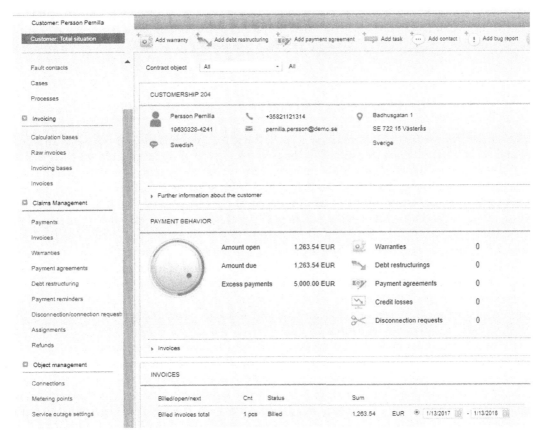

**FIGURE 9.5**   The customer information management environment of EnerimCIS [18].

### 9.3.3 Settlement tools

Making sure trade results are distributed according to contract content and reported to end customers is what is done in reporting and settlement. The overarching settlement system of the energy market, where the local market operates, should always be considered. This is easiest in environments with datahub-enabled settlement such as the Danish environment described in Fig. 9.1.

Local settlement will be dependent on the information streams available to perform it. The local market can make use of a shared energy management system such as EnerimEMS to perform the settlement or decide on a purely transactive scheme that relies on mutually shared information such as a common blockchain, such as EWF.

Essentially, the settlement is an extension of the market implementation itself and thus is typically something that the local market enabler will do and perform with the market platform chosen. Overarching markets typically end up in a competition enabling situation with shared information monopolies existing underneath for neutrality and double investment avoidance. The datahubs are one example of this philosophy and it could be argued that when local markets grow, they could make use of similar developments. Until they do, settlement can be implemented as described earlier.

## 9.4 Community and end customer architecture

Communities and end customers need to be capable of connecting up to the local market and to perform the tasks required of them to be eligible and responsible for their activities. For this there are multiple tools available. The question is which level of activity the community or end customers wish to take. On a large scale end customers regard energy as a tool or a means to an end and are therefore not prepared to focus their activities on energy related processes. This does not mean that they are not interested. It means that any scalable activity should be based on implementing automation while facilitating interaction when customer interest is there or there is a window of opportunity.

Energy is also becoming a statement in personal profiles. This kind of interest is not based on engineering or efficiency, but on the will to be part of the global change and to act in favor of a more sustainable future. This need can be tapped with local community tools, such as the Brooklyn Microgrid App [19]. The app lets participants trade credits based on their solar energy generation.

The local environment as a whole means that the community will need a set of tools that enable basically all of the aspects of running an islanded microgrid and financially distribute the costs of energy and grid access between participants. The decision to make is, whether to use them all, some, or only one of the tools while connecting upward in the value chain to either new or existing service providers. In the Brooklyn example, as with many others, the choice is a mixture of existing infrastructure and new innovation.

Controlling flexible resources and generation assets is at the core of enabling responsible participation in markets. This functionality is enabled by a number of technologies described in the next sections and provided by a number of vendors. A sampling of vendors is included in each technology description below.

### 9.4.1 Smart meter enabled control

A starting point can be tapping into the capabilities provided by smart meters. Depending on the respective networks, smart meter enabled control of simple connected systems such as warm water heaters or storage heating may be good sources of flexibility to start with [20]. It may provide benefits both from an energy savings standpoint and the capability to participate in flexibility trading through a service provider who is eligible for the use of the control interface. These interfaces are now being built in different markets and are most prevalent where previous analog control has already been used to control resources and thus an existing installed base of connected resources already exists. An added benefit is that usually a separate measurement for use of resources when the control is active is provided by the smart meter itself. This aids settlement in many ways.

Most automated smart meter providers have these control relays available and as such have been implemented by networks using meters from Kamstrup, Aidon, L&G, Telvent and several others.

## 9.4.2 Smart device enabled control

Smart devices are devices capable of being remotely controlled in order to either switch on/off or regulate the use or generation of energy at a their location of deployment [21]. Multiple control system variants exist and most are proprietary to their manufacturer while providing an abstract control layer to the user or a connecting system. While these can provide an immense array of capabilities, they share the challenge of verification as there is no secure and trusted way to separate the resulting behavior from the behavior of the metering point under which they are deployed. In many applications this is not an issue as the result being sought is that of the behavior of the metering point in total. This however limits the accountability of multiple devices at the same location.

Many energy companies offer these kinds of smart home or smart control solutions and these are also available as separate systems. Examples include the control system provided by Nokia spin-off There [22], who have been developing their control system and adjoining cloud services for some time. Another example could be the KNX automation system provided by Siemens [23], which is implementable in any location and is meant for general control as well as energy market enabled processes.

In the United States so called smart plugs have been popular. These provide similar capabilities as those offered by more integrated systems but are far simpler to install. Their challenge of accountability in conjunction with the lack of deployment of smart meters is no issue in the United States because of a different market structure from Europe in most states. From a local market standpoint, as long as there is adequate measurement in place for the actual commodity being traded by the market, using either integrated systems or smart plugs for control is possible. The important point to take into consideration there is that if there is trading based on the altered behavior, this behavior is accounted for in the settlement of the responsibilities of the energy position and involved parties.

## 9.4.3 Smart charging control

Electric vehicles (EV) provide a wealth of opportunities for local energy markets as the vehicles contain high-volume batteries, sophisticated control technologies, and ready-made information exchange interfaces for a multitude of functions. In essence the EV is a transportable energy storage and balancing unit capable of valuable interaction with its charging environment. The main commercial challenges lie in ensuring that the vehicle can perform as required in its primary task as a vehicle and will not be hampered by possible add-on or parallel use case implications. Manufacturers are reluctant to take on liability risks linked with having vehicles be part of other energy ecosystems than the vehicle's own system in other roles than being a flexible load while charging. Even that however carries a value and can be enabled through service networks built on smart charging devices or simpler chargers connected to smart control systems.

A sample provider of smart charging services is the Finnish charging service provider Virta. Virta has set up a cloud-based service for managing distributed charging points and can enable both control and reporting of charging on any given network. A local market could therefore implement smart charging enabled trading by utilizing the control and reporting of Virta [24].

### 9.4.4 Microgrid tools

Going beyond the commercial aggregation of local resources requires actual management of networks and the responsibilities that come with it. Managing network safety must be taken seriously and allocated clearly. That all said, microgrids can be implemented by automation on almost any scale allowed by regulation. Regulation is by far the most decisive factor in setting the scope of microgrids as regulation governs energy networks almost everywhere.

In its simplest form, a microgrid can be formed around a single inverter station connected to a solar inverter. With meters installed at load point connections, a small community of houses or huts becomes an automated energy system. From there, next steps could include adding active microgrid controllers and further generation and load equipment. Eventually connecting up to an existing grid could become a possibility with constraints being set on the quality of the connection and associated flows being set, such as for any connection to the grid.

At the heart of the microgrid we have to have the responsible entity who is accountable for the performance and safety of the grid and probably also some or all of the grid equipment. This entity will then take the decisions needed to build and run the grid. Because all microgrids are different, there is no microgrid-in-a-box solution that can be implemented. Multiple component vendors do provide the components and systems needed to run the grid though. A good example is the automation and grid equipment provider ABB who provides solutions from small to large grids and enables scaling through services and connected systems as well.

As local markets are small, these networks tend to be very small and require automation to be managed in a safe and secure way. Multiple vendors provide the solutions needed to manage and run microgrids, that is, microgrid management systems. These vendors include ABB, Schneider, and many small local companies.

## 9.5 Conclusions

The tools presented in this chapter are indicative of the kinds of tools available for creating, running, and developing local energy markets. As described, the local energy market is a multifaceted environment. There is no single silver bullet tool that can implement everything in a flash for an aspiring local market provider. Setting up a local market is a serious affair and one should look at the actual value at play and weigh the challenges against opportunities to scale when choosing how and with whom to set up the market.

The current state of play with the tools introduced here and others that did not make it onto these pages is such that work is needed to fit them into any given local market context. That said, all the needed technologies exist and all the tools are available to go and create something valuable for the local community. The coming years will see even more product-oriented approaches. Actual local market implementations will seek to scale and combine the building blocks that today's tools provide. This will enable easily implementable cloud solutions that can allow most any like-minded group to start sharing their valuable assets and make use of the specific energy that they prefer.

Time will tell whether the local markets of the future will be more on a virtual level, deriving value from commercially created niches of larger markets, or if a more technical approach centering on dedicated and specifically controllable resources will prevail. The market is evolving and we are all part of it, the needed technologies have already come of age.

# References

[1] DOMINOES, Tools for local energy market and end user interaction, 2019. <http://dominoesproject.eu/wp-content/uploads/2020/01/D2.5_DOMINOES_Tools-for-local-energy-market-and-end-user-interaction-vFINAL_1.pdf>.

[2] IER, 2020. <https://www.instituteforenergyresearch.org/history-electricity/>.

[3] Järventausta, et al., Active customer integration, 2008. <https://digital-library.theiet.org/content/conferences/10.1049/ic_20080436>.

[4] Collis, et al., The Role Modelling Guide, Jaron Collis, jaron@info.bt.co.uk Divine Ndumu, ndumudt@info.bt.co.uk, Applied Research and Technology, BT Labs Release 1.01, August 1999.

[5] V. Choudhary, T. Mukhopadhyay, Neutral Versus Biased Marketplaces: A Comparison of Electronic B2B Marketplaces With Different Ownership Structures, 2001, pp. 121–122.

[6] deX, Australian local market, 2020. <https://dex.energy/about-dex/>.

[7] Greensync, 2020. <https://greensync.com/solutions/dex/>.

[8] M. Andoni, V. Robu, D. Flynn, S. Abram, D. Geach, D. Jenkins, et al., Blockchain technology in the energy sector: a systematic review of challenges and opportunities, 2019. <https://www.sciencedirect.com/science/article/pii/S1364032118307184>.

[9] ebXML, 2020. <http://www.ebxml.org/geninfo.htm>.

[10] ebXML, 2020. <ebxml.org/tools>.

[11] Ethereum, 2020. <https://ethereum.org/what-is-ethereum/>.

[12] Ethereum, Ethereum catalogue, 2020. <https://docs.ethhub.io/built-on-ethereum/built-on-ethereum>.

[13] Airswap, 2020. <htltps://www.airswap.io/>.

[14] EWF, Energy Web foundation, 2020. <https://www.energyweb.org/>.

[15] Energinet, Danish retail market, 2020. <https://en.energinet.dk/-/media/Energinet/El-RGD/El-CSI/Dokumenter/ENGELSKE-DOKUMENTER/Danish-electricity-retail-market.pdf>.

[16] O. Kilkki, S. Repo, G. Mendes, J. Rui Ferreira, C. Trocato, Y. Ahmad, et al., Scalable Local Energy Market Architecture, 2019.

[17] Eneginet, Energinet.dk 2020, 2020. <https://en.energinet.dk/Electricity/DataHub#Documents>.

[18] I.M. Empower, Interview, Olli Kilkki, Empower IM Product Manager, 2020.

[19] BMG, Brooklyn Microgrid, 2020. <https://www.brooklyn.energy/about>.

[20] S. Darby, Smart electric storage heating and potential for residential demand response, 2017. <https://link.springer.com/article/10.1007/s12053-017-9550-3>.

[21] Gonzalez, et al., Integrating Building and IoT Data in Demand Response Solutions, 2019. <http://ceur-ws.org/Vol-2389/07paper.pdf>.

[22] There Corporation, 2020. <https://www.there.fi/en/home/>.

[23] Siemens, 2020. Siemens KNX Enabled Products. <https://www.hqs.sbt.siemens.com/cps_product_data/data/produktdb_en.htm>.

[24] Virta, 2020. The Electric Vehicle Charging Company. <https://www.virta.global/company>.

# Distributed energy resource management system

*Quoc Tuan Tran[1], Van Hoa Nguyen[2], Ngoc An Luu[3] and Elvira Amicarelli[4]*

[1]Université Grenoble Alpes, CEA LITEN, Grenoble, France [2]Univ. Grenoble Alpes, Grenoble INP, G2Elab, F-38000, Grenoble, France [3]The University of Danang - University of Science and Technology, Danang, Vietnam [4]ENEL S.p.A., Italia

## 10.1 Energy management system in local energy market

The increase in electricity consumption is an important factor, which impels the judicious planning of the amount and the type of energy resources to use in the future. Legacy and polluting energy generation technologies are to be replaced with modern renewable energy [1]. Besides the slow shifting of policies and the less competitive price, one of the most important obstacles to improve the participation of renewable energies is the difficulty for integration. Many of the renewable energy resources are decentralized, intermittent, and nondispatchable, making them a challenge to integrate into a legacy grid designed for a one-way flow of electricity from centralized generating plants to customer loads. A high penetration rate of renewable energy may also negatively disturb the operating constraints and the power quality of the power system (voltage, frequency) [2].

A microgrid (MG)—consisting of colocated flexible loads, distributed energy resources (DER) and storage technologies, while presenting itself as a single entity to the utility grid via a single point common coupling—can help to remove the perceived challenges to integrating DERs and can help to make widespread DER deployment more manageable [3]. MG can actively maintain local balance of supply and demand, seamlessly island, reconnect to the grid, and improve the resilience and disaster recovery.

In the context of local electricity market, MG deployment also creates an opportunity to deal with energy surplus or deficit due to the intermittent nature of distributed renewable energy sources (DRES). Consumers, becoming prosumers with DRES installation can actively participate in a peer-to-peer or peer-to-operator energy transaction framework.

Such a trading framework benefits both the network operators (flexible peak load management, regulation of voltage/frequency, etc.) and the end users (in terms of energy security and reducing an installation's payback time and carbon footprint). The framework can be implemented at different scales, either among individual prosumers inside a single MG or inter-MG or eventually among distribution system operators (DSO) [4].

Either to support the reliable functionality of the MG, to maximize the penetration of DRES or to achieve optimization of cost and balance in the associated local electricity market, it is primordial to deploy an energy management system (EMS). The EMS is control software that can be embedded to or can act from a higher layer of the microgrid central controller (MGCC) and the individual components controllers. Its role is complementing the control functionalities of the MGCC (voltage, frequency regulation, power sharing, resynchronization and transition between grid-connected and island modes, etc.) with advanced optimizations following predefined objectives (minimize carbon emissions, maximize the financial profit, etc.) based on real-time operating conditions, the MG and the DER, and forecasting information (DRES production, loads).

## 10.2 Architecture and functionalities of EMS

In general, the EMS complements the management and control capabilities of the tertiary and secondary control layer of the MG, taking into account different parameters and various constraints (Fig. 10.1).

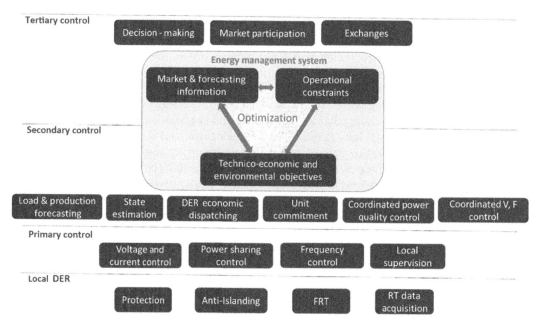

**FIGURE 10.1**    Role and position of an EMS in the control architecture of MG.

The architecture of the EMS is designed in accordance to the considered infrastructure (single MG or multi-MG, grid-connected or isolated), the applied strategy (collaborative or competitive), shared information among the elements in the system, and the method and location of the decision-maker (centralized or distributed). Three popular EMS architectures can be found in the literature: centralized, decentralized, and hybrid [5,6].

In general, the main functionality of the EMS is determining the optimal options for utility power purchases, load dispatch and DER, DRES scheduling, given the load and DER forecasting data, customer preference, energy policy, market price, as well as environmental objectives [7]. The EMS can be installed at the local layer with only a few DERs, at the MGCC for a single MG or eventually in a multi-MG system in a centralized or distributed manner.

## 10.3 EMS design and implementation

In general, the design and implementation of an EMS system varies largely in function of the infrastructure and the power constraints and the functional objectives (Fig. 10.2). The desired outputs of EMS (day-ahead schedule, economic dispatching, and DER dispatching) also influence the data acquisition setup and the choice of optimization software and methods.

### 10.3.1 Production and consumption forecasting

To determine the optimal schedule for MG functionality, it is mandatory to forecast correctly the production and consumption of the MG, the surplus/deficit of energy of nearby MG, or eventually the tendency of market. The objective and output of the EMS

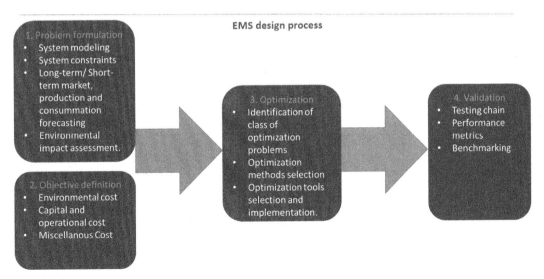

FIGURE 10.2  EMS design and implementation process.

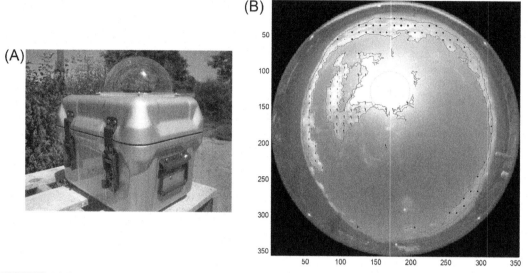

**FIGURE 10.3**   Fisheye camera as acquisition system in CEA INES (A) and cloud detection from the camera image (B).

(day-ahead scheduling, unit commitment, economic dispatching) dictates the required horizon of the forecast (24 hours, short-term or very short-term), and therefore the methods.

Various techniques are employed in the field of photovoltaic (PV) power, load demand, and electricity price forecasting [8–10]. For day-ahead and short-term forecast, historical data and numerical weather predictions can be analyzed with traditional time series (e.g., AutoRegressive Moving Average (ARMA) or Autoregressive Integrated Moving Average (ARIMA)) or newer machine learning (ML) techniques (e.g., Artificial Neural Networks (ANN), k-Nearest-Neighbors (kNNs)) [11]. Time series methods tend to perform better in this case thank to their ability to capture the transition of irradiance over a 24-hour period [12]. The performance of ML techniques can however be improved by dividing the data into different weather regimes and fitting a specific model to a specific weather regime dataset, instead of using one fitted model for all different weather regimes [13]. For short-time PV production forecast, site measurements (power, radiation, temperature) can be combined with the meteorological forecast, and satellite image of the site.

For very short-term forecast (∼5 minutes), the solution is based on direct image of the site sky taken by a local camera (Fig. 10.3A). The principle is to observe the cloud cover from the site, and to predict its evolution in the minutes that follow (Fig. 10.3B). This requires taking periodic images of the sky, then processing these images and integrating them into the forecasting tool. To fully capture the sky, a "fisheye" camera can be used to provide horizon-to-horizon coverage (180 degrees view) [14]. It is however necessary to estimate and take into account the distortion that penalizes the spatial resolution of the images.

## 10.3.2 Optimization problem formulation

The EMS determines the optimal working condition and schedule for the MG by solving an optimization problem on the defined system metrics and constraints, to achieve a specific objective. The correct formulation of this problem is necessary to ensure the functionality of the EMS. In general, the EMS designer needs to consider the following:

- System modeling and metrics: system modeling involves the mathematical or numerical representation of the physical MG, which indicates the system behaviors and the relation among the individual entities (e.g., energy relation, power flow, and loss on line). Systems metrics involve establishing the performance indicators in terms of reliability, or feasibility of system, as well as other economic and environmental impacts (e.g., annualized capital cost, energy payback, battery degradation cost, or $CO_2$ cost) [5,15,16].
- System constraints delineate the physical, operational, and economical requirements (equality) and boundaries (in-equality) of the MG functionalities (e.g., demand response, transformer capacity, energy balancing, physical limits of ESS) [17].
- Objective functions are mathematical or numerical representation of the optimization purposes (minimization of carbon emission, operational cost, maximization of DRES penetration, of energy selling profit, etc.). Several objectives can be combined to a single function for optimization by attributing appropriate weight coefficients. An exhaustive review on popular EMS objective functions can be found in Ref. [15].

## 10.3.3 Optimization methods and solvers

There exist several categories of optimization methods for EMS problems in the literature [17]:

- dynamic programming and rule-based methods (linear, nonlinear, and mixed-integer programming, etc.);
- meta-heuristic approaches (genetic algorithm, particle swarm optimization, artificial bee colony, etc.);
- ML approaches (artificial neural network, reinforced learning based);
- fuzzy logic and multiagent system; and
- the approaches derived from control theory (e.g., model predictive control) [18].

It is also possible to define a hybrid or multistage optimization (e.g., multiagent system combined with a meta-heuristic method). The choice of method varies case by case and depends mostly on the architecture of system (centralized, distributed, hybrid) and the optimization problem formulation (linear, nonlinear, mixed-integer, etc.). Some methods may perform better in one case, but are not suitable for another problem.

This leads to the delicate choice of software/implementation platform for optimization. While there is a wide range of optimization software for EMS [5], some solvers are more adapted to a certain type of problems than others. It is therefore necessary to consider this in the testing chain and to do benchmarking of the EMS finally.

In Sections 10.4 and 10.5, we present two exemplary case studies of design and implementation of EMS in a single MG and in a network multi-MG.

## 10.4 Case study 1: centralized EMS in a single microgrid

In this case study, we consider a centralized EMS for a grid-connected MG consisting of a wind turbine, PV, and battery energy storage system (BESS) (Fig. 10.4). The objective of the EMS is to achieve the minimal operation cost by optimally scheduling DER production and power import from the main grid, respecting the constraints of supply/consumption power balance and the limitations of each DER. The Branch and Bound technique [19,20] is utilized as the optimization method.

The inputs of the EMS are the forecasted profiles of PV and wind production and load/consumption. The EMS will provide the optimal control strategy (minimizing operational cost or maximizing profit) in case of renewable energy surplus (charging BESS or feeding to grid) or deficit (discharging BESS or buying from grid).

Two scenarios are considered: fixed electricity price (feed-in and draw from grid) and dynamic electricity price (feed-in and draw from grid). In the first scenario, the grid price (EgP) is 0.12 €/kWh and the feed-in tariff (FiT) is 0.072 €/kWh. In the second scenario, the prices vary according to the profile in Fig. 10.5A and B.

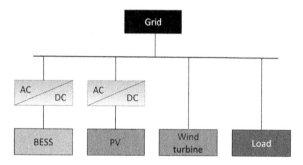

FIGURE 10.4    Configuration of the considered MG.

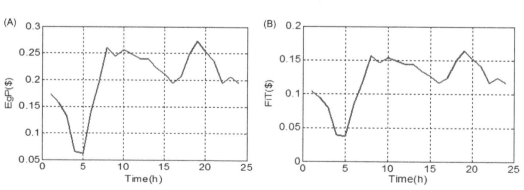

FIGURE 10.5    The grid electricity price (A) and the feed-in tariff profile (B).

### 10.4.1 Formulation

In this MG, the PV, wind power, and load profiles are inputs. We focus mainly on the formulation of the BESS and the constraints:

The State of Charge (SOC) of the BESS is defined as:

$$SOC = \frac{C(t)}{C_{ref}}$$

where $C(t)$ and $C_{ref}(t)$ are the BESS capacity at time $t$ and the reference capacity.

The evolution of SOC is characterized by the relation between $SOC(t)$ and $SOC(t - \Delta t)$:

$$SOC(t) = SOC(t - \Delta t) + \frac{P_{wind}(t) + P_{PV}(t) + P_{grid}(t) - P_L(t)}{C_{ref}} . \Delta t$$

The system is limited by several physical and operational constraints:

- Power balance constraint:

$$P_L(t) = P_{wind} + P_{PV}(t) + P_B(t) + P_{grid}(t)$$

- BESS power output is limited by the physical constraint:

$$P_{Bmin} \leq P_B(t) \leq P_{Bmax}$$

Therefore, SOC variation is also bounded: $SOC_{min} \leq SOC(t) \leq SOC_{max}$
Battery SOC constraint: $SOC_{min} \leq SOC(t) \leq SOC_{max}$

The BESS is also required to be at an acceptable State of Health (SOH) due to environmental reasons and security of supplying capacity: $SOH(t) \geq SOH_{min}$

- Grid power constraint:

$$0 \leq P_{grid}^{max} \leq P_{peakload} \text{ and } P_{grid}^{min} = - P_{grid}^{max}$$

### 10.4.2 Definition of management objectives

In this case study, we consider a financial objective: minimizing the cash flow (CF) from the MG to the grid. The cash flow is expressed as a function of the feed-in profit (cash received—CR) and the consumption price (CP).

The selling profit is determined by the power sold to the main grid $P_{grid}$ and the associated feed-in price at the moment of exchange.

$$CR(t) = P_{grid}(t).FIT(t).t$$

The consumption price is calculated as the cost to buy electricity from the grid and the cost of operation and replacement of the BESS (BrC).

$$CP(t) = (P_{grid}(t).EgP(t).t + BrC(t))$$

In which BrC can be estimated as a function of the SOH of the battery, which is in turn, calculated from amount of charge/discharge (i.e., the variation of SOC) and the aging coefficient $Z$ [19].

$$SOH(x_i, x_j, t) = Z.\left(SOCx_i(t - \Delta t) - SOCx_j(t)\right)$$

Then:

$$BrC(t) = BiC\frac{SOH(t)}{1 - SOH_{min}},$$

With BiC being the BESS investment cost and $SOH_{min}$ being the minimum SOH. The total cash flow is then determined:

$$CF(t) = P_{grid}(t).FIT(t).t + P_{grid}(t).EgP(t).t + BrC(t))$$

The objective function is therefore minimizing:

$$\min \sum_{to}^{T} \left(P_{grid}(t).FIT(t).t\right) + \left(P_{grid}(t).EgP(t).t + BrC(t)\right)$$

### 10.4.3 Optimization method: branch and bound

The evolution of SOC of the battery during 1 day operation can be considered as a multilayer neural network with one root node, one destination node, and 22 hidden layers in which the SOC varies in [$SOC_{min}$, $SOC_{max}$] (Fig. 10.6). $\Delta$SOC is defined as the variation step

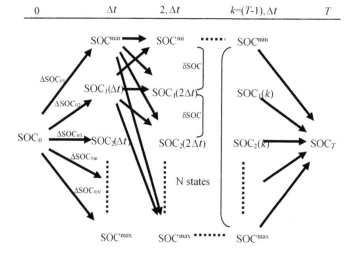

FIGURE 10.6 The battery SOC evolution as a multilayer neural network.

For each layer, the variation interval of $\Delta$SOC is calculated from the forecasted production/consumption and the grid constraints:

$$\Delta SOC_{min}(t) = \frac{P_{PV}(t) + P_{wind}(t) + P_L(t) - P_{grid}^{max}}{C_{ref}}$$

$$\Delta SOC_{max}(t) = \frac{P_{PV}(t) + P_{wind}(t) + P_L(t) + P_{grid}^{max}}{C_{ref}}$$

These $\Delta$SOCare also subjected to the requirement that the next SOC value (i.e., $SOC(t+1) = SOC(t) + \Delta SOC(t)$) stays in the physical constraint of the BESS. The set of nodes that satisfy these constraints at the $i$th layer is denoted as $\psi(i)$. The rest of the nodes are pruned.

The problem now becomes finding the optimal path from the root node to the destination node to minimize the objective function. The Branch and Bound method is applied to find the "leaf" node (i.e., the node having the minimum total cost value from the root node) at every layers.

At every layer, the optimal path to a node $j$ is calculated from the last layer:

$$\min_{i \in \psi(t-1)} CF(i, t-1) + CF(j, t)$$

The process is repeated until we reach the destination node. The minimal cash flow is then determined:

$$CF^* = \min_{i \in \psi(23)} CF(i, 24)$$

### 10.4.4 Implementation results

The designed EMS is applied to the microgrid in Fig. 10.4. The forecasted profiles of consumption and production of DRES (PV and wind) are presented respectively in Fig. 10.7A–C.

The EMS then produces the following power schedules for the MG for both scenarios: fixed prices and dynamic prices (Fig. 10.8A and B).

It can be noted that the fundamental difference is the activity of the BESS. In the fixed price scenario, the EMS aims to minimize the power drawn from the grid by making use of the DRES to supply the load and to charge the battery with the excess power. It also prioritizes discharge of the BESS before demanding the deficit power from grid, when load gets to peak. On the other hand, in the second scenario, the EMS tends to buy energy from the grid when the price is low (i.e., beginning of the day) and sells the excess DRES power to the grid from 9:00 a.m to 5:00 p.m. The BESS, in this case, is charged with low price grid power instead of the DRES.

## 10.5 Case study 2: distributed EMS in a multimicrogrid network

In this case study, we consider a distributed EMS in a multimicrogrid system, which contributes to the active congestion management of a distribution grid. In particular, the massive integration of DRES into the distribution network may cause voltage and power

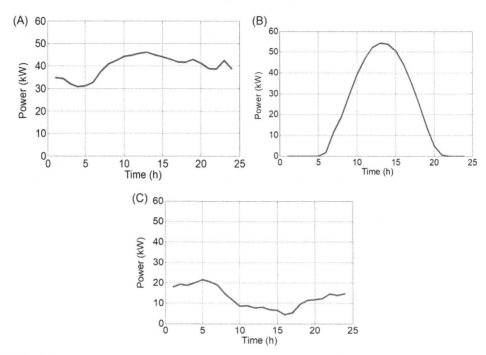

**FIGURE 10.7**    Forecasted profiles of consumption (A), PV production (B), and wind energy production (C).

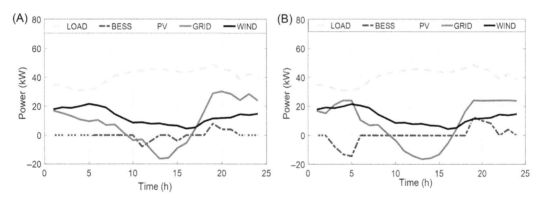

**FIGURE 10.8**    Power schedule of the MG in fixed-price (A) and dynamic price (B) scenarios.

congestion risks. Besides planning a day-ahead schedule, the considered EMS can take into account the flexibilities of users (industrial, commercial, or household customers) in short-term management to mitigate congestion risks and efficiently manage a connected system to optimize the investments.

We consider a system of four MG connected through a MV/LV transformer to the MV network. Each MG is represented to the MV network via a unique aggregator. The considered

MGs consists of communities of users, small-sized residential and commercial activities with inflexible loads, flexible loads, DGs, and energy storage systems (ESSs). The characteristics of all four MGs are given in Tables 10.1–10.4.

## 10.5.1 Day ahead scheduling

The EMS first defines an initial day-ahead planning for each MG, following the available production, the consumption to satisfy, and the storage constraints. This process can

**TABLE 10.1**  Type and nominal data of components in microgrid 1.

| PV | | | Li-ion battery | | |
|---|---|---|---|---|---|
| $n.$ | $P_n$ | LCOE | $n.$ | $P_n$ | $E_n$ |
| 4 | 9 kW | 13 c€/kWh | 1 | 50 kW | 100 kWh |
| **Load** | | | $\eta_{ch}$* | $\eta_{dec}$* | $SOC_{start}$* |
| $n.$ | $P_c$ | **Type** | 0.96 | 0.97 | 0.2 |
| 30 | 6 kW | Inflex. | $SOC_{final}$* | $SOC_{max}$* | $SOC_{min}$* |
| 15 | 3 kW | Inflex. | 0.2 | 0.8 | 0.2 |

**TABLE 10.2**  Type and nominal data of components in microgrid 2.

| PV | | | Li-ion battery | | |
|---|---|---|---|---|---|
| $n.$ | $P_n$ | LCOE | $n.$ | $P_n$ | $E_n$ |
| 2 | 300 kW | 11 c€/kWh | 1 | 100 kW | 200 kWh |
| | | | $\eta_{ch}$* | $\eta_{dch}$* | $SOC_{start}$* |
| | | | 0.98 | 0.98 | 0.2 |
| **Load** | | | $SOC_{final}$* | $SOC_{max}$* | $SOC_{min}$* |
| – | | | 0.2 | 0.9 | 0.1 |

**TABLE 10.3**  Type and nominal data of components in microgrid 3.

| PV | | | Li-ion battery | | |
|---|---|---|---|---|---|
| $n.$ | $P_n$ | LCOE | $n.$ | $P_n$ | $E_n$ |
| 4 | 15 kW | 13 c€/kWh | | | |
| 10 | 25 kW | 12 c€/kWh | 1 | 100 kW | 200 kWh |
| **Load** | | | $\eta_{ch}$* | $\eta_{dch}$* | $SOC_{start}$* |
| $n.$ | $P_c$ | **Type** | 0.97 | 0.98 | 0.4 |
| 30 | 6 kW | Inflex. | $SOC_{final}$* | $SOC_{max}$* | $SOC_{min}$* |
| 5 | 9 kW | Inflex. | 0.4 | 0.9 | 0.1 |

**TABLE 10.4**   Type and nominal data of components in microgrid 4.

| PV | | | Li-ion battery | | |
|---|---|---|---|---|---|
| *n.* | $P_n$ | LCOE | *n.* | $P_n$ | $E_n$ |
| 1 | 25 kW | 12 c€/kWh | | | |
| **Load** | | | 1 | 100 kW | 200 kWh |
| *n.* | $P_c$ | Type | $\eta_{ch}$[a] | $\eta_{dch}$[a] | $SOC_{start}$[a] |
| 30 | 6 kW | Inflex. | 0.97 | 0.98 | 0.4 |
| **Demand response** | | | $SOC_{final}$[a] | $SOC_{max}$[a] | $SOC_{min}$[a] |
| *n.* | Type | $P_n$ | UT | | |
| | | | 0.4 | 0.9 | 0.1 |
| 7 | WM[a] | 3 kW | 6–23 | | |
| 5 | DW[a] | 3 kW | 5–13 | | |

[a]DM, Dishwasher; $\eta_{ch}$, charge efficiency; $\eta_{dch}$, discharge efficiency; SOC, state of charge; UT, utilization time; WM, washing machine.

follow the method described in Section 10.4. Each MG tries to minimize costs for consumers and maximize revenues for producers. The aggregated energy schedule for the four MGs are presented in Fig. 10.9.

## 10.5.2 Flexibilities and problem formulation

The EMS considers two types of flexibilities: downward (production reduction, charging storage systems, and increasing flexible load) and upward (production increase, discharge storage systems, and reduction of flexible load). The EMS can generate two types of offers for flexibility bid:

- Type A: a pair of Quantity–Price (kWh–c€/kWh), that corresponds to a binary response accepted/not accepted.
- Type B: a triple of Maximal Quantity–Minimal Quantity–Price (kWh–kWh–c€/kWh). The accepted amount can be chosen between the minimal and maximal proposed quantity.

Each proposal is referred to a time period (e.g., 0:00 a.m.–1:00 a.m., 1:00 a.m.–2:00 a.m., etc.) and with a bid duration of 1 hour. DGs, ESS, and flexible loads can be used for providing both upward and downward flexibilities using both type A and B models. The bid price could be fixed between zero and the selling price for downward flexibilities. Upward flexibilities price is taken as equal to the one proposed by the user to produce electricity.

In this case study, the downward flexibility is composed of appliances shifting (Fig. 10.10A: washing machines between 8:00–9:00 a.m. and 9:00–10:00 a.m. with a bid proposal of 4.7 kW) and PV reduction (between 10:00 a.m. and 2:00 p.m.) (Table 10.5). Once accepted, the load profile can be modified according to the flexibilities. The upward flexibilities involves using the residual power of the diesel generator (Table 10.5; Fig. 10.10B). These flexibilities are strongly influenced by market prices and DGs costs.

The EMS then considers the combination of all the resources to be submitted to the local energy market while respecting the constraints imposed by the DSO. It requires the

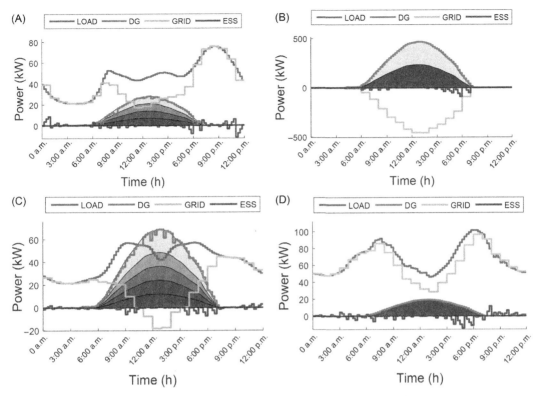

FIGURE 10.9  Initial energy schedules for MG1 (A), MG2 (B), MG3 (C), and MG4 (D).

generation and load forecasted profiles and information about the available flexible appliances (profile, start and end time) to find the final aggregated profiles, which minimize costs to satisfy the required consumption and maximize revenues for selling energy. It determines at each moment $t$ whether or not to activate the flexibilities to respect both the DSO request and the user's schedule.

The objective function is then formed as:

$$\min \sum_{f \in F} C_t^f \cdot PF_t^f \Delta t - CS_t \cdot PS_t \cdot \Delta t + CB_t \cdot PB_t \cdot \Delta t$$

It is optimized under the constraints of:

- Power balance:

$$\sum_{m \in M} PB_t^m - \sum_{m \in M} PS_t^m = -\sum_{gn \in F} P_t^{gn} - \sum_{ln \in F} P_t^{ln} + \sum_{gp \in F} P_t^{gp} + \sum_{lp \in F} P_t^{lp} - PS_t$$

- Flexibilities upper and lower bounds

$$xF^f \cdot PF_{\min}^f \leq PF_t^f \leq xF^f \cdot PF_{m\max}^f$$

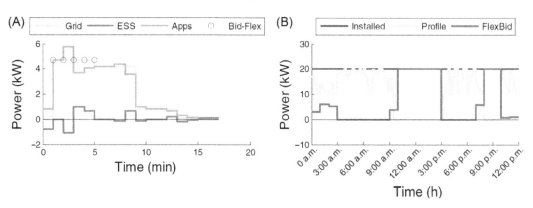

FIGURE 10.10   Flexibility of the washing machines (A) and the diesel generator (B).

TABLE 10.5   Upward and downward DGs available flexibilities.

| Upward | 0:00–1:00 a.m. | 1:00–2:00 a.m. | 2:00–3:00 a.m. | 10:00–11:00 a.m. |
|---|---|---|---|---|
| | 0.00–3.1 kW | 0.0–6.0 kW | 0.0–5.4 kW | 0.0–3.8 kW |
| | 10:00–11:00 a.m. | 11:00–12:00 a.m. | 12:00–1:00 p.m. | 1:00–2:00 p.m. |
| | 10.0–20.0 kW | 10.0–20.0 kW | 10.0–20.0 kW | 10.0–20.0 kW |
| | 2:00–3:00 p.m. | 7:00–8:00 p.m. | 8:00–9:00 p.m. | 9:00–10:00 p.m. |
| | 10.0–20.0 kW | 0.0–5.6 kW | 10.0–20.0 kW | 10.0–20.0 kW |
| | 10:00–11:00 p.m. | 11:00–0:00 p.m. | – | – |
| | 0.0–0.8 kW | 0.0–0.9 kW | | |
| Downward | 10:00–11:00 a.m. | 11:00–12:00 a.m. | 12:00–1:00 p.m. | 1:00–2:00 p.m. |
| | 0.0–4.1 kW | 0.0–24.2 kW | 0.0–37.5 kW | 0.0–31.2 kW |

- DSO required bounds

$$Ps_t \leq Plb_t$$
$$PB_t \leq Pub_t$$

The optimization is then implemented using MATLAB® and the OPTI Toolbox [21].

## 10.5.3 Implementation and results

The MG optimization is updated over the day-ahead initial energy schedule for a time window of 15 minutes. We take a particular attention to MG 4, which has 12 flexible appliances (Table 10.4). Fig. 10.11 represents the initially scheduled operation of these appliances (in green—dishwashers before 12:00 a.m. and washing machines around 12:00 a.m. and in late evening hours). The usage-time flexibility of these appliances is then combined with an ESS to compute a coherent hourly bid. Different bid proposals are aggregated

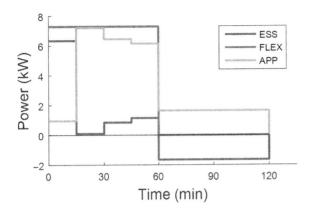

**FIGURE 10.11** Load flexibility in MG4 at $t = 13$.

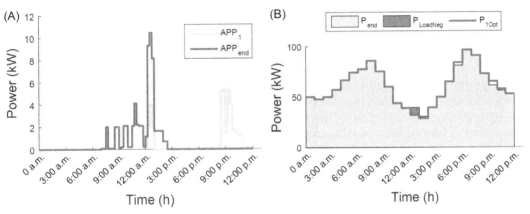

**FIGURE 10.12** (A) Initial and final schedule for flexible appliances in MG4. (B) Initial scheduled profile and accepted flexibilities by aggregator for MG4.

according to the available flexible appliances, for example, at $t = 13$, the aggregated flexibility consists of an on/off bid of 7.3 kW with the price of 5.0 c€/kWh (Fig. 10.11).

The appliances scheduled between 8:30 p.m. and 11:00 p.m. in MG4 are then started at 12:00 a.m. (Fig. 10.12A). It alters the final schedule of the total load/production in MG4 (Fig. 10.12B).

On the contrary, MG2 and MG3 inject energy to the main grid from 11:00 a.m. to 3:00 p.m. and offer downward flexibilities (Table 10.5; Fig. 10.13A). The EMS runs the optimization problem using flexibilities proposed by these MGs to mitigate the energy surplus exceeding the DSO lower limit between 1:00 p.m. and 2:00 p.m. (Fig. 10.13B). The final schedule for MG2 is then presented in Fig. 10.13B, with the reduction of energy by 3.6% and 5.1% of the initial schedule (i.e., 16.4 and 23.4 kW), in time frame 13 and 14, respectively. This energy is stored in the ESS and is sold between 3:00 p.m. and 5:00 p.m.

Detailed implementation of this case study can be found in Ref. [22–24].

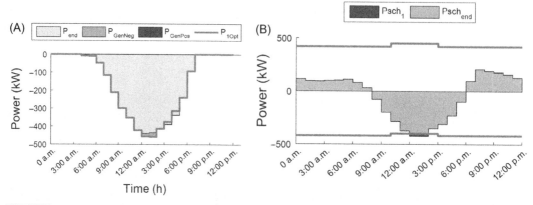

**FIGURE 10.13**   (A) Initial scheduled profile and accepted flexibilities by aggregator for MG2. (B) Final aggregator schedule and DSO upper and lower limits.

## 10.6 Conclusion

The EMS determines the optimal strategy with respect to the local electricity market and helps the MG to function according to the desired strategy. In this chapter, the concept and design of the EMS in a MG was presented along with two exemplary case studies: optimization of cash flow for a single microgrid, and minimizing costs and maximizing revenues for a multimicrogrid network.

## References

[1] International Energy Agency, World Energy Outlook 2018, 2018.
[2] ENTSOE, Dispersed generation impact on CE region security - dynamic study, European Network of Transmission System Operators for Electricity, Report Update, 2014.
[3] A. Hirsch, Y. Parag, J. Guerrero, Microgrids: a review of technologies, key drivers, and outstanding issues, Renew. Sustain. Energy Rev. 90 (2018) 402–411. Available from: https://doi.org/10.1016/j.rser.2018.03.040.
[4] V.H. Nguyen, Y. Besanger, Q.T. Tran, M.T. Le, On the applicability of distributed ledger architectures to peer-to-peer energy trading framework, in: 2018 IEEE International Conference on Environment and Electrical Engineering EEEIC, 2018, pp. 1–5. Available from: https://doi.org/10.1109/EEEIC.2018.8494446.
[5] A.H. Fathima, K. Palanisamy, Optimization in microgrids with hybrid energy systems – a review, Renew. Sustain. Energy Rev. 45 (2015) 431–446. Available from: https://doi.org/10.1016/j.rser.2015.01.059.
[6] L. Meng, E.R. Sanseverino, A. Luna, T. Dragicevic, J.C. Vasquez, J.M. Guerrero, Microgrid supervisory controllers and energy management systems: a literature review, Renew. Sustain. Energy Rev. 60 (2016) 1263–1273. Available from: https://doi.org/10.1016/j.rser.2016.03.003.
[7] W. Su, J. Wang, Energy management systems in microgrid operations, Electr. J. 25 (8) (2012) 45–60. Available from: https://doi.org/10.1016/j.tej.2012.09.010.
[8] T. Hong, P. Pinson, S. Fan, H. Zareipour, A. Troccoli, R.J. Hyndman, Global energy forecasting competition 2014 and beyond, Int. J. Forecast. 32 (3) (2016) 896–913.
[9] J. Antonanzas, N. Osorio, R. Escobar, R. Urraca, F.J. Martinez de Pison, Review of photovoltaic power forecasting, Sol. Energy 136 (2016) 78–111.
[10] R. Inman, H. Pedro, C. Coimbra, Solar forecasting methods for renewable energy integration, Prog. Energy Combust. Sci. 39 (6) (2013) 535–576.

[11] V. Kestylev, A. Pavlovski, Solar power forecasting performance - towards industry standards, presented at the International Workshop on the Integration of Solar Power Into Power Systems, Aarhus, Denmark, 2011.

[12] G. Reikard, Predicting solar radiation at high resolutions: a comparison of time series forecasts, Sol. Energy 83 (3) (2009) 342−349.

[13] H.T.C. Pedro, C.F.M. Coimbra, Assessment of forecasting techniques for solar power production with no exogenous inputs, Sol. Energy 86 (7) (2012) 2017−2028. Available from: https://doi.org/10.1016/j.solener.2012.04.004.

[14] SteadySun, SteadyEye. <https://www.steady-sun.com/technology/steadyeye/>. (accessed 23.04.2020).

[15] A. Ahmad Khan, M. Naeem, M. Iqbal, S. Qaisar, A. Anpalagan, A compendium of optimization objectives, constraints, tools and algorithms for energy management in microgrids, Renew. Sustain. Energy Rev. 58 (2016) 1664−1683. Available from: https://doi.org/10.1016/j.rser.2015.12.259.

[16] A. Omu, A. Rysanek, M. Stettler, R. Choudhary, Economic, climate change, and air quality analysis of distributed energy resource systems, Procedia Comput. Sci. 51 (2015) 2147−2156. Available from: https://doi.org/10.1016/j.procs.2015.05.487.

[17] M.F. Zia, E. Elbouchikhi, M. Benbouzid, Microgrids energy management systems: a critical review on methods, solutions, and prospects, Appl. Energy 222 (2018) 1033−1055. Available from: https://doi.org/10.1016/j.apenergy.2018.04.103.

[18] L.I. Minchala-Avila, L.E. Garza-Castañón, A. Vargas-Martínez, Y. Zhang, A review of optimal control techniques applied to the energy management and control of microgrids, Procedia Comput. Sci. 52 (2015) 780−787. Available from: https://doi.org/10.1016/j.procs.2015.05.133.

[19] N.A. Luu, Control and Management Strategies for a Microgrid (Ph.D. thesis), University Grenoble Alpes, Grenoble, France, 2014.

[20] N.A. Luu, T.M.D. Tran, Q.T. Tran, Optimal energy management for an on-grid microgrid by using Branch and Bound method, presented at the International Conference on Environment and Electrical Engineering, Palermo, Italia, 2018.

[21] J. Currie, D.I. Wilson, OPTI: Lowering the Barrier Between Open Source Optimizers and the Industrial Matlab User, Foundations of Computer-Aided Process Operations, Georgia, United States, 2012.

[22] E. Amicarelli, Q.T. Tran, S. Bacha, Optimization algorithm for microgrid day-ahead scheduling and aggregator proposal, presented at the 2017 IEEE International Conference on Environment and Electrical Engineering, June 2017.

[23] E. Amicarelli, Q.T. Tran, S. Bacha, M.C. Pham, Capacity limit allocation for active congestion management of distribution grids using flexible user's profiles in microgrids, presented at the IECON 2018 - 44th Annual Conference of the IEEE Industrial Electronics Society, October 2018.

[24] E. Amicarelli, Control and Management Strategies of Smart Grids With High Penetration of Renewable Energy (Ph.D. thesis), University Grenoble Alpes, Grenoble, France, 2017.

# Modeling, simulation, and decision support

*Danial Esmaeili Aliabadi[1], Emre Çelebi[2,3], Murat Elhüseyni[4]
and Güvenç Şahin[4]*

[1]Helmholtz Centre for Environmental Research - UFZ, Leipzig, Germany [2]Center for Energy
and Sustainable Development, Kadir Has University, Istanbul, Turkey [3]Industrial Engineering
Department, Yeditepe University, Istanbul, Turkey [4]Faculty of Engineering and Natural
Sciences, Sabanci University, Istanbul, Turkey

## 11.1 Introduction

Determining correct policies requires accurate observations and predictions of future trends. Yet, neither would be achieved without quantitative methods, by which the underlying system, and its components, are represented using mathematical formulations. Energy systems are complex systems that influence, and are affected by, other techno—socio—economic subsystems. In order to study such complicated interactions, researchers rely on two well-known modeling perspectives: the top-down macroeconomic approach and the bottom-up engineering approach, which are synonymous with *aggregated* and *disaggregated* models, respectively [1]. These approaches have their strengths and shortcomings; therefore, choosing a suitable modeling approach is profoundly subject to the problem at hand. For instance, while bottom-up models are better suited for analyzing energy systems in which parameters and technologies may evolve over time, they require an extensive effort in data collection. Hence, researchers should determine a suitable trade-off between the complexity and simplicity.

In this chapter, we categorize various modeling techniques in the context of energy and electricity markets. The advantages and disadvantages of each method are explained to assist researchers in choosing proper methodologies to cope with their problems. Fig. 11.1 depicts an abstract illustration of the chosen pathway in the following sections. As one can perceive, the reviewed papers in this chapter show only one possible path, which is closer to the agenda of this book; thus, it is by no means complete.

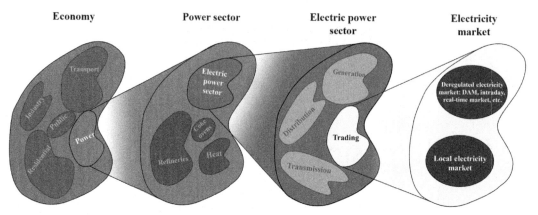

**FIGURE 11.1** The guideline for the following sections. We content ourselves to only a few instances inside each hyperbubble for the sake of clarity.

This chapter unfolds as follows. In the next section, we elaborate on basic modeling approaches through which researchers produce insights. Section 11.3 focuses on electricity markets and examines various modeling techniques with their corresponding pros and cons. Studies that are related to local electricity markets are investigated in Section 11.4. Finally, Section 11.5 concludes the chapter.

## 11.2 Modeling approaches

There are two fundamental modeling approaches to investigate interrelated energy systems [1]:

- *The top-down macroeconomic approach* that emphasizes the possibilities to substitute multiple inputs in order to achieve better outputs. Top-down models focus on economy-wide features.
- *The bottom-up engineering approach*, which is taking into account technological and sectorial details. Bottom-up models can be divided into optimization (including partial equilibrium models), simulation, multiagent, and accounting models [2].

In top-down models, energy, like other inputs, is a production factor and interacts with other factors in the production function to generate economic growth. Top-down models mostly account for macroeconomic feedback and microeconomics realism [3]; however, for the sake of simplicity and tractability, top-down models ignore technical aspects and assume an institutionally, behaviorally, and technologically stable world [4]. We can associate any top-down model to one of the following categories [5–7]:

- *Input-output (IO) models:* these models are characterized through a system of linear equations that describe the financial flow of products among economic sectors for both intermediate and end-use deliveries. IO life cycle assessment (IO LCA) extends economic IO models to include externality costs such as environmental impacts [8,9].

Although these models allow us to study the impacts of structural changes and economic shocks on the whole economy, they assume that prices are provided exogenously [2]. Another drawback of these models is that they often consider sectors in an extremely aggregated form, which may introduce inaccuracy in the results [10,11].

- *Econometric models:* they utilize economic data and statistical inference techniques to examine statistical relations among economic variables with respect to time. Econometric models are ranging from simple linear regression to more rigorous methods in time series analysis [12,13]. While econometric models are used to calculate projections, their ability to project the relation of economic variables into decades ahead is limited, since the correlations among statistical variables may change over time [14].
- *Computable general equilibrium (CGE) models:* CGE models are built upon general equilibrium theory to analyze equilibrium conditions in an economy with rational economic agents. These models can be seen as the general form of the partial equilibrium models, in which interactions between energy markets and the rest of the economy are considered. Models in this category can be conveniently connected to bottom-up models.
- *System dynamics models:* complex nonlinear simulations can be produced by specifying rules to describe different agents' behavior in these models; nonetheless, they often have narrower focus than CGE models.

Top-down models, which are normally used to make projections, become less reliable as the underlying parameters in these models change over time. For example, the past projections of natural gas prices in the United States turned out to be so unreliable that they resulted in billions of wasted dollars in investments in US regasification plants that were constructed to import foreign sourced liquid natural gas (LNG) into the United States. In this case, top-down models failed to anticipate the application of known technologies to the production of natural gas from reserves that were previously thought to be too expensive to produce (i.e., shale gas) [15,16]. Unlike top-down models in which parameters will remain unchanged, the bottom-up models provide specific opportunities to introduce new technologies and understand how they can affect future energy market fundamentals; therefore, bottom-up models serve as the only practical choice to estimate energy trends beyond a few years.

Bottom-up models are qualified to describe the whole energy sector in detail considering different forms of energy and various technologies. In bottom-up models, technologies are characterized based on technical (e.g., availability factor, efficiency), and economic properties (e.g., investment cost, operation and maintenance cost). Parallel technologies compete in the bottom-up frameworks to satisfy demand with low cost and in a sustainable manner. As competing technologies are assumed to be perfect substitutes, it can lead to a market that is dominated by the cheapest technology; therefore, despite the technological explicitness, these models suffer from the lack of behavioral realism [17] and feedback stemmed from economic growth (e.g., income and GDP) [1]. Fig. 11.2 depicts a schematic diagram of a Reference Energy System (RES) in a generic bottom-up model from the supply of primary energy sources, energy conversion, transmission, distribution down to consumption by services.

Major top-down models can be counted as EPPA [18], GEMINI-E3 [19], MACRO [20], GTAP [21], and BaHaMa [22]. MESSAGE [23] and its successor TIMES [24] and LEAP [25]

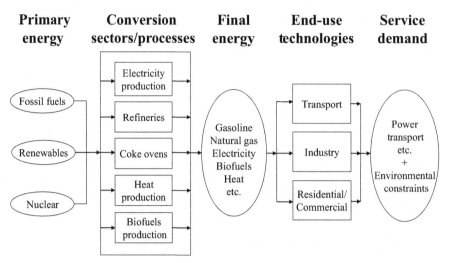

**FIGURE 11.2**  Simplified reference energy system.

are popular software tools for the bottom-up modeling of energy systems.[1] The bottom-up modeling approach has been used in World Energy Outlook (WEO) since 2008 by the International Energy Agency (IEA).

As bottom-up and top-down approaches can complete one another, researchers propose hybrid models, in which they consider the joint impact of macroeconomic factors and technological development [3,26]. These hybrid models can be created by either soft-linking or hard-linking of top-down and bottom-up models. The soft-linking strategy seeks to align these two basic modeling approaches through an iterative process such that the convergence condition of central parameters and variables are fulfilled. However, the hard-linking strategy attempts to unify these two modeling approaches into a single model; thus, the resulted model is often simplified either in the top-down or bottom-up side (e.g., Ref. [17] developed a hybrid CGE model, in which the electric power sector is modeled using bottom-up approach and other sectors are modeled using a CGE model). Fig. 11.3 depicts soft-linking and hard-linking strategies of two approaches with the interactions between them. MARKAL-MACRO [20] and TIMES-MACRO [27] are the known instances of hybrid frameworks.

Since these frameworks are comprehensively covering energy, economy, and environment modules, they require substantial effort on data gathering and organization processes for the region of interest (e.g., city, province, country, or world) based on the desired level of detail. For example, in Ref. [28], the bottom-up model is disaggregated to individual processes instead of employed technologies (see Fig. 11.4), which demands multiple gigabytes of raw data to provide accurate representations. Thus, instead of taking a holistic view, researchers may confine the boundary of their study to a subset of these sectors and consider the interaction with other sectors as exogenous input parameters to simplify the interactions.

---

[1] Please check Table 1 in Ref. [2] for the extensive list of models in each category.

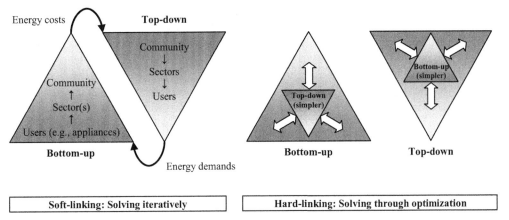

FIGURE 11.3   A simplified demonstration of building hybrid models through hard-or soft-linking. Two models may partially or entirely overlap each other when hard-linking strategy is applied.

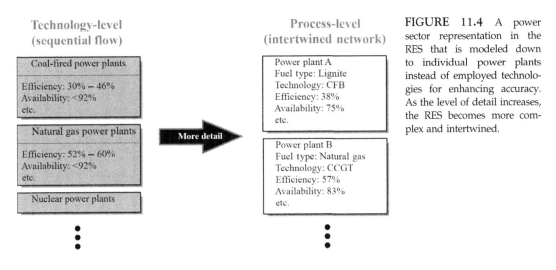

FIGURE 11.4   A power sector representation in the RES that is modeled down to individual power plants instead of employed technologies for enhancing accuracy. As the level of detail increases, the RES becomes more complex and intertwined.

In the next section, we concentrate on various methodologies that replicate and analyze electricity markets.

## 11.3  Modeling electricity markets

Unlike storable commodities, electricity consumption should be instantaneously balanced with generation. This feature introduces technical, managerial, and economical complexity to trading structures. The electricity industry commenced with vertically integrated monopolies; however, unsatisfactory performance of such regulated monopolies as a result of expensive construction and operation costs called for their liberalization [29,30]. Nowadays, power delivery consists of many services, including generation, trading, transmission, and distribution. Trading and generation layers regard electricity as a tradable merchandise, whereas distribution and transmission layers concentrate on resolving technical issues and providing electricity services.

The main objective of deregulated (or liberalized) electricity markets is to maximize social welfare via competition. Nevertheless, designing a perfectly competitive liberalized market is tremendously difficult. In fact, it has been shown that some electricity markets act more like oligopolies [31,32]. A few reasons for this oligopolistic behavior can be reported as:

- restricted number of generators due to entry barriers (e.g., high capital investment) for smaller companies;
- limitations in transmission and distribution networks (e.g., congestion) that isolate certain generators from some consumers; and
- losses in transmission lines that discourage consumers to purchase electricity from remote producers [31].

An unfavorable by-product of oligopolistic markets is the likelihood of participants engaging in collusion. Collusion is an agreement among multiple parties to evade perfect competition. In electricity markets, explicit collusion is prohibited, yet another form of collusion, which is called tacit collusion, exists in the absence of explicit agreement. Market designers and policy makers struggle to mitigate collusion between competitors to attain a competitive market. In general, detecting collusive behavior is of no simple task for regulators [33,34]; nonetheless, researchers in Refs. [35,36,37] have demonstrated that players might have engaged in tacit collusion in Wales and England, California, and Spain, respectively. These negative outcomes can be alleviated by employing innovative concepts such as the Local Electricity Markets (LEMs). In LEMs, the number of small generators is more than conventional markets, and this alone can make the market more competitive, even though there is a growing body of the literature investigating the optimal coalition in LEMs [38]. The LEM also mitigates the congestion in the network. In Section 11.4, we will discuss this concept in more detail.

In the literature, three separate market modeling paradigms exist: equilibrium, optimization, and simulation models. While equilibrium models exhibit the general behavior of markets factoring into account individual participant models, optimization models concentrate on a single entity. In equilibrium models, price-taking behavior relates to perfect competition, whereas strategic behavior pertains to imperfect competition. Equilibrium models that regard imperfect competition are ranging from simple economic models (e.g., Cournot and Bertrand competition) to more sophisticated mathematical models (e.g., Conjectural Variation (CV) and Supply Function Equilibria (SFE)). Finally, when the underlying problem is very complex to be addressed with equilibrium models, simulation models can be used as alternatives to generate insights.

Some researchers prefer equilibrium models, in which the solution can be computed rather straightforwardly using analytical methods. Nonetheless, because of strict simplifying assumptions, there is no guarantee that these outcomes are spotted in practice [39]. One of these streamlining assumptions is associated with the characterization and the length of the time period: most analytical methods are restricted to monitoring the system during a short and fixed time horizon (e.g., Cournot [40,41], CV [42,43], and equilibrium problem with equilibrium constraints (EPEC) [44]). These models implicitly assume a fixed-behavior for the power generation companies (GenCos) over time. In addition, analytically solvable equilibrium models typically do not regard the physical constraints of the transmission network [41,43–45]; for instance, the authors in [44] admit that introducing the physical limitations of transmission lines makes their suggested analytical method intractable; therefore, they demand the employment of a numerical method instead of an analytical one.

On the other side of the spectrum, simulation models enable us to capture the dynamic behavior of participants in a real-life environment. For example, when one GenCo modifies its bidding strategy, other stakeholders observe this new arrangement, and will eventually respond to the new conditions such that serve their interests. A countless number of studies simulated electricity markets using agent-based models [46–51]. Researchers believe that agent-based modeling and simulation (ABMS) is a practical approach that can produce realistic knowledge about players' interactions within complex markets [52].

In the upcoming sections, each method is explained briefly with their advantages and disadvantages. Fig. 11.5 categorizes various modeling techniques in the context of energy markets. Although equilibrium models are shown in top-down and bottom-up methodologies, their level of detail is different: CGE models are more aggregated, focusing on the whole economy; but, equilibrium models that are connected to bottom-up methods are mostly concentrated on the electric power sector. Hybrid models cause *modeling electricity markets* and the *top-down approach* to intersect since the future consumption of demand services are often projected using top-down models.

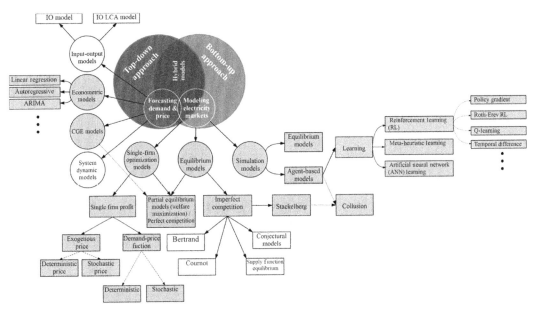

**FIGURE 11.5** The taxonomy of various modeling approaches in the energy market. Blue elements are relevant to both local electricity markets and generic electricity markets (For colors, please check the electronic version).

In Ref. [12,53], the authors provide an extensive review of the forecasting techniques used in the literature with respect to their time frame. The forecasting time frame can be categorized into three groups: short, medium, and long. In the short-term forecasting, researchers consider temporary inputs such as weather and past consumption to calculate projections for the next hour or the next week [54,55]. Beyond a week to a year, researchers use medium-term forecasts [56,57]. In the long-term forecast, socioeconomic parameters such as GDP and population are used to develop projections beyond a year [58].

## 11.3.1 Single firm optimization models

Single firm optimization models mostly focus on the maximization (minimization) of profit (cost) or utility. The price of electricity, which is essential for the profit calculation, can be dictated either exogenously [59–61] or as a function of the demand [62,63].

## 11.3.2 Multiple-firm equilibrium models

A number of equilibrium models have been offered to study the oligopolistic behavior of deregulated markets: Cournot, Bertrand, and SFE are amid the famous models, while others, such as the CV and Stackelberg, have also been employed to analyze electricity markets.

### 11.3.2.1 Cournot competition

In Cournot models, power producers compete over the amount of dispatched power, and the price of electricity is determined via an inverse price-demand function. Pros and cons are as follows:

+ Cournot models are well-established in the literature of microeconomics.
+ With a lower computational burden, it allows researchers to model producers' behavior in electricity markets with adequate detail that represent the real world.
− Each producer presumes that its output can change the market price, but not competitors' production level.
− The cost function of every single GenCo is considered to be known to others; however, in reality, these pieces of information are confidential [64].
− Cournot models are highly sensitive to the demand elasticity.

### 11.3.2.2 Bertrand competition

Unlike Cournot competition where quantities are the strategic decision variables of players, prices are assumed to be the strategic decision variables in Bertrand models. Bertrand models consider no bound on the GenCos' production level, which is not a realistic assumption in electricity markets.

### 11.3.2.3 Supply function equilibrium

Bidding supply curves instead of quantities or prices provide better adaptability for players in dynamic environments. At the equilibrium solution of the supply function game, each GenCo specifies its optimal offer, which is a supply curve, such that maximizes its payoff with respect to other GenCos' reactions concerning developments in market conditions, anticipating their strategies. The quantity and price in any SFE model are bounded by outcomes in Bertrand and Cournot models [65]. Advantages and disadvantages are as follows:

+ By gaming in both quantity and price, SFEs represent realistic pictures of electricity markets; thus, their price predictions are more reliable.
+ SFE prices are not as sensitive as Cournot models to the demand elasticity.
+ Unlike Cournot models, SFE models provide the possibility of developing insights into the bidding behavior.
− Solving SFEs are computationally burdensome; in fact, the equilibrium solution of an SFE model is obtained by solving a system of differential equations, whereas it involves solving a system of algebraic equations in a Cournot model; therefore, one can hardly obtain the closed-form expressions of a solution.
− Proving the existence of a solution, or its uniqueness, is hard except in trivial cases.
− It is not clear which solution better represents GenCos' strategic behavior when multiple SFE solutions exist.
− Transmission lines constraints are regarded solely in simplified SFEs.

### 11.3.2.4 Conjectural variation

The CV method is employed to evaluate players' strategic behavior while reflecting the reactions of others with different competition levels. Many game theoretical bidding strategies and traditional market structures, such as Stackelberg, monopoly, perfect competition, and Cournot, are particular manifestations of CV strategies. Pros and cons are as follows:

+ Similar to SFE models, CV models also overcome the demand elasticity issue.
+ Several market competition levels can be modeled using CV parameters.
− There are discussions opposed to CV models regarding the stability of the conjectures and the likelihood of multiple equilibria.
− The necessity of identifying all rivals' CV parameters makes the method virtually impossible to be utilized in a real-life scenario.

CV models can easily be rendered intractable when transmission networks are being introduced [44].

### 11.3.2.5 Stackelberg and multileader−follower games

The Stackelberg model investigates noncooperative games, in which a dominating leader in the market behaves strategically while followers act upon leader's decision. The multileader−follower game is the extended form of the Stackelberg game, where multiple leaders behave strategically and compete with one another. In a typical electricity pool market, the Independent System Operator (ISO) is considered as a follower while GenCos represent leaders; however, the role of leaders and followers might be different in other contexts. For instance, in Ref. [66], the follower is a virtual entity that determines the unit investment cost of a novel technology based on leaders' (i.e., GenCos) investment decisions.

As the process of decision-making in the multileader—follower and Stackelberg games are sequential, their equilibrium solution may suit better than other oligopolistic models to the long-term investment-decision problem according to microeconomics. Generally, Stackelberg games are modeled using mathematical programming with equilibrium constraints (MPEC) [67,68], whereas multileader—follower games (e.g., Refs. [69,70]) are formulated using EPEC since multiple leaders with dissimilar interests exist. EPEC models are mostly unsolvable by mathematical-based optimization techniques [71], which is why heuristics and metaheuristics algorithms are proposed to resolve the issue [72].

## 11.3.3 Simulation models

Simulation models exploit an alternative approach to calculate the solution of the equilibrium models for the complex problems when they cannot be addressed using conventional frameworks. Typically, simulation models describe agents' strategic decision dynamics using a collection of sequential commands and rules. The main advantage of the simulation approaches is the flexibility to model almost every type of strategic behavior.

### 11.3.3.1 Equilibrium models

Often simulation models are connected to a family of equilibrium models. In this class of simulation models, market players ignore learning and achieve the possible equilibrium by following predetermined rules. For instance, researchers may use Cournot models to support the efficacy of simulation models, in which GenCos' decisions are in the form of quantities. In order to find the CV parameters, Song et al. [45] introduce a simulation model, in which GenCos optimize their bid using the CV method such that it minimizes their perceptual errors about rivals' competition levels at every iteration.

### 11.3.3.2 Agent-based models

In various disciplines, *agent-based modeling* and *multiagent system* have been used interchangeably [73]; nonetheless, despite similarities, they are distinct from one another. Broadly speaking, multiagent systems are emerging in real-world phenomena; however, agent-based models strive to replicate such systems for analytical purposes in simulation frameworks. Thus the nature of agent-based models is to study the collective behaviors of agents who follow certain rules rather than solving a specific engineering problem.

In the agent-based models, modeling learning and intelligence are indispensable to a certain extent as agents have to decide and act autonomously in unknown environments [74]. Learning can assist GenCos in obtaining and enriching helpful information to display desirable performance in the future. Learning is particularly important in the electricity markets as the act of bidding is occurring repetitively.

The current economic theory widely imposes the rational expectations assumption, by which the learning problem is being short-circuited [75]. To the greatest extent, the absence of (dynamic) learning in the game theory-based models may cause miscalculation in results and conclusions [76]. Nonetheless, agent-based models provide adequate flexibility to examine the impact of learning on GenCos' strategic behavior [77]. Various methods are employed to simulate learning in electricity markets; however, researchers emphasize more on reinforcement learning (RL) algorithms, especially model-free RL algorithms such as Q-learning due to the ease of implementation together with acceptable accuracy and convergence. In an environment that can be stated in the form of a finite Markov decision process (MDP), RL algorithms can be utilized to discover an optimal action-selection policy. Krause and Andersson [78] analyzed multiple congestion management mechanisms employing an agent-based model, in which GenCos learn according to the Q-learning algorithm.

Using a generic Q-learning framework, Krause et al. [79,80] studied GenCos' strategic behavior and inferred that in the presence of several Nash equilibria, GenCos' cognitive ability may fail to function properly. Thus to improve Q-learning performance, researchers exploit features of other algorithms. For example, Bakirtzis and Tellidou [81], Tellidou and Bakirtzis [82], and Wang [76] combine simulated annealing with generic Q-learning to adjust exploitation versus exploration. In their method, the exploitation rate increases as time progresses from a low value; simultaneously, the exploration rate decreases to a minimum value at the end of simulation. Fine-tuning Q-Learning parameters using simulated annealing is regarded as a remedy to cope with the slow convergence or divergence.

Roth-Erev learning is a streamlined version of RL when a finite number of pure strategies are played by multiple players [83,84]. Unlike Q-learning, Roth-Erev considers only one state for every agent; therefore, practitioners can avoid the dimensionality curse and process the collected data faster. Veit et al. [51] employed the Roth-Erev RL algorithm to study the impact of several congestion management mechanisms on the German electricity market. Li and Shi [52] used the Roth-Erev RL algorithm in an agent-based simulation framework to analyze the link between weather forecasting and the net earnings of Wind GenCos.

The combined impact of risk sensitivity and learning behavior on GenCos' profits are explored in Ref. [47]. In their work, the authors developed an agent-based simulation model for dynamic electricity markets considering time-varying learning parameters and transmission constraints. They demonstrated that risk aversion to a certain level can increase GenCos' profits, whereas an extreme level of risk aversion can cause intense price competition. In contrast, the model proposed in Ref. [49] focuses on the perspective of a supported player, with the unique aim of supporting its decisions to achieve the maximum profit.

## 11.4 Local electricity markets

In order to create a competitive liberal electricity market, policy makers have envisioned a greater contribution from consumers in generating electricity [85]. In established electricity markets, exorbitant capital investment has prevented end-users from becoming active market players; however, thanks to the development of new technologies such as solar photovoltaic (PV) cells, electricity production at smaller scales is gradually becoming affordable, even without any support mechanism. The steep drop in cost and modularity of solar PV supports the building of an environment, the so-called local electricity market (LEM), in which end users (i.e., prosumers) with limited budgets can generate and trade electricity. However, the higher participation of prosumers corresponds with the following managerial difficulties:

- As the number of active players grows, the organization of the distributed generation (DG) becomes challenging.
- The power delivery is structured unidirectionally to transfer the electricity generated by large-scale fossil-fired power plants to end-users to be consumed instantaneously. Nevertheless, this paradigm is rapidly becoming obsolete since battery storage systems are getting cheaper, and prosumers can feed electricity into the grid. Thus, as the prosumers' share in the market increases, the transmission and distribution networks must transform to enable the bidirectional flow of electricity. To increase the stability of distribution networks, the use of smart grid technology is being advertised, through which aggregators can minimize and manage the reverse flow of power to low- and medium-voltage substations [86].
- Payments to prosumers should be negotiated with electric utilities. These utilities which have already invested heavily in conventional generation systems are reluctant to accommodate their business models to the new market conditions; therefore, prosumers often face resistance [87].

Simulation and optimization models have a proven record of solving managerial problems. Simulation models are mostly employed in disaggregated systems, whereas optimization models are common in centralized decision-making. In Refs. [74,88], the authors show that managing DG is feasible by means of multiagent systems. Pinto et al. [89] use an agent-based model to simulate virtual power plants (VPPs), whose premise is to bring flexibility to DG. However, due to the characteristics of LEMs, mixed models might be a better fit to answer research questions. For instance, Li and Willman [90] utilize a scenario-based analysis to conduct the simulation, in which various aspects of ocean energy penetrating local energy systems in remote areas are analyzed; therefore in Ref. [90] the authors propose a mixed framework that combines optimization methods in submodules with a simulation method that generates scenarios.

In the following sections, we review various modeling approaches in the context of LEMs from different angles: market designs, the integration of renewables and energy storage systems, demand side management, and power system reliability.

## 11.4.1 Market designs

Among multiple network structures, some researchers suggest the peer-to-peer (P2P) market mechanism to facilitate direct and secure trading between prosumers and consumers in the grid [91–94] (Fig. 11.6).

Mengelkamp et al. [94] investigated two market designs using an agent-based model: a direct P2P market and a closed order book market. They conclude that the P2P market design with intelligent agents achieves a lower average electricity price. To facilitate decentralized coordination among nontrusting parties, Münsing et al. [87] propose exercising blockchain technology and smart contracts. Using the alternating direction method of multipliers (ADMM), the authors decompose their decentralized optimization model and implement it in the blockchain.

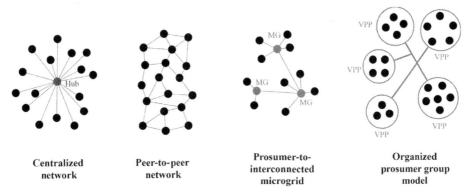

|  Centralized network | Peer-to-peer network | Prosumer-to-interconnected microgrid | Organized prosumer group model |

**FIGURE 11.6**   The network of prosumers in LEMs (for more detail, see Ref. [95]).

## 11.4.2 Renewable and energy storage systems

Integrating an energy system with unpredictable renewables can make the whole system unstable. To prevent instability that can cause electricity outage, researchers have pursued three main approaches: developing better prediction models [96,97], managing consumption, and utilizing storage systems to smooth out volatility in renewables.

In Ref. [93], agent-based simulation is utilized to analyze the impact of community electricity storage in a decentralized P2P local electricity market. Lüth et al. [92] implemented an optimization model to represent P2P interactions in the presence of battery storage for a small community in London. They compare the contribution of centralized storage to that of decentralized storage. Unlike Lüth et al. [92], Xiao et al. [98] modeled a market in which not only electricity but also hydrogen can be stored and traded. A decentralized iterative procedure is developed to clear this market securely. The impacts of centralized and decentralized integration of battery storage and renewables are also studied in Ref. [99] using a Stackelberg game model. In the proposed model, the retailer sets the electricity price, and consumers adjust the consumption level to maximize their surplus accordingly. Their simplified analytic model helped to gain a better understanding of the *death spiral hypothesis* for retail utilities. Also, various control methodologies are studied in Ref. [100] to manage battery storage in residential energy systems based on three control architectures: centralized, decentralized, and distributed control. The objective of their optimization models is to minimize the variation in energy consumption across the network. According to their results, the centralized model offers the minimum variance; therefore, it is used as the basis of comparison to evaluate the performance of other methods.

Unfortunately, the current PV−battery systems are not managed optimally; to mitigate this issue, Klein et al. [101] suggest using scarcity signals to align prosumers with the electricity wholesale market and measure its efficiency using a so-called market alignment indicator (MAI). They illustrate the performance of multiple policy instruments using MAI in a simulation framework, in which technical limitations of battery systems are formulated as a mixed-integer linear programming (MILP) problem. Using agents that exhibit no intelligence in a simulation framework, Ampatzis et al. [102] examined the efficiency of an LEM, in which solar PV and fuel-based generators provide electricity for local consumers and prosumers. In another study Menniti et al. [103] employ a simulation framework to illustrate the performance of their optimization model (i.e., StLM) for managing distributed storage systems.

As experts expect to witness a growing demand for battery electric vehicles (BEVs) and plug-in hybrid electric vehicles (PHEVs) in the market, a body of literature is interested in the idea of exploiting these vehicles as a distributed electricity storage system. Vayá and Andersson [104] introduce a Stackelberg game model, in which the upper-level minimizes the charging cost of electric vehicles, while the lower-level clears the market. They convert their MPEC model to a MILP problem and solve it using CPLEX. Tan et al. [105] propose a distributed optimization algorithm based on the ADMM that solves an optimization problem, by which renewable distributed generators and EVs are integrated. In Ref. [106], a three-tier approach is developed, in which the electricity demand of all PHEVs are aggregated; then an optimization problem is solved to minimize costs for electricity supply; and finally, an incentive signal is created for all PHEVs. Vandael et al. [106] put their method to the test via simulation runs. To the best of our knowledge, utilizing simulation frameworks to validate distributed optimization models is a well accepted approach in the LEM literature.

### 11.4.3 Demand side management

Another subject of study in LEMs is demand-side management. New technologies, such as the smart grid, assist us in applying innovative methodologies to shape demand so as to minimize costs and variation in energy consumption, and omitting network congestion [100,107]. Demand response (DR) can be implemented using load response and price-based programs.

Menniti et al. [86] simulated a DR program for prosumers in a smart network. Ahmadi et al. [107] devised an optimization model to control residential loads which takes into account their necessity, reschedulability, and controllability. In the proposed model, the authors consider both DR and control models. Giordano et al. [108] offer a two-stage optimization model that schedules prosumers' load/storage/production in the first stage and refines the solution by redistributing electricity surplus within the district instead of selling at a cheaper wholesale electricity price to the grid. Similar to Ahmadi et al. [107], electricity loads are divided to schedulable and nonschedulable loads.

### 11.4.4 Power reliability and resilience

Typically, the reliability of the power systems is secured using the $N - k$ contingency constraint [109]; thereby, the system will be able to satisfy the demand, even though $k$ components (out of $N$) fail. However, resilience expands the definition of reliability as the system's capability to predict, prepare for, endure, and efficiently return to normal conditions [110]. Throughout the Iraq war, saboteurs and looters in Baghdad damaged and stole valuable parts of the electrical systems [111]. To cope with threats of this kind and to strengthen national security, experts advise DG technologies by which one can remove targets that are vulnerable to terrorists [112]. Despite the importance of security measures against hostile actors, sound security measures should encompass natural disasters as well. The last example of such a necessity for DG occurred in California, where wildfire caused a protracted blackout [113].

The damage induced by wildfires or earthquakes, even on a massive scale, is no match against bioterrorism and infectious diseases: the unprecedented economic crisis ensuing from the COVID-19 outbreak caused the demand for all energy sources to fall [114]. Partial and full lockdowns reduced the electricity demand by 20% [115]; it also changed the demand shape since most industries were closed. The COVID-19 pandemic damaged many utility-scale power producers (i.e., burning coal, oil, and natural gas) as they are labor-intensive industries. Nevertheless, residential solar PVs, due to their autonomous nature and near-zero marginal cost, can unfold new opportunities to increase the reliability and resilience of communities against natural disasters of such scale [116].

Considering system reliability and supply security, Arefifar et al. [117] suggest a systematic approach to construct an optimal design of microgrids with DG of many different types. The authors formulates their problem as a multiobjective programming model and solved it using Tabu search, graph-theory techniques, and probabilistic power-flow methods. In Ref. [118] the authors developed a MILP model to design a defense strategy for the grid, in which transmission lines are protected against man-made attacks while considering the threshold risks and investment costs. The computation time of the developed model grows exponentially as the number of buses increases from 6 to 57. To improve the resilience of power systems, Lin and Bie [119] proposed a tri-level defender–attacker–defender (DAD) model and solved it using a column-and-row generation method. According to Panteli and Mancarella [120], one can enhance the resilience through hardening and operational measures; therefore the developed model determines the best hardening strategy on the first level. Then, the attacker optimizes its actions to have maximum damage on the second level. Finally, given the damage after hardening plans, the defender finds the best operational plan including DG islanding formation to restore the system on the third level. Although the proposed model is solvable in a matter of hours for two case studies with 33 and 94 buses, the computation efficiency of DAD models should be improved. With some caveats [121], heuristics might be a better fit to solve DAD problems for networks with a huge number of microgrids.

## 11.5 Concluding remarks

Deregulation of electricity markets, although motivated by various reasons in different regions, was considered as revolutionary. However, the challenges came along. To begin with, it requires an elaborate analysis to understand whether deregulation had served the purpose of intention while the outcomes heavily rely on various external parameters. While the biggest challenge lies in recognizing and interpreting the behaviors of players in the market, imperfect conditions are usually detected to result in unwanted outcomes. In this respect, while the top-down macroeconomic approaches are only conducive in constructing hypotheses on principals of fundamental market design issues, the bottom-up engineering approaches bear more potential in predicting behaviors of market players and prescribing solutions to prevent undesirable outcomes. Nonetheless, to address certain questions that may arise in analyzing energy and climate policies, scientists suggest combining these basic approaches to benefit from technological explicitness and behavioral realism, simultaneously.

A more recent milestone in deregulation and liberalization of markets has emerged as the advancement in DG technologies; their deployment throughout distribution systems have led to a rapid increase in new players, that is, prosumers, in local supply of electricity. While this development dictates reconsideration of market designs in an unstructured manner, renewables and energy storage systems as well as demand side management have also risen together with LEMs. There are many new challenges ahead for integrating LEMs into existing power grids originally designed for centralized generation. However, both the technical and market-wise integration of LEMs into the conventional grid would provide many benefits including the following aspects:

- Current wholesale competition among generation companies would be spread over all players in the market; hence, small consumers would benefit by participating in the market directly as well as through increasing energy autonomy.
- Incorporation of renewable and DG would facilitate decarbonization of existing power sector.
- DG, especially when it is equipped with the rooftop solar PV systems, can improve the network security.
- Transmission system losses due to a centralized power system would diminish due to local generation.
- Energy for remote areas would be easily and conveniently provided.
- These new resources may provide ancillary services for the power system.
- Last but not least, integration of new players in the market would eliminate opportunities for market power and collusive behavior of players.

On the other hand, there are also some very likely threats that come along with this new era. Uncertain and intermittent nature, near-zero marginal cost, strong site specificity of DG, and its penetration in LEMs create new challenges in day-ahead scheduling, real-time dispatch, and network security for system operators. Furthermore, bidirectional power flows introduced by these resources in the power system that is designed for unidirectional flows creates new technical challenges, that is, emerging system configuration and setup.

Both the expected benefits and the impacts of upcoming threats are to be reflected in the analytical models including and not limited to optimization, simulation, and decision-making tools. The existence of DG through LEMs changes the span and scope of such tools. Integration of these new agents into the existing grid system requires new methodologies and approaches. In addition, both the degree of uncertainty and the degree of freedom in decisions are magnified; it will naturally require more sophisticated analytical modeling tools not only to analyze and understand the market but also to provide decision support for various players in the market. While such developments are inexorable from the point of view of policy makers and practitioners, novel propositions are to emerge in research immediately.

# References

[1] H. Farzaneh, Energy Systems Modeling: Principles and Applications, Springer, 2019.
[2] P.I. Helgesen, Top-down and bottom-up: combining energy system models and macroeconomic general equilibrium models, Cent. Sustain. Energy Stud. Work. Pap. 30 (2013).

[3] K.S. Andersen, L.B. Termansen, Bottom-up and top-down modelling approach, IntEract. Model. (2013).

[4] D. Wilson, J. Swisher, Exploring the gap: top-down versus bottom-up analyses of the cost of mitigating global warming, Energy Policy 21 (3) (1993) 249–263.

[5] S. Firouzi Alizade, Analysis of $CO_2$ Emissions and Economic Impacts for Turkey Using TIMES-MACRO Energy Modeling System (Master's thesis), Boğaziçi University, 2018.

[6] M. Herbst, F. Toro, F. Reitze, J. Eberhard, Bridging Macroeconomic and Bottom Up Energy Models – The Case of Efficiency in Industry, ECEE, The Netherlands, 2012.

[7] A.S.R. Subramanian, T. Gundersen, T.A. Adams, Modeling and simulation of energy systems: a review, Processes 6 (12) (2018) 238.

[8] C.T. Hendrickson, L.B. Lave, H.S. Matthews, A. Horvath, Environmental Life Cycle Assessment of Goods and Services: An Input-Output Approach, Resources for the Future, 2006.

[9] H.S. Matthews, M.J. Small, Extending the boundaries of life-cycle assessment through environmental economic input-output models, J. Ind. Ecol. 4 (3) (2000) 7–10.

[10] Y. Chang, R.J. Ries, Q. Man, et al., IO LCA model for building product chain energy quantification: a case from China, Energy Build. 72 (2014) 212–221.

[11] A.T. Murray, Minimizing aggregation error in input-output models, Environ. Plan. A 30 (6) (1998) 1125–1128.

[12] S.K. Aggarwal, L.M. Saini, A. Kumar, Electricity price forecasting in deregulated markets: a review and evaluation, Int. J. Electr. Power Energy Syst. 31 (1) (2009) 13–22.

[13] C. Deb, F. Zhang, J. Yang, S.E. Lee, K.W. Shah, A review on time series forecasting techniques for building energy consumption, Renew. Sustain. Energy Rev. 74 (2017) 902–924.

[14] R.E. Lucas, Econometric policy evaluation: a critique, in: Carnegie-Rochester Conference Series on Public Policy, 1976, pp. 19–46.

[15] USEIA, Annual Energy Outlook 2004 with projections to 2025, US DOE, 2004.

[16] USEIA, Annual Energy Outlook 2018 with projections to 2050, 2018.

[17] C. Bohringer, A. Loschel, Promoting renewable energy in Europe: a hybrid computable general equilibrium approach, Energy J. (2006). Hybrid Modeling(Special Issue# 2).

[18] S. Paltsev, J.M. Reilly, H.D. Jacoby, R.S. Eckaus, J.R. McFarland, M.C. Sarofim, et al., The MIT emissions prediction and policy analysis (EPPA) model: version 4. Technical report, MIT Joint Program on the Science and Policy of Global Change, 2005.

[19] A. Bernard, M. Vielle, GEMINI-E3, a general equilibrium model of international–national interactions between economy, energy and the environment, Comput. Manag. Sci. 5 (3) (2008) 173–206.

[20] A.S. Manne, C.-O. Wene, MARKAL-MACRO: a linked model for energy-economy analysis, Technical report, Brookhaven National Lab., Upton, NY, 1992.

[21] T.W. Hertel, Global Trade Analysis: Modeling and Applications, Cambridge university press, 1997.

[22] O. Bahn, A. Haurie, R. Malhamé, A stochastic control model for optimal timing of climate policies, Automatica 44 (6) (2008) 1545–1558.

[23] S. Messner, User's guide for the matrix generator of MESSAGE II, in: International Institute for Applied Systems Analysis (IIASA), 1984.

[24] R. Loulou, U. Remne, A. Kanudia, A. Lehtila, G. Goldstein, Documentation for the TIMES model, Part I: Energy technology systems analysis programme, 2005.

[25] C.G. Heaps, Long-Range Energy Alternatives Planning (LEAP) System, Stockholm Environment Institute, Somerville, MA, 2012.

[26] I.S. Wing, et al., The synthesis of bottom-up and top-down approaches to climate policy modeling: electric power technology detail in a social accounting framework, Energy Econ. 30 (2) (2008) 547–573.

[27] S. Kypreos, A. Lehtila, TIMES-MACRO: decomposition into hard-linked LP and NLP problems, ETSAP TIMES (2014).

[28] D. Esmaeili Aliabadi, IICEC-Sabanci university TIMES energy model: overview, Technical report, Istanbul International Center for Energy and Climate (IICEC), 2019.

[29] P.L. Joskow, Electricity sectors in transition, Energy J. (1998) 25–52.

[30] P.L. Joskow, Deregulating and Regulatory Reform in the US Electric Power Sector, DSpace@MIT, 2000.

[31] A.K. David, F. Wen, Market power in electricity supply, IEEE Trans. Energy Convers. 16 (4) (2001) 352–360.

[32] Z. Younes, M. Ilic, Generation strategies for gaming transmission constraints: will the deregulated electric power market be an oligopoly? Decis. Support. Syst. 24 (3–4) (1999) 207–222.

[33] D.E. Aliabadi, M. Kaya, G. Şahin, Determining collusion opportunities in deregulated electricity markets, Electr. Power Syst. Res. 141 (2016) 432−441.

[34] E. Çelebi, G. Şahin, D.E. Aliabadi, Reformulations of a bilevel model for detection of tacit collusion in deregulated electricity markets, in: 2019 16th International Conference on the European Energy Market (EEM), IEEE, 2019, pp. 1−6.

[35] A. Sweeting, Market power in the England and Wales wholesale electricity market, Econ. J. 117 (520) (2007) 654−685.

[36] X. Guan, Y.-C. Ho, D.L. Pepyne, Gaming and price spikes in electric power markets, IEEE Trans. Power Syst. 16 (3) (2001) 402−408.

[37] N. Fabra, J. Toro, Price wars and collusion in the Spanish electricity market, Int. J. Ind. Organ. 23 (2005) 155−181.

[38] Z. Li, L. Chen, G. Nan, Small-scale renewable energy source trading: a contract theory approach, IEEE Trans. Ind. Inform. 14 (4) (2017) 1491−1500.

[39] A.K. David, F. Wen, Strategic bidding in competitive electricity markets: a literature survey, Power Engineering Society Summer Meeting, vol. 4, IEEE, 2000, pp. 2168−2173.

[40] A.R. Kian, J.B. Cruz, Bidding strategies in dynamic electricity markets, Decis. Support. Syst. 40 (3) (2005) 543−551.

[41] C. Ruiz, A. Conejo, R. Garca-Bertrand, Some analytical results pertaining to Cournot models for short-term electricity markets, Electr. Power Syst. Res. 78 (10) (2008) 1672−1678.

[42] C.A. Daz, J. Villar, F.A. Campos, J. Reneses, Electricity market equilibrium based on conjectural variations, Electr. Power Syst. Res. 80 (12) (2010) 1572−1579.

[43] C. Ruiz, A.J. Conejo, R. Arcos, Some analytical results on conjectural variation models for short-term electricity markets, IET Gener. Transm. Distrib. 4 (2) (2010) 257−267.

[44] C. Ruiz, S.J. Kazempour, A.J. Conejo, Equilibria in futures and spot electricity markets, Electr. Power Syst. Res. 84 (1) (2012) 1−9.

[45] Y. Song, Y. Ni, F. Wen, F.F. Wu, Conjectural variation based learning model of strategic bidding in spot market, Int. J. Electr. Power Energy Syst. 26 (10) (2004) 797−804.

[46] D.E. Aliabadi, M. Kaya, G. Şahin, An agent-based simulation of power generation company behavior in electricity markets under different market-clearing mechanisms, Energy Policy 100 (2017) 191−205.

[47] D.E. Aliabadi, M. Kaya, G. Sahin, Competition, risk and learning in electricity markets: an agent-based simulation study, Appl. Energy 195 (2017) 1000−1011.

[48] D.W. Bunn, F.S. Oliveira, Agent-based simulation-an application to the new electricity trading arrangements of England and Wales, IEEE Trans. Evolut. Comput. 5 (5) (2001) 493−503.

[49] T. Pinto, F. Falcão-Reis, Strategic participation in competitive electricity markets: internal versus sectorial data analysis, Int. J. Electr. Power Energy Syst. 108 (2019) 432−444.

[50] T. Sueyoshi, G.R. Tadiparthi, Agent-based approach to handle business complexity in us wholesale power trading, IEEE Trans. Power Syst. 22 (2) (2007) 532−543.

[51] D.J. Veit, A. Weidlich, J.A. Krafft, An agent-based analysis of the german electricity market with transmission capacity constraints, Energy Policy 37 (10) (2009) 4132−4144.

[52] G. Li, J. Shi, Agent-based modeling for trading wind power with uncertainty in the day-ahead wholesale electricity markets of single-sided auctions, Appl. Energy 99 (2012) 13−22.

[53] K.B. Debnath, M. Mourshed, Forecasting methods in energy planning models, Renew. Sustain. Energy Rev. 88 (2018) 297−325.

[54] J. Fan, J. McDonald, A real-time implementation of short-term load forecasting for distribution power systems, IEEE Trans. Power Syst. 9 (2) (1994) 988−994.

[55] M.T. Hagan, S.M. Behr, The time series approach to short term load forecasting, IEEE Trans. Power Syst. 2 (3) (1987) 785−791.

[56] S. Barak, S.S. Sadegh, Forecasting energy consumption using ensemble ARIMA−ANFIS hybrid algorithm, Int. J. Electr. Power Energy Syst. 82 (2016) 92−104.

[57] E. Barakat, Modeling of nonstationary time-series data. Part II: Dynamic periodic trends, Int. J. Electr. Power Energy Syst. 23 (1) (2001) 63−68.

[58] C. Yuan, S. Liu, Z. Fang, Comparison of China's primary energy consumption forecasting by using ARIMA (the autoregressive integrated moving average) model and GM (1, 1) model, Energy 100 (2016) 384−390.

[59] S.-E. Fleten, S.W. Wallace, W.T. Ziemba, Hedging electricity portfolios via stochastic programming, Decision Making Under Uncertainty, Springer, 2002, pp. 71−93.

[60] M. Pereira, et al., Methods and tools for contracts in a competitive framework, Task. Force 38.05 9 (2001).

[61] G. Unger, Hedging Strategy and Electricity Contract Engineering (Ph.D. thesis), Royal Institute of Technology, 2002.

[62] E.J. Anderson, A.B. Philpott, Optimal offer construction in electricity markets, Mathematics Oper. Res. 27 (1) (2002) 82−100.

[63] A. Baillo, A methodology to develop optimal schedules and offering strategies for a generation company operating in a short-term electricity market, Dep. Ind. Organ. 102 (2002).

[64] C. Peng, H. Sun, J. Guo, G. Liu, Multi-objective optimal strategy for generating and bidding in the power market, Energy Convers. Manag. 57 (2012) 13−22.

[65] P.D. Klemperer, M.A. Meyer, Supply function equilibria in oligopoly under uncertainty, Econometrica: J. Econometric Soc. (1989) 1243−1277.

[66] D. Esmaeili Aliabadi, Decarbonizing existing coal-fired power stations considering endogenous technology learning: a Turkish case study, J. Clean. Prod. 261 (2020) 121100.

[67] Y. Chen, B.F. Hobbs, S. Leyffer, T.S. Munson, Leader-follower equilibria for electric power and $NO_x$ allowances markets, Comput. Manag. Sci. 3 (4) (2006) 307−330.

[68] M. Ventosa, R. Denis, C. Redondo, Expansion planning in electricity markets. Two different approaches, in: Proceedings of the 14th Power Systems Computation Conference (PSCC), Seville, 2002.

[69] S. Leyffer, T. Munson, Solving multi-leader-follower games, Prepr. ANL/MCS-P1243-0405 4 (2005) 4.

[70] F. Murphy, Y. Smeers, Capacity expansion in imperfectly competitive restructured electricity market, C. Chambolle et E. Giraud-Héraud E. Giraud-Héraud, LG Soler et H. Tanguy (2001).

[71] D. Ralph, Y. Smeers, EPECs as models for electricity markets, in: 2006 IEEE PES Power Systems Conference and Exposition, IEEE, 2006, pp. 74−80.

[72] B. Bahmani-Firouzi, S. Sharifinia, R. Azizipanah-Abarghooee, T. Niknam, Scenario-based optimal bidding strategies of gencos in the incomplete information electricity market using a new improved prey-predator optimization algorithm, IEEE Syst. J. 9 (4) (2015) 1485−1495.

[73] M. Niazi, A. Hussain, Agent-based computing from multi-agent systems to agent-based models: a visual survey, Scientometrics 89 (2) (2011) 479.

[74] G. Rohbogner, U. Hahnel, P. Benoit, S. Fey, Multi-agent systems' asset for smart grid applications, Comput. Sci. Inf. Syst. 10 (4) (2013) 1799−1822.

[75] L. Tesfatsion, Agent-based computational economics: a constructive approach to economic theory, Handb. Comput. Econ. 2 (2006) 831−880.

[76] J. Wang, Conjectural variation-based bidding strategies with Q-learning in electricity markets, in: 42nd Hawaii International Conference on System Sciences IEEE, 2009, pp. 1−10.

[77] K. Poplavskaya, J. Lago, L. de Vries, Effect of market design on strategic bidding behavior: model-based analysis of European electricity balancing markets, Appl. Energy 270 (2020) 115130.

[78] T. Krause, G. Andersson, Evaluating congestion management schemes in liberalized electricity markets using an agent-based simulator, in: Power Engineering Society General Meeting, IEEE, 2006, p. 8.

[79] T. Krause, G. Andersson, D. Ernst, E. Vdovina-Beck, R. Cherkaoui, A. Germond, Nash equilibria and reinforcement learning for active decision maker modelling in power markets, in: Proceedings of the 6th IAEE European Conference: Modelling in Energy Economics and Policy, 2004.

[80] T. Krause, E.V. Beck, R. Cherkaoui, A. Germond, G. Andersson, D. Ernst, A comparison of nash equilibria analysis and agent-based modelling for power markets, Int. J. Electr. Power Energy Syst. 28 (9) (2006) 599−607.

[81] A.G. Bakirtzis, A.C. Tellidou, Agent-based simulation of power markets under uniform and pay-as-bid pricing rules using reinforcement learning, in: Power Systems Conference and Exposition, IEEE, 2006, pp. 1168−1173.

[82] A.C. Tellidou, A.G. Bakirtzis, Agent-based analysis of capacity withholding and tacit collusion in electricity markets, IEEE Trans. Power Syst. 22 (4) (2007) 1735−1742.

[83] I. Erev, A.E. Roth, Predicting how people play games: reinforcement learning in experimental games with unique, mixed strategy equilibria, Econ. Rev. 88 (4) (1998) 848−881.

[84] A.E. Roth, I. Erev, Learning in extensive-form games: experimental data and simple dynamic models in the intermediate term, Games Econ. Behav. 8 (1) (1995) 164−212.

[85] E. Commission, The Strategic Energy Technology (SET) Plan, Publications Office, Luxembourg, 2017.

[86] D. Menniti, A. Pinnarelli, N. Sorrentino, A. Burgio, G. Brusco, Demand response program implementation in an energy district of domestic prosumers, 2013 Africon, IEEE, 2013, pp. 1–7.

[87] E. Münsing, J. Mather, S. Moura, Blockchains for decentralized optimization of energy resources in micro-grid networks, 2017 IEEE Conference on Control Technology and Applications (CCTA), IEEE, 2017, pp. 2164–2171.

[88] M. Hommelberg, C. Warmer, I. Kamphuis, J. Kok, G. Schaeffer, Distributed control concepts using multi-agent technology and automatic markets: an indispensable feature of smart power grids, 2007 IEEE Power Engineering Society General Meeting, IEEE, 2007, pp. 1–7.

[89] T. Pinto, Z. Vale, H. Morais, I. Praça, C. Ramos, Multi-agent based electricity market simulator with VPP: conceptual and implementation issues, 2009 IEEE Power & Energy Society General Meeting, IEEE, 2009, pp. 1–9.

[90] Y. Li, L. Willman, Feasibility analysis of offshore renewables penetrating local energy systems in remote oceanic areas–a case study of emissions from an electricity system with tidal power in Southern Alaska, Appl. Energy 117 (2014) 42–53.

[91] K. Chen, J. Lin, Y. Song, Trading strategy optimization for a prosumer in continuous double auction-based peer-to-peer market: a prediction-integration model, Appl. Energy 242 (2019) 1121–1133.

[92] A. Lüth, J.M. Zepter, P.C. del Granado, R. Egging, Local electricity market designs for peer-to-peer trading: the role of battery flexibility, Appl. Energy 229 (2018) 1233–1243.

[93] E. Mengelkamp, J. Garttner, C. Weinhardt, The role of energy storage in local energy markets, 2017 14th International Conference on the European Energy Market (EEM), IEEE, 2017, pp. 1–6.

[94] E. Mengelkamp, P. Staudt, J. Garttner, C. Weinhardt, Trading on local energy markets: a comparison of market designs and bidding strategies, 2017 14th International Conference on the European Energy Market (EEM), IEEE, 2017, pp. 1–6.

[95] Y. Parag, B.K. Sovacool, Electricity market design for the prosumer era, Nat. Energy 1 (4) (2016) 16032.

[96] A. Patel, H.K. Nunna, S. Doolla, et al., Multi-agent-based forecast update methods for profit enhancement of intermittent distributed generators in a smart microgrid, Electr. Power Compon. Syst. 46 (16–17) (2018) 1782–1794.

[97] H. Wang, Z. Lei, X. Zhang, B. Zhou, J. Peng, A review of deep learning for renewable energy forecasting, Energy Convers. Manag. 198 (2019) 111799.

[98] Y. Xiao, X. Wang, P. Pinson, X. Wang, A local energy market for electricity and hydrogen, IEEE Trans. Power Syst. 33 (4) (2017) 3898–3908.

[99] L. Jia, L. Tong, Renewables and storage in distribution systems: centralized vs. decentralized integration, IEEE J. Sel. Areas Commun. 34 (3) (2016) 665–674.

[100] K. Worthmann, C.M. Kellett, P. Braun, L. Grüne, S.R. Weller, Distributed and decentralized control of residential energy systems incorporating battery storage, IEEE Trans. Smart Grid 6 (4) (2015) 1914–1923.

[101] M. Klein, A. Ziade, L. De Vries, Aligning prosumers with the electricity wholesale market–the impact of time-varying price signals and fixed network charges on solar self-consumption, Energy Policy 134 (2019) 110901.

[102] M. Ampatzis, P. Nguyen, W. Kling, Implementation and evaluation of an electricity market operated at district level, 2015 IEEE Eindhoven PowerTech, IEEE, 2015, pp. 1–6.

[103] D. Menniti, N. Sorrentino, A. Pinnarelli, G. Belli, A. Burgio, P. Vizza, Local electricity market involving end-user distributed storage system, 2015 IEEE 15th International Conference on Environment and Electrical Engineering (EEEIC), IEEE, 2015, pp. 384–388.

[104] M.G. Vayá, G. Andersson, Optimal bidding strategy of a plug-in electric vehicle aggregator in day-ahead electricity markets under uncertainty, IEEE Trans. Power Syst. 30 (5) (2014) 2375–2385.

[105] Z. Tan, P. Yang, A. Nehorai, An optimal and distributed demand response strategy with electric vehicles in the smart grid, IEEE Trans. Smart Grid 5 (2) (2014) 861–869.

[106] S. Vandael, B. Claessens, M. Hommelberg, T. Holvoet, G. Deconinck, A scalable three-step approach for demand side management of plug-in hybrid vehicles, IEEE Trans. Smart Grid 4 (2) (2012) 720–728.

[107] M. Ahmadi, J.M. Rosenberger, W.-J. Lee, A. Kulvanitchaiyanunt, Optimizing load control in a collaborative residential microgrid environment, IEEE Trans. Smart Grid 6 (3) (2015) 1196–1207.

[108] A. Giordano, C. Mastroianni, D. Menniti, A. Pinnarelli, L. Scarcello, N. Sorrentino, A two-stage approach for efficient power sharing within energy districts, IEEE Trans. Syst. Man Cybern. Syst. (2019).

[109] D. Bienstock, A. Verma, The $n - k$ problem in power grids: new models, formulations, and numerical experiments, SIAM J. Optim. 20 (5) (2010) 2352–2380.

[110] Z. Bie, Y. Lin, A. Qiu, Concept and research prospects of power system resilience, Autom. Electr. Syst. 39 (22) (2015) 1–9.

[111] L. Milford, R. O'Meara, Distributed power generation for homeland security: proposal for a new federal and state partnership, Clean. Energy Group. (2003) 1–19.

[112] P. Asmus, The war against terrorism helps build the case for distributed renewables, Electr. J. 14 (10) (2001) 75–80.

[113] L. Miranda, As California burns, generator companies make a power grab. [Online]. <https://www.nbcnews.com/business/business-news/california-burns-generator-companies-make-power-grab-n1076611> November 2019 (accessed 02.05.2020).

[114] IEA, *Global Energy Review 2020: The Impacts of the COVID-19 Crisis on Global Energy Demand and* $CO_2$ *Emissions*, IEA, Paris, 2020.

[115] E. Şengül, COVID-19 represents biggest shock to energy system: IEA. [Online]. <https://www.aa.com.tr/en/energy/electricity/covid-19-represents-biggest-shock-to-energy-system-iea/29117> April 2020 (accessed 03.05.2020).

[116] F.M. Cordova, F.F. Yanine, Homeostatic control of sustainable energy grid applied to natural disasters, Int. J. Comput. Commun. Control. 8 (1) (2013) 50–60.

[117] S.A. Arefifar, A.-R.M. Yasser, T.H. El-Fouly, Optimum microgrid design for enhancing reliability and supply-security, IEEE Trans. Smart Grid 4 (3) (2013) 1567–1575.

[118] N. Nezamoddini, S. Mousavian, M. Erol-Kantarci, A risk optimization model for enhanced power grid resilience against physical attacks, Electr. Power Syst. Res. 143 (2017) 329–338.

[119] Y. Lin, Z. Bie, Tri-level optimal hardening plan for a resilient distribution system considering reconfiguration and dg islanding, Appl. Energy 210 (2018) 1266–1279.

[120] M. Panteli, P. Mancarella, The grid: Stronger, bigger, smarter? Presenting a conceptual framework of power system resilience, IEEE Power Energy Mag. 13 (3) (2015) 58–66.

[121] G. Brown, M. Carlyle, J. Salmerón, K. Wood, Defending critical infrastructure, Interfaces 36 (6) (2006) 530–544.

# 12

# Blockchain as messaging infrastructure for smart grids

*Iván S. Razo-Zapata*[1,2,3] *and Mihail Mihaylov*[4]

[1]Instituto Tecnológico Autónomo de México (ITAM), Mexico City, Mexico
[2]COCOA Collaborative Innovation, Delft, The Netherlands [3]LICORE, Queretaro, Mexico
[4]i.LECO, Geel, Belgium

## 12.1 Introduction

The smart grid (SG) offers a new way to make more efficient and effective the electric system [1]. At its core, the SG combines information and communication technology (ICT) with traditional and new forms to generate electricity, such as photovoltaic panels and wind turbines [2].

Among other objectives, the SG aims to optimize the joint operation of power flows, economic flows, and information flows. The optimization of power flows mostly deals with maximizing the use of the electricity infrastructure while also improving the integration of renewable sources [1]. Likewise, optimizing economic flows aims to provide cost-effective ways for operating the overall infrastructure. Regarding the optimization of information flows, one of the main goals is to offer secure, trustworthy, and resilient ways to communicate among different actors within the SG [1,3]. Such communication infrastructure can be offered either through wired (e.g., Ethernet) or wireless technologies (e.g., WiMax,[1] 5G,[2] or Wi-Fi 6[3]). Regardless of being wired or wireless, the offered infrastructure usually relies on (standardized) communication protocols, which facilitate the exchange of messages/information among SG actors [1,3].

The remainder of this chapter is structured as follows. Section 12.2 presents an overview on (1) current and emerging protocols for SG communication, and (2) a description

---

[1] Worldwide Interoperability for Microwave Access.

[2] Fifth generation wireless technology for digital cellular networks.

[3] Highly efficient wireless based on IEEE 802.11ax.

on how blockchain-based mechanisms can be used for communication in SGs. Section 12.3 presents a case study in which an energy community in The Netherlands pilots a novel blockchain-based mechanism for local energy trade. Finally, Section 12.4 provides some discussion on benefits and challenges related to blockchain messaging mechanisms in SGs, whereas Section 12.5 presents a brief summary and future research work.

## 12.2 Current, emerging, and blockchain-based communication protocols for smart grids

This section presents an overall description on different application layer protocols that will potentially be used for communication purposes among actors in smart (electricity) grids. Section 12.2.1 presents emerging standards for communications in SGs, whereas Section 12.2.2 describes what blockchain-based solutions can offer for communication purposes within SGs.

### 12.2.1 Emerging standards for smart grids

#### 12.2.1.1 IEC 61850

Initially conceived as a communication standard for automation of electrical substations, the IEC 61850 standard is currently being investigated as a potential solution for communication between distributed energy resources [4,5]. The IEC 61850 standard defines (1) a device information model, which is a hierarchical structure to organize the elements composing a given device (e.g., smart meter, inverter, substation, etc.), and (2) an abstract communication service interface (ACSI), which defines communication functionality/services that can be provided by such devices (e.g., get, set, report).

On the one hand, following a tree-like structure (see Fig. 12.1A), the device information model represents electrical devices as logical devices (LDs) that encapsulate logical nodes (LNs) (functionality associated to the device, e.g., AC or DC metering), which can be described by means of data objects being composed of specific data attributes. For example, a physical device such as a smart inverter can be represented as a LD that encapsulates LNs such as metering and switch control, which are described by means of so-called data objects like total watts per hour (TotWh) and information on the health of the device. For instance, Fig. 12.1B shows an example representation of a smart inverter (LD) composed of metering and switching capabilities (LNs). Note the use of a hierarchical structure for describing the overall device.

On the other hand, a physical device defines the communication services being supported through a specific communication service mapping (SCSM), which the device manufacturer must specify within an ACSI service conformance statement and can freely define/choose using alternatives such as MMS, Web service, or XMPP [5].

#### 12.2.1.2 Modbus

Modbus provides a communication protocol, which is currently used across different industries that implement so-called supervisory control and data acquisition (SCADA)

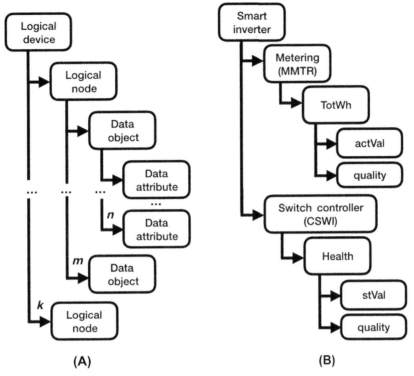

**FIGURE 12.1** (A) Abstraction of the tree-like structure of the IEC 61850 data information model. (B) Example describing some elements of a smart inverter.

systems [6]. Likewise, Modbus follows a client—server model (CSM) for communicating and controlling devices connected on buses or networks.[4] Since ModBus is an application layer (i.e., sitting at the top of the Open Systems Interconnection (OSI) reference model), it can be (physically) implemented using options such as serial transmission (based on either radio, fiber, RS-232, or RS-485) or TCP/IP over Ethernet [7]. Fig. 12.2 shows a possible ModBus configuration in which TCP/IP-based and serial lines are used to communicate different devices through a shared communication bus.

Through a CSM, ModBus supports different connection-related services (i.e., requesting, establishing, and managing connections), which operate in a request—reply fashion using an application data unit (ADU) and a protocol data unit (PDU) as basic messaging components to exchange information among devices. An ADU encapsulates a PDU and includes an address and error check fields, whereas a PDU is composed of a function code (e.g., commands/actions to be performed by the client or server) and a data field containing pieces of information [6]. Likewise, due to its simplicity, flexibility, and industrial maturity, ModBus is being currently investigated as a potential solution for communicating devices in SGs [8].

[4] http://www.modbus.org/docs/Modbus_Application_Protocol_V1_1b3.pdf.

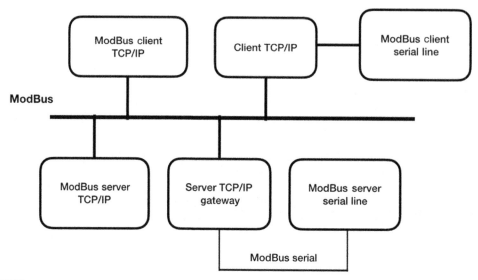

**FIGURE 12.2**    Example of the ModBus protocol running on top of TCP/IP. Source: *The example has been adapted from S. Automation, Modbus Messaging on TCP/IP Implementation Guide v1. 0b, MODBUS Organization, 2006 (last accessed 30.06.15).*

### *12.2.1.3 OPC UA*

The Open Platform Communications—Unified Architecture (OPC UA) is a standard that offers a structured approach to facilitate the communication among different devices, which must follow either a CSM or a publisher—subscriber model (PSM) [9]. First, similar to IEC 61850, the CSM allows devices acting as servers to provide different services[5] (e.g., setting up communication channels, monitoring items, or managing nodes) to clients, that is, other devices. Second, the PSM provides a message-oriented middleware to exchange data between publishers (data producers) and subscribers (data consumers). Furthermore, PSM supports broker-based and broker-less mechanisms for exchanging messages between the middleware and both publishers and subscribers.

Fig. 12.3 depicts the CSM and PSM models as defined in Ref. [9]. Both models, however, rely on a common infrastructure model that defines (1) an information model, which specifies structural, behavioral, and semantic elements to represent a given communication solution (i.e., a network of devices communicating with each other); (2) a message model, which prescribe how messaging is realized between clients and servers (through service request messages) as well as among publishers, subscribers, and a middleware (through *NetworkMessages*); (3) a communication model, which provides mechanisms to transfer data among devices; and (4) a conformance model, which provides interoperability between systems through the use of OPC profiles.

Finally, OPC UA also specifies the possibility of using different protocols for CSM and PSM. Although CSM and PSM can operate using TCP/IP-based transport protocols, they

---

[5] Considered as abstract Remote Procedure Calls (RPCs).

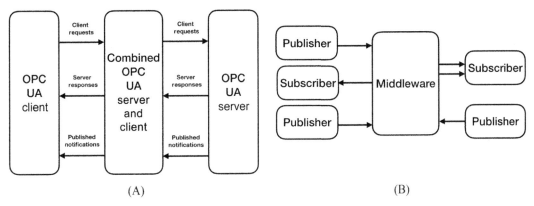

**FIGURE 12.3**    OPC UA models. (A) Client−server model. (B) Publisher−subscriber model.

rely on extra protocols. In this way, CSM needs either OPC UA TCP (a simple TCP-based protocol designed for server devices lacking the resources to implement XML[6] encoding), HTTP (HyperText Transfer Protocol), or SOAP (Simple Object Access Protocol), whereas PSM needs either MQTT (Message Queue Telemetry Transport) or AMQP (Advanced Message Queuing Protocol).

### 12.2.1.4 DNP3

The Distributed Network Protocol 3 (DNP3) emerged as a communication protocol intended for SCADA applications. DNP3 is based on early work around the IEC 60870-5 protocol and is currently maintained by the DNP user group and IEEE as part of the IEEE Standard 1379 [10].

DNP3 provides mechanisms for data transmission between masters (e.g., servers in control centers) and outstations (e.g., computer clients located in the field), which help support the operation of public or private infrastructures (e.g., water or electricity networks).

Likewise, based on these basic components (masters and outstations), DNP3 supports different system architectures such as: (1) point-to-point, (2) point-to-multipoint, and (3) hierarchical structures. Fig. 12.4 illustrates the abstract operation of such architectures.

DNP3 uses a specific data fragment to exchange messages between masters and outstations. The DNP3 data fragment encapsulates messages using three DNP3 layers: application, pseudo-transport, and data. Such DNP3 fragments (containing messages) can then be forwarded either using serial or TCP/IP-based communications.

## 12.2.2 Blockchain as a communication protocol

The set of blockchain-based applications in the electricity sector has been growing in recent years [11]. The range of application spans over numerous areas, such as (1) metering and billing, (2) decentralized grid and energy management, (3) green certificates, (4) electric (sustainable) mobility, and (5) security [11]. Note that all these areas rely on

---

[6] Extensible Markup Language.

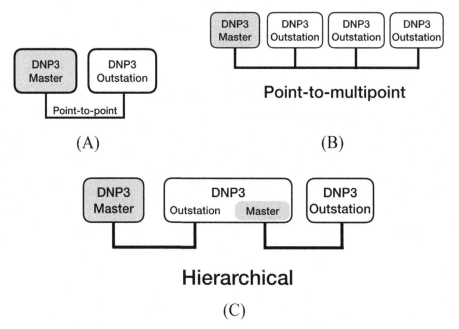

**FIGURE 12.4**    A few DNP3 supported architectures. (A) Point-to-point. (B) Point-to-multipoint. (C) Hierarchical.

communication between devices. Since the blockchain technology is built on messaging and storage of information, it already offers an existing communication infrastructure which SG applications can utilize.

The main idea is to use blockchain transactions to exchange and record (encrypted) messages among different devices. The main advantages of using a blockchain-based messaging mechanism can be categorized along with the following nonexhaustive features:

- *Openness*: unlike protocols described in the previous section, device manufacturers do not need to develop several application programming interfaces (APIs) to communicate with equipment from other manufacturers as they only need to communicate with a smart contract running on the blockchain. Likewise, this way of working also avoids "lock-in effects" since devices do not need to meet communication requirements from a single manufacturer.
- *Transparency*: if the smart contract code governing message exchange between applications is public, the communication protocols are clear and transparent to all stakeholders.
- *Immutability*: once messages are time-stamped and recorded, they remain forever in the blockchain without any possibility to be modified by malicious actors. This feature becomes relevant for settings such as settlements among different parties.
- *Data provenance*: taking advantage of immutability, a blockchain-based mechanism also allows to keep track of properties around transactions (i.e., messages) such as *when* was generated, *who* was involved, and *what* was exchanged.

- *Data governance*: If permissioned (whether public or private), a blockchain-based mechanism allows to determine *who* can create and get access to *what* types of information and *when*.

Due to its novel technology, however, a blockchain-based messaging infrastructure faces also some challenges, which can be categorized as follows:

- *Economic*: Both the cost of developing and deploying the infrastructure (CapEx) as well as the operational costs (OpEx) can be high. For instance, depending on the technology used, the costs of transactions can become costly per each message.
- *Technical*:
  - *Real-time:* the messaging infrastructure lacks support for real-time applications. Currently, many of the well-established blockchain implementations suffer from low scalability, low throughput (i.e., few messages can be processed), and considerable latency (i.e., messages are processed within minutes or hours). There is often a trade-off between latency and resilience, as well as between throughput, scalability, and decentralization. Numerous initiatives (among which Ethereum 2.0 and IOTA) are attempting to address these limitations, but the technology is still in its infancy and has yet to be validated.
  - *Fast changes:* as the technology is still under constant development incorporating new features, there is the risk of lacking interoperability between devices. For instance, due to chain reorganization, rollbacks, or forks messaging functionalities can be broken.
- *Regulatory*: since messages exchanged among stakeholders remain recorded in the blockchain history forever, the infrastructure could face issues with regulations such as the General Data Protection Regulation (GDPR) in Europe. To alleviate this, precautions must be taken such as (1) encrypting data and (2) carefully defining who can access what and for how long, provided the blockchain implementation allows.

Besides performing a trade-off between advantages and current challenges, when designing a blockchain-based application, one also needs to consider functionality offered by current communication protocols. In this way, Table 12.1 summarizes some of the features offered by the protocols described in this section. Table 12.1 aims to provide a quick

**TABLE 12.1**    Summary of popular communication protocols for smart grids.

| Protocol | Supported topologies | Main messaging component | Support protocols |
| --- | --- | --- | --- |
| IEC 61850 | Tree-like client−server | Services defined by SCSM | TCP/IP |
| Modbus | Tree-like client−server | ADU & PDU | TCP/IP |
| OPC UA | Tree-like client−server and ad hoc based on PSM | CSM: service request messages; PSM: *NetworkMessages* | TCP/IP, AMQP, MQTT |
| DNP3 | Tree-like client−server | DNP3 data fragment | TCP/IP |
| Blockchain-based alternative | Ad hoc | Transactions | TCP/IP |

way of comparing different messaging options. The next section provides examples of current blockchain initiatives that use blockchain as a messaging infrastructure.

## 12.3 Use case: renewable energy community pilots in the Netherlands

As discussed by Andoni et al. [11], blockchain technology can have a positive impact on the traditional operations and business processes within energy companies. Overall, blockchain-based solutions can help controlling and managing complex and decentralized applications that are part of modern energy systems and microgrids [11].

Likewise, Andoni et al. have also classified different blockchain initiatives into eight application groups: (1) metering/billing and security; (2) cryptocurrencies, tokens, and investment; (3) decentralized energy trading; (4) green certificates and carbon trading; (5) grid management; (6) IoT, smart devices, automation, and asset management; (7) electric e-mobility; and (8) general purpose initiatives and consortia [11].

Following on from this, it becomes clear that communication and messaging infrastructure is a key component to support the operation of these application groups [11,12].

This section outlines a particular use case where blockchain technology was applied as part of pilot studies in real homes. The aim of these studies is to demonstrate the effectiveness of a novel solution to a variety of existing and future energy system problems resulting from the energy transition. This novel solution is called a "layered energy system"[7] (LES) [13], where households and enterprises can trade energy with each other in different layers: locally among houses of one community, between communities of the country, as well as with the cross-country wholesale electricity market.

It is relevant to note that many components of the LES concept (such as the market process, message structure, etc.) are designed based on the USEF framework.[8] However, a novel aspect in LES is that its logical components (see below) communicate to each other using blockchain technology. In particular, LES uses Energy Web Foundation's Volta blockchain[9]—a consortium blockchain which is public to read, but only a permissioned set of nodes can validate transactions. The Volta blockchain platform uses a virtual machine that is identical to public Ethereum, but was specifically designed for the energy sector. However, LES is not tied to that particular blockchain implementation and can use any other blockchain that supports smart contracts.

A detailed description of the LES concept and its benefits is out of the scope of this chapter, but the interested reader is referred to the LES white paper [13]. Here we briefly outline the main logical components or "apps" that the LES ecosystem consists of and how they interact with the blockchain:

- Market Service app—an application without a user interface that collects bids from the blockchain for buying and selling electricity, computes a "market program" for the next day or intraday, and publishes it back to the blockchain.

---

[7] https://ileco.energy/wp-content/uploads/2019/08/layered-energy-system-white-paper.pdf.

[8] https://www.usef.energy/.

[9] https://www.energyweb.org/technology/energy-web-chain/.

- Grid Safety app—an application without a user interface that collects the market program from the blockchain, analyzes it for congestions based on grid specifications, and publishes the result back to the blockchain. This app also publishes requests for flexibility that will mitigate any potential congestions.
- Billing Service App—an application without a user interface that collects the above information from the blockchain, computes the provisional costs of residents, and publishes them back to the blockchain. This information can be used by settlement providers to create the final invoices and settle all payments—either using on-blockchain currency, or via traditional banking.
- Prosumer App—a web-based application with a user interface that is the single interaction point of end users to the LES ecosystem. It fetches all of the above information from the blockchain, and displays it on the interface. In addition, this app allows residents to submit day-ahead and intraday bids for buying and selling electricity, as well as offers for flexibility, among other functionalities.

All sensitive information that needs to be published to the blockchain is first encrypted with the cryptographic key for that user and is decrypted upon fetching the encrypted data from the blockchain. In the event that a given user wishes to exercise their right to be forgotten (Art. 17 of GDPR), this can be achieved by destroying the cryptographic key assigned to that user. In that scenario, all user's information that has been encrypted with a key that is subsequently destroyed, remains encrypted and is unrecoverable, therefore it can be considered as "forgotten." Fig. 12.5 shows an overview of how the above LES apps communicate with each other securely via the blockchain. Note that each app can be owned by a different stakeholder and hosted in a different location, for example, a cloud platform or an enterprise server.

Likewise, Fig. 12.6 depicts the message flow between LES apps in more detail. Technically, when a given app needs to share information with another (e.g., the Prosumer App sending day-ahead bids to the Market App), it sends this information via its blockchain node (i.e., Node 1 in Fig. 12.6) as a transaction in the form of a function call to the respective LES smart contract (i.e., Bids in this case). This smart contract function then "broadcasts" the information by attaching it to the receipt of that same transaction. Once this transaction is validated, the information sent by the LES app becomes permanently stored in the blockchain and can be seen by any other app via their blockchain node (i.e., Node 2). Readers interested in the technical details of how the Ethereum blockchain, Virtual Machine, or smart contracts work are referred to Ref. [14]. Note that the above process resembles that of the PSM shown in Fig. 12.3B.

Thus the purpose of blockchain technology in LES is twofold:

- LES relies on blockchain as a communication medium. Therefore there is no need to maintain custom APIs between apps. All communication between apps takes place through the blockchain using smart contracts. The latter serve as public communication protocols in LES. An added advantage is that no single party can change this

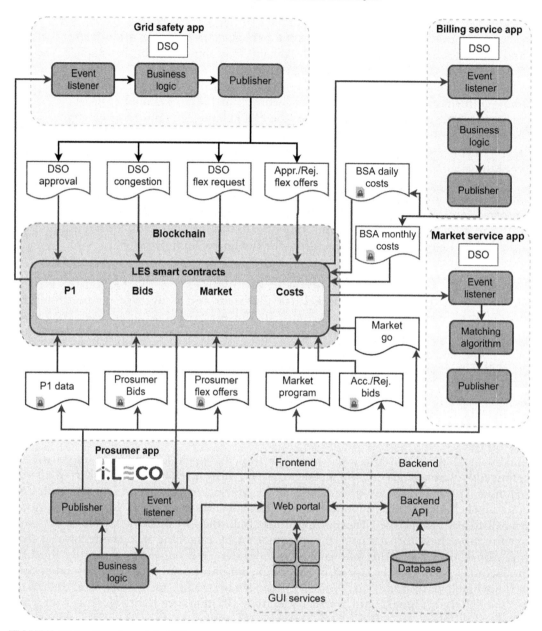

**FIGURE 12.5**  Overview of the LES ecosystem. White boxes denote messages (information) and the lock symbol means that the message is encrypted. The company logo denotes intellectual property ownership.

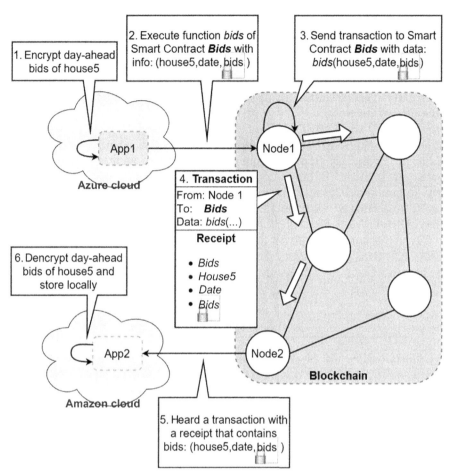

**FIGURE 12.6** Depiction of the message flow between apps and blockchain nodes. The lock symbol signifies encrypted data.

communication protocol unilaterally, as it is stored in the blockchain. If all app owners agree to use a different or upgraded communication protocol, then a new smart contract can be deployed.

- LES relies on blockchain as a database. Thus no single party is entrusted with holding all data, which also means no tampering and no single point of failure. Blocks and transactions can be seen publicly through the Volta blockchain explorer.[10] Nevertheless, all private information is placed encrypted on the blockchain in order to comply with GDPR.

The combination of the above use cases for blockchain prevents a "lock-in effect" where a single privileged (commercial) provider becomes too expensive or difficult to be replaced by another. At any time any party can offer better services in the form of LES apps, for

[10] https://volta-explorer.energyweb.org.

example, a more user-friendly Prosumer App, more efficient Market Service App, more accurate billing or settlement services, etc. Existing stakeholders, for example, end customers, or grid operators, can then freely decide the services of which provider to connect to in LES. Since all data is kept in the decentralized database and all communication takes place via the decentralized blockchain network, any LES party is replaceable at any time by any other party. This promotes healthy competition and the need for constant improvements of the offered services.

In September 2019 the LES implementation was released to one community in Hoog Dalem—a small city in South Holland, the Netherlands. Each end user was given login credentials to the Prosumer App. Residents do not need to have any knowledge of how blockchain works, because all transactions are handled by the Prosumer App.

Similarly, in January 2020, residents of Woerden, the Netherlands were the second pilot community to experiment with the LES concept. This deployment was part of the EU H2020 project DRIvE.[11] It is foreseen that in mid-2020 LES will be deployed in a third community in the Netherlands, namely in the city of Eemnes, as part of the EU H2020 project RENAISSANCE.[12] Dialogues are ongoing with several other communities who are interested to be first movers in the energy transition by becoming pilot communities of LES.

## 12.4 Discussion

We have illustrated the use of a blockchain-based messaging mechanism in the context of the above real-world pilots. They help to illustrate not only the feasibility of blockchain-based communication but also the potential benefits and challenges associated with this communication approach. In this way, we first elaborate on the benefits and later on describe challenges.

### 12.4.1 Benefits

Transparency, immutability, and data provenance are a few of the main benefits offered by a blockchain-based messaging infrastructure. As illustrated in Section 12.3, all exchanged messages (i.e., blockchain transactions) can be seen by any node that is part of the blockchain consortium. Consequently, all involved actors can always check what has been happening within the energy community.

Likewise, immutability helps to keep all messages unchanged, which prevents both intended and unintended attempts to modify recorded messages (information).

Lastly, all the recorded messages within the blockchain also help to support data governance and provenance. The former defines *who* can record and access to *what* kind of messages, whereas the later helps tracking *when* and *who* sent a given message to *whom*. Both mechanisms, data provenance and data governance, help dealing with settlement issues such as who (e.g., an energy consumer) owes what (e.g., a given amount of electricity) to whom (e.g., an energy prosumer).

[11] https://www.h2020-drive.eu.

[12] https://www.renaissance-h2020.eu.

## 12.4.2  Challenges

In a similar vein, adding to the general blockchain-related challenges described in Section 12.2.2, the economic, technical, sociotechnical aspects are among the main challenges faced by blockchain-based messaging solutions.

### 12.4.2.1  Economic aspects

Since the profitability of blockchain-based solutions is an important concern, it is required to ensure a sustainable flow of revenues that helps covering both initial investments (i.e., CapEx) and operational costs (i.e., OpEx). For blockchain-based messaging, CapEx can be related to required software development (i.e., service Apps) as well as new equipment required for operationalizing a given solution (i.e., smart meters and communication devices). Moreover, OpEX can be related to the cost of validating and storing transactions (messages among all participants) within the blockchain (i.e., maintaining computer servers to store copies of the blockchain database).

### 12.4.2.2  Technical aspects

Broadly speaking, the influence of IT-legacy systems should also be carefully considered. For instance, utilities and system operators need to have a clear picture not only on who is allowed to access information on a blockchain-based solution (i.e., public vs private) and on who can add information to the blockchain (i.e., permissioned vs permissionless) but also on the IT systems that need to interact with a blockchain-based solution (e.g., energy billing systems) [15].

Furthermore, a blockchain-based messaging mechanism as the one described in Section 12.3 is mostly suitable for one-to-many communication, which may not be ideal for applications such as real-time demand side management that might require a more ad hoc or many-to-many communication paradigm.

### 12.4.2.3  Sociotechnical aspects

Privacy aspects are of utmost importance to blockchain-based messaging mechanisms since such mechanisms could disclose sensitive people-related information to a group of nodes within the blockchain that could misuse this information. This issue, nonetheless, could be addressed through the implementation of off-chain encryption (i.e., sensitive people-related information/messages are encrypted prior to being stored in the blockchain, as in the example in Section 12.3). This feature would also help addressing issues related to GDPR within a European context.

Distributed governance is also a relevant issue as smart contracts offer a new way to automatically rule the interactions among stakeholders. Examples of projects that apply distributed governance without a central coordinator are Bisq[13] and Colony.[14] For these kinds of novel decentralized organizations, smart contracts must ideally be defined in close collaboration with all stakeholders [16]. Traditional organizations, on the other hand, must be prepared to (1) invest in (re)training human resources, as well as (2) supporting

---

[13] https://bisq.network/dao/.

[14] https://colony.io/.

new forms of interaction within and outside the organization [16]. In the first case for instance, utilities will probably require employees acquiring new skills in areas such as contract management and cryptography (e.g., to handle contracts with customers and system operators). In the second case, it would be possible to explore more flexible ways of interacting with electricity utilities, for instance pay-as-you-go services or democratized models in which consumers are fully empowered so that they can responsibly manage their electricity demand [11].

## 12.5 Conclusions and future work

The chapter has presented an overview on how a blockchain-based communication mechanism can be used to offer messaging infrastructure for SGs. It has also described current and emergent ICT protocols that can potentially be used for the same purpose without the need to deploy a blockchain. Along the same line, however, the chapter also described (1) current pilots that use blockchain as messaging infrastructure, and (2) a discussion on some of the benefits of using such an infrastructure as well as the associated challenges.

Finally, this chapter mostly looks at application-layer communication protocols. Readers interested in technologies related to lower OSI layers should explore the work done in Refs. [17,18].

## Acknowledgments

The work in this chapter was funded in part by the European Union's Horizon 2020 research and innovation program under grant agreement number 774431, project DRIvE. Likewise, financial support from the Asociación Mexicana de Cultura A.C. is gratefully acknowledged.

## References

[1] S.F. Bush, Smart Grid: Communication-Enabled Intelligence for the Electric Power Grid, IEEE Press, 2014.
[2] IEA, Next Generation Wind and Solar Power - From Cost to Value, Technology Report, International Energy Agency (IEA), 2016.
[3] A.M. Annaswamy, M. Amin, IEEE Vision for Smart Grid Controls: 2030 and Beyond, IEEE, 2013.
[4] S. Feuerhahn, R. Kohrs, C. Wittwer, Remote control of distributed energy resources using iec 61850 as application-layer protocol standard, Energy Technol. 2 (1) (2014) 77−82.
[5] S. Feuerhahn, Analysis and Evaluation of the IEC 61850 Communication Standard for Monitoring and Control of Distributed Energy Resources (Ph.D. thesis), TU Dortmund, 2017.
[6] I. Modbus, Modbus Application Protocol Specification v1.1b3.
[7] S. Automation, Modbus Messaging on TCP/IP Implementation Guide v1. 0b, MODBUS Organization, 2006 (last accessed 30.06.15).
[8] S. Kenner, R. Thaler, M. Kucera, K. Volbert, T. Waas, Comparison of smart grid architectures for monitoring and analyzing power grid data via modbus and rest, EURASIP J. Embedded Syst. 2017 (1) (2017) 12.
[9] OPC UA specifications and information models. <https://reference.opcfoundation.org/v104/> (accessed 10.01.2020).
[10] The distribute network protocol DNP3. <https://www.dnp.org/> (accessed 10.01.2020).
[11] M. Andoni, V. Robu, D. Flynn, S. Abram, D. Geach, D. Jenkins, et al., Blockchain technology in the energy sector: a systematic review of challenges and opportunities, Renew. Sustain. Energy Rev. 100 (2019) 143−174.

[12] C. Burger, J. Weinmann, A. Kuhlmann, P. Richard, Blockchain in the energy transition. A survey among decision-makers in the german energy industry. <https://esmt.berlin/knowledge/blockchain-energy-transition-survey-among-decision-makers-german-energy-industry> (accessed 10.01.2020).

[13] Energy21, Stedin, Layered energy system, Tech. rep., Whitepaper, 2018.

[14] A.M. Antonopoulos, G. Wood, Mastering Ethereum: Building Smart Contracts and Dapps, O'reilly Media, 2018.

[15] M. Walport, Distributed ledger technology: beyond block chain (a report by the UK government chief scientific adviser), UK Government.

[16] J. Mendling, I. Weber, W.V.D. Aalst, J.V. Brocke, C. Cabanillas, F. Daniel, et al., Blockchains for business process management - challenges and opportunities, ACM Trans. Manage. Inf. Syst. 9 (1) (2018) 4:1−4:16. Available from: https://doi.org/10.1145/3183367.

[17] V.C. Gungor, D. Sahin, T. Kocak, S. Ergut, C. Buccella, C. Cecati, et al., Smart grid technologies: communication technologies and standards, IEEE Trans. Ind. Inform. 7 (4) (2011) 529−539. Available from: https://doi.org/10.1109/TII.2011.2166794.

[18] A. Mahmood, N. Javaid, S. Razzaq, A review of wireless communications for smart grid, Renew. Sustain. Energy Rev. 41 (2015) 248−260. Available from: https://doi.org/10.1016/j.rser.2014.08.036. <http://www.sciencedirect.com/science/article/pii/S1364032114007126>.

# Load profiling revisited: prosumer profiling for local energy markets

*Gianfranco Chicco and Andrea Mazza*

Dipartimento Energia "Galileo Ferraris," Politecnico di Torino, Torino, Italy

## 13.1 Introduction

This chapter discusses some challenging aspects concerning how to extract knowledge from the data referring to the usage of electricity from local groups of users, and how to make this knowledge useful in local energy markets. The key challenges come from the rapid changes in progress in the current evolution of the energy systems and markets. The roles of the operators are rapidly changing as well, and formerly consolidated practices need to be revisited. Load profiling is an example. Load profiling is a practice followed for many years to understand the characteristics of the electricity consumption of the users, in particular, when there was no possibility of gathering measured data from every user.

To give a clearer view of the concepts, let us start from the definitions. In 2001 the International Energy Agency provided the following definition of load profiling [1]: "The study of the consumption habits of consumers to estimate the amount of power they use at various times of the day and for which they are billed. Load profiling is an alternative to precise metering."

After about two decades, various aspects have to be discussed. Today, the situation is changing from different prospects, which are introducing many provoking questions, for example:

1. The number of consumers that can be metered is growing fast, due to the installation of smart meters. Does the use of load profiling still make sense when all or most consumers could be equipped with interval meters?
2. The local users are not only consumers of electricity, but some of them have become local producers, and as such the view on profiling has to be extended to cover the more general category of producers/consumers (called *prosumers*). In addition, the development of local energy storage is changing the way to consume energy. With all these changes, the

electricity usage seen from the prosumer's supply point may reach virtually any shape. Which is the most effective way to construct and deal with the prosumer profiles of electricity usage?

3. The electricity markets are evolving toward smaller market sizes, also depending on the increasing interest in the creation of local energy communities. For these communities, even the supply point could be different from the one currently considered, up to the concept of distributed supply. Could prosumer profiling provide useful information in the case of local energy communities?

4. The time intervals of interest to categorize the consumption (or local production) are becoming progressively shorter, to encompass the possibility of introducing short-term demand response programs or flexible energy services. Is there any room for developing short-term prosumer profiling?

5. Dealing with smaller prosumer groups, the uncertainty on the local generation and demand becomes higher, and the local generation and demand patterns tend to be less similar to the ones found for large prosumer groups (e.g., at nationwide level). In this context, is stochastic prosumer profiling a viable option?

Further definitions of interest are:

• *Load pattern* (or *load curve*): evolution in time of the average power determined in a given time step from an individual consumer or an aggregation of consumers. It represents the shape of electricity usage during the time period of interest.

• *Load profile*: load pattern that describes how electricity is used by a group of consumers having similar characteristics. The similarity is determined by using a specified criterion depending on a given notion of distance. The load profile is typically defined by using a reference power and a pattern in time normalized to the reference power.

Both definitions refer strictly to the load and the consumer, without taking into account the possible presence of prosumers. According to a further definition from Elexon [2], the load profile is a "pattern of electricity usage for a customer segment of the electricity supply market."

The presence of prosumers changes the prospect for the analysis of electricity usage. The terminology itself has to be adapted. In principle, the term "load" could be still used in an extensive way, considering the "net load" given by the sum between the (positive) demand and the (negative) local generation at the supply point. Therefore, for the sake of increasing clarity, the following terminology could be preferable:

• *Net load curve* (or net load pattern, or net average power curve): evolution in time of the average power determined in a given time step from an individual prosumer or an aggregation of prosumers.

• *Net load profile*: net load pattern that represents how electricity is used by a group of prosumers having similar characteristics.

The notion itself of "load profiling" has to be revisited, by introducing the updated view of *"prosumer profiling,"* in which each prosumer profile is associated with a net load profile. The term *prosumer profiling* is used in this chapter to reflect this updated view.

In the past, the adoption of load profiling was motivated by the reduction of the costs associated with the estimation of the aggregate demand of several consumers not equipped with interval meters. This load profiling was obtained by avoiding the installation of individual meters with sufficiently high granularity (e.g., with time intervals of 15 minutes, 30 minutes, or 1 hour) for all the load points [1]. In fact, it was virtually impossible or at least highly impractical to meter all consumers individually. In particular, small-size consumers (below a given threshold on the contract power) remained without interval metering installed, using only fiscal meters for billing purposes. For these consumers, the load profiles were defined based on specific measurement campaigns, carried out on a statistically significant set of consumers, then using statistical methods or unsupervised clustering approaches to identify the consumer groups based on the shape of their consumption [3].

During the years, the roll-out of smart meters has progressively increased. Correspondingly, the wide use of standard load profiles has been seen as an obstacle to the advantages offered by using real-time smart metered data in the liberalized market context [4]. In particular, real-time metering supports the application of advanced pricing schemes, as well as the development of programs for demand response and flexibility enhancement. The benefits appear when peak loads can be reduced, or the demand can become more predictable. However, the determination of the peak load also depends on the measurement recording time interval [5]. For the same purposes, also energy metering schemes going beyond the classical interval metering, such as event-driven energy metering [6], could be practical to set up tariffs or contracts for flexibility. A remarkable advantage of event-driven energy metering is to detect the peak loads more effectively [7]. On these bases, the role of load profiling or prosumer profiling in today's context has to be discussed with a critical appraisal.

The next sections of this chapter point out various aspects involved in the definition and application of prosumer profiling and are organized as follows. Section 13.2 recalls the load profiling principles, objectives, and types, recalling the main aspects of their usage for customers without interval meters. Section 13.3 deals with the timing and amplitude aspects of the electricity usage patterns. Section 13.4 discusses new needs and opportunities to apply prosumer profiling for local energy markets. The last section contains the conclusions.

## 13.2 Load profiling principles

### 13.2.1 Basic principles

As already mentioned, the traditional usage of load profiling is to address consumers without interval meters [2]. Since the cost of a measurement campaign and the accuracy of the representation of the population increase when the percentage of measured units grows, a statistically significant sample of the consumers is determined by using an appropriate technique such as the stratified sampling approach [8], in which the budget limit for the measurement campaign determines the maximum number of users to monitor, and the stratified sampling outcomes provide the partitioning of these users among the predefined users' macrocategories (e.g., households, industry, commerce, traction, other services, and lighting). The stratified sampling approach is applicable only when the information used to characterize the users is available from every user. Typically, the data used are the contract power or the annual energy consumption taken from the billing information. The information on energy consumption is, in general, more useful than the contract power to partition the consumers based on their actual usage of electricity. As such, the availability of energy consumption data for periods shorter than 1 year could be advantageous. However, it is often difficult to get data for all the consumers belonging to all the macrocategories.

The main principles of load profiling are:

- The information available has to cover the whole population of consumers.
- The time interval of interest can be partitioned into consistent time periods (CTPs) in such a way that the entire time interval is covered by a sequence of nonoverlapping CTPs without loss of continuity.
- Load profiling has to categorize all consumers into nonoverlapping groups.
- Each user has to be uniquely associated with one group.
- Each group has to be characterized by a load profile in the corresponding CTP.
- The load profiles have to be sufficiently different from each other.
- The time granularity (i.e., the duration of the time step) used to define the load profiles has to be sufficiently significant to represent the main features of the load profile.
- The load profile is generally defined by using a normalized (dimensionless) load curve and is associated with a reference amplitude (e.g., in kW), in such a way to reconstruct the load pattern in actual quantities by multiplying the load profile by the reference amplitude. In the case the load profile corresponds to an aggregation of users, the reference amplitude of the aggregation is the sum of the reference amplitudes of the individuals that form the aggregation.
- The load profiles have to be energetically equivalent to the real energy consumption in the time period under analysis. For this purpose, after a preliminary definition of the initial load profiles based on computational considerations, the final load profiles of each group are rescaled by multiplying the initial load profiles by the *load profile scaling factor*. This factor is defined as the ratio between the total energy of the actual set of users that compose the group, and the total energy determined for that group from the preliminary load profiles in the time period of analysis.

## 13.2.2 Scopes of load profiling

Load profiling can be used with different scopes, some of which are summarized below.

### 13.2.2.1 Settlement

In the electricity market, the term *settlement* can indicate two different concepts: (1) the process that reconciles the difference between the amount of energy used by the user served by the supplier and the energy previously bought by the supplier (i.e., balancing) [9,10]; and (2) the process that leads to recognizing the entity responsible for the unbalance and to settle the cost of it to the responsible entity [2]. In the second case, the settlement can be done only by knowing the load pattern of every consumer served by the supplier. However, not all the customers are equipped with interval meters having an adequate resolution in time (i.e., suitable for measuring energy at time steps not longer than the settlement time interval). For this reason, it is necessary to profile these customers by including them in categories characterized by proper load profiles, modified according to external conditions, such as temperature, season, and so on. The energy losses can be included in the settlement process based on appropriate regulatory provisions.

### 13.2.2.2 Tariff setting

One of the traditional ways to use the load profiles is to formulate tariff structures with different price levels, which may be applied to different types of day and different hours of the day, for each macrocategory of consumers [11]. For example, to reduce the peaks of the total demand, higher rates can be defined in the time intervals at which the peaks occur, for the macrocategories of consumers that mostly contribute to these peaks.

### 13.2.2.3 Forecasting

The exploitation of load profiles could be useful for load forecasting. Some results presented in Ref. [12] indicate that the effectiveness of using load profiles in neural network-based load forecasting requires appropriate segmentation of the consumers, accurate determination of the load profiles that represent each consumer class, and proper selection of the inputs for the neural network. From another point of view, the load profiles can result from the application of forecasting tools. In Espinoza et al. [13] the load profiles are obtained from a periodic autoregressive model that satisfies the convergence conditions, that is, a model that exhibits stationary behavior.

### 13.2.2.4 Demand side management, demand response and flexibility

The incorporation of specific features, such as the trends of variation of the load patterns, makes the load profiling procedures appropriate to be used for demand response applications. The procedure presented in Lin et al. [14] is defined to select the loads suitable for demand response programs by considering a fast load increase, which occurs when a large portion of the load is switched on in short time duration. Some concepts to apply clustering techniques in demand response studies are reviewed in Wang et al. [15]. Utilization and availability factors taken as load profiling attributes are successfully exploited in Konda et al. [16] within a demand response exchange mechanism.

### 13.2.2.5 Energy not served

Following a service interruption with a given duration, the *energy not served* (also called *energy not supplied*) is the estimated energy that conceptually would have been used by the consumers without the service interruption. The load profiles may be used to establish a *conventional* way to partition the energy not served among previously established consumer classes. In this way, it is possible to make a distinction among the interruptions that could occur at different times during the day. This approach has been used in Heggset et al. [17], where the energy not served is estimated by using information concerning the external temperature, the annual energy consumption, and (if available) the load that has been measured hour before the interruption. On the other hand, the load profiles are used to estimate the load curves that would have been followed if no interruption had occurred, for each consumer class. Then, the sums of the integrals of these load curves during the period of interruption provide the total amount of profiled energy. The sharing coefficients among the consumer classes are calculated based on the profiled energy and are then used to partition the estimated energy not served among the consumer classes.

### 13.2.2.6 *Aggregate load modeling, simulations, and benchmarking*

The load profiles may be used for creating different aggregations of consumer classes, to assess different characteristics of the grid operation. These characteristics include the power flows and the potential reverse power flow to the supply point, the possible over-loading conditions of one or more network branches, the voltage profile at the network nodes, the network losses, and the losses allocated to the different nodes (taking into account the net power load at each node). Further quantities depend on the network operation, such as the possible greenhouse gas emission reduction due to the generation scheduled to serve the demand. Moreover, the load profiles may be used to construct different *scenarios*, in which the amount of load given by different consumer classes could change.

## 13.2.3 Types of load profiles

The classical distinction among load profiles depends on how these profiles are created, with reference to consumer classes and time periods (daily, monthly, or seasonal) generally specified a priori:

- *Deemed* load profiles are defined based on *engineering knowledge*, without resorting to measured data, typically for load patterns that can be easily predicted.
- *Static* load profiles are created based on *historical data*; a load profile is defined for each consumer class at each time period.
- *Dynamic* load profiles are created based on *metered data*, with the aim of better representing the actual energy usage by the consumers.

In the context of local energy markets, deemed load profiles are not detailed enough to represent the actual variability of the consumption patterns. Static load profiles could be insufficient, as they cannot be easily adapted to daily variations of the external conditions (e.g., ambient-related variables). Correction coefficients on static load profiles could be determined (e.g., for ambient conditions), but additional analyses would be needed by using local data to validate the results. In this case, it would not be simple to gather these local data. Dynamic load profiles depend on the availability of a significant number of data. For this purpose, a statistically significant number of samples have to be determined for each consumer class, to gather and elaborate on the metered data of the selected consumers. For local energy systems with extensive smart metering, data gathering from a significant number of consumers could become viable. The issues, in this case, are shifted to data elaboration, considering timing aspects (discussed in the next section), and the consumer categorization tools indicated below.

## 13.2.4 How to obtain the load profiles

The procedure for determining the load profiles is based on a series of steps, represented in Fig. 13.1.

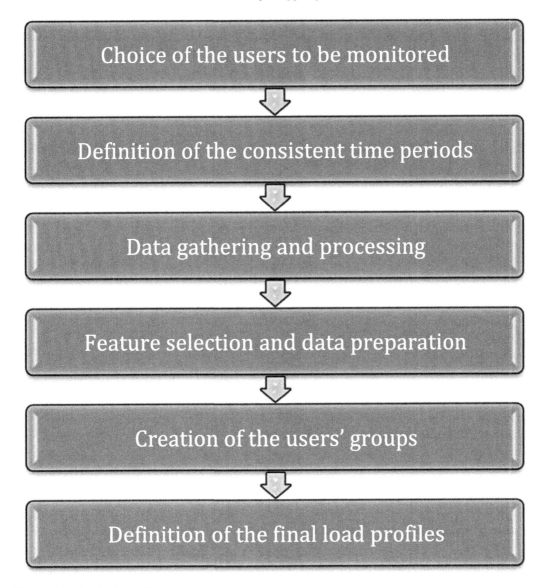

**FIGURE 13.1**   Load profiling procedure.

1. *Choice of the users to be monitored* (called *Sample design* and *Sample selection* in Elexon [2]). It is impractical to gather data from the whole population of users. Hence, engineering knowledge is applied to set up a preliminary macroclassification (e.g., residential, industrial, commercial users, and so forth). A representative variable, such as contract power or energy consumption in a given period (the latter being generally more representative of the actual way to use energy) is considered. The representative

variable has to be known in advance (e.g., from the supply company's database). Statistical techniques such as the stratified sampling approach [8] are applied for each macrocategory, to determine the partitioning of the users to be monitored for a given total number of users depending on the cost of the sampling campaign. Then, a second-level partitioning of the users into internal layers for each macrocategory is possible based on the same representative variable or another variable. In this case, the internal layers and the number of users to be monitored are chosen to minimize the variance. In this way, more homogeneous groups are found inside each layer, thus leading to a reduction of the sample size needed to represent the internal distribution of the relevant variable accurately. After that, the selection of the users to be monitored inside each internal layer is carried out in a random way.

2. *Definition of the CTPs.* The CTPs (also called *loading conditions* in Chicco et al. [11]) define the partitioning of the overall time period of analysis (e.g., 1 year) into time intervals (also noncontiguous) in which the user's load patterns are expected to exhibit similar characteristics. A classical partitioning is obtained by dividing the seasons, and for each season weekdays from weekends (and, in case, also Saturdays from Sundays). If there is a historical database of recorded entries, more refined solutions are found by looking for similarities among the days that change from one macrocategory to another [3].

3. *Data gathering and processing* [18]. If the data metering/monitoring equipment is not installed for the selected users, the first phase will be to install it. Then, data are gathered for a sufficiently long time period. The data have to be checked to avoid the presence of bad data (due to missing values, incorrectly transmitted values, or misleading entries. The latter entries depend, for example, on short interruptions that make the data useless for load profiling—in which only data representing normal conditions are relevant). Different data cleansing procedures can be used for this purpose [19]. Replacement of missing or bad data may be provided through data imputation procedures [20], even though it could not be strictly needed (for example, in an averaging process unreliable data occurring at a given time step could be simply ignored). For each CTP, a *Representative Load Pattern* (RLP) is defined for each user by averaging the time series of the clean data.

4. *Feature selection and data preparation.* Each load pattern is initially represented as a series of *points* in the time domain. To make the calculation tools more efficient, from these points, it is also possible to construct other *representative features* [18]. This process could involve data normalization and the choice of one (or more) specific features. The features can be defined in different domains. In the time domain, it is possible to change the averaging time step or to transform the values into symbols referring to irregular time steps [21]. In other domains, the features can be defined in the frequency domain [22], in the wavelet-based domain [23], in the domain of appropriately defined shape factors [11], as a given set of entries from the principal component analysis, curvilinear component analysis or Sammon maps [24], or from duration curves [25]. Once the data have been transformed into their representative features, data preparation is carried out by forming, for each CTP, a matrix whose rows indicate the users and the columns contain the representative features. If desired (and preferably upon the availability of features defined in similar ranges of values), multiple features can be cascaded in the matrix columns.

5. *Creation of the users' groups.* Unsupervised *clustering* algorithms are generally used to create the users' groups based on the shape of their RLPs. A full review of the clustering algorithms is outside the scope of this chapter. The interested reader may find details in various literature references (e.g., [18,24,26]). The clustering output is a vector that contains the number of the cluster to which each user has been assigned. Clustering validity indicators can also be used to quantify the effectiveness of the clustering results, again by using a number of indicators defined on the basis of specific notions of distances (mainly with the Euclidean distance). The general aim is to declare good clustering results when the groups are well separated with each other and are relatively compact internally. If the clustering validity indicators have to be used to compare the results, they have to be computed by using the same type of data (e.g., in the time domain at a given time step), regardless of the features used to execute the clustering procedure. Moreover, some clustering algorithms are more suitable to isolate uncommon patterns (outliers), while other clustering algorithms tend to create more uniform partitions. The definition of the number of clusters is another critical issue. In some cases, the number of clusters is a user-defined input value, in other cases, the input information is a threshold, and the number of clusters is not fixed a priori. The expertise of the data analyst in the energy domain is essential to choose the method and parameters, as well as to interpret the results. Clustering methods based on random number extractions should be executed more times, making a statistical analysis of the outcomes.

6. *Definition of the final load profiles.* The creation of the groups is not sufficient to determine the final profiles. The first reason is that the composition of the groups could be modified if some significant properties of the data grouped in the same cluster, which characterize the cluster entries, are discovered. In this case, all the users that share the same significant properties could be added to the cluster, even though they were assigned to other groups after clustering. For the same reason, some users could be excluded from the group (and have to be reassigned to different groups based on a notion of distance, or depending on other characteristics of the user). Once the groups are formed in their final composition, the load profiles have to be established for each group and in each CTP. Let us call *tentative load profiles* the ones determined from the sum of the contributions of the users belonging to the same user's group. If there are data available from the field, a validation phase is then needed. For example, if real measurements are available at the substation level, and the number of users served by the substation for each group is known, it is possible to reconstruct the load pattern at the substation level based on the sum of the contributions to the load pattern of each group. This procedure can be followed for different substations (if data are available) to validate the load profiles. The tentative load profiles of the groups are then rescaled to obtain for each group energetically equivalent load patterns, such that the total energy represented by the patterns (plus a reasonable estimation of the energy due to the network losses) matches as much as possible the total energy measured at the substation level, for each substation. The final load profiles are then obtained from the rescaled tentative load profiles. In addition, classification of the load profiles can be carried out by using decision trees or rule-based methods, in such a way to obtain the information needed to assign a new consumer to one of the classes defined [27,28].

## 13.3 Timing and amplitude aspects of the electricity usage patterns

With the evolution of the communication systems, the metered data could become available with a precision (in time) better than in the past. Under the ubiquitous connection paradigm (i.e., any device is connected at any time and any place), smart meters and data logger tend to be seen as a part of the Internet of Things (IoT) framework [29]. The extended amounts of data could be possibly handled with hierarchical levels of computing, to pass to the higher level only the information needed for the high-level process. The conjugation of the capability of efficiently handling many data, with the precision in time of the data representation, is a crucial point for present and future applications. This point also enables exploiting the potential of real-time simulators, in which the latency of the communication channels is taken into account when processing data from different locations. In this way, any reduction of the local causes of misplacements in time of the data should be avoided.

### 13.3.1 Timescales

The definition of the timescales of interest is crucial to obtain meaningful data representations for the problem under analysis. For load profiling, the timescales may be defined by considering:

- The *time horizon* chosen for the study (e.g., 1 year or multiyear, also depending on the data availability)
- The partitioning of the time horizon into *time periods* that exhibit similar characteristics (e.g., seasons, or the distinction between weekdays and weekend days, or other types of partitioning, based for example on the identifications of CTPs in which the demand exhibits similar characteristics).
- The *time step* chosen for the data representation inside the time horizon and inside each time period.

The above definition of the timescales is representation based and goes from the longest to the shortest timescale. However, looking at how data is gathered, the process of data construction goes in a reverse way, with:

- The internal timing of the data measurement, depending on the data sampling occurring inside the meter or data logger. This timing is not disclosed to the load profiling applications, as there is no need to enter in these details.
- The elementary time interval is the shortest duration for which the data provided by the meter or data logger is of interest for the problem under analysis. The elementary time interval is not necessarily the shortest duration for which the meter or data logger may provide an output (see Section 13.3.2). This elementary time interval can be seen as the constant time step for interval metering, in which data are represented at a regular cadence, or as the minimum duration considered in nonregular schemes such as event-driven metering [6].
- The time period is a multiple of the elementary time interval and is considered for separate analyses of the load patterns. For example, if weekdays are partitioned with

respect to the weekend days (also including bank holidays), each partition corresponds to a different time period.

- The time horizon includes all the time periods used for the analysis.

For load pattern analysis and profiling, the quantity that corresponds to the elementary time interval (or time step) is generally the *average power* referring to the duration of the time step. For this reason, it is also possible to use the terms *averaging time interval* or *averaging time step* [30]. The data representation for the time periods inside the time horizon is then a sequence of average power levels. In this way, the most appropriate way to represent a demand (or generation) pattern is a sequence of constant levels, each one equal to the average power in the corresponding time step. If time steps of 1 hour are used, the load pattern may be represented with the same number for either the average demand (kW) or the total consumption (kWh) at a given time step. Otherwise, it has to be clearly specified whether the numbers that represent the load pattern refer to the average power or to the energy consumption.

From the previous discussion, it is also evident that the time series corresponding to the load patterns contain information having an *integral* meaning (i.e., referring to energy or average power, which can be only determined after a certain time has elapsed [31]). Thereby, these time series are not aimed at reproducing the actual waveform by using data sampling as it happens in signal processing. For this reason, the word sampling will not be used here to indicate the data gathering that gives as an outcome the sequence of average power values.

## 13.3.2 Horizontal and vertical resolutions

With today's smart metering technologies, the elementary time interval could become very small, conceptually up to one cycle of the alternating current (20 ms for 50 Hz systems). However, gathering data with such a fast cadence would provide a number of data excessively high for energy-based analyses such as load profiling, in which the main scope is to categorize the consumption trends.

The way to gather data can also be discussed in terms of the resolution required to collect the data points of the time series that represents the demand pattern. In particular, it is possible to identify two types of resolutions, whose combined effect determines the effectiveness of data representation [30]:

- *Vertical resolution*, referring to the discretization step of the vertical axis. It depends on the number of digits used to indicate the output, or on the number of bits for digital instruments. This issue was more relevant when the measuring instruments were less accurate. With the smart meters and data loggers used today, this issue has become less relevant. However, it may still happen when a measuring instrument targeted at measuring relatively high values is used to measure demand patterns with low amplitudes. An example is reported in Fig. 13.2, where the average power is measured for an aggregation of households with a vertical resolution of 0.75 kW. One of the ways to mitigate the effect of vertical resolution issues is to gather data at lower time steps (e.g., 1 minute) and average them at longer time steps (e.g., 15 minutes). In this way, the averaged values also reach intermediate values with respect to the levels imposed by the vertical resolution.

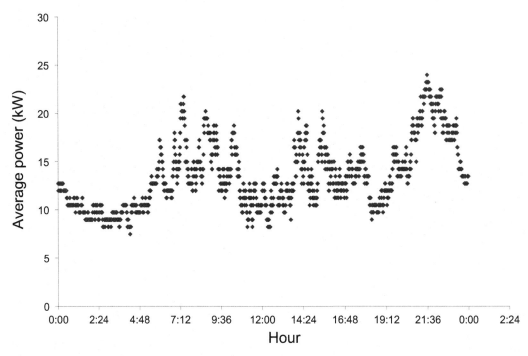

**FIGURE 13.2**　Example of data with low vertical resolution.

- *Horizontal resolution*, referring to the discretization step of the horizontal (time) axis. The determination of the most appropriate time step is a challenging issue and depends on the problem under analysis. Shorter time steps could be useful when it is essential to identify specific issues in the demand patterns, or to gather more information to be used in data disaggregation procedures aimed at understanding which devices are used during the time period of interest [32]. For load profiling purposes, fast averaging time steps are, in general, not useful. In fact, the number of data points to be processed would increase considerably (also increasing the computation time for the calculation of the distance between patterns), without providing real advantages in the comparison among the shapes of different demand patterns.

Some hints for the determination of the averaging time step that increases the information which can be extracted from the demand pattern may come from the analysis of the *entropy* contained in the time series [33]. The entropy information is relevant, as in conditions of high entropy the global properties of the time series can be represented by using less information, while the representation of the local features would need more information. As such, conditions of high entropy are of interest for load profiling, which depends more specifically on the global properties of the demand patterns. Additional information may come from further indicators adopted in the information theory domain (whose analysis is beyond the scope of this chapter).

As an example, let us consider a demand pattern in which the average power values are taken at time steps of 1 second (assumed here as the elementary time interval), for a time

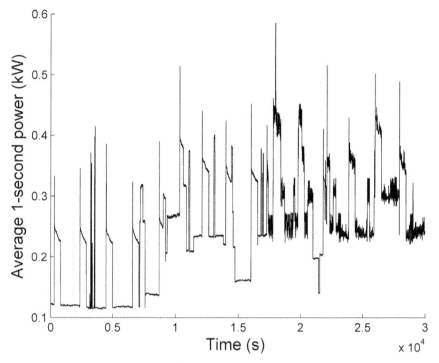

**FIGURE 13.3**   Demand pattern with 1 s time step.

interval of 12 hours, from a household consumer with low energy consumption (Fig. 13.3). In addition to the demand pattern, the demand *variation* pattern is determined for a given time step as the point-by-point difference between two successive values. The entropy of the demand pattern and of the demand variation pattern is then calculated. This process is repeated by aggregating the data points for multiples of the elementary time interval. Fig. 13.4 shows the result. From these calculations, the maximum entropy of the demand pattern is located at about 600 seconds (i.e., 10 minutes). The maximum entropy of the demand variation pattern appears from about 600 to 1200 seconds (that is, from 10 to 20 minutes). For very small values of the averaging time step (e.g., below 1 minute) the demand variations between successive values and their entropy are relatively low. These results are consistent with the use of traditional time steps, such as 15 minutes for data gathering.

## 13.3.3   Timing issues for net power analysis

For a prosumer that manages both local generation and local demand, the *net power demand* is the difference between the local demand and local generation patterns. The net power demand is of interest when the metering is carried out at the supply point of the prosumer, without separating the contributions of local demand and generation (and of local storage, if any). This kind of net metering is not widely practiced today. However, it could be applied more extensively with the diffusion of local energy markets, in which the

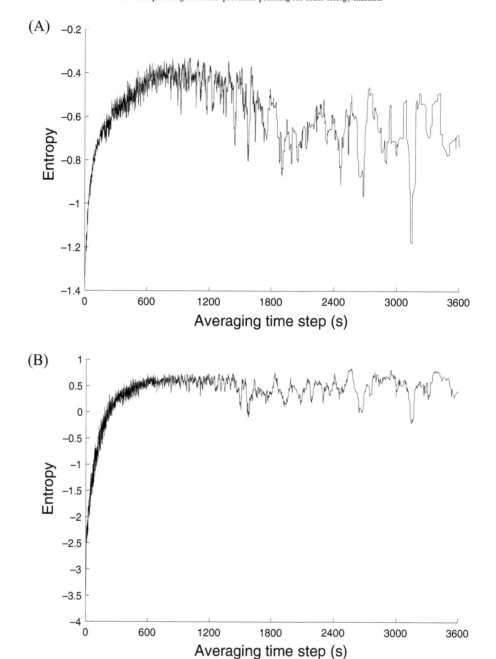

**FIGURE 13.4** Entropy determined for the demand values and the demand variations: (A) with demand values; (B) with demand variations.

separate outcomes of the local demand, generation, and storage could be considered as private information of the prosumers.

In order to assess some specific aspects concerning net metering, let us define with $\tau_0$ the base averaging time step, and with $\mathbf{p}_{T,\tau_0}$ and $\mathbf{g}_{T,\tau_0}$ the column vectors that contain the demand and generation pattern data, respectively [31], with the data defined for a given time period $T$. For further generalization, for the demand let us consider the averaging time steps $m\tau_0$, multiple of $\tau_0$ with the multiplier $m = 1,\ldots, M$, and for the generation the averaging time steps $v\tau_0$, multiple of $\tau_0$ with the multiplier $v = 1,\ldots, M$. In this way, the net demand pattern is defined in its general form as

$$\mathbf{d}_{T,m\tau_0,v\tau_0} = \mathbf{p}_{T,m\tau_0} - \mathbf{g}_{T,v\tau_0} = \left\{ d_{T,m\tau_0,v\tau_0}(k), \; k = 1, \ldots, \frac{T}{\tau_0} \right\} \tag{13.1}$$

where all the patterns are composed of the same number of data points in order to enable their comparison. In particular, the data points are the same as in the case with the averaging time step $\tau_0$, and constant values are assumed in all time steps of duration $m\tau_0$ and $v\tau_0$ for the generation.

Fig. 13.5 shows an example with data gathered from the field, in which $\tau_0 = 5$ min, and the generation (from a photovoltaic system) and demand (from an aggregation of loads) have the same time step ($m = v = 1$). The net demand assumes both positive and negative values at different time steps. If the averaging time step changes, assuming the same change for both patterns, the net demand pattern reaches different shapes, losing the highest variations (Fig. 13.6). If these variations occurred close to the null net demand, in the smoothed net demand curve some positive (or negative) entries could be missing. On the purely technical point of view, this corresponds to a lack of detail in the representation of the net demand pattern. However, if the sign of the net power demand is associated with different economic rates, the costs or revenues related to the positive or negative net energy changes, introducing a possible weak point in the treatment of the prosumers, linked to the averaging time step [31]. This aspect may be crucial in a local energy market,

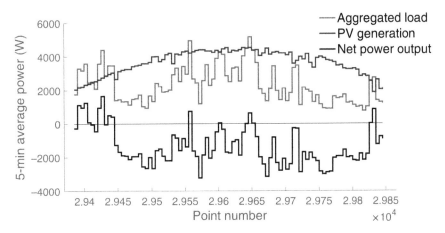

**FIGURE 13.5** Average power data gathered at 5-min time step.

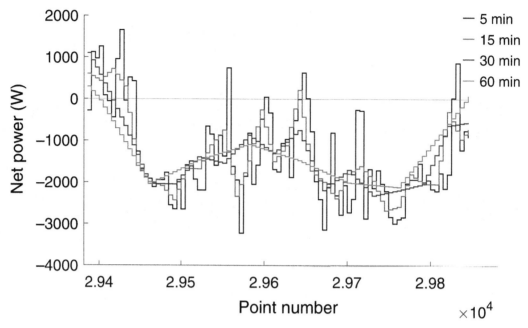

**FIGURE 13.6** Changes in the net load pattern for different averaging time steps (with equal time steps for demand and generation patterns).

as well in the prospect of the energy exchanges inside a local energy community, also owing to the possible application of community energy storage solutions [34].

Further considerations refer to the variety of representations of the demand and generation patterns. If the demand and generation patterns constructed for different averaging time steps have the same energy of the corresponding patterns represented at the base time step $\tau_o$ for the same time period $T$, the total net energy resulting in the time period $T$ is the same in all cases and is calculated as:

$$W_{T,\tau_o}^{(m,v)} = \sum_{k=1}^{T/\tau_o} d_{T,m\tau_o,v\tau_o}(k) \tag{13.2}$$

For the sake of comparison with other situations, it is possible to form the matrix $\mathbf{W}_{T,\tau_o} = \{W_{T,\tau_o}^{(m,v)}\} \in \mathcal{R}^{M,M}$, with all equal entries. The other situations of interest are the ones in which only the positive (or the negative) values of net energy are calculated, taking into account the different time steps between demand and generation. For this purpose, the following matrices are defined, for $m = 1, \ldots, M$ and $v = 1, \ldots, M$:

- The matrix with energy calculated from positive net demand components only

$$\hat{\mathbf{W}}_{T,\tau_o} = \left\{\hat{W}_{T,\tau_o}^{(m,v)}\right\} \in \mathcal{R}^{M,M}, \text{ with } \hat{W}_{T,\tau_o}^{(m,v)} = \sum_{k=1}^{T/\tau_o} \max\{d_{T,m\tau_o,v\tau_o}(k), 0\} \tag{13.3}$$

• The matrix with energy calculated from negative net demand components only

$$\check{\mathbf{W}}_{T,\tau_o} = \left\{ \check{W}_{T,\tau_o}^{(m,v)} \right\} \in \mathcal{R}^{M,M}, \text{ with } \check{W}_{T,\tau_o}^{(m,v)} = \sum_{k=1}^{T/\tau_o} \min\left\{ d_{T,m\tau_o,v\tau_o}(k), 0 \right\} \qquad (13.4)$$

By representing the entries of these matrices, it is possible to observe the different values of total positive (or negative) net energy demand resulting in the different combinations of the averaging time steps. To exemplify, let us consider the demand and generation patterns shown in Fig. 13.7, both gathered at the base averaging time step $\tau_o = 5$ min. The total energy calculated from positive (and negative, reported in absolute value) net power variations for different combinations of the averaging time steps is shown in Fig. 13.8. It can be seen that there is a trend of reduction of the total energy in both cases. From these results, it can be seen how much the averaging time step impacts the net energy values. The reduction of the averaging time step (when possible for both demand and generation patterns) allows capturing more detailed situations. This reduction is particularly effective when the net power pattern exhibits fluctuations around the null value.

The accuracy of the net power analysis outcomes is then conditioned by the data set with the longest averaging time step duration. The differences depending on the averaging time steps may have economic relevance if the positive and negative net power values are given a different economic treatment [31]. Furthermore, the averaging

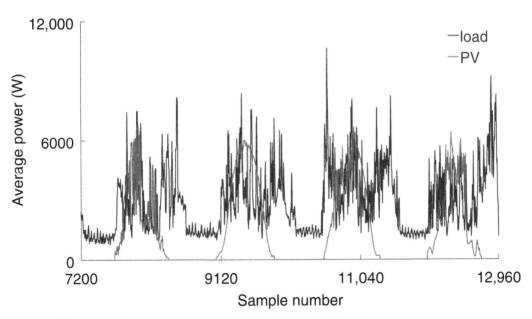

FIGURE 13.7    Demand and generation patterns at 5-min averaging time step.

(A)

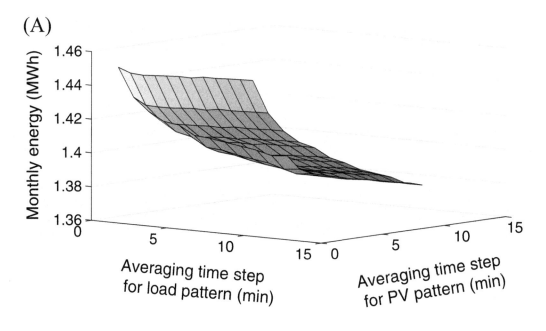

(B)

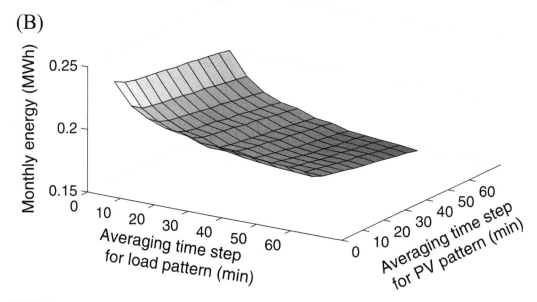

FIGURE 13.8    Total energy calculated from positive and negative net power variations for different combinations of the averaging time steps: (A) positive net power variations; (B) (absolute) negative net power variations.

time step also impacts on self-consumption and self-sufficiency of a local generation system.[1] Progressive overestimations when the averaging time step increases are reported for photovoltaic systems in Beck et al. [35] and Ayala-Gilardón et al. [36]. These aspects and their impact on the local energy market operation deserve further studies.

### 13.3.4 Creation of time series with the same time step

When streams of data come from different sources, different data representations could, in general, be used. In particular, the pattern types could have regular or irregular cadence. In the latter case, the data could be represented only after the occurrence of an event (e.g., linked to rapid variations [37]). In addition, the data could be considered with a stepwise structure, or as the interpolation of successive points. In any case, the total energy corresponding to the pattern in the period of analysis has to be preserved. The tool presented in Chicco et al. [31] transforms an input pattern $\mathbf{p}_{T,\tau_i}^T$ (with a regular time step $\tau_i$ in minutes, or with irregular cadence), into a preliminary output pattern $\tilde{\mathbf{p}}_{T,\tau_o}^T$ with a regular time step $\tau_o$ (in minutes) within a total period of analysis $T$. For a regular input pattern, the time steps $\tau_i$ and $\tau_o$ may be generally different and even not multiples of each other. If the time steps $\tau_i$ and $\tau_o$ are multiples of each other, the total energy of the input and preliminary output patterns is the same, so that the final output pattern is $\mathbf{p}_{T,\tau_o}^T = \tilde{\mathbf{p}}_{T,\tau_o}^T$. Otherwise, a correction factor is needed to reproduce the total energy of the input pattern. This correction factor is determined as the ratio $\chi = W_i/W_o$, where $W_i$ and $W_o$ are the total energy of the input pattern and of the preliminary output patterns, respectively. The final output pattern is then $\mathbf{p}_{T,\tau_o}^T = \chi \tilde{\mathbf{p}}_{T,\tau_o}^T$.

### 13.3.5 Horizontal and vertical normalization

Normalization is the process with which different time series are made comparable by rescaling them in such a way that the *shape* of the pattern becomes the main aspect used to distinguish the time series. There are different types of normalizations, which can be denoted as:

- *Vertical normalization*, with which the pattern is rescaled only in its amplitude, without changing the location of the points in the horizontal (time) axis. This normalization is carried out by using some classical techniques that are used in time series analysis, for example:
  - Normalization with respect to the *maximum value* of the time series: considering the amplitude $y$ of a given data in a time series having the maximum amplitude $y_{max}$, the normalized amplitude is $y_{normMax} = y/y_{max}$. This normalization leads to giving similar importance to all patterns regardless of their actual amplitude.

[1] Self-consumption is defined as the electrical energy generated by the photovoltaic system that is consumed locally by the local user, divided by the total electrical energy generation of the photovoltaic system (the excess of which is exported to the grid). Self-sufficiency is defined as the self-consumed electrical energy divided by the total electrical energy consumption of the local user.

- *Min−max normalization*: considering the amplitude $y$ of a given data in a time series having minimum amplitude $y_{min}$ and maximum amplitude $y_{max}$, the normalized amplitude is $y_{norm_{MinMax}} = y - y_{min}/y_{max} - y_{min}$. The normalized values belong to the range $(0,1)$; namely, the minimum value corresponds to 0 and the maximum value to 1. This normalization has the drawback that if the time series contains outliers (e.g., a point much higher than all the others), the resulting normalized time series will have many points represented in a restricted vertical range.
- Normalization with respect to the *average value* of the time series: the normalized amplitude of a given data $y$ belonging to a time series with mean value $\mu$ is $y_{norm_{Average}} = y/\mu$. With this normalization, the normalized values obtained have not precisely the same vertical scale. For its application to electrical load patterns, the average power is proportional to the energy used in the time interval of the time series.
- *Z-score normalization*: the normalized amplitude of a given data $y$ belonging to a time series with mean value $\mu$ and standard deviation $\sigma$ is $y_{norm_{Zscore}} = y - \mu/\sigma$. This normalization is useful to handle the outliers better, but the normalized values obtained have not precisely the same vertical scale.

For the specific application to load profiles, a significant normalization is performed by dividing the amplitude of the time series data by the contract power $P_{contract}$ of the user [11]. The normalized amplitude of a given data $y$ is $y_{norm_{Contract}} = y/P_{contract}$. In this way, a benefit is that the normalization does not depend on the actual data, and it is possible to create differences between load patterns that evolve more or less closer to the contract power.

- *Horizontal normalization*, with which some operations on the time intervals are made in order to make the patterns comparable. This type of normalization is generally used only in particular cases. For example, for photovoltaic systems, it is crucial to take into account that the solar irradiance patterns obtained in different days or seasons are not directly comparable, because the duration of the daylight period (from sunrise to sunset) changes every day [38]. Moreover, the maximum solar irradiance that can be found on a bright day varies day by day. The power generation from the photovoltaic system, if it is not directly measured on-site, is determined by knowing the solar irradiance, the temperature, other ambient variables, and the model of the photovoltaic system, including the internal losses [39]. Thereby, a comparison among the power generation patterns during the days of the year cannot be carried out in a meaningful way by considering just the absolute values. For this purpose, the power generation patterns are rescaled both in time and amplitude, by defining a normalized horizontal scale in which the period of time from sunrise to sunset is mapped onto the interval $(0,1)$. Likewise, the amplitude is defined for each day from 0 to 1, where the value 1 corresponds to the power generation that could occur in the clear sky conditions determined from an appropriate model, for example, the Moon−Spencer model [40]. Finally, in order to obtain comparable patterns, the power generation points determined during the days have to be represented with the same number of points on the horizontal axis. For this purpose, the tool for the creation of time series with the same time step indicated in the previous section is used to calculate the normalized amplitudes for the same (user-defined) number of points in the interval $(0,1)$. The resulting patterns can be sent as inputs to the solution methods (e.g., clustering algorithms) for grouping the days (e.g., clear sky, quasi-clear sky, quasi-cloudy, and cloudy [38]), or the photovoltaic sites in a given time period.

### 13.3.6 Data alignment and synchronization

A crucial aspect of comparing multiple patterns is to ensure that the data are perfectly aligned in time. There are various issues concerning data alignment, for example:

- The average power data refer to a given time step. However, the data have to be associated with a given point in time. It has to be clear whether this point in time is the initial, the final, or the central time instant, and all the data have to be represented with the same time references.
- The clocks used to gather the data have to be precise. Otherwise, their lagging or leading times have a direct impact on the data representation.
- The synchronization of the data collected has to be ensured as much as possible, through time stamp mechanisms that provide clear and robust time representation of the timing information [29]. Time-stamping is essential for both data gathering at a regular cadence, and event-driven energy metering with asynchronous events. Time-stamping can be enhanced by exploiting the mechanisms used in the blockchain technology [41].

## 13.4 Trends and opportunities for local energy systems and markets

### 13.4.1 From large-scale to local load profiling

The classical view of load profiling was to create estimated profiles for a large number of nonmetered consumers. Today, with the increasing diffusion of smart meters, the availability of clouds and big data techniques, it may seem that load profiling is losing its role. Indeed, in the future different load profiles are expected with respect to the current ones [42]. Prosumer profiling remains important to define the baselines for demand response analysis and the formulation of demand response programs [43]. In any case, the overall interest in load profiling will remain, as 100% diffusion of smart metering will not be reached soon in every jurisdiction and for each type of user. A particular situation reported in Council of European Energy Regulators (CEER) [9] indicates that a legal exemption from having a smart meter installed is possible in Norway for clients who obtain a medical certificate from a medical doctor or psychologist.

Various aspects require particular attention in the present evolution of the energy systems and markets, namely:

1. *Metered data are available to larger extents*. The nature of the clustering analysis is changing, and more opportunities are becoming available. To summarize the situation, four types of studies can be defined taking into account the (static or dynamic) nature of the data used and the (again, static or dynamic) nature of the definition of the classes in the clustering procedure adopted, as indicated in Table 13.1 [44]. The attention is more and more focused on Type 3 and Type 4 analyses, in which the availability of

**TABLE 13.1** Definition of the types of analysis.

| Type of analysis | Data | Clustering | Notes |
|---|---|---|---|
| Type 1 | Static | Static | Fixed data |
| | | | Fixed classes |
| Type 2 | Static | Dynamic | Fixed data |
| | | | Classes updated during the execution of the procedure |
| Type 3 | Dynamic | Static | Variable data during time |
| | | | Fixed classes |
| Type 4 | Dynamic | Dynamic | Variable data during time |
| | | | Classes updated during the execution of the procedure |

updated data streams gives the possibility of confirming or uploading the structure of the classes in time. The changing conditions in which the users are called to operate today include participation in demand response programs, local smoothing of renewable energy production by using storage or demand side management. With the growing impact of local generation and energy storage, long-term historical load profiles are less useful, unless suitable weighting factors are introduced to give more relevance to the present situation. This dynamic view of the present energy systems reflects in the application of a Type 4 analysis, for example, using online adaptive clustering methods like the one presented in Ref. [45]. An advantage of this approach is the reduction of the number of data used in the computational procedures, as only recent data are assumed to have practical relevance.

2. *The succession of data points during time is no longer the main factor.* The increased flexibility of energy use calls for the availability of time-adjustable power patterns. At the same time, the presence of local generation and storage makes it possible to obtain a compensation between production and demand at any time, thus reducing the distinction between periods of peak pricing and low-rate periods, and increasing the diffusion of flat-rate tariffs. Moreover, the effects of programs to shift the demand from peak to off-peak periods are the progressive flattening of the load pattern, reducing the need to provide additional flexibility [46]. In this context, it can be considered that there is no specific synchronization of the load patterns based on the users' behavior [7]. Therefore, the calculation of the distances between the average power values at individual points in time (which is at the basis of many clustering procedures) becomes less relevant. In this case, the features can be defined by using formulations that do not depend on the individual points in time (e.g., with duration curves [25]). In this way, the representations can be adapted to enable the use of comparison techniques such as dynamic time warping and shape-based distance [47]. Flexibility and demand response analyses are more and more based on switchable profiles. Limits to switch the users' contribution to the power patterns could arise from the stiffness appearing in some time periods in which the users' willingness to change their consumption behavior is lower. The time periods

with higher stiffness of the users' behavior may be detected by looking at the load pattern variations for a given aggregation of users, for example during the early morning for an aggregation of residential users [48].

3. *The purpose of the users' grouping is changing.* The classical load profiling aimed to a large extent at defining tariffs and pricing mechanisms. Today, the energy markets are becoming smaller, both in terms of timing (intraday markets) and of size (local energy markets, microgrids, and energy communities) [49]. The definition of energy services is becoming increasingly significant. The evolution from load profiling to prosumer profiling requires the identification of the relevant features with which it is important to categorize the users in local energy systems. Some of these features depend on the load pattern *variations*, together with the shape of the load pattern. Other features come from net metering analysis, looking at the changes with respect to the "neutral" conditions in which the user does not exchange energy with the grid, as well as from the exposure of the user to average power variations during time. Further features can be defined to represent self-production and self-sufficiency of the local energy system, to account for the thermal response of the users [50], or to take into account the correlations among different types of energy demands.

4. *Uncertainty plays a major role in small energy systems.* Load profiling carried out by using the data of a large energy company has the effect of forming user groups that are similar across the whole set of users. The characteristics of local portions of the energy system, in general, do not appear, with notable exceptions. For example, exceptions are found when there is a portion of the territory with cold temperatures and the use of electric heating, while in other areas the climate is different and electric heating is not used. When local energy systems are considered, the load profiles depend more on local characteristics, and general-purpose load profiles are less useful. In addition, the uncertainty of the local energy production plays a significant role, and the number of users is reduced so that the aggregate power patterns are more variable, and the definition of the prosumer profiles could be less stable. For the analyses, it will be more and more appropriate to resort to probabilistic load profiling techniques [51].

## 13.4.2 Market opportunities

The evolution of the energy markets (daily and intraday) and of the markets to procure ancillary services has to take into account the current trends to promote the development of local energy systems, microgrids, and new local energy communities. Prosumer profiling is a key tool for an aggregator willing to create a portfolio of prosumers and to trade predefined bands of energy consumption (or even local production) at given time steps with the manager of the relevant market [52]. The crucial aspect is the definition of the penalties the aggregator will incur if the actual consumption (or production) deviates from the predefined band. Mechanisms like the share balancing responsible model [53] or the scheduled high voltage/medium voltage exchange profile market model [54] establish the roles and responsibilities of the market players, in some cases with the distribution system operator (DSO) acting to guarantee the programmed profile during time at the interface with the grid. In an energy system in which any prosumer can buy and sell any amount of electricity, it is essential to have a responsible balance entity that takes over the risk of having differences during the successive time steps between the measured energy and the scheduled energy. The discussion on how to set the right incentive structures for DSOs is in progress [55].

### 13.4.3 Not only electrical load profiling

The energy systems are evolving toward integrated solutions in which the synergy among different energy sources to supply multienergy loads provides overall benefits to energy efficiency, environmental aspects, and energy sustainability. The notion of load profiling, as used in the electrical domain, has been applied to the thermal energy systems, especially to study the energy management in buildings [56]. The averaging time step needed for the thermal load patterns is generally lower than the one for the electrical load patterns, as the time constants for thermal energy systems are longer. In many studies the load profiles of electrical and thermal loads have been constructed in an almost independent way, depending on the availability of measured data, taking into account individually the possible dependence on environmental variables [57]. Profiling for gas stations has been addressed in Mihai et al. [58]. One of the directions for future studies is to consider multienergy loads [59], with extensions to integrated profiling of multienergy outcomes. Furthermore, in multienergy systems, it is essential to take into account the simultaneous presence of heating and cooling loads, which could impact on the use of equipment such as heat pumps, able to provide both heating and cooling outputs [60]. In addition to battery storage, also thermal energy storage has the role of decoupling the electrical and thermal demand over time [61,62]. It will be quite likely to identify in the future different prosumer profiles for prosumers with the local photovoltaic generation with storage, without storage, and profiles in which there is a remarkable contribution of electric vehicles.

Finally, in the determination of the load profiles, some approaches have introduced social and psychological components to represent the behavior of the users, in households [63], and for the assessment of incentive-based demand response [64].

## 13.5 Conclusions

This chapter has discussed various challenges concerning the possible role of users' profiling practices in the evolving context of local energy systems and markets. Some open questions have been indicated in the Introduction. Some remarks referring to these questions are summarized as follows:

1. Does it still make sense to use load profiling when all consumers could be equipped with interval meters? On the one hand, load profiling will continue to be useful for settlement or tariff definition purposes, as 100% diffusion of interval meters will not be reached in every jurisdiction. On the other hand, the use of profiling will change from the traditional load profiling to a broader view on prosumer profiling, able to guarantee appropriate segmentation of the average power curves over time, to respond to different energy services and flexibility needs.
2. Which is the most effective way to construct and deal with the prosumer profiles of electricity usage? In the evolving context in which more metered data are available, and the focus is set on local energy systems, general-purpose load profiles are less useful.

The incorporations of local characteristics and the possible synergies within multienergy system operations become essential. Both the purpose and the feature selection for prosumer profiling have to be adapted. Finding the most suitable ways to exploit the outcomes from net power analysis is a big challenge.

3. Could prosumer profiling provide useful information in the case of local energy communities? The focus on local energy communities introduces novel aspects for the development of prosumer profiling, at two different levels. The first one is a local level for individual users or groups of users supplied from the same point, in which the individual data will have to be considered as private information. The second one is a community level, in which the aggregation of multipoint users may benefit from the heterogeneity among the different users, to compensate for the deviations within the energy community and offer smoother average power patterns to the external system.

4. Is there any room for developing short-term prosumer profiling? In the short term, it becomes crucial to capture the peaks of the average power patterns. For this purpose, fine-grained metering or event-driven energy metering can provide better data. In the latter case, appropriate clustering procedures have to be developed.

5. Is stochastic prosumer profiling a viable option? For local energy systems, short-term dynamic profiling may be addressed by considering uncertainty and taking into account the correlations between different types of energy loads, with their possible dependence on common variables such as temperature.

# References

[1] IEA - International Energy Agency, Competition in Electricity Markets, IEA Publication, 2001. ISBN 92-64-185593, Paris, France, February 2001.

[2] Elexon, Load Profiles and Their Use in Electricity Settlement, Version 3.0, October 25, 2018. Available from: <https://www.elexon.co.uk/documents/training-guidance/bsc-guidance-notes/load-profiles/>. (accessed 29.12.19).

[3] D. Labate, P. Giubbini, G. Chicco, M. Ettorre, SHAPE: A New Business Analytics Web Platform for Getting Insights on Electrical Load Patterns, CIRED Workshop - Rome, June 11–12, 2014, Paper 0354.

[4] D. Balmert, D. Grote, K. Petrov, Development of Best Practice Recommendations for Smart Meters Rollout in the Energy Community, KEMA International B.V. Final Report, Bonn, Germany, February 2012.

[5] A. Wright, S. Firth, The nature of domestic electricity-loads and effects of time averaging on statistics and on-site generation calculations, Appl. Energy 84 (4) (2007) 389–403.

[6] M. Simonov, G. Chicco, G. Zanetto, Event-driven energy metering: principles and applications, IEEE Trans. Ind. Appl. 53 (4) (2017) 3217–3227.

[7] G. Chicco, A. Mazza, New insights for setting up contractual options for demand side flexibility, J. Eng. Sci. Innov. 4 (4) (2019) 381–398.

[8] J. Neyman, On the two different aspects of the representative method: the method of stratified sampling and the method of purposive selection, J. R. Stat. Soc. Part IV (1934) 558–606.

[9] CEER - Council of European Energy Regulators, Implementing Technology that Benefits Consumers in the Clean Energy for All Europeans Package - Selected Case Studies, Ref: C19-IRM-16-04, July 22, 2019. Available from: <https://www.ceer.eu/documents/104400/-/-/bd457593-900f-f995-eac4-ed989255b26f>. (accessed 29.12.19).

[10] OFGEM, Elective Half-Hourly Settlement: Conclusions Paper, London, May 26, 2016.

[11] G. Chicco, R. Napoli, P. Postolache, M. Scutariu, C. Toader, Customer characterisation options for improving the tariff offer, IEEE Trans. Pow. Syst. 18 (1) (2003) 381–387.

[12] J.C. Sousa, L.P. Neves, H.M. Jorge, Assessing the relevance of load profiling information in electrical load forecasting based on neural network models, Electr. Power Energy Syst. 40 (2012) 85–93.

[13] M. Espinoza, C. Joye, R. Belmans, B. De Moor, Short-term load forecasting, profile identification and customer segmentation: a methodology based on periodic time series, IEEE Trans. Power Syst. 20 (3) (2005) 1622–1630.

[14] S. Lin, F. Li, E. Tian, Y. Fu, D. Li, Clustering load profiles for demand response applications, IEEE Trans. Smart Grid 10 (2) (2019) 1599–1607.

[15] Y. Wang, Q. Chen, C. Kang, M. Zhang, K. Wang, Y. Zhao, Load profiling and its application to demand response: a review, Tsinghua Sci. Technol. 20 (2) (2015) 117–129.

[16] S.R. Konda, L.K. Panwar, B.K. Panigrahi, R. Kumar, Investigating the impact of load profile attributes on demand response exchange, IEEE Trans. Ind. Inform. 14 (4) (2018) 1382–1391.

[17] J. Heggset, G.H. Kjolle, F. Trengereid, H.O. Ween, Quality of supply in the deregulated Norwegian power system, in: Proc. 2001 IEEE Porto Power Tech, Porto, Portugal, 2001.

[18] G. Chicco, Overview and performance assessment of the clustering methods for electrical load pattern grouping, Energy 42 (1) (2012) 68–80.

[19] M. Martinez-Luengo, M. Shafiee, A. Kolios, Data management for structural integrity assessment of offshore wind turbine support structures: data cleansing and missing data imputation, Ocean. Eng. 173 (2019) 867–883.

[20] N. Bokde, M.W. Beck, F. Martínez Álvarez, K. Kulat, A novel imputation methodology for time series based on pattern sequence forecasting, Pattern Recognit. Lett. 116 (2018) 88–96.

[21] A. Notaristefano, G. Chicco, F. Piglione, Data size reduction with symbolic aggregate approximation for electrical load pattern grouping, IET Gener. Transm. Distrib. 7 (2) (2013) 108–117.

[22] E. Carpaneto, G. Chicco, R. Napoli, M. Scutariu, Electricity customer classification using frequency-domain load pattern data, Int. J. Electr. Power Energy Syst. 28 (1) (2006) 13–20.

[23] M. Petrescu, M. Scutariu, Load diagram characterisation by means of wavelet packet transform, in: Proc. 2nd Balkan Power Conference, Belgrade, Yugoslavia, June 19–21, 2002, pp. 15–19.

[24] G. Chicco, R. Napoli, F. Piglione, Comparisons among clustering techniques for electricity customer classification, IEEE Trans. Power Syst. 21 (2) (2006) 933–940.

[25] T. Cerquitelli, G. Chicco, E. Di Corso, F. Ventura, G. Montesano, M. Armiento, et al., Clustering-based assessment of residential consumers from hourly-metered data, in: International Conference on Smart Energy Systems and Technologies (SEST 2018), Seville, Spain, September 10–12, 2018.

[26] G.J. Tsekouras, P.B. Kotoulas, C.D. Tsirekis, E.N. Dialynas, N.D. Hatziargyriou, A pattern recognition methodology for evaluation of load profiles and typical days of large electricity customers, Electr. Power Syst. Res. 78 (9) (2008) 1494–1510.

[27] V. Figueiredo, F. Rodrigues, Z. Vale, J.B. Gouveia, An electric energy consumer characterization framework based on data mining techniques, IEEE Trans. Pow. Syst. 20 (2) (2005) 596–602.

[28] S. Verdu, M. Garcia, C. Senabre, A. Marin, F. Franco, Classification, filtering, and identification of electrical customer load patterns through the use of self-organizing maps, IEEE Trans. Power Syst. 21 (4) (2006) 1672–1682.

[29] M.S. Omar, S.A.R. Naqvi, S.H. Kabir, S.A. Hassan, An experimental evaluation of a cooperative communication-based smart metering data acquisition system, IEEE Trans. Ind. Inform. 13 (1) (2017) 399–408.

[30] G. Chicco, Challenges for smart distribution systems: data representation and optimization objectives, in: Proc. 12th International Conference on Optimization of Electrical and Electronic Equipment (OPTIM 2010), Brasov, Romania, May 20–22, 2010, pp. 1236–1244.

[31] G. Chicco, V. Cocina, A. Mazza, F. Spertino, Data pre-processing and representation for energy calculations in net metering conditions, in: Proc. IEEE Energycon 2014, Dubrovnik, Croatia, May 13–16, 2014, Paper 262.

[32] K.C. Armel, A. Gupta, G. Shrimali, A. Albert, Is disaggregation the holy grail of energy efficiency? The case of electricity, Energy Policy 52 (2013) 213–234.

[33] R. Moddemeijer, On estimation of entropy and mutual information of continuous distributions, Signal. Process. 16 (3) (1989) 233–248.

[34] E. Barbour, D. Parra, Z. Awwad, M.C. González, Community energy storage: a smart choice for the smart grid? Appl. Energy 212 (2018) 489–497.

[35] T. Beck, H. Kondziella, G. Huard, T. Bruckner, Assessing the influence of the temporal resolution of electrical load and PV generation profiles on self-consumption and sizing of PV-battery systems, Appl. Energy 173 (2016) 331–342.

[36] A. Ayala-Gilardón, M. Sidrach-de-Cardona, L. Mora-López, Influence of time resolution in the estimation of self-consumption and self-sufficiency of photovoltaic facilities, Appl. Energy 229 (2018) 990−997.

[37] M. Simonov, G. Chicco, G. Zanetto, Real-time event-based energy metering, IEEE Trans. Ind. Inform. 13 (6) (2017) 2813−2823.

[38] G. Chicco, V. Cocina, F. Spertino, Characterization of solar irradiance profiles for photovoltaic system studies through data rescaling in time and amplitude, in: 49th International Universities' Power Engineering Conference (UPEC 2014), Cluj-Napoca, Romania, September 2−5, 2014, Paper 52.

[39] F. Spertino, A. Ciocia, P. Di Leo, R. Tommasini, I. Berardone, M. Corrado, et al., A power and energy procedure in operating photovoltaic systems to quantify the losses according to the causes, Sol. Energy 118 (2015) 313−326.

[40] P. Moon, D.E. Spencer, Illumination from a non uniform sky, Trans. Illumination Eng. Soc. 37 (12) (1942) 707−726.

[41] O. Van Cutsem, D.H. Dac, P. Boudou, M. Kayal, Cooperative energy management of a community of smart-buildings: a blockchain approach, Int. J. Electr. Power Energy Syst. 117 (2020) 105643.

[42] J. Haakana, J. Haapaniemi, R. Härmä, M. Ryhänen, J. Lassila, J. Partanen, Electricity demand profile for residential customer 2030, in: 25th International Conference on Electricity Distribution Madrid, Spain, June 3−6, 2019, Paper 1785.

[43] M. Sun, Y. Wang, F. Teng, Y. Ye, G. Strbac, C. Kang, Clustering-based residential baseline estimation: a probabilistic perspective, IEEE Trans. Smart Grid 10 (6) (2019) 6014−6028.

[44] I. Benítez, A. Quijano, J.L. Díez, I. Delgado, Dynamic clustering segmentation applied to load profiles of energy consumption from Spanish customers, Int. J. Electr. Power Energy Syst. 55 (2014) 437−448.

[45] G. Le Ray, P. Pinson, Online adaptive clustering algorithm for load profiling, Sustain. Energy Grids Netw. 17 (2019) 100181.

[46] F. Abbaspourtorbati, A.J. Conejo, J. Wang, R. Cherkaoui, Is being flexible advantageous for demands? IEEE Trans. Power Syst. 32 (3) (2017) 2337−2345.

[47] H. Teichgraeber, A.R. Brandt, Clustering methods to find representative periods for the optimization of energy systems: an initial framework and comparison, Appl. Energy 239 (2019) 1283−1293.

[48] I.A. Sajjad, G. Chicco, R. Napoli, Definitions of demand flexibility for aggregate residential loads, IEEE Trans. Smart Grid 7 (6) (2016) 2633−2643.

[49] F. Lezama, J. Soares, P. Hernandez-Leal, M. Kaisers, T. Pinto, Z. Vale, Local energy markets: paving the path toward fully transactive energy systems, IEEE Trans. Power Syst. 34 (5) (2019) 4081−4088.

[50] A. Albert, R. Rajagopal, Thermal profiling of residential energy use, IEEE Trans. Power Syst. 30 (2) (2015) 602−611.

[51] Z.A. Khan, D. Jayaweera, M.S. Alvarez-Alvarado, A novel approach for load profiling in smart power grids using smart meter data, Electr. Power Syst. Res. 165 (2018) 191−198.

[52] S.R. Konda, A.S. Al-Sumaiti, L.K. Panwar, B.K. Panigrahi, R. Kumar, Impact of load profile on dynamic interactions between energy markets: a case study of power exchange and demand response exchange, IEEE Trans. Ind. Inform. 15 (11) (2019) 5855−5866.

[53] H. Gerard, E.I.R. Puente, D. Six, Coordination between transmission and distribution system operators in the electricity sector: a conceptual framework, Uti. Policy 50 (2018) 40−48.

[54] N. Natale, F. Pilo, G. Pisano, G.G. Soma, Scheduled profile at TSO/DSO interface for reducing balancing costs, in: 1st International Conference on Energy Transition in the Mediterranean Area SyNERGY MED, Cagliari, Italy, May 28−30, 2019.

[55] Smart Energy Europe, Design Principles for (Local) Markets for Electricity System Services, SmartEn Position Paper, September 2019. Available: <https://www.smarten.eu/wp-content/uploads/2019/09/20190903-smartEn-Flexibility-Markets-Position-Paper-Final.pdf>. (accessed 29.12.19).

[56] M.S. Piscitelli, S. Brandi, A. Capozzoli, Recognition and classification of typical load profiles in buildings with non-intrusive learning approach, Appl. Energy 255 (2019) 113727.

[57] K.B. Lindberg, S.J. Bakker, I. Sartori, Modelling electric and heat load profiles of non-residential buildings for use in long-term aggregate load forecasts, Uti. Policy 58 (2019) 63−88.

[58] C. Mihai, D. Ilea, P.M. Mircea, Use of load profile curves for the energy market, in: 13th International Conference on Development and Application Systems, Suceava, Romania, May 19−21, 2016.

[59] F. Lombardi, S. Balderrama, S. Quoilin, E. Colombo, Generating high-resolution multi-energy load profiles for remote areas with an open-source stochastic model, Energy 177 (2019) 433–444.

[60] R. Ghoubali, P. Byrne, J. Miriel, F. Bazantay, Simulation study of a heat pump for simultaneous heating and cooling coupled to buildings, Energy Build. 72 (2014) 141–149.

[61] D. Enescu, G. Chicco, R. Porumb, G. Seritan, Thermal energy storage for grid applications: current status and emerging trends, Energies 13 (2) (2020) 340.

[62] D. Patteeuw, K. Bruninx, A. Arteconi, E. Delarue, L. Helsen, Integrated modeling of active demand response with electric heating systems coupled to thermal energy storage systems, Appl. Energy 151 (2015) 306–319.

[63] N. Pflugradt, U. Muntwyler, Synthesizing residential load profiles using behavior simulation, Energy Procedia 122 (2017) 655–660.

[64] Q. Shi, C.F. Chen, A. Mammoli, F. Li, Estimating the profile of incentive-based demand response (IBDR) by integrating technical models and social-behavioral factors, IEEE Trans. Smart Grid 11 (1) (2020) 171–183.

# Forecasting

*Elena Mocanu[1], Decebal Constantin Mocanu[1,2], Nikolaos G. Paterakis[3] and Madeleine Gibescu[2,4]*

[1]Department of Computer Science, EEMCS, University of Twente, Enschede, The Netherlands
[2]Department of Mathematics and Computer Science, Eindhoven University of Technology, Eindhoven, The Netherlands [3]Department of Electrical Engineering, Eindhoven University of Technology, Eindhoven, The Netherlands [4]Copernicus Institute of Sustainable Development, Utrecht University, Utrecht, The Netherlands

## 14.1 Introduction

As prediction developed, different subfields were created. The electrical forecasting problem can be regarded as a nonlinear time series prediction problem depending on many complex factors since it is required at various aggregation levels and high resolution [1]. Furthermore, the electrical forecasting accuracy and the resulting errors will be reflected in the performance of the local energy market. In this context, a variety of forecasts are necessary at national level, regional level, or specific to the type of consumers (residential, industrial). Worldwide, residential buildings have one of the highest energy consumption rates, on average they consume around 40% of the global primary energy and contribute to over 30% of $CO_2$ emission. Within Europe, residential energy usage grows at an annual rate of 1.5%. This is higher than the industrial and transportation sector energy consumption increase rate [2].

Consequently, the current growth of urbanization and electricity demands introduce new requirements for future power grids and keep the electricity market under pressure. To satisfy these demands, future power grids will need to predict, learn, schedule, make decisions, and monitor local energy production and consumption. Following this, to improve the flow of energy requires energy predictions over various time horizons [3,4].

As both the aggregation level and the prediction horizon are decreasing more and more, the fluctuations are increasing in the electrical patterns. To solve these challenging problems, various time series and machine learning (ML) approaches have been proposed in the literature. These range from heuristic based approaches to mathematically grounded ones such as those residing in the realm of ML.

When analyzing the local energy market impact, it is imperative to not only predict the electrical pattern, but also to consider a deeper range of factors. This allows the decomposition of demand and price forecasting, to not only help identify consumption and generation trends, detect faults, or predict savings, but it allows for better decision-making strategies to control and schedule loads to off-peak times [3,4]. The choice of a high-performance method depends on the special characteristics that electrical patterns have.

## 14.2 Energy prediction: particularities

The complexity of the consumers' energy producing and consuming technologies and the uncertainty in the influencing factors yield frequent fluctuations. Nowadays, commercial, industrial, and residential buildings represent a tremendous amount of the global energy used. Moreover, urbanization and electrification trends show that the total energy demand will increase in the future, and the penetration of energy from renewable sources is increasing as well.

Therefore the electrical demand forecasting problem, at various aggregation levels, can be regarded as a highly nonlinear and nonstationary time series prediction problem [5–7]. In Tang et al. [8] a comprehensive list of data characteristics of energy time series are summarized, such as stationarity (and nonstationarity), linearity (and nonlinearity), complexity, chaotic property, fractality, regularity (and irregularity), cyclicity, seasonality, saltation (or mutability), and randomness. Therein, the characteristics are split in two, that is, the *nature* and *pattern* characteristics, analyzing energy time series data from different perspectives. The first type, given by the nature, refers to a series of components, that is, trend, cyclical, seasonal, saltatory, and noisy patterns. The pattern refers to the ability of a prediction method to extract coexisting hidden patterns from data. In this chapter the following three energy pattern characteristics are considered.

### 14.2.1 Prediction horizon and resolution

Prediction of temporal energy consumption enables building operators to schedule the energy usage over time, shift energy usage to off-peak periods, and make more effective energy purchase plans. From this perspective, demand forecasting can be considered to fall into three categories:

1. Short-term forecasts are usually applied to intervals ranging from 1 hour to 1 week. Traditionally, the short-term forecast is performed using data with 1-hour and 15-minute resolutions, but higher resolutions make the problem even more complicated.
2. Medium-term forecasts are usually from 1 week to 1 year.
3. Long-term forecasts are for ranges longer than 1 year.

The predictions performed in this chapter are restricted to the short-term intervals, with a special focus on the day-ahead energy prediction with various resolutions.

### 14.2.2 Level of aggregation

In the local energy market context it is important to predict not only aggregated data, but to go deep into the individual building level, so that distributed generation resources can be deployed based on the local forecast. Decomposition of demand forecasting helps analyze energy consumption patterns and identify the prime targets for energy conservation. As the level of aggregation decreases, the electricity patterns are becoming harder to forecast due to a broad range of influencing factors inside the local communities.

### 14.2.3 Influencing factors

Without presenting an exhaustive list of explanatory variables, one should consider countries and regional tendencies, socioeconomic indicators, as well as climate and meteorological data in order to perform a forecast. Additional variables, which proved to have a significant impact on energy systems as the Covid-19 pandemic, should be considered. Furthermore, as the role of end-users progresses in the local energy markets, the features selection step starts to increasingly account for multimodal data recorded at low-level aggregation (e.g., residential and industrial). The complexity of building energy behavior and the uncertainty of the influencing factors, such as more fluctuations in demand, make energy prediction a hard problem. These fluctuations are given by weather conditions, the building construction and thermal properties of the physical materials used, the occupants and their behavior, sublevel systems components lighting or HVAC (heating, ventilating, and air-conditioning).

## 14.3 Energy prediction: methods

The highly nonlinear behavior of energy consumption is difficult to predict, due to the uncertainty of inconsistent fluctuations. These fluctuations are influenced by a range of factors from climate, building envelope, systems, control and maintenance, indoor environment to occupant behavior, and can even extend as far as socioeconomic factors [9].

Given training vectors $x_i \in \mathcal{R}^n, \forall i = 1, ..n$, and a label vector $y_i \in \mathcal{R}^d$ the forecasting problem aims to find at any moment in time an estimated $\tilde{y}_i$, such that the difference between the expected and actual value is as small as possible. In an attempt to provide an accurate estimate over the years, many methods have been developed.

*Linear Models.* In this chapter, we have chosen three linear models to study: Ordinary Least Squares (OLS), Ridge Regression (RR), and Bayesian Ridge Regression (BRR). Each of these linear models minimizes a loss function. In the case of RR and BRR, $\alpha$ is the $L_2$ regularization penalty. A summary of these loss functions is depicted in Table 14.1. The BRR uses the same loss function, the $L_2$ regularization, and assumes the output to be a Gaussian distribution around $Xw$, such that $p(y|X, w, \alpha) = \mathcal{N}(y|X, w, \alpha)$.

**TABLE 14.1**  Cost function of the linear models used in this chapter.

| Linear models | Cost function |
| --- | --- |
| Ordinary least squares | $\min_w \|Xw - y\|_2^2$ |
| Ridge regression | $\min_w \|Xw - y\|_2^2 + \alpha \|w\|_2^2$ |
| Bayesian ridge regression | $\min_w \|Xw - y\|_2^2 + \alpha \|w\|_2^2$ where $p(y|X, w, \alpha) = \mathcal{N}(y|X, w, \alpha)$ |

*Kernel Models.* Furthermore, we have chosen two kernel-based methods, Kernel Ridge Regression (KRR) and Support Vector Regression (SVR). The first one, KRR, uses the kernel in combination with the ride regression loss function. SVR solves the primal problem defined by:

$$\underset{w,b,\zeta,\zeta^*}{\text{minimize}} \quad \frac{1}{2} w^T w + C \sum_{i=0}^{n} (\zeta + \zeta^*)$$

$$\text{subject to} \quad y_i - w^T \phi(x_i) - b \leq \varepsilon + \zeta_i$$

$$w^T \phi(x_i) + b - y_i \leq \varepsilon + \zeta_i^*$$

$$\zeta_i, \zeta_i^* \leq 0, \forall i = 1..n$$

Here, we are penalizing samples whose prediction is at least $\varepsilon$ away from their true target. These samples penalize the objective by $\zeta_i$ or $\zeta_i^*$ depending on whether their predictions lie above or below the $\varepsilon$ tube, as discussed in Ref. [10].

*Tree-based methods.* The tree-based methods showed superior performance on tabular data in various applications. As all data in the local energy market context are tabular data, three tree-based methods were chosen to be compared in this chapter, that is, Decision Trees (DT), AdaBoost (AB), and Random Trees (RT). We consider them in a binary splitting format, where for each node $n$ a specific threshold is considered (i.e., $\tau_n$). For DT the optimal placement of the loss is calculated using as an impurity measure the Gini index. Consequently, for each node, the minimization criteria used to identify the locations for future splits was Mean Squared Error, which minimizes the $L2$ error using mean values at terminal nodes.

RT uses a similar procedure in a recursive setting for a large number of randomized trees. Finally, the AB method produces a committee as a result of a linear combination of week splittings. Fore more details regarding the use of tree-based methods for electricity forecast we refer to Refs. [11,12] for random forest and to Ref. [13] for the use of boosting based algorithms, for example, XGBoost.

### 14.3.1 From machine learning to neural networks

Besides the methods discussed above, a few more popular ML methods were used for comparison, based on their ability to perform very well on specific tasks, such as Nearest Neighbors Regression (NNR), and Gaussian Process (GP). For example, in Ref. [14] k-Nearest Neighbors provide the most accurate results for load forecasting.

For almost three decades, Artificial Neural Networks were used for electricity forecasting [15,16], while recently more and more advanced neural network models, ML, and artificial intelligence methods are advancing the field. A simple neural network has three layers of neurons: input layer, hidden layer, and output layer. In the continuous search for powerful methods with better generalization capabilities, deep learning methods — as a general umbrella for artificial neural networks with many hidden layers — appear to be more popular.

*Deep Learning.* In addition to the traditional statistical learning approaches, deep learning for supervised energy prediction was used from 2014 onwards when methods, such as Conditional Restricted Boltzmann Machine (CRBM) [3], and Factored Conditional Restricted Boltzmann Machine (FCRBM) [4] were introduced. Besides these works, based on our best knowledge, there are only a few publications that extend the use of these types of methods for building energy prediction applications. For example, the Long Short-Term Memory (LSTM) method was used for building energy prediction in Refs. [17,18]. In Ref. [6] a hybrid method suitable for a price-responsive context based on FCRBM with a Gaussian Restricted Boltzmann Machine (GRBM) added to perform feature selection was proposed for energy prediction at the aggregated level. Another related work is that of Marino et al. [19], which is followed by the work of Manic et al. [20]. In 2017 a Multilayer Perceptron (MLP) enhanced with deep learning capabilities was introduced in Ref. [12] and showed a good accuracy level in comparison with other seven ML methods. Later on, many other deep learning methods have been investigated. In Ref. [21], XGBoost was used for feature selection in order to improve the accuracy of a LSTM model.

Given the fact that the number of deep learning algorithms has increased a lot in the last 2−3 years we can observe a mirroring effect in the electricity forecast area. Yet, deep learning methods for load and price forecasting are in an incipient phase where new advanced neural networks methods shall be explored and extended.

## 14.3.2 From neural networks to sparse neural networks

Despite their success, the dense deep learning models are computationally very expensive. In general, we all know that there are too many connections between neurons (parameters) to be optimized in a deep neural network. The traditional way to reduce the number of parameters in neural networks follows the *dense-to-sparse* learning paradigm, which is widely known as pruning and was started in the early 1990s [22−24]. The typical approach is to perform cycles of training and pruning starting from a dense (or fully connected) network. The traditional procedure for any pruning method is:

- Step 1: Start with a fully connected neural network;
- Step 2: Train the network until some convergence criteria is satisfied;
- Step 3: Identify unimportant connections based on some saliency criteria;
- Step 4: Prune those unimportant connections;
- Step 5: Repeat from Step 2 until the neural network satisfies the target trade-off sparsity/accuracy.

The most used in practice saliency criteria are magnitude (connections with values close to zero) or various information-theoretic criteria. However, there is a considerable amount of work in the last years on this topic and for more details we refer to a recent review [25].

Still, all pruning methods need large fully connected layers to start from. In 2016 we introduced a new learning paradigm [26] to obtain sparse neural networks which starts from sparsely connected layers and offers computational advantages also in the training phase, while making use of the gradient-based optimization method benefits. Lately, this paradigm has started to be known as *sparse-to-sparse* training or *sparse training*. A schematic architecture of a sparse neural network and its dense counterpart is depicted in Fig. 14.1.

Within the sparse training context, while in Ref. [26] we used static sparse connectivity, in 2017 [27,28] we have introduced the idea of adaptive sparse connectivity to fit the data distribution and the Sparse Evolutionary Training (SET) algorithm which fructifies the idea of adapting the sparse connectivity to the data. SET is able to quadratically reduce

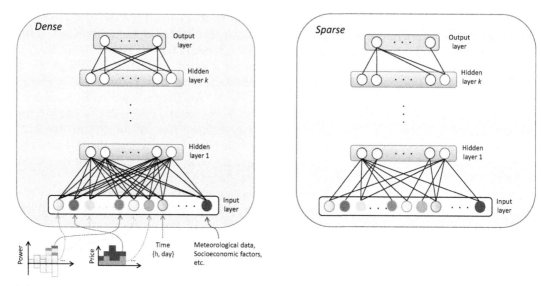

**FIGURE 14.1**   A visual representation of a fully connected neural network and its sparsified version.

the number of parameters before training. The basic idea is that instead of starting with fully connected layers, we start the training procedure with sparse connected layers. On short, the schematic procedure for SET is:

- Step 1: Start with a sparsely connected network.
- Step 2: Train the model on all data for one epoch.
- Step 3: Identify a percentage, $\zeta$, of unimportant connections using magnitude;
- Step 4: Remove those unimportant connections.
- Step 5: Add $\zeta$ random connections to fit the data distribution and to keep a constant sparsity level.
- Step 6: Repeat from Step 2 until the neural network reaches the targeted performance (e.g., accuracy).

In 2019 Peterson et al. [29] found that sparse training outperformed dense neural networks and ensemble methods on tabular data using SET. To summarize, we can distinguish between (1) static sparse training where the aim is to find a sparse subnetwork at the beginning and then train it without changing the topology; and (2) dynamic (adaptive) sparse training where the initial sparse subnetwork change its topology while training. In this chapter, we focus on the latter case, that is, dynamic (adaptive) sparse training as it is more flexible and has the capacity of outperforming dense neural networks.

## 14.4 Experiments and results: country level

In this section, we analyze the ability of all the previously mentioned algorithms to estimate the day-ahead total load in five countries, as it can be seen in Fig. 14.2. We use online data retrieved from the ENTSO-E Transparency Platform [30] with 1-hour and 15-minute

**Electricity forecast**

- France
- Germany
- Netherlands
- Portugal
- Spain

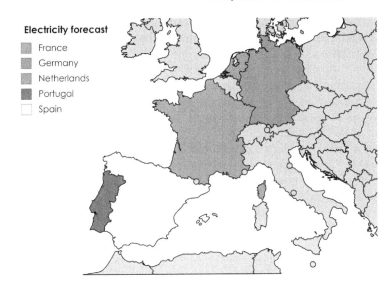

**FIGURE 14.2** The experiments show the load forecasting for five EU countries, using data with hourly resolution for France, Portugal, and Spain and with 15-min resolution in the case of the Netherlands and Germany.

resolutions. The training dataset was considered from November 2018 to October 2019, while the testing was performed successively for 30 days using all data from November 2019. Overall, no major preprocessing steps were necessary, with one exception, given by four duplicate extra values reported on October 27, 2019, between 2:15 a.m. and 3:00 a.m. Consequently, all four data have been removed. In all cases, we have considered a time window of 7 days to predict the next day.

In both scenarios discussed further, we perform a comparison between the following methods: OLS, RR, BRR, KRR, SVR, NNR, GP, DT, AB, RT, MLP, and sparse MLP trained with Sparse Evolutionary Training (SET-MLP).

## 14.4.1 Implementation details

*One output prediction methods.* With the exception of the neural network models, that is, MLP and SET-MLP, all the other methods can predict just one output. In this case, we have created a model for each necessary prediction. This means that we trained for each method 24 models for each country in the case of day-ahead forecasting with 1-hour resolution, and 96 models for each country in the case of day-ahead forecasting with 15-minute resolution. All these methods have been implemented using the Scikit-learn library [31] in Python using their default hyperparameters.

*Multiple output prediction methods.* As MLP and SET-MLP can predict multiple outputs, we have implemented just one model for each method for each country. Consequently, both have 24 output neurons for day-ahead forecasting with 1-hour resolution, and 96 output neurons for day-ahead forecasting with 15-minute resolution. The data have been normalized in [0,1] intervals. MLP has been implemented using the Scikit-learn library, while SET-MLP has been implemented based on the Refs. [28,32]

and their corresponding code for the truly sparse implementations was downloaded from Github.[1] The code has been adapted to perform regression. For the case with 15-minute resolution both MLP and SET-MLP have five layers (one input, three hidden, and one output layer) with the following numbers of neurons per layer 672−250−250−250−96; while for the case with 1-hour resolution the number of neurons per layer is 168−120−80−40−24. The hidden layers have ReLU activation function. To train the dense MLP models we used Adam, while for SET-MLP models we used backpropagation with a stochastic gradient descent, momentum of 0.9, learning rate of 0.01, weight decay of 0.0002, and a batch size of 20. For SET-MLP, Sigmoid has been used as activation function for the output layer and mean squared error as loss function, while for MLP the unspecified hyperparameters have been set to the default ones. For the 15-minute resolution scenario (Section 14.4.4), MLP had in total 317,000 connections (parameters to optimize), while SET-MLP had 29,635 connections and approximately a sparsity level of 90%. For the 1-hour resolution scenario (Section 14.4.3), MLP had 33,920 connections, while SET-MLP had 8701 connections and a sparsity level of about 75%.

### 14.4.2 Metrics used for accuracy assessment

To assess the performance of any prediction model, various metrics are used to evaluate the error between the predicted (output) and the measured (ground truth) values. The root mean square error (RMSE) is used to assess the error, as follows:

$$\text{RMSE} = \sqrt{\frac{1}{N} \sum_{i=1}^{N} (y_i - \hat{y}_i)^2}, \tag{14.1}$$

where $N$ is the amount of data samples, $y_i$ is the output data, and $\hat{y}_i$ is the predicted output data. We used the RMSE values to understand the magnitude of the deviation between the real value and the predicted one. The RMSE is then normalized to put the error into a percentage unit. The normalized root mean square error (NRMSE) is given by

$$\text{NRMSE\%} = \frac{\sqrt{\frac{1}{N} \sum_{i=1}^{N} (y_i - \hat{y}_i)^2}}{(y_{max} - y_{min})} \cdot 100 \tag{14.2}$$

More specifically, one should look to the NRMSE values when the prediction is performed on another dataset (e.g., from a different country or, more locally, at the building level) and we want to understand what is the absolute difference between the values predicted with one algorithm and another. Lastly, the Pearson correlation coefficient (PCC) is used to evaluate the similarity between $y_i$ and $\hat{y}_i$, given by the following:

$$\text{PCC} = \frac{E[(y - \mu_y)(\hat{y} - \mu_{\hat{y}})]}{\sigma_y \cdot \sigma_{\hat{y}}}, \tag{14.3}$$

[1] https://github.com/dcmocanu/sparse-evolutionary-artificial-neural-networks https://github.com/SelimaC/Tutorial-SCADS-Summer-School-2020-Scalable-Deep-Learning.

where $\mu_y$, $\mu_{\hat{y}}$ is the mean of each data set and $\sigma_y$, $\sigma_{\hat{y}}$ is the variance of each data set respectively. A value of one represents the perfect correlation and zero means no correlation; the entire output range of PCC is between $[-1,1]$.

### 14.4.3 Total load forecast with 1-hour resolution

In this set of experiments, we perform a forecast with hourly resolution for the total load of France, Portugal, and Spain. The results are summarized in Tables 14.2–14.4, respectively. For all three assessment metrics, the results are provided as an average value over 30 days, and the corresponding standard deviation. Furthermore, the last row in each table is calculated using the reference forecast values available on the ENTSO-E Transparency Platform [30]. The best performance in the case of the total load forecast of France and Spain is obtained using the BRR method, while for Portugal the Sparse Evolutionary Training (SET-MLP) algorithm provides the most accurate results. In this set of experiments, it is worth noting that when BRR is the best performer, SET-MLP is the second best performer; while when SET-MLP is the first performer, BRR is the second best performer.

### 14.4.4 Total load forecast with 15-minute resolution

Complementary, in the case of Germany and the Netherlands, we use data with 15-minute resolution. As discussed above, this is a more difficult scenario due to the data

**TABLE 14.2** Day-ahead total load forecast with 1-h resolution, averaged over 1 month, using all 12 methods in teerms of root mean square error (RMSE), normalized RMSE, and Pearson correlation coefficient.

|  | Method | RMSE | NRMSE (%) | PCC |
|---|---|---|---|---|
| France Total load forecast (MW) | OLS | 2446.24 ± 1308.93 | 6.12 ± 3.27 | 0.96 ± 0.05 |
|  | RR | 2446.24 ± 1308.93 | 6.12 ± 3.27 | 0.96 ± 0.05 |
|  | BRR | 1795.94 ± 1255.48 | 4.49 ± 3.14 | 0.97 ± 0.04 |
|  | KRR | 2459.23 ± 1351.78 | 6.15 ± 3.38 | 0.96 ± 0.05 |
|  | SVR | 10,978.18 ± 4854.93 | 27.45 ± 12.14 | 0.87 ± 0.07 |
|  | NNR | 3718.02 ± 2350.41 | 9.30 ± 5.88 | 0.94 ± 0.07 |
|  | GP | 60,315.54 ± 6676.15 | 150.83 ± 16.69 | nan ± nan |
|  | DT | 3987.89 ± 1573.50 | 9.97 ± 3.93 | 0.79 ± 0.13 |
|  | AB | 2977.35 ± 1285.04 | 7.45 ± 3.21 | 0.93 ± 0.06 |
|  | RT | 2650.57 ± 1733.39 | 6.63 ± 4.33 | 0.92 ± 0.08 |
|  | MLP | 2675.71 ± 1601.89 | 6.69 ± 4.01 | 0.95 ± 0.05 |
|  | SET-MLP | 2090.84 ± 1531.40 | 5.23 ± 3.83 | 0.96 ± 0.05 |
|  | TSO-FR [30] | 1229.58 ± 270.95 | 3.07 ± 0.68 | 0.98 ± 0.02 |

**TABLE 14.3** Day-ahead total load forecast with 1-h resolution, averaged over 1 month, using all 12 methods in terms of root mean square error (RMSE), normalized RMSE, and Pearson correlation coefficient.

| | Method | RMSE | NRMSE (%) | PCC ± |
|---|---|---|---|---|
| Portugal Total load forecast (MW) | OLS | 281.03 ± 176.80 | 6.63 ± 4.17 | 0.98 ± 0.02 |
| | RR | 281.03 ± 176.80 | 6.63 ± 4.17 | 0.98 ± 0.02 |
| | BRR | 203.29 ± 139.31 | 4.80 ± 3.29 | 0.99 ± 0.02 |
| | KRR | 278.29 ± 169.74 | 6.56 ± 4.00 | 0.98 ± 0.02 |
| | SVR | 609.05 ± 215.27 | 14.37 ± 5.08 | 0.95 ± 0.05 |
| | NNR | 218.38 ± 137.53 | 5.15 ± 3.24 | 0.98 ± 0.03 |
| | GP | 6099.11 ± 526.67 | 143.88 ± 12.42 | nan ± nan |
| | DT | 355.51 ± 162.87 | 8.39 ± 3.84 | 0.92 ± 0.09 |
| | AB | 250.36 ± 111.03 | 5.91 ± 2.62 | 0.98 ± 0.02 |
| | RT | 216.04 ± 152.23 | 5.10 ± 3.59 | 0.98 ± 0.03 |
| | MLP | 257.58 ± 143.67 | 5.88 ± 4.11 | 0.98 ± 0.03 |
| | SET-MLP | 188.53 ± 177.32 | 4.45 ± 4.18 | 0.99 ± 0.03 |
| | TSO-PT [30] | 164.07 ± 82.02 | 3.87 ± 1.93 | 0.99 ± 0.01 |

**TABLE 14.4** Day-ahead total load forecast with 1-h resolution, averaged over 1 month, using all 12 methods in terms of root mean square error (RMSE), normalized RMSE, and Pearson correlation coefficient.

| | Method | RMSE | NRMSE (%) | PCC |
|---|---|---|---|---|
| Spain Total load forecast (MW) | OLS | 842.11 ± 447.41 | 4.58 ± 2.43 | 0.99 ± 0.01 |
| | RR | 842.11 ± 447.41 | 4.58 ± 2.43 | 0.99 ± 0.01 |
| | BRR | 645.78 ± 360.40 | 3.51 ± 1.96 | 0.99 ± 0.01 |
| | KRR | 866.00 ± 471.97 | 4.71 ± 2.57 | 0.99 ± 0.01 |
| | SVR | 2782.17 ± 1295.12 | 15.14 ± 7.05 | 0.95 ± 0.05 |
| | NNR | 1126.52 ± 816.85 | 6.13 ± 4.45 | 0.98 ± 0.03 |
| | GP | 29,250.31 ± 2778.41 | 159.19 ± 15.12 | nan ± nan |
| | DT | 1414.54 ± 984.22 | 7.70 ± 5.36 | 0.94 ± 0.06 |
| | AB | 1138.50 ± 545.64 | 6.20 ± 2.97 | 0.98 ± 0.01 |
| | RT | 958.96 ± 852.61 | 5.22 ± 4.64 | 0.98 ± 0.04 |
| | MLP | 1193.23 ± 891.19 | 6.49 ± 4.85 | 0.98 ± 0.04 |
| | SET-MLP | 793.43 ± 607.41 | 4.32 ± 3.31 | 0.98 ± 0.03 |
| | TSO-ES [30] | 356.52 ± 162.52 | 1.94 ± 0.88 | 1.00 ± 0.00 |

fluctuations. A summary of these experiments can be seen in Table 14.5. SET-MLP outperformed all the other models, including here the reference forecast values available on the ENTSO-E Platform. It is worth highlighting that SET-MLPs made better predictions than the dense MLP models, while having about 10 times fewer parameters.

**TABLE 14.5** Day-ahead total load forecast using data with 15-min resolution, averaged over 1 month, corresponding to the 12 methods studied, in terms of root mean square error (RMSE), normalized RMSE, and Pearson correlation coefficient.

|  | Method | RMSE | NRMSE (%) | PCC |
|---|---|---|---|---|
| The Netherlands Total load forecast (MW) | OLS | 1019.41 ± 519.32 | 13.69 ± 6.97 | 0.92 ± 0.09 |
|  | RR | 1019.79 ± 519.83 | 13.70 ± 6.98 | 0.92 ± 0.09 |
|  | BRR | 382.23 ± 222.44 | 5.13 ± 2.99 | 0.98 ± 0.02 |
|  | KRR | 1049.43 ± 549.24 | 14.09 ± 7.38 | 0.92 ± 0.10 |
|  | SVR | 1368.74 ± 487.48 | 18.38 ± 6.55 | 0.94 ± 0.05 |
|  | NNR | 426.47 ± 193.08 | 5.73 ± 2.59 | 0.98 ± 0.01 |
|  | GP | 13,991.52 ± 934.43 | 187.91 ± 12.55 | 0.00 |
|  | DT | 795.84 ± 237.75 | 10.69 ± 3.19 | 0.92 ± 0.05 |
|  | AB | 515.82 ± 234.36 | 6.93 ± 3.15 | 0.98 ± 0.02 |
|  | RT | 462.26 ± 244.52 | 6.21 ± 3.28 | 0.98 ± 0.02 |
|  | MLP | 498.93 ± 240.01 | 6.70 ± 3.22 | 0.98 ± 0.03 |
|  | SET-MLP | 364.47 ± 173.42 | 4.89 ± 2.33 | 0.98 ± 0.02 |
|  | TSO-NL [30] | 2304.27 ± 311.63 | 30.95 ± 4.19 | 0.97 ± 0.03 |
| Germany Total load forecast (MW) | OLS | 3103.66 ± 1812.25 | 8.52 ± 4.98 | 0.98 ± 0.02 |
|  | RR | 3102.26 ± 1812.64 | 8.52 ± 4.98 | 0.98 ± 0.02 |
|  | BRR | 2065.33 ± 1163.67 | 5.67 ± 3.20 | 0.99 ± 0.01 |
|  | KRR | 3038.35 ± 1669.97 | 8.35 ± 4.59 | 0.98 ± 0.02 |
|  | SVR | 6339.94 ± 2089.72 | 17.41 ± 5.74 | 0.94 ± 0.08 |
|  | NNR | 2150.20 ± 1806.20 | 5.91 ± 4.96 | 0.98 ± 0.02 |
|  | GP | 59,550.84 ± 6052.62 | 163.57 ± 16.62 | 0 |
|  | DT | 3394.26 ± 2031.05 | 9.32 ± 5.58 | 0.93 ± 0.07 |
|  | AB | 2426.06 ± 1654.39 | 6.66 ± 4.54 | 0.98 ± 0.01 |
|  | RT | 2039.49 ± 1457.70 | 5.60 ± 4.00 | 0.99 ± 0.02 |
|  | MLP | 2650.41 ± 1388.62 | 7.28 ± 3.81 | 0.97 ± 0.02 |
|  | SET-MLP | 1492.10 ± 1126.51 | 4.10 ± 3.09 | 0.99 ± 0.01 |
|  | TSO-GE [30] | 1962.46 ± 977.24 | 5.39 ± 2.68 | 0.99 ± 0.00 |

## 14.5 Forecasting: a glimpse into the future

Nowadays, deep learning methods for power system data analysis are more and more popular, given their ability to cope with highly nonlinear and highly dimensional time series. Furthermore, the power system transition toward the big data era encourages the use of deep learning, as the most advanced solutions for large-scale applications. In this chapter, we have provided an example using statistical learning and deep learning methods in supervised learning settings. By doing so, we assume that we have historical data available from the quantity which we would like to forecast. However, sometimes it is often impossible to have historical data [33]. Also, a large number of open questions are related to the performance of the deep learning models based on the architecture choice and hyperparameters optimization, robustness, adaptivity, scalability, etc. In recent years, several solutions have been developed to overcome these limitations, and some preliminary answers could be considered as a starting point for further research:

1. *How we can overcome data limitation problems?* There are plenty of concepts, more and more popular in data science and artificial intelligence which have never been, to the best of our knowledge, investigated in the local energy market context. For example, few-shot learning methods related with the problem of learning from few data [34], advanced augmentation techniques based on the ability of artificially increasing the number of data points using, for instance, generative replay [35], multitask learning, adaptive learning, collaborative learning, and so on.

   Another approach is to use transfer learning and unsupervised learning. In Ref. [33], we proposed a method able to perform a cross-building transfer that can target the new behavior of existing buildings (due to changes in their structure, installations, and energy price), as well as completely new types of buildings.

2. *How can we optimally choose the parameters of a deep learning model?* There are a few attempts to address this problem. Bilevel random optimization techniques or evolutionary techniques were the first attempts aiming to start the neural network training with a better configuration. Later on, both model-based and model-free methods, as well as gradient-based methods, were develop for hyperparameters optimization (e.g., self-tuning networks [36]). These methods have the potential to increase prediction accuracy and should be considered as an integral part of a trading mechanism.

3. *How can we make the forecasting methods scalable?* Traditional deep learning approaches make use of cloud computing facilities and do not scale well to autonomous agents with low computational resources. Even in the cloud they suffer from computational and memory limitations and cannot be used to model properly large physical worlds for agents which assume networks with billion of neurons. These issues were addressed in the last few years by the emerging topics of scalable and efficient deep learning and sparse training methods (e.g., Refs. [26,28,37−41]). All these methods can play a role in the local energy market.

4. *Further possible improvements in the model accuracy.* Artificial intelligence-based approaches bring important new options, enabling efficient individual and aggregated energy management. Such approaches can provide different players aiming to

accomplish individual and common goals in the frame of a market-driven environment with advanced decision support and automated solutions [42]. One possible improvement is to further investigate multitask learning, as a bridge between forecasting and local market strategies.

5. *What if two or three consecutive steps in the electricity market design are performed simultaneously by one algorithm?* For example, energy flexibility detection (classification) and prediction tasks [43,44] or building energy forecasting and scheduling tasks [45].

Acting at the local level, metering and submetering data could be used to strengthen the market role. These data are coming with fine granularity and are expected to bring many benefits, as well as challenges. The first attempt one can make is to adapt the existing forecasting methods with the aim to obtain a more adaptive and scalable solution. Toward this goal, all three ML paradigms can be explored:

1. Supervised learning (e.g., for load regression task)
2. Unsupervised learning (e.g., for customers clustering)
3. Reinforcement learning (e.g., for control, adaptive and active learning)

All three paradigms can be successfully used to cope with the range of data sets available. The day-ahead load forecast problem at the country level, as typical in an energy market context, may be hard to be adapted through the ongoing transition toward local energy markets. To cope with the high level of uncertainty at the building and aggregated level, recently more advanced AI methods have been proposed. We can now distinguish various possibilities in local energy markets:

- Adapt the day-ahead forecasting methods to local energy markets, as a supervised learning solution. These should be the preferred option when historical data are available [42].
- Use adaptive AI methods (e.g., deep reinforcement learning [45]), when a direct integration of your forecast model is embedded in a multiagent local energy market.
- Perform multitask learning (e.g., costumer clustering and load forecasting [43]).

All of these highlight some open questions as well as new possible applications, which are expected to bring benefits for the better planning and operation of the smart grid, by helping customers to adopt energy conserving behaviors and by easing their transition from a passive to an active role.

# References

[1] E. Mocanu, Machine Learning Applied to Smart Grids (Ph.D. thesis), Technische Universiteit Eindhoven, 2017.
[2] A. Costa, M.M. Keane, J.I. Torrens, E. Corry, Building operation and energy performance: monitoring, analysis and optimisation toolkit, Appl. Energy 101 (2013) 310–316.
[3] E. Mocanu, P.H. Nguyen, M. Gibescu, W.L. Kling. Comparison of machine learning methods for estimating energy consumption in buildings, in: Proceedings of the 13th International Conference on Probabilistic Methods Applied to Power Systems, Durham, 2014.
[4] E. Mocanu, P.H. Nguyen, M. Gibescu, W.L. Kling, Deep learning for estimating building energy consumption, Sustain. Energy, Grids Netw. 6 (2016) 91–99.

[5] H. Madsen, Time Series Analysis, 2008.

[6] E. Mocanu, E. Mahler Larsen, P.H. Nguyen, P. Pinson, M. Gibescu, Demand forecasting at low aggregation levels using factored conditional restricted boltzmann machine, in: IEEE Power Systems Computation Conference (PSCC), 2016.

[7] J. Wang, D. Chi, J. Wu, Hy Lu, Chaotic time series method combined with particle swarm optimization and trend adjustment for electricity demand forecasting, Expert. Syst. Appl. 38 (7) (2011) 8419–8429.

[8] L. Tang, C. Wang, S. Wang, Energy time series data analysis based on a novel integrated data characteristic testing approach, Procedia Computer Sci. 17 (2013) 759–769.

[9] H.R. Khosravani, M. Del Mar Castilla, M. Berenguel, A.E. Ruano, P.M. Ferreira, A comparison of energy consumption prediction models based on neural networks of a bioclimatic building, Energies 9 (57) (2016).

[10] B.-J. Chen, M.-W. Chang, C.-J. lin, Load forecasting using support vector machines: a study on eunite competition 2001, IEEE Trans. Power Syst. 19 (4) (2004) 1821–1830.

[11] G. Dudek, Short-term load forecasting using random forests, in: Intelligent Systems'2014, Springer International Publishing, Cham, 2015, pp. 821–828.

[12] N.G. Paterakis, E. Mocanu, M. Gibescu, B. Stappers, W. van Alst, Deep learning versus traditional machine learning methods for aggregated energy demand prediction, in: 2017 IEEE PES Innovative Smart Grid Technologies Conference Europe (ISGT-Europe), September 2017, pp. 1–6.

[13] S.B. Taieb, R.J. Hyndman, A gradient boosting approach to the kaggle load forecasting competition, Int. J. Forecast. 30 (2) (2014) 382–394.

[14] R. Zhang, Y. Xu, Z.Y. Dong, W. Kong, K.P. Wong. A composite k-nearest neighbor model for day-ahead load forecasting with limited temperature forecasts, in: 2016 IEEE Power and Energy Society General Meeting (PESGM), 2016, pp. 1–5.

[15] A.G. Bakirtzis, V. Petridis, S.J. Kiartzis, M.C. Alexiadis, A.H. Maissis, A neural network short term load forecasting model for the greek power system, IEEE Trans. Power Syst. 11 (2) (1996) 858–863.

[16] D.C. Park, M.A. El-Sharkawi, R.J. Marks, L.E. Atlas, M.J. Damborg, Electric load forecasting using an artificial neural network, IEEE Trans. Power Syst. 6 (2) (1991) 442–449.

[17] S. Ryu, J. Noh, H. Kim, Deep neural network based demand side short term load forecasting, in: 2016 IEEE International Conference on Smart Grid Communications (SmartGridComm), November 2016, pp. 308–313.

[18] W. Kong, Z.Y. Dong, Y. Jia, D.J. Hill, Y. Xu, Y. Zhang, Short-term residential load forecasting based on lstm recurrent neural network, IEEE Trans. Smart Grid 10 (1) (2017) 841–851.

[19] D.L. Marino, K. Amarasinghe, M. Manic, Building energy load forecasting using deep neural networks, in: Proceedings of the 42nd Annual Conference of the IEEE Industrial Electronics Society (IECON), 2016.

[20] M. Manic, K. Amarasinghe, J.J. Rodriguez-Andina, C. Rieger, Intelligent buildings of the future: Cyberaware, deep learning powered, and human interacting, IEEE Ind. Electron. Mag. 10 (4) (2016) 32–49.

[21] H. Zheng, J. Yuan, L. Chen, Short-term load forecasting using emd-lstm neural networks with a xgboost algorithm for feature importance evaluation, Energies 10 (8) (2017).

[22] S. Han, J. Pool, J. Tran, W. Dally, Learning both weights and connections for efficient neural network, in: C. Cortes, N.D. Lawrence, D.D. Lee, M. Sugiyama, R. Garnett (Eds.), Advances in Neural Information Processing Systems 28, Curran Associates, Inc., 2015, pp. 1135–1143.

[23] Y. LeCun, J.S. Denker, S.A. Solla, Optimal brain damage, in: D.S. Touretzky (Ed.), Advances in Neural Information Processing Systems 2, Morgan-Kaufmann, 1990, pp. 598–605.

[24] M.C. Mozer, P. Smolensky, Using relevance to reduce network size automatically, Connect. Sci. 1 (1) (1989) 3–16.

[25] T. Gale, E. Elsen, S. Hooker, The State of Sparsity in Deep Neural Networks, CoRR, abs/1902.09574, 2019.

[26] D.C. Mocanu, E. Mocanu, P.H. Nguyen, M. Gibescu, A. Liotta, A topological insight into restricted boltzmann machines, Mach. Learn. 104 (2) (2016) 243–270.

[27] D.C. Mocanu, Network Computations in Artificial Intelligence (Ph.D. thesis), Technische Universiteit Eindhoven, 2017.

[28] D.C. Mocanu, E. Mocanu, P. Stone, P.H. Nguyen, M. Gibescu, A. Liotta, Scalable training of artificial neural networks with adaptive sparse connectivity inspired by network science, Nat. Commun. 9 (2383) (2018).

[29] D.D. Bourgin, J.C. Peterson, D. Reichman, T.L. Griffiths, S.J. Russell, Cognitive model priors for predicting human decisions, in: International Conference on Machine Learning (ICML), 2019, pp. 5133–5141.

[30] ENTSO-E Transparency, Electricity Market Transparency, 2020.

[31] F. Pedregosa, G. Varoquaux, A. Gramfort, V. Michel, B. Thirion, O. Grisel, et al., Scikit-learn: machine learning in Python, J. Mach. Learn. Res. 12 (2011) 2825–2830.

[32] S. Liu, D.C. Mocanu, A. Reddy Ramapuram Matavalam, Y. Pei, M. Pechenizkiy, Sparse evolutionary deep learning with over one million artificial neurons on commodity hardware, Neural Comput. Appl. (2020).

[33] E. Mocanu, P.H. Nguyen, W.L. Kling, M. Gibescu, Unsupervised energy prediction in a smart grid context using reinforcement cross-building transfer learning, Energy Build. 116 (2016) 646–655.

[34] D.C. Mocanu, E. Mocanu, One-shot learning using mixture of variational autoencoders: a generalization learning approach, in: International Conference on Autonomous Agents and Multi-Agent Systems (AAMAS), 2018.

[35] D.C. Mocanu, M.T. Vega, E. Eaton, P. Stone, A. Liotta. Online Contrastive Divergence With Generative Replay: Experience Replay Without Storing Data, CoRR, abs/1610.05555, 2016.

[36] M. Mackay, P. Vicol, J. Lorraine, D. Duvenaud, R. Grosse, Self-tuning networks: Bilevel optimization of hyperparameters using structured best-response functions, in: International Conference on Learning Representations, 2019.

[37] T. Dettmers, L. Zettlemoyer, Sparse networks from scratch: faster training without losing performance, arXiv preprint arXiv:1907.04840 (2019).

[38] U. Evci, T. Gale, J. Menick, P.S. Castro, E. Elsen, Rigging the lottery: making all tickets winners, in: Proceedings of Machine Learning and Systems 2020, 2020, pp. 471–481.

[39] S. Liu, T. Van der Lee, A. Yaman, Z. Atashgahi, D. Ferraro, G. Sokar, et al., Topological insights into sparse neural networks, in: European Conference on Machine Learning and Principles and Practice of Knowledge Discovery in Databases (ECMLPKDD), 2020.

[40] H. Mostafa, X. Wang, Parameter efficient training of deep convolutional neural networks by dynamic sparse reparameterization, arXiv preprint arXiv:1902.05967 (2019).

[41] H. Zhu, Y. Jin, Multi-objective evolutionary federated learning, in: IEEE Transactions on Neural Networks and Learning Systems, 2019.

[42] L.A. Hurtado, E. Mocanu, P.H. Nguyen, M. Gibescu, R.I.G. Kamphuis, Enabling cooperative behavior for building demand response based on extended joint action learning, IEEE Trans. Ind. Inform. 14 (1) (2018) 127–136.

[43] D.C. Mocanu, E. Mocanu, P.H. Nguyen, M. Gibescu, A. Liotta, Big IoT data mining for real-time energy disaggregation in buildings, in: 2016 IEEE International Conference on Systems, Man, and Cybernetics (SMC), 2016, pp. 3765–3769.

[44] D.C. Mocanu, H.B. Ammar, D. Lowet, K. Driessens, A. Liotta, G. Weiss, et al., Factored four way conditional restricted boltzmann machines for activity recognition, Pattern Recognit. Lett. 66 (2015) 100–108.

[45] E. Mocanu, D.C. Mocanu, P.H. Nguyen, A. Liotta, M.E. Webber, M. Gibescu, et al., On-line building energy optimization using deep reinforcement learning, IEEE Trans. Smart Grid 10 (4) (2019) 3698–3708.

# Mathematical models and optimization techniques to support local electricity markets

*John Fredy Franco*[1], *Leonardo H. Macedo*[2],
*Nataly Bañol Arias*[3], *Alejandra Tabares*[2], *Rubén Romero*[2] and
*João Soares*[4]

[1]School of Energy Engineering, São Paulo State University, Rosana, Brazil [2]Department of
Electrical Engineering, São Paulo State University, Ilha Solteira, Brazil [3]Department of Energy
Systems, School of Electrical and Computer Engineering, University of Campinas, Campinas,
Brazil [4]GECAD Research Center, School of Engineering of Polytechnic of Porto, Porto, Portugal

## 15.1 Introduction

The management of distributed energy resources (DERs) requires the development of optimization tools so they can be properly integrated into distribution networks under future local electricity markets (LEMs) [1,2]. Extensive research has been done to overcome the challenges of energy management in a smart grid context [3]. This chapter presents mathematical formulations that can be used to represent optimization problems in LEMs including DERs. Furthermore, some metaheuristic and decomposition methods are introduced as efficient alternatives to solve complex optimization problems in the context of distribution networks with DERs.

## 15.2 Mathematical models for power flow and distributed energy resources

This section presents a mathematical formulation for the operation of distribution networks, as well as to model DERs (renewable distributed generation, energy storage devices, and electric vehicles) and voltage control devices (capacitor banks and voltage regulators).

### 15.2.1 Power flow representation in unbalanced distribution networks

The operation of unbalanced distribution networks can be represented using current or power-based formulations [4]. Here, the latter is adopted to illustrate the mathematical expressions that can model the AC steady-state operation in a distribution network based on the branch flow equations [5], in which mutual effects among phases and unbalanced loads are included.

Considering Fig. 15.1, the active/reactive power balances for a given bus $i$, phase $\phi$, are represented by (15.1) and (15.2), which are written in terms of the active/reactive power flows ($P_{ki}^\phi/Q_{ki}^\phi$ arriving at bus $i$ through line $ki$), the complex power losses ($S_{ki}^{L,\phi}$), the active/reactive powers demanded ($P_{i,\phi}^D$ and $Q_{i,\phi}^D$), and the power supplied ($P_{i,\phi}^S$ and $Q_{i,\phi}^S$). The voltage drop of line $ij$ in phase $\phi$ is represented using the square of the voltage at each terminal (variable $V_{i,\phi}^2$ for bus $i$) as shown in (15.3). The complex power losses are defined by (15.4).

The parameter $\hat{Z}_{ki}^{\phi,\psi}$ is an equivalent impedance that takes into account the phase angle difference between phases; it is calculated by multiplying the impedance of line $ki$ between phases $\phi$ and $\psi$, $Z_{ki}^{\phi,\psi}$, by the factor $1\angle\left(\theta_\psi - \theta_\phi\right)$ [4]; $\hat{R}_{ki}^{\phi,\psi}$ and $\hat{X}_{ki}^{\phi,\psi}$ are the real and imaginary parts of the impedance $\hat{Z}_{ki}^{\phi,\psi}$.

$$\sum_{ki}\left[P_{ki}^\phi + \mathrm{Re}\left\{S_{ki}^{L,\phi}\right\}\right] - \sum_{ij}P_{ij}^\phi + P_{i,\phi}^G = \sum_i P_{i,\phi}^D \qquad \forall i,\phi \tag{15.1}$$

$$\sum_{ki}\left[Q_{ki}^\phi + \mathrm{Im}\left\{S_{ki}^{L,\phi}\right\}\right] - \sum_{ij}Q_{ij}^\phi + Q_{i,\phi}^G = \sum_i Q_{i,\phi}^D \qquad \forall i,\phi \tag{15.2}$$

$$V_{i,\phi}^2 - V_{j,\phi}^2 = 2\sum_\psi\left(\mathrm{Re}\left\{\hat{Z}_{ij}^{\phi,\psi}\right\}P_{ij}^\psi + \mathrm{Im}\left\{\hat{Z}_{ij}^{\phi,\psi}\right\}Q_{ij}^\psi\right) + \sum_\psi\left|\hat{Z}_{ij}^{\phi,\psi}\right|^2 \frac{\tilde{P}_{ki}\,\psi P_{ij}^\psi + \tilde{Q}_{ki}\,\psi Q_{ij}^\psi}{\tilde{V}_i\,\phi\tilde{V}_i\,\psi} \qquad \forall ij,\phi \tag{15.3}$$

$$S_{ki}^{L,\phi} = \sum_\psi\left(\frac{\hat{Z}_{ki}^{\phi,\psi}S_{ki}^{\psi*}}{V_{j,\phi}V_{j,\psi}}\right)S_{ki}^\phi \qquad \forall ki,\phi \tag{15.4}$$

Furthermore, the complex power losses of line $ki$ and phase $\phi$ can be separated into active and reactive power losses, $P_{ki,\phi}^L = \mathrm{Re}\{S_{ki}^{L,\phi}\}$ and $Q_{ki,\phi}^L = \mathrm{Im}\{S_{ki}^{L,\phi}\}$, shown in (15.5) and

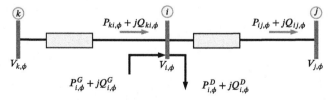

**FIGURE 15.1** Active and reactive power balances in a distribution network. *Source: Adapted from J.F. Franco, L.F. Ochoa, R. Romero, AC OPF for smart distribution networks: an efficient and robust quadratic approach, IEEE Trans. Smart Grid 9 (2018) 4613–4623. https://doi.org/10.1109/TSG.2017.2665559 [6].*

(15.6), and the square of the magnitude of the current through phase $\phi$ of line $ki$ ($I_{ki,\phi}^2$) is defined by (15.7).

$$P_{ki,\phi}^L = \sum_{\psi} \left[ \hat{R}_{ki}^{\phi,\psi} \frac{(P_{ki,\phi}P_{ki,\psi} + Q_{ki,\phi}Q_{ki,\psi})}{V_{k,\phi}V_{i,\psi}} + \hat{X}_{ki}^{\phi,\psi} \frac{(-Q_{ki,\phi}P_{ki,\psi} + P_{ki,\phi}Q_{ki,\psi})}{V_{k,\phi}V_{i,\psi}} \right] \quad \forall ki, \phi \quad (15.5)$$

$$Q_{ki,\phi}^L = \sum_{\psi} \left[ \hat{X}_{ki}^{\phi,\psi} \frac{(P_{ki,\phi}P_{ki,\psi} + Q_{ki,\phi}Q_{ki,\psi})}{V_{k,\phi}V_{i,\psi}} + \hat{R}_{ki}^{\phi,\psi} \frac{(Q_{ki,\phi}P_{ki,\psi} - P_{ki,\phi}Q_{ki,\psi})}{V_{k,\phi}V_{i,\psi}} \right] \quad \forall ki, \phi \quad (15.6)$$

$$I_{ki,\phi}^2 = \frac{P_{ki,\phi}^2 + Q_{ki,\phi}^2}{V_{i,\phi}^2} \quad \forall ki, \phi \quad (15.7)$$

The set of Eqs. (15.1)–(15.7) is a basic formulation to represent the state of unbalanced distribution networks, which can be extended to incorporate different DERs as discussed below. Moreover, regarding the nonlinearities present on those equations, linearization techniques can be applied to transform the formulation into convex or linear models that make possible the use of off-the-shelf solvers to find optimal solutions of optimization problems.

### 15.2.2 Representation of distributed energy resources

DERs such as renewable distributed generation, energy storage devices, and electric vehicles can both affect and benefit the distribution network. Thus, they can be included within optimization processes to enhance LEMs. Mathematical formulations representing those resources are presented below.

#### 15.2.2.1 Electric vehicles

Uncontrolled charging of electric vehicles can lead to the overloading of distribution network assets as well as low-voltage issues and an increase in power losses. Therefore their impact on the network should be represented, as well as the possibility of implementing a coordinated charging control. Moreover, the ability of injecting power into the network (vehicle-to-grid technology) could be taken into account in order to enable the provision of services, such as frequency support and reserve power [7].

The power demanded/injected ($P_e^{EV}$) by the electric vehicle $e$ depends on the states of the binary variables $y_e$ and $z_e$ and the maximum charging/discharging rates ($\overline{P}_e^{EV+}/\overline{P}_e^{EV-}$) as discussed in Ref. [4] and defined in (15.8).

$$P_e^{EV} = \overline{P}_e^{EV+} y_e - \overline{P}_e^{EV-} z_e \quad \forall e \quad (15.8)$$

$$y_e + z_e \leq 1 \quad \forall e \quad (15.9)$$

$$\min\left\{E_e^{EVi}, DoD \cdot \overline{E}_e^{EV}\right\} \leq E_e^{EVi} + \Delta t \left(\overline{P}_e^{EV+} y_e \eta_e^{EV+} - \overline{P}_e^{EV-} z_e/\eta_e^{EV-}\right) \leq \overline{E}_e^{EV} \quad \forall e \quad (15.10)$$

Only one state for the battery is permitted (idle, charging, or discharging), which is guaranteed by (15.9). Moreover, the energy state of the battery during a time interval $\Delta t$ should satisfy its lower and upper limits according to the energy capacity of the battery ($\overline{E}_e^{EV}$), the maximum depth of discharge ($DoD$), the charging/discharging state, the initial state of charge ($E_e^{EVi}$), and the charging/discharging efficiencies ($\eta_e^{EV+}/\eta_e^{EV-}$), as given by (15.10).

### 15.2.2.2 Energy storage devices

A stationary energy storage device $s$ can be represented through expressions and variables like those related to electric vehicles, as defined by (15.11)–(15.13).

$$P_s^{SD} = \overline{P}_s^{SD+} y_s - \overline{P}_s^{SD-} z_s \qquad \forall s \tag{15.11}$$

$$y_s + z_s \leq 1 \qquad \forall s \tag{15.12}$$

$$\min\left\{E_s^{SDi}, DoD \cdot \overline{E}_s^{SD}\right\} \leq E_s^{SDi} + \Delta t\left(\overline{P}_s^{SD+} y_s \eta_s^{SD+} - \overline{P}_s^{SD-} z_s / \eta_s^{SD-}\right) \leq \overline{E}_s^{SD} \qquad \forall s \tag{15.13}$$

The influence of both electric vehicles and energy storage devices is represented in the operation state of the network by their corresponding active powers; this can be done by adding $P_e^{EV}$ and/or $P_s^{SD}$ to the right-hand side of (15.1) if the electric vehicle $e$ and/or energy storage device $s$ are connected to the bus $i$.

### 15.2.2.3 Renewable distributed generation

The integration of renewable distributed generation brings benefits to society and has the potential to improve the operation of distribution networks [8,9]. The power injection of renewable distributed generation can be represented considering the primary resource (e.g., wind or solar irradiation) and the capacity of the corresponding coupling elements (e.g., doubly-fed induction generators or inverters) [7].

The active power that can be generated ($P_i^{DG}$) is limited by the availability of the primary resource ($P_i^{prim}$), a nonlinear function $f(P_i^{prim}, \overline{P}_i^{DG})$ representing the energy conversion, that is, wind or solar irradiation to electrical power, and the capacity of the distributed generator unit ($\overline{P}_i^{DG}$), according to (15.14). The combination of active and reactive powers injected by the distributed generator connected at bus $i$ ($P_i^{DG}/Q_i^{DG}$) should not overpass its apparent power limit ($\overline{S}_i^{DG}$) defined by (15.15). Moreover, a limit for the operating power factor ($pf_i^{DG}$) is also included in the model. Additional constraints limiting the active and reactive powers according to the capability curves of doubly-fed induction generators can also be included [4]. The active/reactive powers of distributed generation units should be added to the left-hand side of (15.1) and (15.2).

$$P_i^{DG} \leq f\left(P_i^{prim}, \overline{P}_i^{DG}\right) \qquad \forall i \tag{15.14}$$

$$\left(P_i^{DG}\right)^2 + \left(Q_i^{DG}\right)^2 \leq \left(\overline{S}_i^{DG}\right)^2 \qquad \forall i \tag{15.15}$$

$$\left|Q_i^{DG}\right| \leq P_i^{DG}\tan\left(\cos^{-1}\left(pf_i^{DG}\right)\right) \qquad \forall i \tag{15.16}$$

## 15.2.3 Representation of voltage control devices

### 15.2.3.1 Capacitor banks

Capacitor banks inject reactive power ($Q_{i,\phi}^C$) and could be managed to correct the power factor and mitigate voltage variations [10]. They are formed by modules (the number of modules is $\overline{n}_i^C$ for bus $i$) with a power $Q_i^{mod}$, which could be switched to change the injected reactive power according to the integer state variable $n_i^C$, limited by (15.17). The reactive power injected by a capacitor bank, defined by (15.18), should be added to the left-hand side of (15.2).

$$0 \leq n_i^C \leq \overline{n}_i^C \qquad \forall i \tag{15.17}$$

$$Q_{i,\phi}^C = Q_i^{mod} \cdot n_i^C \qquad \forall i, \phi \tag{15.18}$$

### 15.2.3.2 Voltage regulators

The control provided by a voltage regulator at the end of line $ij$ can be represented through the nonregulated voltage ($\tilde{V}_{i,\phi}$), the controlled voltage ($V_{i,\phi}$), the regulation percentage of the device ($R\%$), and the tap position ($nt_{ij}$), as shown in (15.19) [10]. The latter is limited by the number of tap positions ($\overline{nt}_{ij}$) according to (15.20).

$$V_{i,\phi} = \left(1 + R\% \cdot \frac{nt_{ij}}{\overline{nt}_{ij}}\right) \cdot \tilde{V}_{i,\phi} \qquad \forall ij \tag{15.19}$$

$$-\overline{nt}_{ij} \leq nt_{ij} \leq \overline{nt}_{ij} \qquad \forall ij \tag{15.20}$$

## 15.3 Mathematical programming models: mixed-integer linear, stochastic, and robust optimization

The purpose of optimization is to provide relevant information for decision-making in order to achieve one or several objectives within a process, which is restricted by the availability of resources and constraints of physical, technical, and economical types. All optimization problems have three elements in common:

- Decision variables: numerical values representing the choices to be optimized.
- Objective functions: mathematical expressions used to assess the value of decision variables. Usually, the objective functions are maximized or minimized, the choice depends on the nature of the optimization problem.
- Restrictions: establish relationships between available resources and decision variables, limiting in a physical, operational, or economic way the decision-making process.

Grouping all the three previous elements, optimization problems can be generically written, as presented in (15.21) and (15.22).

$$\text{minimize}_{x_1,\ldots,x_n} \qquad f(x_1, \ldots, x_n) \tag{15.21}$$

$$\text{subject to}: \quad (x_1, \ldots, x_n) \in \beta \tag{15.22}$$

in which $x_1, \ldots, x_n$ represent the $n$ decision variables of the problem and the objective function is $f(x_1, \ldots, x_n)$. The constraints are represented by (15.22), while $\beta$ denotes the feasible region. A decision variable vector $x_1, \ldots, x_n$ in $\beta$, that is, $(x_1, \ldots, x_n) \in \beta$, is called a feasible solution. One or several feasible solutions optimizing the objective function are known as optimal solutions.

## 15.3.1 Mixed-integer linear programming

If the objective function and constraints from an optimization problem are represented by a linear function and linear constraints, the formulation is known as a linear programming problem. This kind of problem is one of the most studied and preferred by researchers, given its convergence characteristics and well-defined properties of feasibility and optimality. Nonetheless, in the mathematical representation of the decision-making process, it is common that some decision variables take binary or integer values (see Eqs. (15.8)–(15.13), (15.17)–(15.20)). This particular linear optimization problem is called a mixed-integer linear programming (MILP) problem, of which three features can be highlighted:

- The set of decision variables is divided into two subsets. One for variables that must take integer values and another for those that may take real values, that is, $\mathbf{x} = \{x_i \in \mathbb{Z}\} \cup \{x_i \in \mathbb{R}\}$ where $\mathbf{x} = x_1, \ldots, x_n$.
- The objective function is a linear combination of those two subsets of decision variables, that is, $f(x_1, \ldots, x_n) = \mathbf{c}^T \mathbf{x}$ where $c_i \in \mathbb{R}$.
- Constraints are linear combinations of the decision variables. The constraints may be equal-to, greater-than-or-equal-to, or less-than-or-equal-to.

Considering the above features, a MILP problem can be written in a generic form as shown in (15.23)–(15.26).

$$\text{minimize}_x \quad \mathbf{c}^T \mathbf{x} \tag{15.23}$$

$$\text{subject to}: \quad \mathbf{A}\mathbf{x} = \mathbf{b} \tag{15.24}$$

$$x_i \in \mathbb{Z} \quad \forall i = 1, \ldots, n - r \tag{15.25}$$

$$x_i \in \mathbb{R} \quad \forall i = 1, \ldots, r \tag{15.26}$$

in which (15.24) represents any type of constraint written in the standard form. The matrix $\mathbf{A}$ and the vector $\mathbf{b}$ on the right-hand side of (15.24) include only real numbers. Constraints (15.25) and (15.26) represent the mixed nature of the decision variables, where some of them should take integer values and the others can be continuous. Analogously to linear programming, MILP presents necessary and sufficient theoretical conditions that guarantee whether a given feasible solution is optimal or not.

In recent years, MILP commercial solvers such as CPLEX [11] and its competitors (XPRESS, GUROBI, etc.) have become extremely efficient as a result of the advancement of solving techniques based on branch-and-bound algorithms. Consequently, the possibility of using those solvers has endorsed the development of mathematical models of optimization problems that require MILP representations. Currently, it is more common using linearization techniques to recast nonlinear restrictions such as those presented in (15.1)–(15.7), (15.10), (15.15), and (15.21) to be used later within MILP formulations.

## 15.3.2 Stochastic programming

Numerous optimization problems define input data in a deterministic way based on expected values to be solved by exact techniques. Although most of the time input data is uncertain and it can be described by using probability functions. This behavior is common in electricity market problems in which a lack of perfect information is present in electricity prices, energy needed to supply customers' loads, and future electricity demands [12−14]. Frequently, stochastic programming addresses these kinds of optimization problems, in which one or several input data presents a random behavior, by creating tools for decision-making when perfect input information is not available. Important characteristics of stochastic models are:

- Random variables: represent the random behavior of the input data. Usually, they are represented by a finite set of realizations called scenarios [15].
- Stochastic processes: related to the dependence of the value of random variables and time [16].
- Scenarios: represent discrete values of the realizations of uncertainty, used to circumvent the intrinsic difficulty of working with probability distributions that characterize stochastic processes.

### 15.3.2.1 Recourse models

A logical way to see the stochastic programming problems is to require a decision to be made here-and-now, while minimizing the expected future costs of these decisions. This paradigm is called *recourse model*, where the aim is not to make an arbitrary correction (recourse) for the here-and-now decision but to make the best correction considering the possible futures scenarios. Fig. 15.2 shows this decision process based on scenarios for the two-stage and multistage stochastic programming approach.

For two stages, the mathematical model can be written as shown in (15.27)−(15.30).

$$\text{minimize}_x \quad \mathbf{c}^T\mathbf{x} + \varepsilon\{Q(\omega)\} \tag{15.27}$$

$$\text{subject to}: \quad \mathbf{Ax} = \mathbf{b} \tag{15.28}$$

$$\mathbf{x} \in X \tag{15.29}$$

$$Q(\omega) = \left\{\text{minimize}_{\mathbf{y}(\omega)} \ \mathbf{q}(\omega)^T\mathbf{y}(\omega) \text{ subject to}: \ \mathbf{T}(\omega)\mathbf{x} + \mathbf{W}(\omega)\mathbf{y}(\omega) = \mathbf{h}(\omega); \mathbf{y}(\omega) \in Y\right\} \quad \forall \omega \in \Omega \tag{15.30}$$

where $\mathbf{x}$ and $\mathbf{y}(\omega)$ are the first- and second-stage decision variable vectors, respectively, and $\mathbf{q}(\omega)$, $\mathbf{h}(\omega)$, and $\mathbf{T}(\omega)$ are known vectors and matrices of appropriate sizes.

In the specialized literature, one of the most common ways to extend the recourse models is by including more stochastic stages, as shown at the bottom of Fig. 15.2. This optimization problem is known as a multistage stochastic problem, in which each stage represents the realization of uncertain data. Under this recourse model, the decision-maker takes one decision now, waits for some uncertainty to be realized, and then makes another decision based on the new data. This procedure is repeated until the stochastic stages are exhausted in order to minimize the expected future costs of all decisions taken.

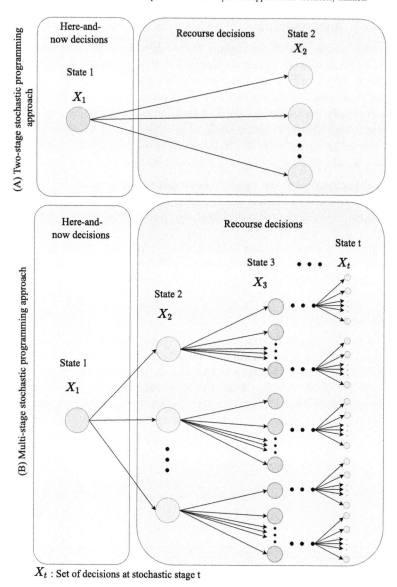

FIGURE 15.2   Representation of stochastic problems with recourse. *Source: Adapted from A.J. Conejo, M. Carrion, J.M. Morales, Decision Making Under Uncertainty in Electricity Markets, Springer, 2010 [17].*

### 15.3.2.2 Probabilistically constrained models

Some optimization problems under uncertainty do not have the ability to take corrective actions (recourse) after the uncertainty has been realized or when the cost and benefits of second-stage decisions are difficult to assess. In those cases, it is preferable to seek a set of first-stage decisions, which ensures to hold a set of constraints expressed in terms of probabilistic statements with a certain level of probability. Again, assume that $\mathbf{x}$ is a vector of decisions. These probabilistic, or chance constraints take the form shown in (15.31).

$$P\left\{ \mathbf{A}^i(\omega)\mathbf{x} \leq \mathbf{h}^i(\omega) \right\} \geq \alpha^i \qquad (15.31)$$

where $0 < \alpha^i < 1$ and $i = 1, \ldots, m$ is an index of the set of restrictions that must be hold.

In chance-constrained programming [18], the objective is often an expected value of the objective function, or its variance, or the probability of some occurrence (such as the satisfaction of restrictions).

### 15.3.3 Robust programming

Robust programming is an alternative to address optimization problems under uncertainty when input data cannot be described through probability distributions or when information about its behavior is insufficient. This approach can be considered as an alternative to stochastic programming aiming to minimize the sensitivity of the solution to the uncertainties concerning parameters. Since the user defines the uncertainty limits and the critical value of the objective function would be calculated based on this uncertain predefined set, robust programming has a low computational cost when compared to other risk management methods [19,20]. A general formulation for robust programming is presented in (15.32)–(15.35).

$$\text{minimize}_{\mathbf{x}} \quad \mathbf{c}^T\mathbf{x} \qquad (15.32)$$

$$\text{subject to}: \quad \mathbf{A}\mathbf{x} = \mathbf{b} \qquad (15.33)$$

$$\mathbf{c}, \mathbf{A}, \mathbf{b} \in \Phi \qquad (15.34)$$

$$\mathbf{x} \in X \qquad (15.35)$$

where $\Phi$ denotes the user-specified uncertainty set.

Robust programming is a deterministic and set-based method that does not require the probabilistic description of uncertainty. It is categorized in the methods of interval optimization and models the worst case of uncertain parameters [21,22]. The objective is to define a feasible solution for any realization of the uncertainty of a certain set. Thus robust programming aims at immunizing the decision-maker from the worst results, which makes it a conservative approach [23,24]. The main characteristics of a robust programming model are:

- All entries in the decision vector represent here-and-now decisions, i.e., the decisions should be taken before the uncertainty is realized.
- The decision-maker evaluates the consequences of the taken decisions when the actual data is within the prespecified uncertainty set $\Phi$.
- The restrictions are inflexible, that is, no violations are permitted when the data is in $\Phi$.

Robust programming creates several strategies to find good solutions. The main strategies solve the robust counterpart and use adjustable robust optimization. Under special conditions related to the uncertain parameter behavior, one of those strategies can be used to guarantee the global optimal results [25].

## 15.4 Metaheuristic and decomposition methods

An alternative approach for solving optimization problems, which does not require an explicit mathematical formulation of the problem, is the adoption of a metaheuristic algorithm [26]. Although this type of approach does not ensure optimality, the obtained solutions are usually near-optimal, the computational times are acceptable, and the processing and memory requirements are low [27].

Decomposition methods are also an option to solve especially large-scale optimization problems, allowing their solution in a decentralized or distributed manner. In contrast to metaheuristic algorithms, decomposition methods require a complete mathematical formulation of the problem and only work successfully when the problem has the appropriated structure, for example, complicating constraints or complicating variables. There are various types of decomposition methods such as the Dantzig–Wolfe algorithm, Lagrangian relaxation, augmented Lagrangian, optimality conditions, and Benders' decomposition, and their usage depends on the nature of the problem, for example, linear, nonlinear, continuous, etc. Furthermore, according to the problem nature, decomposition methods may or may not guarantee optimality [28].

This section presents and discusses four metaheuristics and the Benders' decomposition approach with a focus on the optimal management of DERs.

### 15.4.1 Genetic algorithm

The genetic algorithm (GA) is a population-based optimization technique that was originally formulated using mechanisms from Darwin's theory of evolution [26]. It was first proposed by Holland in the 1970s [29] and can be viewed as a class of evolutionary algorithms [27].

The first GAs used binary codification for the variables [26]. More recent implementations propose to codify each type of variable using a natural representation, that is, to represent discrete variables using integer and binary numbers and real variables using floating-point numbers, usually calculated by a subsidiary model [30]. Fig. 15.3 presents the structure of a basic GA.

At each iteration of the basic GA of Fig. 15.3, a new population is generated based on the previous population. Each solution in the new population is obtained by applying the selection, recombination, and mutation operators. The selection operator chooses the fittest (best) solutions of the current population to go through recombination, when the solutions are combined to generate new solutions. Finally, the variables of the recombined solutions are slightly changed by applying the mutation operator. The recombination and mutation rates define the intensity of each one of these operators [27].

| |
|---|
| **Initialization:** Generate a random initial population after choosing its size and the type of codification; define the recombination and mutation rates; choose a stopping criterion. <br> **Repeat** the following steps until the stopping criterion is met: <br>    i.   Evaluate the fitness function of each solution in the population. Update the incumbent solution (best solution found in the process); <br>   ii.   Implement the selection operator; <br>  iii.   Implement the recombination operator; <br>  iv.   Implement the mutation operator and generate the new population. |

FIGURE 15.3 Structure of the basic GA [30].

The GA can be used to solve the optimal management of DERs problem, with both continuous and discrete variables. The control variables of the problem can be represented in each individual of the population, and the quality of each solution can be evaluated by a subsidiary problem, which may be (1) a modified version of the optimization problem with only continuous variables or (2) a power flow algorithm that provides the state of the system. Depending on the strategy chosen, the continuous control variables of the problem (power injected by the distributed generators and energy storage devices) can be directly represented in a solution proposal (for strategy (2)) or can be provided by the subsidiary optimization model (for strategy (1)). The violations in the technical and operational limits of the network can be penalized in the fitness function in both approaches.

## 15.4.2 Particle swarm optimization

The particle swarm optimization (PSO) algorithm was proposed by Kennedy and Eberhart in 1995 [31] and is a branch of the swarm intelligence field. As the GA, it is also a population-based technique, which is based on the collective behavior of decentralized and self-organized natural systems [27]. Fig. 15.4 presents the steps of the original PSO algorithm.

In the algorithm of Fig. 15.4, the population has particles representing the variables of the problem. At each iteration, the best position of each particle and the position of the particle that represents the incumbent solution are identified, the velocity of each particle is calculated, and their positions are updated using (15.36) [27]. In this algorithm, $\mathbf{rnd}(0, \gamma_i)$ is a vector of random numbers uniformly distributed in $[0, \gamma_i]$, generated at each iteration for each particle, $\circ$ is the Hadamard product (element-wise multiplication); each component of $\mathbf{v}_i \in [-\overline{v}, \overline{v}]$ [32].

Since the PSO was originally designed to solve continuous and unconstrained optimization problems, for solving mixed-integer nonlinear programming problems, such as the optimal management of DERs, (15.36) should be modified to round the discrete variables to their nearest values, or other more sophisticated alternatives should be considered [33]. Besides that, the limits of bounded variables should also be considered. The constraints of the problem can be penalized in the evaluation function. Other strategies may also be

**Initialization:** Generate an initial population with random positions and velocities; let $\mathbf{x}_i$ and $\mathbf{v}_i$ be the current position and velocity of the $i^{th}$ particle, respectively; define the parameters $\gamma_1$ and $\gamma_2$; choose a stopping criterion.
**Repeat** the following steps until the stopping criterion is met:
   i.    Calculate the evaluation function of each particle $\mathbf{x}_i$;
   ii.   Compare the value of the evaluation function of each particle $\mathbf{x}_i$ with $p_i^{best}$ (the evaluation function of the best position of particle $i$, $\mathbf{p}_i$, found so far in the process). If the current value is better than $p_i^{best}$, then let $\mathbf{p}_i \leftarrow \mathbf{x}_i$ and update $p_i^{best}$;
   iii.  Identify the best particle $\mathbf{p}_i$ in the population and let $g \leftarrow i$;
   iv.  Update the velocity and position of each particle using (15.36):

$$\begin{cases} \mathbf{v}_i \leftarrow \mathbf{v}_i + \mathbf{rnd}(0, \gamma_1) \circ (\mathbf{p}_i - \mathbf{x}_i) + \mathbf{rnd}(0, \gamma_2) \circ (\mathbf{p}_g - \mathbf{x}_i) \\ \mathbf{x}_i \leftarrow \mathbf{x}_i + \mathbf{v}_i \end{cases} \tag{15.36}$$

**FIGURE 15.4** Structure of the original PSO algorithm [32].

used, including strategies for updating the values of $\gamma_i$ during the execution of the algorithm. Since the modified PSO algorithm can handle both continuous and discrete variables, the quality of a solution proposal can be evaluated by a power flow algorithm, with the control variables represented in each particle of the population.

### 15.4.3 Variable neighborhood search

Differently from the GA and the PSO algorithm, the search strategies of the variable neighborhood search (VNS) algorithm [27] are not based on natural phenomena. Indeed, the VNS algorithm is a generalization of the local search method [34], that changes the neighborhood of the current solution as a means for escaping local optima.

Several versions of the VNS algorithm are available in the literature, including the basic variable neighborhood search (BVNS) [34], variable neighborhood descent (VND) [35], reduced variable neighborhood search (RVNS) [35], and general variable neighborhood search (GVNS) [36]. This section discusses the VND algorithm, which is described in Fig. 15.5, for the problem of optimal management of DERs.

For the VND algorithm shown in Fig. 15.5, no arbitrary parameter must be tuned. The only thing that must be defined is the set of neighborhood structures and the sequence in which they will be applied to the solution proposal. The algorithm considers a single initial solution **s**, that may be generated randomly or using some optimization strategy, and its neighborhood $N_1$ is analyzed. If the best neighbor solution of $\mathbf{s} \in N_1$ is better than the current solution, then the current solution is updated, and the search is performed in the neighborhood $N_1$ of the new current solution. If no improvement is achieved, the algorithm evaluates the neighborhood $N_2$ of the current solution. If an improvement is achieved, the algorithm returns to $N_1$, and if no improvement is achieved in the last neighborhood, $N_{\bar{l}}$, the algorithm stops. One way of defining the neighborhoods is with growing size and complexity, with $N_1$ being the smallest and simplest one [35].

For the problem of optimal management of DERs, an efficient approach is to represent only the discrete variables (related, e.g., to the operation of the capacitor banks and voltage regulators) in the solution vector. For each solution proposal, the resultant linear programming, quadratic programming, or nonlinear programming model (with the discrete variables provided by the VNS algorithm) is solved, providing the values of the continuous decision variables and evaluation function of the problem.

The neighborhoods can then be defined according to their sizes, for example, $N_1$: change the state of a capacitor bank; $N_2$: change the state of two capacitor banks; $N_3$: change the tap position of one voltage regulator, etc.

---

**Initialization:** Define the neighborhood structures $N_l$, $l = 1, \cdots, \bar{l}$. Generate an initial solution **s**. Let $l \leftarrow 1$.

**Repeat** while $l \leq \bar{l}$:
   i.   Find the best neighboring solution $\mathbf{s}'$ in $N_l(\mathbf{s})$;
   ii.  If $\mathbf{s}'$ is better than **s**:
          Let $\mathbf{s} \leftarrow \mathbf{s}'$;
          Let $l \leftarrow 1$;
     Else:
          Let $l \leftarrow l + 1$.

**FIGURE 15.5** Structure of the VND algorithm [35].

## 15.4.4 Greedy randomized adaptive search procedure

Greedy randomized adaptive search procedure (GRASP) is a heuristic algorithm proposed by Feo and Resende [37], as an extension of classic constructive heuristic algorithms. GRASP is a heuristic inspired in operational research concepts and works based on a random and adaptative component. This technique solves the problem in an iterative manner selecting at each iteration, one random element from a restricted candidate list (RCL) that contains the best components classified by a greedy function. The GRASP algorithm is composed of two phases: a constructive and a local search phase.

The first phase consists of creating an initial solution (not necessarily feasible nor locally optimal), iteratively selecting one element from the RCL, at a time. Soon after, the RCL is sorted, such that the next element is chosen according to a greedy function that measures the benefit of choosing each element. The adaptative component of the GRASP is here reflected since, at each iteration, the contribution of each element to the benefit is updated considering the changes resulting from the selection of former elements. The probabilistic component of the GRASP remains in the random choice among the best elements in the RCL, not necessarily the best.

The RCL is constructed as described in (15.37); it is sorted in an increasing or decreasing interval according to the objective function nature: minimization or maximization. In (15.37), $\mathscr{S}$ represents the set of indices of the elements that can be added to the solution, $f_i$ is the objective function value related to the element $i$, $f_{min}$ and $f_{max}$ are the minimum and maximum objective function values, and $\alpha$ is a parameter within the interval $[0,1]$ that indicates how random or how greedy the construction of the solution is. For instance, in minimization problems, if $\alpha = 0$, the construction of the solution will be totally greedy. On the other hand, if $\alpha = 1$, the choosing process will be totally random.

$$RCL = \left\{ i \in \mathscr{S} \mid f_{min} \leq f_i \leq f_{min} + \alpha \left( f_{max} - f_{min} \right) \right\} \tag{15.37}$$

The local search phase intends to improve the solution provided by the constructive phase (since it might not be locally optimal), exploring solutions close to it (neighboring solutions). Sequential substitutions of the current solution by a better solution in the current solution's neighborhood are carried out in an iterative manner. This process is repeated until no better solutions are found in the current solution's neighborhood. The quality of the solutions here provided highly depends on a suitable neighborhood structure and the quality of the initial solution. The general structure of the GRASP algorithm is described in Figs. 15.6 and 15.7.

The construction phase of the GRASP shown in Fig. 15.6 initiates with an empty solution **s**, and at each iteration, an element is added until the stopping criterion is achieved.

**Initialization:** Read the input data and set $\alpha$; set $\mathbf{S} \leftarrow \{\,\}$; define a stopping criterion.
**Repeat** the following sequence until the stopping criterion is achieved:
  i.    *Create the RCL;*
  ii.   *Select a random element from the RCL:* find $\kappa \in RLC$
  iii.  $\mathbf{S} = \mathbf{S} \cup \{\kappa\}$;
  iv.   *Adapt the greedy function:* Update $f$;

**FIGURE 15.6**  Structure of GRASP algorithm—Phase I [38].

> **Initialization:** Read the initial solution **s**, provided in Phase I; select a neighborhood structure $N(\mathbf{x})$; define a stopping criterion.
> **Repeat** the following sequence until the stopping criterion is achieved.
>   i.   *Exploration of the neighborhood:* find the best neighbor **s′** of **s**;
>   ii.  *Move or not:* if **s′** is better than **s**, set $\mathbf{s} \leftarrow \mathbf{s}'$.

**FIGURE 15.7**   Structure of the GRASP algorithm—Phase II [38].

The element to be added to **s** from the RCL could be randomly chosen or chosen by using a linear probability distribution function with proportional selection [39]. In the problem of optimal management of DERs, the stopping criterion at this phase could be that all the DERs have attributed a control action (for instance, tap position of voltage regulators or charging of electric vehicles at specific periods of time, etc.).

The main input for the local search phase shown in Fig. 15.7 is the solution constructed in phase I. After defining a neighborhood structure for this current solution, the neighborhood is explored. If the best neighbor solution of $\mathbf{s} \in N(\mathbf{s})$ is better than the current solution, then the latter is updated, and the search is performed in the neighborhood $N(\mathbf{s})$ of the new current solution. The stopping criterion at this phase could be a maximum number of iterations without any improvement of the current solution or a maximum execution time of the algorithm.

Similar to the VNS algorithm, the GRASP algorithm is able to solve problems with both continuous or integer/binary variables, which can be represented by a vector **s** or matrix **S**. Thus the problem of optimal management of DERs can be solved via the GRASP algorithm, following the same approach proposed for the VNS in the previous section.

### 15.4.4.1 *Benders' decomposition*

Benders' decomposition is a solving technique for linear programming problems with complicating variables through a distributed and iterative process. It was first proposed to deal with large-scale MILP problems, in which after fixing the integer variables the resulting problem is linear, easier to solve, and for which duality theory can be applied [40]. Benders' decomposition can guarantee optimality only for convex problems. Despite that, this approach has also been explored to solve nonconvex, bilevel, and other kinds of problems [41]. The main idea of Benders' decomposition is dividing the original optimization problem into two smaller problems: the master problem and the subproblem(s), each of them only including part of the decision variables. The two smaller problems are then synchronized through an iterative mechanism so that the solutions defined are consistent with the original problem. These solutions can be achieved by parameterizing the problem as a function of the complicating variables. A generalized form of the optimization problem (15.21)–(15.22) with complicating variables is given by (15.38)–(15.41), in which $\mathbf{d}, \mathbf{e}, \mathbf{f}, \mathbf{g}, \mathbf{h}$ are constants, $\mathbf{z}$ and $\mathbf{w}$ are variables, and $\bar{\mathbf{z}}$ and $\overline{\mathbf{w}}$ are their maximum limits.

$$\text{minimize} \quad \mathbf{d}^{\mathrm{T}}\mathbf{z} + \mathbf{e}^{\mathrm{T}}\mathbf{w} \tag{15.38}$$

$$\text{subject to:} \quad \mathbf{f}^{\mathrm{T}}\mathbf{z} + \mathbf{g}^{\mathrm{T}}\mathbf{w} \leq \mathbf{h} \tag{15.39}$$

$$0 \leq \mathbf{z} \leq \bar{\mathbf{z}} \tag{15.40}$$

$$0 \leq \mathbf{w} \leq \overline{\mathbf{w}} \tag{15.41}$$

Assuming that $\mathbf{z}$ is the set of complicating variables, the original problem can be divided into the subproblem represented by (15.42)–(15.45), and the master problem represented by (15.46)–(15.49). The subproblem is an instance of the original problem, obtained after fixing the values of the complicating variables $\mathbf{z}$. Then, the solution consists of finding the optimal values for the set of variables $\mathbf{w}$, along with the dual variables $\boldsymbol{\lambda}^{(v)}$ associated with the constraints that fix the values of the complicating variables (15.46), within an iterative process $v$.

$$\text{minimize} \quad \mathbf{e}^T\mathbf{w} \tag{15.42}$$

$$\text{subject to}: \quad \mathbf{f}^T\mathbf{z} + \mathbf{g}^T\mathbf{w} \le \mathbf{h} \tag{15.43}$$

$$0 \le \mathbf{w} \le \overline{\mathbf{w}} \tag{15.44}$$

$$\mathbf{z} = \mathbf{z}^{(v)} : \boldsymbol{\lambda}^{(v)} \tag{15.45}$$

After finding the optimal solution of the subproblem, it is possible to solve the master problem. Here, the objective function is composed by the set of complicating variables plus an auxiliary variable $\alpha^v$ that represents the objective function of the subproblem(s) within the objective function of the original problem. Constraint (15.47) allows the Benders' cuts, adding one new cut per iteration, and (15.49) is an additional constraint that allows feasibility in the first iteration when no cuts have been added, using a large negative constant $\alpha_{down}$.

$$\text{minimize} \quad \mathbf{c}^T\mathbf{x} + \alpha^v \tag{15.46}$$

$$\text{subject to}: \quad \alpha^v \ge \mathbf{g}^T\mathbf{w}^{(k)} + \boldsymbol{\lambda}^{(k)}\left(\mathbf{z} - \mathbf{z}^{(k)}\right) \forall k = 1 \ldots v - 1 \tag{15.47}$$

$$0 \le \mathbf{z} \le \overline{\mathbf{z}} \tag{15.48}$$

$$\alpha^v \ge \alpha_{down} \tag{15.49}$$

The general structure of the Benders' decomposition algorithm is described in Fig. 15.8, where $v$ is the current Benders' iteration, $\mathbf{z}^{\text{fixed}}$ represents an initial solution ($\mathbf{z}^{\text{initial}}$) for the set of complicating variables $\mathbf{z}$, and $LB$ and $UB$ are the lower and upper bound. $LB$ corresponds to the objective function of the master problem. In step $i$, given initial values for complicating variables $\mathbf{z}$, the subproblem provides the optimal solution for the other set of variables (e.g., $\mathbf{w}$), along with the dual variables $\boldsymbol{\lambda}$. If the convergence criterion is not

---

**Initialization:** Initialize the iteration counter $j \leftarrow 1$; set a small tolerance value $\epsilon$ to control convergence; $\mathbf{z}^{\text{fixed}(1)} \leftarrow \mathbf{z}^{\text{initial}}$; $LB^1 \leftarrow -\infty$.
**Repeat** the following sequence until convergence is achieved:
  i.    *Solve the subproblem(s):* Obtain the optimal values of primal variables $\mathbf{w}^v$, all dual variables $\boldsymbol{\lambda}^v$, and the objective function of the original problem, which corresponds to the upper bound ($UB^v$);
  ii.   *Convergence check:* if $|UB^v - LB^v| \le \epsilon$, an optimal solution within an error $\epsilon$ is obtained, otherwise $j \leftarrow j + 1$;
  iii.  *Solve the master problem:* Obtain the updated values for the complicating variables $\mathbf{z}^v$ and $\alpha^v$; update $LB^v$, and set $\mathbf{z}^{\text{fixed}}$ as the updated values of $\mathbf{z}^v$.

**FIGURE 15.8**  Structure of Benders' decomposition algorithm [42].

satisfied (step *ii*), a new iteration is required and based on the dual variables provided by the subproblem(s), the master problem is solved again in step *iii* to obtain updated values for complicating variables **z** and $\alpha$, and so on.

In the context of the optimal management of DERs problem, Benders' decomposition technique is a powerful tool due to the inherent characteristics of the problem. Integer/ binary variables representing control actions of storage devices, electric vehicles, distributed generators, among others, can be treated as complicating variables in the subproblem, and operating variables (continuous) representing power injections/consumptions can be regarded in the master problem. Additionally, in this problem, it is common to find many levels of optimization due to the existence of various stakeholders with different interests, for example, distribution system operators want to reduce power losses while owners of electric vehicle fleets or solar parks want to maximize revenue. In those cases, Benders' decomposition allows solving the problem in a decentralized manner such that individual and collective interests are achieved. Moreover, naturally, DERs present high levels of uncertainty for which most of the mathematical representations result in stochastic models that can be more tractable and easier to solve through decomposition techniques [43].

## 15.5 Conclusion

Mathematical models (MILP, stochastic programming, and robust optimization) to represent the operation of distribution networks with DERs have been presented in this chapter. Metaheuristic and decomposition methods have been also discussed as optimization techniques that allow finding good-quality solutions and reducing the computational effort. Those deterministic and approximate approaches are a stack of optimization techniques that can be adopted to solve optimization problems in LEMs such as feasible and economic operation, integration of DERs, and demand response.

## Acknowledgments

This work was supported by the Coordination for the Improvement of Higher Education Personnel (CAPES)— Finance Code 001, the Brazilian National Council for Scientific, the Technological Development (CNPq), under grants 305852/2017-5 and 305318/2016-0, and the São Paulo Research Foundation (FAPESP), under grants 2015/ 21972-6, 2017/02831-8, 2018/08008-4, 2018/20355-1, and 2018/23617-7. The Portuguese partner received funding from FEDER funds, through COMPETE 2020, project POCI-01−0145-FEDER-028983; and by National Funds through FCT under project PTDC/EEI-EEE/28983/2017, UIDB/00760/2020 and CEECIND/02814/2017 grant.

## References

[1] F. Lezama, J. Soares, P. Hernandez-Leal, M. Kaisers, T. Pinto, Z. Vale, Local energy markets: paving the path toward fully transactive energy systems, IEEE Trans. Power Syst. 34 (2019) 4081−4088. Available from: https://doi.org/10.1109/TPWRS.2018.2833959.

[2] F. Teotia, R. Bhakar, Local energy markets: concept, design and operation, 2016 National Power Systems Conference (NPSC), IEEE, 2016, pp. 1−6. Available from: https://doi.org/10.1109/NPSC.2016.7858975.

[3] J. Soares, T. Pinto, F. Lezama, H. Morais, Survey on complex optimization and simulation for the new power systems paradigm, Complexity 2018 (2018) 1−32. Available from: https://doi.org/10.1155/2018/2340628.

[4] C.F. Sabillon-Antunez, J.F. Franco, M.J. Rider, R. Romero, Mathematical optimization of unbalanced networks with smart grid devices, Electric Distribution Network Planning, Springer, 2018, pp. 65−114. Available from: https://doi.org/10.1007/978-981-10-7056-3_3.

[5] J.F. Franco, A.T. Procopiou, J. Quirós-Tortós, L.F. Ochoa, Advanced control of OLTC-enabled LV networks with PV systems and EVs, IET Generation, Transm. Distrib. 13 (2019) 2967−2975. Available from: https://doi.org/10.1049/iet-gtd.2019.0208.

[6] J.F. Franco, L.F. Ochoa, R. Romero, AC OPF for smart distribution networks: an efficient and robust quadratic approach, IEEE Trans. Smart Grid 9 (2018) 4613−4623. Available from: https://doi.org/10.1109/TSG.2017.2665559.

[7] A.T. Al-Awami, E. Sortomme, Coordinating vehicle-to-grid services with energy trading, IEEE Trans. Smart Grid 3 (2012) 453−462. Available from: https://doi.org/10.1109/TSG.2011.2167992.

[8] Y.M. Atwa, E.F. El-Saadany, M.M.A. Salama, R. Seethapathy, Optimal renewable resources mix for distribution system energy loss minimization, IEEE Trans. Power Syst. 25 (2010) 360−370. Available from: https://doi.org/10.1109/TPWRS.2009.2030276.

[9] J.A.P. Lopes, N. Hatziargyriou, J. Mutale, P. Djapic, N. Jenkins, Integrating distributed generation into electric power systems: a review of drivers, challenges and opportunities, Electr. Power Syst. Res. 77 (2007) 1189−1203. Available from: https://doi.org/10.1016/j.epsr.2006.08.016.

[10] J.F. Franco, M.J. Rider, M. Lavorato, R. Romero, A mixed-integer LP model for the optimal allocation of voltage regulators and capacitors in radial distribution systems, Int. J. Electr. Power Energy Syst. 48 (2013) 123−130. Available from: https://doi.org/10.1016/j.ijepes.2012.11.027.

[11] CPLEX Optimizer | IBM, n.d. <https://www.ibm.com/analytics/cplex-optimizer> (accessed 07.01.2020).

[12] Z. Geng, A.J. Conejo, Q. Chen, C. Kang, Power generation scheduling considering stochastic emission limits, Int. J. Electr. Power Energy Syst. 95 (2018) 374−383. Available from: https://doi.org/10.1016/j.ijepes.2017.08.039.

[13] S. Nojavan, H.A. Aalami, Stochastic energy procurement of large electricity consumer considering photovoltaic, wind-turbine, micro-turbines, energy storage system in the presence of demand response program, Energy Convers. Manag. 103 (2015) 1008−1018. Available from: https://doi.org/10.1016/j.enconman.2015.07.018.

[14] S. Nojavan, B. Mohammadi-Ivatloo, K. Zare, Optimal bidding strategy of electricity retailers using robust optimisation approach considering time-of-use rate demand response programs under market price uncertainties, IET Gener. Transm. Distrib. 9 (2015) 328−338. Available from: https://doi.org/10.1049/iet-gtd.2014.0548.

[15] J.R. Birge, F. Louveaux, Introduction to Stochastic Programming, Springer, 2011.

[16] A. Prékopa, Stochastic Programming, Kluwer Academic Publishers, 1995.

[17] A.J. Conejo, M. Carrion, J.M. Morales, Decision Making Under Uncertainty in Electricity Markets, Springer, 2010.

[18] A. Charnes, W.W. Cooper, Deterministic equivalents for optimizing and satisficing under chance constraints, Oper. Res. 11 (1963) 18−39. Available from: https://doi.org/10.1287/opre.11.1.18.

[19] Y. Zhang, J. Tang, A robust optimization approach for itinerary planning with deadline, Transport. Res. Part E: Logist. Transport. Rev. 113 (2018) 56−74. Available from: https://doi.org/10.1016/j.tre.2018.01.016.

[20] R.M. Lima, A.Q. Novais, A.J. Conejo, Weekly self-scheduling, forward contracting, and pool involvement for an electricity producer. An adaptive robust optimization approach, Eur. J. Oper. Res. 240 (2015) 457−475. Available from: https://doi.org/10.1016/j.ejor.2014.07.013.

[21] I. Ahmadian, O. Abedinia, N. Ghadimi, Fuzzy stochastic long-term model with consideration of uncertainties for deployment of distributed energy resources using interactive honey bee mating optimization, Front. Energy 8 (2014) 412−425. Available from: https://doi.org/10.1007/s11708-014-0315-9.

[22] S. Nojavan, B. Mohammadi-Ivatloo, K. Zare, Robust optimization based price-taker retailer bidding strategy under pool market price uncertainty, Int. J. Electr. Power Energy Syst. 73 (2015) 955−963. Available from: https://doi.org/10.1016/j.ijepes.2015.06.025.

[23] Z. Haider, H. Charkhgard, C. Kwon, A robust optimization approach for solving problems in conservation planning, Ecol. Model. 368 (2018) 288−297. Available from: https://doi.org/10.1016/j.ecolmodel.2017.12.006.

[24] A. Noruzi, T. Banki, O. Abedinia, N. Ghadimi, A new method for probabilistic assessments in power systems, combining monte carlo and stochastic-algebraic methods, Complexity 21 (2015) 100−110. Available from: https://doi.org/10.1002/cplx.21582.

[25] A. Najafi-Ghalelou, S. Nojavan, K. Zare, Heating and power hub models for robust performance of smart building using information gap decision theory, Int. J. Electr. Power Energy Syst. 98 (2018) 23−35. Available from: https://doi.org/10.1016/j.ijepes.2017.11.030.

[26] M. Gendreau, J.-Y. Potvin, Handbook of Metaheuristics, vol. 57, second ed., Springer, New York, 2003. Available from: https://doi.org/10.1007/b101874.

[27] second ed., M. Gendreau, J.-Y. Potvin (Eds.), Handbook of Metaheuristics, vol. 146, Springer US, Boston, MA, 2010. Available from: https://doi.org/10.1007/978-1-4419-1665-5.

[28] A.J. Conejo, E. Castillo, R. Mínguez, R. García-Bertrand, Decomposition Techniques in Mathematical Programming: Engineering and Science Applications, Springer, 2006. Available from: https://doi.org/10.1007/3-540-27686-6.

[29] J.H. Holland, Adaptation in Natural and Artificial Systems, first ed., University of Michigan, Ann Arbor, MI, 1975.

[30] Z. Michalewicz, Genetic Algorithms + Data Structures = Evolution Programs, third ed., Springer-Verlag, Heidelberg, 2013. Available from: https://doi.org/10.1007/978-3-662-03315-9.

[31] J. Kennedy, R. Eberhart, Particle swarm optimization, Proceedings of ICNN'95 - International Conference on Neural Networks, vol. 4, IEEE, Perth, 1995, pp. 1942−1948. Available from: https://doi.org/10.1109/ICNN.1995.488968.

[32] R. Poli, J. Kennedy, T. Blackwell, Particle swarm optimization: an overview, Swarm Intell. 1 (2007) 33−57. Available from: https://doi.org/10.1007/s11721-007-0002-0.

[33] M. Clerc, Discrete particle swarm optimization, illustrated by the traveling salesman problem, New Optimization Techniques in Engineering: Studies in Fuzziness and Soft Computing, Springer, Heidelberg, 2004, pp. 219−239. Available from: https://doi.org/10.1007/978-3-540-39930-8_8.

[34] N. Mladenović, P. Hansen, Variable neighborhood search, Comput. Oper. Res. 24 (1997) 1097−1100. Available from: https://doi.org/10.1016/S0305-0548(97)00031-2.

[35] P. Hansen, N. Mladenović, A Tutorial on Variable Neighborhood Search, Groupe d'Études et de Recherche en Analyse des Décisions; Montreal, 2003.

[36] P. Hansen, N. Mladenović, Variable neighborhood search: principles and applications, Eur. J. Oper. Res. 130 (2001) 449−467. Available from: https://doi.org/10.1016/S0377-2217(00)00100-4.

[37] A.F. Thomas, R. Mauricio, Greedy randomized adaptive search procedures, J. Glob. Optim. 6 (1995) 109−134.

[38] M.G.C. Resende, C.C. Ribeiro, Optimization by GRASP, Springer, New York, 2016. Available from: https://doi.org/10.1007/978-1-4939-6530-4.

[39] F. Glover, G.A. Kochenberger, Handbook of Metaheuristics, Kluwer Academic Publishers, 2003.

[40] J.F. Benders, Partitioning procedures for solving mixed-variables programming problems, Numerische Mathematik (1962). Available from: https://doi.org/10.1007/BF01386316.

[41] R. Rahmaniani, T.G. Crainic, M. Gendreau, W. Rei, The Benders decomposition algorithm: a literature review, Eur. J. Oper. Res. (2017) 801−817. Available from: https://doi.org/10.1016/j.ejor.2016.12.005.

[42] J. Kazempour, Decomposition techniques for optimization problems with complicating variables, Lecture of the DTU CEE Summer School 2017 "Modern Challenges in Power System Operation and Electricity Markets: An Optimization Perspective", 2017, pp. 1-77. Available from: https://energy-markets-school.dk/summer-school-2017/

[43] J. Soares, B. Canizes, M.A.F. Ghazvini, Z. Vale, G.K. Venayagamoorthy, Two-stage stochastic model using benders' decomposition for large-scale energy resource management in smart grids, IEEE Trans. Ind. Appl. 53 (2017) 5905−5914. Available from: https://doi.org/10.1109/TIA.2017.2723339.

# PART IV

# Regulatory framework: Current trends and future perspectives

# An economic analysis of market design: Local energy markets for energy and grid services

*L. Lynne Kiesling*[1,2,3]

[1]Institute for Regulatory Law & Economics, Carnegie Mellon University, Pittsburgh, PA, United States [2]Engineering & Public Policy, Carnegie Mellon Universitys, Pittsburgh, PA, United States [3]Wilton E. Scott Institute for Energy Innovation, Carnegie Mellon University, Pittsburgh, PA, United States

## 16.1 Introduction

Two converging areas of innovation shape today's dynamic energy systems. Catalyzed by policies to promote low-carbon energy technologies, the production of distributed energy resources (DERs) and renewable energy sources (RESs) has increased, and their energy efficiency has grown while their production and installation costs have fallen [1]. DERs range from battery storage to solar photovoltaics to electric vehicles (EVs), among other technologies, and range in size from a car battery up to a resource that can power an office building or neighborhood.

At the same time, digitization of production and consumption of many goods and services has created new markets and value propositions in a wave of Schumpeterian dynamism [2]. Digitization has also led to the emergence of new platform business models that challenge existing industries and firms to rethink their strategies [3].

As DERs and digitization converge in the electricity industry, they reveal new ways to create economic and environmental value. But the existing distribution grid infrastructure and regulatory institutions are ill-suited to these new value propositions ([4], p. 4082). A policy impetus for decarbonization has led to policies like feed-in tariffs that uncouple consumer and prosumer incentives from market prices, creating network congestion without a decentralized feedback mechanism to alleviate it [1,5].

Using market processes to coordinate physical and economic flows enables parties to benefit from reducing congestion, with prices as the decentralized coordination signals (control signals in an engineering sense). Digitization of the distribution grid facilitates these opportunities. This economic logic from transactive energy (TE) provides a conceptual framework for local energy markets (LEMs).

A LEM is a "...platform on which individual consumers and prosumers trade energy supporting regional scopes such as a neighborhood environment" ([4], p. 4082). LEMs reflect the idea that congestion and imbalances are local and short-term in nature. The fundamental aspect of LEMs is decentralized coordination. Coordination of consumption and production occurs at an individual and a local level, and the distribution system operator (DSO) aligns that balance with system-level operational coordination [6].

For example, EVs connected through LEMs act as a large distributed battery resource that can respond to local changes by responding to price signals in the LEMs for grid services. An EV owner in a LEM could use her home energy management system to automate LEM participation by establishing trigger prices ($p_{bid}$ and $p_{offer}$): if the market price is below $p_{bid}$, buy energy and charge the battery, if the market price is above $p_{offer}$, sell energy and discharge the battery.[1] As a dual-nature resource EVs can both take power and supply power when necessary for voltage management, and the signal of which action is valuable is the price signal in the LEM. EVs with trigger prices [both willingness to pay (WTP) to buy and willingness to accept (WTA) to sell] reflecting their owner's or operator's preferences can respond autonomously when the market price changes. A different example of a potential LEM-enabled EV resource is a fleet of school buses—the buses operate during a few hours of the day during specific times of the year, and in their idle hours they can provide either power or storage as needed, creating an additional revenue stream for the fleet owner.

These examples also illustrate shifting the focus from a resource's physical *capacity* to an emphasis on its *capability* to perform a particular service at a particular time. With diverse resources and conceptualizing flexibility as a service, what a resource can do and how quickly it can do it matters more than the standard measurement of physical capacity. What matters is not the capacity quantity, but rather the time and place at which the resource can offer the service and how firmly its owner-operator can make that commitment.

Grid services include voltage management, system reserves (spinning and nonspinning response periods), and frequency support [7]. These grid services can be provided locally to address local operational objectives as well as support bulk system issues [8]. Doing so requires designing LEMs thoughtfully. This chapter synthesizes exchange theory, transaction cost economics, mechanism design, and auction theory into an economic framework to suggest market design principles for why and how LEMs enable DERs to provide energy and grid services in a decentralized architecture. It then provides a brief analysis of the institutional context in the United States, and how the institutional diversity and experimentation in the United States may affect LEM development and adoption.

[1] The algorithm for implementing this simple strategy is likely to be more complicated, due to the different potential states of charge of the battery and how those states may influence $p_{bid}$ and $p_{offer}$, which will vary depending on the charge state.

## 16.2 DERs for energy and grid services

DERs are possible suppliers of grid services because of their smaller scale, their spatial distribution throughout the network, and the ability to use information and communications technology (ICT) to interconnect them and automate their operations. In the transition from conceptualizing electricity as a commodity to energy as a service [9], several activities that used to be necessarily bundled with generation and delivery can now be provided in a decentralized manner using resources owned by parties other than a vertically integrated distribution utility.

DERs change resource ownership and operation patterns along with the change in scale and architecture. With DERs, various parties in the energy ecosystem can provide grid services, and can consume grid services provided by parties other than the incumbent utility/wires company/DSO. The evolution to a decentralized architecture with DERs means that prosumers can provide grid services through LEMs, aggregators and retail energy providers can combine prosumer and consumer resources and demand response and operate them to participate in LEMs, and consumers can purchase grid services from them through LEMs to enable the delivery of their purchased energy. The DSO delivers energy and coordinates the grid services needed to do so with safety, reliability, and resilience. These heterogeneous DER capabilities and a portfolio of LEMs for grid services enable prosumers, consumers, aggregators, and retail energy providers to participate in these markets, and to benefit from providing energy and grid services ([6], p. 4). Digital technologies that enable LEMs also reduce the cost of DER integration, and the innovation incentives that markets provide will induce further technology evolution that reduces DER costs. Over time the costs of using LEMs to enable DERs to provide energy and grid services will fall (Fig. 16.1).

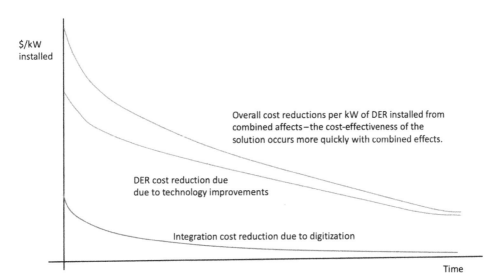

FIGURE 16.1    Combined effects of digitization and falling DER costs.

Some grid services are better suited to decentralized LEMs, such as voltage management, frequency regulation, and fast frequency response.[2] Where the relative value of flexibility on short time frames is high, connecting smaller DER resources together into an LEM can enable local decentralized coordination. In this decentralized architecture, the system optimization framework would have to change from a centralized optimal power flow calculation to a layered decomposition optimization [10].[3]

Other grid services are better suited to more centralized procurement, to the extent that they serve primarily to enable the wires company/DSO to engage in the planning they do to maintain reliability and resilience. DSO would want to continue to procure reserves. For example, procurement contracting for reserves as an insurance service (which does not necessarily have to be in capacity terms). Again, here the shift is to think in terms of payment for a service rather than fixed capacity payments.

In the short run, imbalance, scarcity, and congestion in the distribution grid are local phenomena and short-term in nature [4]. The intermittency of RES may exacerbate these system issues, and wholesale markets lack the ability to predict or counteract such intermittency. For this reason, resource flexibility is a capability of growing importance and value. A resource's ability to take or provide, buy or sell power, potentially quickly and in a contractible/dispatchable way, will be a valuable capability in a distribution network with high DER penetration [4,5]. High-penetration DER networks will benefit from markets for flexibility as a service. A portfolio of LEMs for grid services provides a contracting framework that matches the locality of the need for grid services with the DERs capable of supplying them.

DERs participating in LEMs can provide grid services like voltage management, reserves, and frequency support. In wholesale power markets these services are transacted as ancillary services at the larger, bulk power level. LEMs open the possibility of doing so at a smaller, more decentralized scale in the distribution network. The capabilities of DERs to bring flexibility as a service to a distribution system highlight the potential value that LEMs can contribute to an energy system. Designing LEMs that do so effectively requires an understanding of economics and of market design.

## 16.3  An economic framework for LEM grid services

Market processes coordinate the actions and plans of distributed, independent actors in complex adaptive systems. As social processes that enable coordination and thus cooperation, markets operate in an institutional context that is a mix of formal rules, informal norms, and behavioral heuristics. In some contexts, these rules emerge through experience, while in others they are more deliberately designed based on both experience and

---

[2] The specific details from a power system engineering perspective matter greatly in these three examples. Voltage management is perhaps easiest of the three to coordinate in a decentralized manner, while frequency regulation and fast frequency response still require some level of system coordination even if supplied in a decentralized manner.

[3] A discussion of optimal power flow is beyond the scope of this analysis, but the development and potential value of LEMs and local balancing indicate the importance of rethinking grid optimization when grid architecture becomes more decentralized.

economic theory. Here I take that general context as a starting point for applying economic theory from transaction cost economics, mechanism design, and auction theory to LEMs, and for developing basic market design principles.

## 16.3.1 A theoretical framework for LEM economics

An economic analysis of LEMs starts from the observation that the demand for energy is a derived demand. Consumer preferences for consuming energy arise because energy is an input into other products, services, and activities that they value. They thus have preferences for energy as an input, and conceptualize their WTP accordingly. Similarly, grid services are inputs that are complements to energy consumption, and thus to the consumption into which energy is an input. This characteristic of consumer preferences applies to both the conventional view of electricity as a physical commodity and the evolving view of energy as a service.

As more consumers purchase DERs and become prosumers, they will also form preferences over their WTA to sell energy and grid services. Digital technologies and a TE system (including LEMs) enables them to automate bids and offers reflecting these preferences and communicating them to others through LEMs.

These individual preferences and perceived opportunity costs are subjective, meaning that each person forms their own judgment about what they value, how much they are willing to pay to buy/willing to accept to sell, and the tradeoffs they face across alternatives (known as opportunity costs). Market processes create value because they use a parsimonious signal—price—to coordinate actions and plans among these diverse people with subjective preferences that can only be known to them. A system of prices that emerge from market processes acts as a system of signals, turning some of that diffuse private knowledge into public information in the form of market prices [11]. The *emergence* of prices is essential. Administratively determined or cost-only-based "prices" are not accurate prices, but are rather cost-based estimates that fail to reflect the distributed knowledge and subjective preferences of the people in the network.

From financial commodity markets to the local farmer's market, economic models incorporate the importance of prices as signals of relative value, relative scarcity, and opportunities for investment and innovation. Prices also signal local conditions, an essential capability for energy systems. How those price signals operate, and how well they enable distributed agents to communicate diffuse private knowledge, depends on a market's institutional framework. By communicating relative scarcity to participants and enabling them to change their behavior, prices enable flexibility and can make flexibility profitable. Prices make markets adaptable frameworks for coordination in the face of change. This economic concept forms the foundation of the LEM value proposition.

Within this context, consider the effects of DERs. Ownership and/or operation of DERs expands the choice set of people, especially people who used to be solely consumers. It also increases heterogeneity in the network because of their nonuniform ownership decisions as well as higher technological diversity. DER innovation reduces the opportunity costs of local energy compared to traditional vertical integration with central generation. Opportunity costs are subjective, and to the extent that people value the low-carbon and/or local characteristics of DERs, those preferences mean that they perceive the opportunity costs of DERs as lower now than they were before.

The heterogeneous nature of DERs enlarges the choice set for people and enables the satisfaction of a larger range of individual preferences, including preferences for low-carbon energy or local exchange of energy among neighbors. Examples of projects that demonstrate this phenomenon include the Brooklyn Microgrid project [8] and the Cornwall Local Energy Market [12].

DER and digital innovation are also inducing a transition from thinking of power as a commodity with the distribution grid as its delivery system to thinking of electricity provision and delivery as a bundle of services. In the traditional vertically integrated monopoly utility, a single firm owned and operated all supply chain assets from generation to the meter, internalizing all of the steps in this complicated supply chain. With large-scale electromechanical technologies comprising the physical network and with analog communication technologies, the 20th century distribution utility was characterized by economies of scale and scope, meaning (among other things) that the least-cost way to provide the services required for reliable delivery was for the utility to provide them alongside generation, transmission, distribution, and metering.[4]

Digital and DER innovation change those economics. Economies of scale no longer dominate either generation or the retail relationship with the end-use consumer, as innovations in generation and ICT have made rivalrous wholesale and retail markets feasible while also reducing costs and increasing reliability in distribution systems. Economies of scale do continue to dominate the economics of the transmission and distribution wires network infrastructure, and as long as network participants can benefit enough from being interconnected on this network rather than relying solely on self-supply, the grid will continue to have net value.

The complementary relationship between DERs and digitization also contributes to the economic foundation of LEMs. Economic complements are products and services that increase value when consumed together. DERs and digital technologies are complements because digitization reduces the costs of interconnecting DERs into networks, both the distribution grid network and a network of markets. Digital technologies are DER enablers and make flexibility as a service possible.

As an important complement to DERs, digitization reduces transaction costs, the administrative and organizational costs of making a transaction happen: acquiring, assembling, monitoring, organizing, and repurposing resources. By reducing transaction costs, digital innovations make decentralized coordination using prices through markets possible instead of centralized control. This phenomenon is now common in the modern economy—digitization reduces transaction costs and makes market exchange more value-creating, which is the reason for so much digital platform innovation in the past decade [3]. Lower transaction costs make market platforms feasible and shift the margin at which transactions occur more profitably through markets than through vertical integration. In the analog communications world of the 20th century, many markets did not exist that exist now (e.g., Airbnb for home rentals,

---

[4] Economies of scale occur when the long-run average cost of production declines as the quantity produced increases. Economies of scope occur when the total cost of using a given set of capital assets to produce two goods $X$ and $Y$ is less than the standalone cost of producing each good separately (i.e., $C(X + Y) < C(X) + C(Y)$). Economies of scale and scope combined are known as cost subadditivity, which is the defining characteristic of a natural monopoly, because they imply that for a given demand, a single firm is the least-cost method of producing those goods (see Chapter 4 in Ref. [13]).

TaskRabbit and Upwork for small jobs, Uber and Lyft for ride sharing). Higher transaction costs mean that gains from exchange are lower.

One manifestation of high transaction costs is vertical integration in industries, where individual firms own and operate multiple related steps in the supply chain. Vertical integration is an organizational response to high transaction costs, a way to create gains from exchange in a high transaction cost environment. In the electricity industry, vertical integration originated as a consequence of technological necessity, due to the invention and architecture of the network as a system and the high cost of transacting in markets for each separate transaction in the supply chain [14].

Falling transaction costs due to digitization mean that vertical integration may no longer be the least-cost or most beneficial way to organize transactions. Market exchange that is now feasible and cheaper may be the more efficient way to organize exchange at the local level as well as the bulk/wholesale level.

This transaction cost economics analysis underlies the economic argument for LEMs: falling transaction costs make LEMs feasible, and they can serve as digital market platforms for creating new value streams by innovating new markets and services within the distribution network. They also make it possible for parties other than the distribution utility to provide those services and transact for them in markets.

## 16.3.2 Market design principles for LEMs

The analysis thus far establishes a general economic foundation for LEMs for energy and grid services. Such a novel proposal requires attention to market design, which establishes market rules and thus shapes both the incentives to participate and the trading incentives and strategies of participants. To align individual and system objectives and to enable participants to maximize value (i.e., to enable economic efficiency), LEM market design should focus on maximizing economic value subject to the physical constraints of the network ([5], p. 7). Here I offer a framework for thinking about the economics of design choices for LEMs using mechanism design and auction theory, and offer some economic design principles to consider.

Market design principles draw on a field of economics called mechanism design. This game-theoretic approach examines situations in which individuals make decisions and interact when they have private knowledge that could affect the decisions they make and the ultimate outcomes.

> The mechanism design problem is to design a mechanism so that when individuals interact through the mechanism, they have incentives to choose messages as a function of their private information that leads to socially desirable outcomes ([15], p. 8).

In other words, each agent in a system has a unique and private input into achieving a shared objective, and a mechanism is an institution for achieving that objective by aligning individual and shared incentives across the agents in the system.[5]

---

[5] Seminal work in mechanism design includes Myerson [16], Myerson and Mark [17], D'Aspremont and Gérard-Varet [18], and Laffont and Tirole [19]. Jackson [15] provides a concise introduction to the theory. Börgers [20] provides an extensive survey of the literature.

Generally, then, mechanism design is relevant to a variety of choice settings, including designing rules for LEMs. The range of various formal and informal mechanisms that have emerged or have been created shows how important institutions are to shaping incentives in different contexts and environments.

Mechanism design focuses on eliciting truthful revelation of some information about an agent's private knowledge/type. Agents may have incentives to misrepresent their preferences (e.g., a buyer understating value to lower price, a seller overstating cost to raise price, or a community member understating value of a public project and engaging in "free riding"). Depending on the environment and the objective function of the principal, different mechanisms can perform better or worse at truthful revelation.

The general structure of mechanism design is to maximize some objective function across all individuals subject to two constraints: individual rationality or participation (IR) and incentive compatibility (IC). The objective function varies depending on the specific context, but represents profit, surplus, or welfare maximization. The IR or participation constraint requires that people participate voluntarily, and only when their expected benefit exceeds the expected cost they will incur. The IC constraint ensures that each agent is better off truthfully revealing their type in the chosen mechanism's outcome compared to other alternative outcomes. IC requires that the mechanism induces truthful reporting of each person's WTP (as a buyer) or WTA (as a seller).[6]

Market design also uses a related field of economics—auction theory—to identify incentive compatible market institutions. Auctions are used in practice to allocate a wide range of different goods and services in different settings. Governments use auction mechanisms for procurement and for policy objectives, most notably since the 1990s for the allocation of use of the radio spectrum for mobile communications, and auctions are the predominant market design in wholesale power markets.[7] Lin et al. [21] and Mengelkamp et al. [28] use auction theory to analyze market design and bidding strategies in peer-to-peer energy markets.

In auctions the participants must determine their bidding strategies, and economists model that process as net value maximization. In a private-value English auction, for example, a bidder's dominant strategy is to bid until the price reaches her value, the point where she is indifferent between winning and not winning the auction.[8]

---

[6] In discussing auction mechanisms for peer-to-peer energy markets, Lin et al. [21] point to the Myerson—Satterthwaite theorem (1983): an ideal auction mechanism satisfies IR and IC, is efficient (maximizes surplus), and has a balanced budget/does not require subsidies, but the class of feasible mechanisms satisfies at most three of those four conditions.

[7] Vickrey [22] is the seminal paper in auction theory, notable for demonstrating the efficiency properties of the second-price sealed-bid auction. McAfee and McMillan [23], Klemperer [24], Klemperer's [25] edited volume, and Menezes and Monteiro [26] provide valuable surveys of a large and diverse auction theory literature. Wilson [27] provides an overview of models of strategic behavior in auctions.

[8] A well-known problem with the ascending English auction is the winner's curse, in which the winning bid is above the winner's value. The winner's curse arises in common value auctions, where the value of the good is the same across bidders but their information varies, rather than private-value auctions. The second-price auction, where the winner pays the second-highest bid, emerged as a response to the winner's curse. Vickrey [22] demonstrated that in a second-price sealed-bid auction it is optimal for a bidder to bid her true value.

In many auction designs the design must specify a market-clearing mechanism when there are multiple participants. Two common market-clearing mechanism are uniform price or discriminatory price, the choice of which yields different price and revenue outcomes. In a uniform price auction all bidders pay the market-clearing bid, while in a discriminatory auction all bidders pay their stated bids. Discriminatory auctions create an incentive to understate bids (a behavior called "bid shading") but can yield higher prices, while uniform price auctions elicit truthful bidding but can have more price volatility and more opportunities for tacit collusion [29]. The comparison of uniform price and discriminatory designs has been extensive in analyses of wholesale power markets [30–33].

## 16.3.3 Summary of economic design principles and challenges

This economic foundation of exchange theory, transaction cost economics, mechanism design, and auction theory suggests some implications and design principles for LEM contexts. The primary emphasis should be on individual participation and IC. Transparent rules enable participants to learn and discover new means of value creation by acting on those incentives. LEM design challenges to consider also include minimizing the potential exercise of market power, and potentially anticompetitive effects of market operator and DSO participation in market transactions.[9]

The individual participation constraint means that people will participate voluntarily if benefits exceed costs. Individual benefits and individual opportunity costs are subjective and thus beyond the control of the market designer. One factor market design can influence is market entry barriers and entry cost. Low-cost entry and exit takes the form of transparent rules, standard interconnection, and widespread interoperability standards.

Some people may find that participation costs exceed benefits for them, but would participate if an aggregator or retail energy provider did so on their behalf. One design question is whether LEMs should be open to aggregators and retail energy providers, or only to individual consumers/prosumers. Many individuals may own DERs yet be uninterested in active participation in LEMs (even if their participation is automated in a TE system), so aggregators and retail energy providers could be important entities for increasing participation, customer satisfaction and value creation, and liquidity in LEMs. By aggregating across consumers and prosumers they would also play a valuable risk management role and contribute to system resilience while also mitigating price risks facing consumers/prosumers. Aggregators and retail energy providers reduce entry costs for some consumers and prosumers, at a possible tradeoff of an increased ability to exercise market power.

As a new type of market, LEMs will also be experiential platforms—without much precedent for comparison, parties will learn how much they value the platform and the grid services that various DERs can provide at different times and in different contexts and conditions, and they learn by experiencing it. This learning process is another reason to prioritize low-cost entry and exit.

[9] Two other important market design challenge for LEMs are the equitable distribution of participation opportunities and of the surplus created through exchange and the role and functions of the DSO. These topics, worthy of separate analysis, are beyond the scope of this chapter. For discussion of the role of the DSO in LEMs as market operator, see Mengelkamp et al. [8], p. 872 and Küster et al. [5], p. 2.

The IC constraint focuses on incentives for truthful revelation of type/preferences/ WTP/WTA. The most important IC issue in market design is market power, so the market designer should focus on minimizing the ability of a resource owner/operator to exercise local market power and on rules that induce them to bid/offer their true underlying subjective value/opportunity cost.

Local market power in LEMs will be an issue due to the spatially distributed nature of many grid services and the potential for lower liquidity and lower volume of transactions. Mitigating the exercise of market power is another reason to focus on low entry and exit costs, because potential entry (also called contestability) can be a potent force against market power.

Market design can mitigate local market power in several ways. Automated demand response is a capability that already exists, and by introducing elasticity into the market demand curve can mitigate market power ([6], p. 5). In specific grid services markets, such as voltage regulation, auction designs like the uniform price double auction decrease bid shading. For other grid services that are more like single-buyer procurement markets, such as reserves, a second-price auction design can decrease the winner's curse while still eliciting low-cost participation. Another market design that may be useful for grid services LEMs is a call option market, in which participants are paid some amount to be available (dispatchable) and then paid the market-clearing price for actual services rendered. Call option markets harness the benefits of contestability to mitigate market power.

A novel aspect of LEMs (and of TE more generally), worthy of further research, is how dual-nature resources like batteries and EVs potentially participate on both sides of this market, and how this flexibility and contestability can mitigate local market power. Consider this example: suppose a person owns several resources including one that is the marginal resource in the market going into peak periods. If that owner realizes the unit is marginal and thus marks the difference between a lower and a higher market-clearing price, s/he has an incentive to withhold that resource, which would shift the supply curve to the left and increase the market-clearing price, resulting in higher revenues earned by the other resources in the market. If demand is sufficiently inelastic, the revenues earned will more than offset the revenues lost by not operating the withheld resource. Such withholding is the core of market power, especially in illiquid markets.

The participation of batteries on both sides of the market contributes to mitigating such market power in two important ways. On the supply side, a larger number of smaller suppliers reduces the extent to which any single resource is likely to be the marginal unit on a regular basis. The ease of entry and exit of batteries and their varying offer prices as their charge levels change also make it more difficult for resource owners to discern whether or not they have the marginal unit in any particular market period. On the demand side, ease of entry and exit for batteries make them a flexible resource that adds responsiveness and elasticity to market demand, even in a small LEM. That elasticity disciplines the ability of any marginal supplier to exercise market power through withholding.

This concern about market power is particularly relevant in a market design in which aggregators and retail energy providers participate in LEMs. They are more likely to operate several resources and have control over a larger portion of the overall portfolio in an LEM, which means they are more likely to control a marginal resource and know when it is marginal. Ensuring low-cost entry and exit, particularly of dual-nature resources like batteries, reduces the likelihood of any single aggregator having sufficient market power in any LEM to be able to raise prices and reduce consumer surplus.

## 16.4 Economics of LEMs in the US regulatory context

The prospect for implementing LEMs, whether for energy or for grid services, depends on the institutional framework in a jurisdiction. In particular, it depends on the legislative mandate that empowers regulation and how regulatory authorities implement that mandate. LEMs have started to gain traction in Europe, and have not done so in the United States, which has a more complicated institutional framework.[10]

Regulatory institutions in the United States grew out of the electricity industry's evolution and out of the technologies themselves. The industry started with local distribution companies in cities like New York and Chicago (and with direct current distribution). The pioneering work of Tesla and Westinghouse on alternating current technologies for building a system that could generate and transmit current over long distances had two important effects on the economics of electricity: it amplified the economies of scale and scope in the cost structure of the system, and it changed the spatial nature of the location of different system elements and the architecture of the system itself. All of these consequences emerged in a bottom-up fashion as technologies evolved, local distribution systems started and grew, and larger electric utilities emerged through growth and mergers to exploit economies of scale and scope.

By the first decade of the 20th century, electricity's future as a large and capital-intensive industry that would transform peoples' daily lives was apparent. Electricity was also a vertically integrated industry, with all transactions in the supply chain (generation–transmission–distribution–retail) taking place between a single firm and the end-use consumer. In the United States this was the Progressive Era, a time of skepticism about large companies and their ability to exercise market power. As investor-owned electric utilities consolidated and grew these concerns mounted. From 1907 through the mid-1930s, most of the states in the United States established state public utility commissions (PUCs) with missions to regulate the investment and pricing decisions of electric utilities "in the public interest" (a regulatory standard dating to the *Munn v. Illinois* case at the US Supreme Court in 1887). Implementation of this regulation took the form of rate-of-return regulation, in which the regulated vertically integrated utility estimated its capital costs and operating costs, applied a financial market-based rate of return to its capital costs, and thus generated a revenue requirement. They then constructed a retail rate tariff that would enable them to earn that revenue in expectation, and submitted these proposals to the state PUC in a rate case, which regulators could approve, amend, or reject. As electric utilities grew into the 1930s they began to formulate interconnection and emergency supply agreements with adjoining utilities, often in other states. Transactions crossing state lines invoke the Constitution's commerce clause, introducing federal regulatory jurisdiction. In 1935 the Federal Power Act established a role for federal regulation to ensure that the prices

---

[10] Much of the analysis in this section draws on [14], which examines U.S. regulatory institutions in greater detail.

at which these interstate transactions occurred were "just and reasonable," a role that persists today as the guiding mission of the Federal Energy Regulatory Commission.

This split jurisdiction continues today, although it evolved considerably in the 1990s as a consequence primarily of cost overruns in nuclear plant construction and of the invention of the combined cycle gas turbine (CCGT). The CCGT transformed generation and broke down enough of the economies of scale to make wholesale power markets possible, and potentially competitive. Thus in the United States the 1990s saw both the development of organized wholesale power markets in organizations that would evolve into Regional Transmission Organizations (RTOs) and the restructuring of state-level PUC regulation in several states. Restructuring required vertically integrated utilities to sell their generation assets (i.e., unbundle generation) and, to varying degrees in different restructured states, to allow retail competition and retail choice. As a result, some states retain the vertically integrated industry structure and traditional PUC regulation of rates and some states are restructured. Some RTOs have wholesale markets whose participants are entirely in restructured states, and some have a mix of restructured and vertically integrated participants. The notable outlier is Texas, which in the late 1990s unbundled both generation and retail from the regulated utility, leaving the footprint of the regulated utility to be solely the transmission and distribution wires [34,35]. Texas is the only state in the United States with both fully competitive wholesale and retail markets without the supply participation of regulated incumbent utilities.

Another complicating dimension of utility industry structure in the United States is the diversity of ownership structures. In addition to regulated investor-owned utilities, some areas have municipal utilities or public utilities that are operated by the local government. Other areas, particularly rural areas, have electric cooperatives in which the customers of the utility are also members that can participate to a certain extent in business decision-making and strategy. Finally, regional federal power administrations operate hydroelectric power facilities and sell power to distribution utilities. In general, municipal utilities, cooperatives, and federal power administrations are exempt from state and federal economic regulation.

This regulatory patchwork offers both benefits and costs for the development of LEMs. The costs arise from a lack of uniformity and from the perspective of lowering the costs of widespread LEM development. That perspective assumes, though, that consensus would emerge in favor of developing LEMs in a more uniform regulatory environment, an assumption that seems to be borne out in the EU context but may not fare as well in a more bottom-up federal structure like the United States. The main benefits associated with the US regulatory patchwork arise from viewing the states as platforms for experimentation, the trial-and-error process through which discovery of beneficial institutional changes occurs [35]. Viewing states as institutional competitors offering natural experiments on different institutional arrangements goes back to the founding of the United States, reflecting James Madison's cautions in *Federalist 10* against the potential political power of special interest factions. While a more uniform approach may seem more efficient, if the point is to discover the effects of new institutional arrangements and new market designs, then a framework that allows for multiple parallel experiments is more likely to achieve the objective (although perhaps not as quickly as some may hope).

In the traditionally regulated, vertically integrated US states, LEMs would have to be established within the conventional structure of the regulated utility and with regulatory approval. Doing so entails bringing LEMs into the utility's regulated tariff structure and determining a specific tariff for LEM transactions. Increasing shares of DERs and digitization have made regulated utility rate design even more complicated than it was before. In the analog electromechanical system with large-scale central generation, utilities built generation and wires capacity to meet peak demand and designed rates with a fixed charge and a variable or volumetric charge. Now with DERs (and particularly with behind-the-meter DERs that are harder to observe), forecasting that daily peak demand is more challenging, and designing a rate structure that approximately matches rates to costs (a principle called cost-causation) is more difficult. Regulated rates increasingly have added "demand charges" for users based on their individual peak demand, using the argument that system costs arise from having to build to accommodate peak and the most substantial costs in distribution systems are not energy-based variable costs. Thus rate design philosophy has evolved into a combination of cost-causation and the objective of flattening individual load duration to the extent possible, to manage system peaks more effectively. Figuring out a regulated rate tariff for LEM transactions in this context will be theoretically and computationally tricky, in addition to the political economy challenges of whether or not politically powerful regulated utilities would accept LEMs, despite their being an effective mechanism for coordinating individual actions.

In states that have retail competition, and particularly in Texas, a path forward for LEMs could be more straightforward. For example, retail energy providers could either enable their customers to bid/offer into LEMs, or aggregate them and make bids/offers into LEMs. Or a coalition of retail energy providers could collaborate to provide an LEM platform around the distribution edge and interconnect to the distribution grid, in much the same way that microgrids around the distribution edge can operate. Microgrids essentially operate as LEMs for their members, so microgrid owners could also become LEM operators and allow participation connecting microgrids to each other, or allowing participation from parties outside of the physical microgrid network. The LEM opportunity and architecture in cases with retail competition might be the LEM serving as another platform around the distribution edge that would benefit from the distribution utility having open-access interconnection standards. Another configuration could be having the distribution utility operate the LEM platform as part of its DSO functions, in much the same way as would happen with a vertically integrated utility, and be compensated with a regulator-approved cost-based service fee.

The US regulatory context and the utility industry structure differ from the DSO-based structure that is evolving in Europe. Those differences may enable more diverse experimentation that could contribute meaningfully to the evolution of LEMs in different models and different contexts.

## 16.5  Conclusion

DERs are capable of providing energy and a wide range of grid services to others due to their technological heterogeneity and spatial diversity. Digitization and DER innovation are complements, creating the potential for DERs technically capable of providing energy and grid services to do so through markets, using price signals to coordinate and automation to respond. Economies of scale and scope are no longer the predominant cost driver in the energy system (although still an important driver in the wires network) as DERs become more energy-efficient generators and their production and installation costs fall.

As DER penetration increases and intermittency and unpredictability change the nature of forecasting and optimization in distribution networks. LEMs for energy and grid services will be useful in that context because they promote flexibility and enable the valuation of flexibility as a service. They are also adaptable to spatial changes in the way people use space, and use energy in those spaces, another dimension of flexibility. Designing LEMs to harness these benefits must be grounded in fundamental economic principles.

The economic theory of exchange and transaction cost economics combine to provide an economic foundation for LEMs. Price signals that emerge from market interactions convey information about the preferences and opportunity costs of market participants, information that would not otherwise be accessible. Digital innovation reduces transaction costs and increases the range of activities over which decentralized coordination through markets is feasible. Such innovation also changes the available value propositions, and LEMs enable people to experience them and learn how much they value them. Automation through TE makes DER participation in LEMs user-friendly. This process enables decentralized coordination of independent actors in complex systems like the distribution system. A transparent and thoughtful market design is important for achieving such decentralized coordination.

Mechanism design and auction theory combine to inform market design for LEMs. In an energy market context, low-cost entry and exit are paramount for enabling voluntary participation, which entails clear and transparent market rules, DER interconnection standards, and widespread interoperability standards. Low-cost entry and exit increase rivalry, contestability, and liquidity, which are aspects of market processes that discipline an agent's ability to misrepresent their true preferences.

Market design is part of a jurisdiction's institutional framework, and institutions will affect incentives for the development and adoption of LEMs. In the US regulatory context, federalism along with its heterogeneity means that multiple parallel institutional experiments, rather than a uniform adoption rule, will inform the growth of LEMs and the specifics of their market design.

## References

[1] E. Mengelkamp, T. Schönland, J. Huber, C. Weinhardt, The value of local electricity-a choice experiment among German residential customers, Energy Policy 130 (2019) (2019) 294–303.
[2] J. Schumpeter, Capitalism, Socialism, and Democracy, Harper Perennial, New York, 1942.
[3] G. Parker, M. Van Alstyne, S. Choudary, Platform Revolution, W.W. Norton, New York, 2016.
[4] F. Lezama, J. Soares, P. Hernandez-Leal, M. Kaisers, T. Pinto, Z. Vale, Local energy markets: paving the path toward fully transactive energy systems, IEEE Trans. Power Syst. 34 (5) (2019) 4081–4088.
[5] K.K. Küster, A.R. Aoki, G. Lambert-Torres, Transaction-based operation of electric distribution systems: a review, Int. Trans. Electr. Energy Syst. (2019) e12194.
[6] F. Teotia, R. Bhakar, Local energy markets: concept, design and operation, in: 2016 National Power Systems Conference (NPSC), IEEE, 2016, pp. 1–6.
[7] S. Widergren, Grid Services, Pacific Northwest National Laboratory, February 27, 2018.
[8] E. Mengelkamp, J. Gärttner, K. Rock, S. Kessler, L. Orsini, C. Weinhardt, Designing microgrid energy markets: a case study: The Brooklyn Microgrid, Appl. Energy 210 (2018) 870–880.

[9] P. Lacy, J. Rutqvist, The product as a service business model: performance over ownership, Waste to Wealth, Palgrave Macmillan, London, 2015, pp. 99–114.

[10] R. Ghorani, H. Farzin, M. Fotuhi-Firuzabad, F. Wang, Market design for integration of renewables into transactive energy systems, IET Renew. Power Gener. 13 (14) (2019) 2491–2500.

[11] L. Kiesling, Knowledge problem, in: P. Boettke, C. Coyne (Eds.), Oxford Handbook of Austrian Economics, Oxford University Press, Oxford, 2015.

[12] Centrica, Cornwall Local Energy Market. Available at: https://www.centrica.com/innovation/cornwall-local-energy-market, 2018.

[13] W. Sharkey, Natural monopoly and subadditivity of costs, The Theory of Natural Monopoly, Cambridge University Press, Cambridge, 1982 (Chapter 4).

[14] L. Kiesling, Implications of smart grid innovation for organizational models in electricity distribution, Smart Grid Handbook, Wiley, 2016, pp. 1–15.

[15] M. Jackson, Mechanism Theory, Stanford University, 2003.

[16] R.B. Myerson, Optimal auction design, Mathematics Oper. Res. 6 (1981) 58–63.

[17] R.B. Myerson, A.S. Mark, Efficient mechanisms for bilateral trade, J. Econ. Theory 29 (1983) 265–281.

[18] C. D'Aspremont, L.A. Gérard-Varet, Incentives and incomplete information, J. Public Econ. 11 (1979) 25–45.

[19] J.-J. Laffont, J. Tirole, Using cost observation to regulate firms, J. Polit. Econ. 94 (3) (1986) 614–641. Part 1.

[20] T. Börgers, An Introduction to the Theory of Mechanism Design, Oxford University Press, Oxford, 2015.

[21] J. Lin, M. Pipattanasomporn, S. Rahman, Comparative analysis of auction mechanisms and bidding strategies for P2P solar transactive energy markets, Appl. Energy 255 (2019) 113687.

[22] W. Vickrey, Counterspeculation, auctions, and sealed tenders, J. Financ. 16 (1961) 8–37.

[23] P. McAfee, J. McMillan, Auctions and bidding, J. Econ. Lit. 25 (1987) 699–738.

[24] P. Klemperer, Auctions: Theory and Practice, Princeton University Press, Princeton, 2004.

[25] P. Klemperer (Ed.), The Economic Theory of Auctions, Edward Elgar, Cheltenham, 2000.

[26] F. Menezes, P. Monteiro, An Introduction to Auction Theory, Oxford University Press, Oxford, 2005.

[27] R. Wilson, Strategic analysis of auctions, in: R. Aumann, S. Hart (Eds.), The Handbook of Game Theory, North-Holland/Elsevier Science Publishers, Amsterdam, 1991.

[28] E. Mengelkamp, P. Staudt, J. Garttner, C. Weinhardt, Trading on local energy markets: a comparison of market designs and bidding strategies, in: 2017 14th International Conference on the European Energy Market (EEM), IEEE, 2017, pp. 1–6.

[29] N. Fabra, Tacit collusion in repeated auctions: uniform versus discriminatory, J. Ind. Econ. 51 (3) (2003) 271–293.

[30] S.J. Rassenti, L.S. Vernon, B.J. Wilson, Discriminatory price auctions in electricity markets: low volatility at the expense of high price levels, J. Regul. Econ. 23 (2) (2003) 109–123.

[31] N. Fabra, N.-H. von der Fehr, D. Harbord, Designing electricity auctions, RAND J. Econ. 37 (1) (2006) 23–46.

[32] G. Federico, D. Rahman, Bidding in an electricity pay-as-bid auction, J. Regul. Econ. 24 (2) (2003) 175–211.

[33] Y. Son, R.B. Seok, K.-H. Lee, S. Siddiqi, Short-term electricity market auction game analysis: uniform and pay-as-bid pricing, IEEE Trans. Power Syst. 19 (4) (2004) 1990–1998.

[34] L. Kiesling, A.N. Kleit (Eds.), Electricity Restructuring: The Texas Experience, AEI Press, Washington, DC, 2009.

[35] L. Kiesling, Incumbent vertical market power, experimentation, and institutional design in the deregulating electricity industry, Indep. Rev. 19 (2) (2014) 239–264.

# South American Markets—regulatory framework: current trends and future prospects in South America

*Rubipiara C. Fernandes, Edison A.C. Aranha Neto and*
*Fabrício Y.K. Takigawa*

Federal Institute of Santa Catarina, Florianópolis, Brazil

## 17.1 Introduction

The objective of the reforms implemented in the electric power industry organization in South American countries, following a trend that initiated in the 1990s in Europe, was to create a competitive environment and increase private investments in the energy sector with the end of the public financing model. This reform had an important impact on the electric power industry in these countries given that it was anchored in the model based on competition, with the separation of the old, verticalized monopolies in companies specialized in power generation, commercialization, transmission, and distribution.

An important characteristic of the reform in the power sector was the search for incentive mechanisms for competition and productive efficiency, with emphasis given to energy commercialization which allowed for an increase in flexibility within the industry. As a result, the development can be seen as an electricity spot market which has the increase in transaction flexibility as its main purpose. This market allows for a balance between the offer of energy and meeting demand, based on defining a short-term price which will then serve as a reference for negotiating long-term contracts.

In contrast, the establishment of a free market functions as a second stage in which the consumers can buy energy directly from generators and sellers. This causes an unregulated impact on prices and conditions. However, this free market is restricted to a few segments with large-scale consumers since there is a concern among governments in relation to the remaining consumers, given the possibility for price fluctuations. Consequently, these consumers should have access to a regulated and stable tariff.

In South America, advances in the liberalization process of the electric power industry lost strength after the initial wave of privatizations. The proposed reform displayed significant flaws both in planning and execution. The implementation of the new sector design changed in large part due to growing governmental intervention. For instance, in Brazil, following the supply crisis due to the lack of investment in generation expansion; in Argentina, following the currency devaluation crisis; and in Bolivia, following the reestablishment of the sector.

Additionally, other countries adopted only the most basic aspect of the market model structure—the liberalization of generation—such as Uruguay. It is also important to highlight other successful cases, such as Chile, which is one of the oldest examples worldwide and Colombia, who adopted a hybrid model, with compatibilization of generation competition in a next-day market and a reasonably developed derivative market.

## 17.2 Brazil

Brazil is one of the most populous countries in the world with a total area of approximately 8.5 million $km^2$, making it almost as large as Europe. Brazil has long been considered a global player and has the eighth largest economy in the world. Thus it is not surprising that the Brazilian electricity sector is by far the largest in South America, characterized by a large interconnected system with installed capacity of 167 GW and hydrothermal generation basis (60% hydro and 40% thermal/wind/complementary) [1]. Although renewable energies make a large contribution to the energy share, this scenario of hydropower dependence makes the grid very sensitive to drought periods where the reduction of stored water in the reservoirs increases the electricity costs and risks to the system operation.

### 17.2.1 Electrical system

The Brazilian Interconnected System (SIN) can be considered unique in the world due to its size and characteristic. The generation and transmission of the Brazilian system is considered a hydrothermal system with hydroelectric predominance. Less than 1% of requested energy in Brazil is not connected to SIN. It is a small and isolated system mainly in the Amazon region [2].

### 17.2.2 Institutional agents

The creation of a new market structure by itself is not enough to ensure the success of the Brazilian Power Sector (SEB) without the simultaneous creation of agents responsible for sector operation.

These agents are shown in Fig. 17.1 and listed with a brief description and their functions [3].

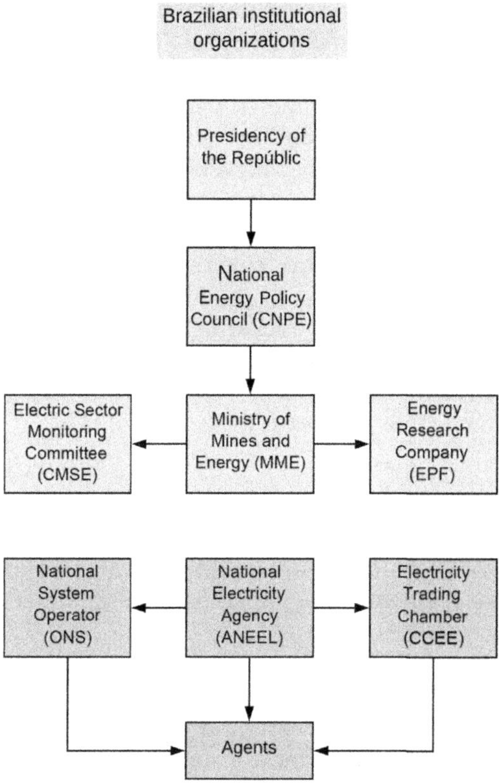

Brazilian institutional
organizations

**FIGURE 17.1** Brazilian institutional organizations.

*National Energy Policy Council (CNPE)*: Advisory body for the President to formulate national policies and guidelines for energy. Responsible for energy policies and promoting rational use of energy resources.

*Ministry of Mines and Energy (MME)*: Has competence in the areas of geology, mineral, and energy resources; use of hydraulic energy; mining and metallurgy; and oil, fuel, and electricity, including nuclear. It is the main element of the CNPE together with the Coordinator.

*Energy Research Company (EPE)*: Coordinates studies and research to support the planning of the energy sector. This company is linked to MME.

*Sector Agents*: Companies of generation, transmission, distribution, trading, and free consumers.

*Electric Energy National Agency (ANEEL)*: Electric system regulator, linked to MME, with the purpose of regulating and monitoring the generation, transmission, distribution, and trending of electric energy.

*Electric System National Operator (ONS)*: Conducts activities to coordinate and control the operation of electric energy generation and transmission in the SIN, under the regulation and supervision of ANEEL. Responsible for the elaboration of operational planning studies.

*Power Commercialization Chamber (CCEE)*: Its purpose is to carry out the wholesale transactions and commercialization of electric power within the Brazilian National Interconnected System, under the regulation and supervision of ANEEL. Main attributions: promotes auctions to sell and buy electric energy; keeps a record of sales commercialization agreements of electricity; investigates the Differences Settlement Price (PLD); and investigates the failure of electric power contracting limits and other infractions and applies penalties.

*Electric Power Sector Monitoring Committee (CMSE)*: Monitors and assesses the continuity and safety of the electric energy supply throughout the country. Tracks the development of generation, transmission, distribution, sale, import and export of electricity, natural gas, and oil and its derivatives.

The companies that act in generation, distribution, and trading (including free and special consumers) of electric energy are called Agents of the Chamber of Electric Energy Commercialization (CCEE), whose creation was authorized by decree n° 5.177/2004 to enable the commercialization of electric energy on the National Interconnected System (SIN) [4,5].

### 17.2.3 Energy market

On SEB, energy transmission and distribution are considered a monopoly because competition would not bring savings earnings to these segments. The generation and commercialization segments are based on competition.

Every energy contract must be registered with the CCEE. Energy trading in Brazil is realized in two different ways, according to Law no 10.848/2004. In the first, named Free Market (ACL), the generators (public service, self-generators, independent producers, marketers, importers, and exporters) and the free and special consumers. Free consumers (consumers with a load that is greater than or equal to 2000 kW) can freely negotiate their energy purchases. Special consumers (consumers with a load that is greater than or equal to 500 kW) can also freely negotiate their energy purchases but with the condition that at least 51% of the energy is from Alternative Sources. These are free to negotiate energy purchases, quantities, prices, and supply delivery dates. The second, called Captive Market (ACR), is an environment where the distribution utilities participate and purchase their energy by auctions arranged by the CCEE, and by the delegation of National Electricity Regulatory Agency (ANEEL). The energy buyers and sellers that are participating in the auction formalize their business relationship through registered contracts within ACR [2]. In this environment, the captive consumers do not have the option to choose their energy supplier. They buy energy only from the utility that they are connected to and pay the regulated tariffs defined by ANEEL. The regulated auctions for energy generation and transmission are a fundamental component of the Brazilian Electrical Sector's new legislation, introduced by Law no 10.848/2004. This law regulated energy trading for consumers through auctions within ACR [6].

## 17.3 Chile

Chile occupies a long, narrow strip along the coast between the Andes and the Pacific Ocean. It shares its northern border with Peru, its northeastern border with Bolivia, its western border with Argentina and the Drake Passage is the southernmost tip of the country. Chile has more than 18 million inhabitants in an area of 756,000 km$^2$.

### 17.3.1 Electrical system

Due to the geographical characteristics of the country, Chile has a unique system because of its length, practically 3100 km, running from the city of Arica in the north to the Island of Chile in the south [7]. Generation in Chile is 40% renewable and 60% thermoelectric [8].

The electrical system in Chile is currently divided into three electrical systems: The National Electrical System (SEN), the Aysén System (SEA), and the Magallanes System (SEM).

The SEN represents the merging of the old systems, Interconectado Central (SIC) and Interconectado do Norte Grande, and the largest installed capacity in Chile can be found here. It is important to highlight that each system has a distinct energy matrix, for example, the SEN is almost proportionately shared between renewable sources (hydroelectric, solar, wind, biomass, and geothermal). and thermoelectric sources (coal, natural gas, and petroleum), while at SEN nearly all sources are thermoelectric (natural gas and diesel) [9,10].

The three independent systems of the Chilean electrical system [8]:

*SEN:* 46% of the installed capacity corresponds to renewable sources (30% hydraulic, 8% solar, 6% wind, 2% biomass, and 0.2% geothermal), while 54% corresponds to thermal sources (21% coal, 20% natural gas, and 13% oil). The generation of renewable energy has increased significantly in recent years, from 35% in 2011 to 42% in 2017, respectively. Similarly, the market penetration of solar and wind technologies increased dramatically from 1% in 2011 to 10% in 2017. The system has a net installed capacity of 22.369 MW.

*SEA:* System that produces electricity to supply the Aysén region of General Carlos Ibañez del Campo. In December 2017, its net installed capacity was 62 MW, with 57% diesel, 37% hydraulic, and 6% wind.

*SEM:* System that produces electricity to supply the Magalhães and Chilean Antarctic regions. It has a net installed capacity of 104 MW, with 82% natural gas, 15% diesel, and 3% wind power.

### 17.3.2 Institutional agents

The Chilean electricity sector is composed of the institutional agents, as shown in Fig. 17.2 [11]:

*The Ministry of Energy:* The Ministry of Energy is the highest authority in the field. It is responsible for planning, policies and standards for the development of the energy sector and for the promotion of energy efficiency. The institutional framework of the Ministry of Energy was established by Law 20.402.

*National Energy Commission (CNE):* The CNE is a public organ, under the Ministry of Energy, responsible for the elaboration and coordination of planning, policies, and norms related to the development of the national energy market as well as ensuring

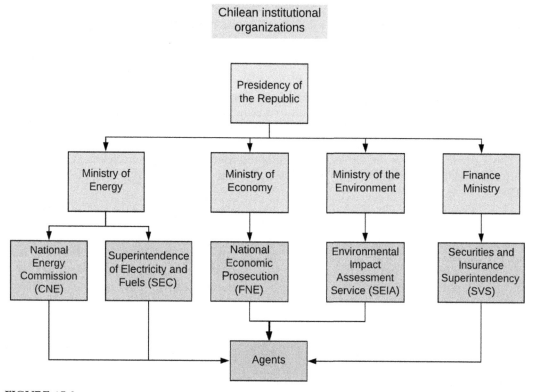

FIGURE 17.2    Chilean institutional organizations.

compliance of these. The CNE's institutional landmark is the Law Decree 2.224, modified by Law 20.402.

*Superintendency of Electricity and Fuels (SEC)*: The SEC is a public organ, under the Ministry of Energy, responsible for monitoring the level of electricity, gas, and fuel services in terms of safety, quality, and cost. Its tasks include monitoring compliance with laws, regulations, and technical norms in the sector of liquid fuels, gas, and electricity to ensure the quality of services provided to users is adequate and that the operation and use is safe for people and artifacts. SEC's institutional framework is Law 18.410, modified by Law 20.402. The CNE, SEC, and the Chilean Nuclear Energy Commission are dependent on the Ministry of Energy.

*Economic Cargo Dispatch Center—SIC (CDEC-SIC)*: The CDEC-SIC is the organ responsible for coordinating and determining SIC installation operations including generation plants, transmission system lines, and substations, and connection of free customer consumption. Their work includes ensuring service safety in the power system, ensuring the most economical operation of all electrical system installations, and guaranteeing the right of servitude over the transmission systems established by means of an electrical concession decree. The CDEC-SIC is composed of generators, transmitters, and free customers which operate in the SIC and it is financed by its members. The institutional framework of the CDEC is Decree 291.

*Economic Cargo Dispatch Center−SING (CDEC-SING)*: The CDEC-SING is the organ responsible for coordinating and determining SING installations operations. It is similar to the CDEC-SIC.

*Ministry of the Environment*: The Ministry of the Environment is the public organ responsible for proposing environmental policies, ensuring compliance of current environmental regulation, developing quality and emission standards, and maintaining a National System of Environmental Information (SINIA), among others. The Environmental Evaluation Service, under the Ministry, is responsible for the administration of the Evaluation of Environmental Impact Service (SEIA). The Ministry is decentralized through the Ministry Regional Secretaries of the Environment (SEREMI). Each region also has an Evaluation Committee, responsible for assessing projects submitted to the SEIA. Transregional projects, however, are directly qualified by the Ministry. The institutional framework of the Ministry of the Environment is Law 20.417.

## 17.3.3 Energy market

Generation, transmission, and distribution activities are identified in the energy market and developed by private companies. This authority plays the role of regulator and inspector, seeking to establish criteria that favor an economically efficient expansion of the electrical system. The electricity sector in Chile is governed by the general law for electrical services. The public organ responsible for the sector is the Ministry of Energy. It is in charge of planning, policies, and standards for the development of the energy sector. It also grants concessions to hydroelectric plants, transmission lines, substations, and electric energy distribution areas. It depends on the CNE, the technical body responsible for analyzing prices, tariffs, and technical norms which the companies should adhere to, calculating the tariffs based on the technical reports of learnability, and generating an operating plan, an indicative 10-year guide for the expansion of the system attached to the technical report. The SEC defines the technical standards and monitors compliance to these. In Chile there are two big interconnected systems, the Central Interconnected System (SIC) and the Interconnected System of Norte Grande (SING), in addition to the smaller systems of Aisén and Magallanes. The generator companies should coordinate their plant operations through the CDEC-SIC and CDEC-SING, respectively. CDEC's main function is to ensure the system is safe and program dispatch of the plants to meet demand at all times, at the lowest cost possible, and subject to safety restrictions [12].

Generating companies are paid for energy. Energy refers to effective consumption and is paid both at the marginal cost of system production as well as the amount agreed upon in the case of free contracts. The concept of energy compensates the generator for providing capacity to the system. Company power is paid for by the marginal cost of the system expansion. The firm power is the maximum power the generator is capable of injecting subject to its probable unavailability and is determined by CDEC for each plant. The sale of energy occurs through financial contracts. Supply contracts with distribution companies and free clients are signed and positions in the spot market are adjusted at marginal cost determined by the CNE every hour. This means that the companies whose production fails to meet obligations go to the spot market to balance their position. In this regard, the modifications introduced in the DFL-4 (Decree with force of law) through the Law Curta II determine that distribution companies should support consumption by their customers regulated by long-term supply contracts. The same law also sets out incentives for renewable energy. Transmission installations are paid through transmission tariffs paid by the generation companies proportionate to use. The tariffs are planned to offer transmission companies a real annual return of 10% on top of the value of their transmission installations. To these, additional costs are added for operations, maintenance, and administration. The transmission companies operate on an open access basis which guarantees access to the existing system by new actors who wish to participate in the expansion of the transmission system. The distributors are obliged to transfer to their regulated customers the node price the energy was bought for, added to the value added tax [12].

## 17.4 Colombia

Colombia is located in the northeast of South America and shares its eastern border with Venezuela and Brazil, its southern border with Ecuador and Peru, the North with the Caribbean Sea, the northeast with Panama, and the west with the Pacific Ocean. It has a population of more than 50 million people and its territory stretches over 1.1 million km$^2$.

### 17.4.1 Electrical system

The electrical sector has been grouped together into a network of generation, transmission, distribution, and commercialization since reforms in the sector in 1994. Almost half of the generation capacity is private. Private participation in electricity distribution is much smaller.

The Mesoamerica Project, the former Pueblo Panama Plan, includes an electrical interconnection project between Colombia and Panama, which would enable Colombia to become part of Central America. This project carried out by Interconexiones Eléctricas SA (ISA) in Colombia and Empresa de Transmissão Elétrica SA (ETESA) in Panama, involves the construction of a 300 MW transmission line (3% installed capacity) from Colombia to Panama and reverse capacity of 200 MW. Since 2011 the project has suffered all kinds of setbacks, including environmental, political, and social. In addition, the Colombian government signed an agreement with the national governments of the Dominican Republic and the state government of Puerto Rico to supply electricity through an underwater network that would link the north of Colombia to the Dominican Republic at a cost of approximately US$ 4–5 million and is currently undergoing an economic prefeasibility assessment.

The Colombian generator park has a net installed capacity of 17,366.43 MW and another 1072.29 MW from ChP, divided into 66 producing agents. The electricity sector in Colombia is mainly dominated by the generation of hydraulic (66% of production) and thermal (33%) energy [13].

Currently, Colombia has 12 market agents present in energy transmission, with a total of 24,912 km of extension lines in its national interconnected system.

## 17.4.2 Institutional agents

The Colombian electric sector is composed by the institutional agents, as can be seen in Fig. 17.3.

*Ministry of Mines and Energy (MME):* Some of its functions with regard to public service companies are [14]: establish the technical requirements the companies should meet; prepare a plan for the expansion of public service coverage to be checked by the Ministry

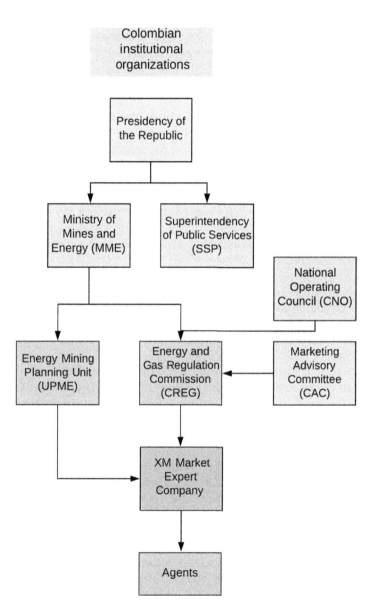

FIGURE 17.3   Colombian institutional organizations.

every 5 years; identify the amount of subsidies the Nation should give to respective public service; collect information on new technologies and management systems in the sector; promote, under the Direction of the President of the Republic, and in coordination with the Ministry of Foreign Affairs, international negotiations related to the relevant public service; and develop and maintain an adequate system of sector data for the use of authorities and public service in general.

*Energy Mining Planning Unit (UPME):* Organized as a Special Administrative Unit linked to the Ministry of Mining and Energy, its main functions are to determine the energy needs of the population and economic agents of the country. These are based on the demand projections which take into consideration the most probable evolution of the demographic and economic variables and costs of energy resources and develop the National Energy Plan and the Expansion of the Electric Sector Plan, according to the National Development Plan [15].

*Energy and Gas Regulation Commission (CREG):* Organized as a Special Administrative Unit of the Ministry of Mines and comprises the Ministry of Mines and Energy, which is presided over by the Minister of Finance and Public Credit, director of the National Department of Planning, five specialists exclusively dedicated to energy matters (nominated by the President of the Republic for 4-year periods), and the Superintendent of Home Public Services, with a voice but without a vote [16].

*Superintendency of Public Domestic Services (SSPD):* Technical organ linked to the Ministry of Economic Development with legal status and administrative and economic autonomy. Conducts specific functions of control and surveillance independent of the Service Commissions and with the immediate collaboration of the delegated superintendents. The superintendent and his delegates are appointed and freely dismissed by the president of the Republic [17].

*Commercial Exchange System Administrator (ASIC):* Unit responsible for the registration of long-term energy contracts, liquidation, collection and payment of the amount of acts or energy contracts on the stock exchange by generators and traders, maintenance of data systems and required computer programs, and the fulfilment of the tasks necessary for the proper functioning of the Commercial Exchange System (SIC).

*Liquidator and Account Manager (LAC):* Entity responsible for liquidating and invoicing charges for the use of the National Interconnected System, determining regulated revenue to carriers and administering the accounts which, due to use of the networks, are caused to wholesale market agents, according to current regulations.

*National Dispatch Center (CND):* It is the unit responsible for the planning, supervision, and control of the integrated operation of the generation, interconnection, and transmission resources of the National Interconnected System. It is also responsible for giving instructions to the regional expedition centers to coordinate a maneuver of the installations to ensure a safe and reliable operation and to ensure that these are consistent with the operational regulations and agreements stipulated by the National Operations Council.

*National Operational Council (CNO):* Its main function is to agree technical aspects to ensure the integrated operation of the National Interconnected System is safe, reliable, and economical and it is the executive operation regulation body. Decisions by the National Operational Council can be appealed with the Energy and Gas Regulation Commission.

### 17.4.3 Energy market

The electricity sector is based on the fact that commercial companies and large consumers acquire energy and power in a market of large blocks of energy. In order to promote competition among generators, the participation of economic, public, and private agents is allowed, which should be integrated into the interconnected system to participate in the wholesale energy market. On the other hand, traders and large consumers act by signing electricity contracts with the generators. The price of electricity in this market is agreed upon by both contracting parties, without state intervention. The operation and administration of the market is undertaken by the XM, which is responsible for the National Dispatch Center (CNN), Commercial Exchange System Administrator (ASIC), and Liquidator and Account Manager for charges for SIN-LAC Network Use [13].

In terms of the market, the regulatory framework established the separation of users into two categories: regulated and unregulated users. The basic difference between the two is related to the handling of prices or tariffs applicable to the sale of electricity. While in the first case the tariffs are set by the CREG through a tariff formula, in the second, the prices of sales are free and agreed between both parties.

## 17.5 South America countries comparison

It is noteworthy that Brazil, Chile, and Colombia are similar in many dimensions such as culture, language, and macroeconomic development. They are also all depending, to a large extent, on hydrogenerated electricity. All three countries have implemented different deregulated systems, allowing a unique possibility to compare the performance of different implementations of deregulation on one continent. Brazil, Chile, and Colombia have already moved to major changes in their electricity system structure, accompanied by the restructuring of their energy markets as they have faced severe power deficits. In a short time, they pushed aggressively for reforms, opened up their energy sectors, built renewable energy capacities to reduce their dependence on fossil fuels, and made a serious effort to lower their carbon footprint.

It can be observed that there is no doubt that technology will be an important driver going forward. Power plants will have to strike the right balance between storage, thermal energy generation, and renewables that are intermittent in nature.

Importantly, the significant role of hydropower in several South American countries, meanwhile, creates opportunities to harness complementarities and scale up other renewable energy technologies. In a deeper analysis, it can be verified that the mechanisms through which hydropower and other renewables can complement each other will result in the improvement of the the cost-efficiency and reliability of power systems.

With electricity demand in the region expected to rise more than 70% by 2030, regional integration could play an important role in increasing efficiencies and bringing down the price of energy.

The following tables show different aspects of the liberalization policies of the energy market, the ways the generation and consumer agents can access the free market, the ways these market prices are set, and the ways generation sources are dispatched to meet demand. It also highlights the insertion of new renewable sources (such as wind, solar, and biomass) in the composition and diversification of the energy network in South American countries.

| Country | Electricity market structures | Electricity generation by source | Market models |
|---|---|---|---|
| | I—Vertically integrated regulated monopoly<br>II—Vertically integrated utility + IPPs<br>III—Wholesale market<br>IV—Wholesale market + retail competition | Diversification in the region in South America power sector | Pool/based on offers/bilateral. |
| Argentina | III—Wholesale market | 3.8% Renewable (data 2018) | Dispatch based on marginal costs. |
| Bolivia | I—Generation, transmission and distribution are by the State | 4.5% Biomass; 0.8% hydro | Cost dispatch. |
| Brazil | III—Wholesale market | 6.6% Hydro; 8.6% natural gas; 8.5% biomass; 7.6% wind; 3.2% coal; 2.5% nuclear; 2.4% petroleum; 0.5% solar (data 2018) | Dispatch based on marginal costs. |
| Chile | III—Wholesale market | SING—51% coal; 35% natural gas; 9% diesel; 4% hydro; 5% renewables<br>SIC—41% hydro; 21% diesel; 12% natural gas; 15% coal; 13% renewables | Cost dispatch. |
| Colombia | IV—Oligopoly in the generation, transmission and distribution | 68.4% Hydro; 13.3% natural gas; 9.5% mineral coal; 7.8% liquid petroleum derivatives; 1% Intermittent Sources of Renewable Energy (data 2018) | Loose pool. |
| Ecuador | I—Migration of state model to private | 58% Hydraulic; 39% thermal; 3% others (photovoltaic, wind, biogas biomass) | Cost dispatch. |
| Paraguay | I—State monopoly | 99% Hydro; 1% thermal for isolated systems | Cost dispatch. |
| Peru | III—Deverticalized GTDC | 4% (2% wind; 2% solar) | Gross Pool Model with presence of free contracts. |
| Uruguay | II—Vertically integrated utility + IPPs | 97% Renewable (3% solar; 38% wind; 7% biomass, 49% hydro), and 3% thermal | Cost dispatch. |
| Venezuela | I—State monopoly | 62% Hydro; 35% thermal (oil and gas); 3% DG (conventional sources) Fundamentally in diesel generation for isolated systems (data 2018) | The country has been experiencing a continuous process of energy rationing over the last few years. This indicates the absence of options for energy sources and the dispatch of all available sources. |

IV. Regulatory framework: Current trends and future perspectives

| Country | Energy pricing | Energy market aspects | Interconnections | Interconnected system |
|---|---|---|---|---|
| | CMO/day ahead/intra-day (hourly). | Captive/free/criteria to be free. Commercialization:<br><br>• Contract markets/Horizon.<br>• Short-term market (spot price). | Interconnection with neighboring countries. | Interconnected system/isolated systems. |
| Argentina | Hourly pricing (marginal pricing of the system or market pricing, and represents the economic cost of generating the next kWh). | Demand greater than 30 kW to enter the free market. Corresponds to 15% of energy (data 2018). | Interconnection with Brazil, Chile, Uruguay, and Paraguay. | Totally interconnected with exception of Tierra del Fuego. |
| Bolivia | Hourly and nodal pricing. | Has only four unregulated consumers (free), which can have free bilateral contracts but supervised by the CNDC. | No interconnection with neighboring countries. | System interconnected through 69, 115, and 203 kV systems. Responsible for 90% of the country's electrical energy. |
| Brazil | Hourly and zonal pricing. | Demand greater than 500 kW (special consumer) and greater than 2 MW (free consumer) to enter the free market. Corresponds to 28% of energy (data 2019). | Interconnection with Argentina, Uruguay, Paraguay, and Venezuela. | Totally interconnected with exception of small parts of Amazon forest. |
| Chile | Pricing is composed by marginal pricing and remuneration to agents for the ability to meet peak demand. | Demand greater than 500 kW.<br>SING—89% in free, 11% in captive.<br>SIC—69% in captive, 31% in free. | Although plans exist, there is no connection with other countries. | The country is divided into four regions which are not interconnected: SING, SIC, SEA, and SEM. |
| Colombia | Hourly offers defined in the day-ahead market. As there is a centralized operator, there is no need for an intra-day market of the agents, the operator is the unit responsible for liquidating and invoice energy surpluses. | All consumers, whether residential (67%) or commercial (13%), are competitive and free to choose their energy supplier. The industrial consumers (21%) are also free to negotiate directly with the generators and participate in the settlement of surpluses made by the operator. | Interconnection with Ecuador and Venezuela. Interconnection under construction with Panama (and, indirectly, all of Central America). | Eastern region is interconnected. Only some isolated systems in the western region (Amazon forest). |

| Country | Energy pricing | Energy market aspects | Interconnections | Interconnected system |
|---|---|---|---|---|
| Ecuador | Weekly pricing. All plants are "subsidiaries" of one national energy company. Each agent is paid for its contribution. | Captive consumers. Large clients should have a demand greater than 675 kW in the last 6 months. | Interconnection with Colombia and Peru. | Interconnected system in the mountains and on the coast. Isolated systems in the Amazon forest and on the islands. |
| Paraguay | Pricing formation by cost but with tariffs defined by ANDE. | Natural monopoly. | Interconnection with Brazil and Argentina. Two large isolated plants. | Interconnected systems. Isolated systems are supplied by thermoelectrics. |
| Peru | Day ahead (by generation costs) with difference resolution during dispatch day. Price per bus is defined for 15-min intervals. | Captive market of the distributors and free market for consumers with demand greater than 1000 kW. Free term contracts. Short-term market. | Interconnection with Ecuador and project with Chile. | SEIN 99% T/90% G, SSAA (isolated) 0.1% T/10% G (installed power). |
| Uruguay | Nodal and hourly pricing. | Captive market of the distributors and free market for consumers with demand greater than 250 kW. | Interconnection with Argentina and Brazil. | Totally interconnected with exception of small parts. |
| Venezuela | The prices are defined by the only state administrative company and are subsidized mainly for the residential class. | 100% Captive Regulated market (only autoproduction of energy is allowed). | Interconnection with Colombia and Brazil. | The electric system is divided in six regions interconnected. |

As a result of the data comparing the South American countries shown in the tables above, it can be seen that the regulatory policies were decisive and led to a series of pioneering solutions being implemented to regulate the electric sector, driving the market liberalization process to new investors and removing total power from the state in terms of expansion and establishment of the free energy market. It is also important to highlight that in countries such as Brazil, Chile, and Colombia, this opening occurred significantly with the existence of a free energy market and the attraction of private investments to promote the expansion of the electric sector in an adequate and sustainable way to meet the growing demand. Meanwhile, other countries in the region have a vertical integration of activities in the form of a regulated monopoly or by allowing the insertion of independent generators only in the sector.

All these factors contributed so that South American countries have some of the highest and lowest energy prices in the world. This can be explained by several market fundamentals and characteristics and pricing policies, which impact the mix of energy in each country, affecting the insertion and renewable energy incentive policies. This occurs with diversification with regard to nonhydroelectric renewable energies, which are reflected in the mix of generation, where the participation of hydroelectric energy in total renewable generation has decreased in these countries in recent years.

## 17.6 Conclusion

This chapter began with considerations of the reforms in the energy sector in the search for mechanisms to promote competition and productive efficiency with an emphasis on the commercialization of energy, to allow and increase flexibility in the sector. This was followed by a general overview of the implementation of these reforms in the South American countries, highlighting Brazil, Chile, and Colombia, where this liberation process is more assertively configured by the establishment of a liberalized energy market. Then, a comparative analysis between the countries was carried out with the aim of acquiring a general view of this process and the current policies in place to reach this objective. The trends of changes in the power matrix with the alterations and propositions of the regional political landscape for renewable energies, including institutional structure and specific policies in the energy sector, were outlined.

# References

[1] EPE, Balanço Energético Nacional – BEN 2019 - Ano Base 2018, Empresa de Pesquisa Energética – Rio de Janeiro: EPE, 2019.
[2] ONS, O que é o SIN? < http://www.ons.org.br/paginas/sobre-o-sin/o-que-e-o-sin > .
[3] E. Nery, Mercados e Regulação de Energia Elétrica, Interciência, Rio de Janeiro, 2012.
[4] CCEE, Ambiente livre e ambiente regulado. < http://www.ccee.org.br/portal/faces/pages_publico/como-participar/ambiente-livre-ambiente-regulado > .
[5] CCEE, Quem participa. < http://www.ccee.org.br/portal/faces/pages_publico/quem-participa > .
[6] F.Y.K. Takigawa, et al., Energy management by the consumer with photovoltaic generation: Brazilian Market, IEEE Latin America, 2016.
[7] Coordinador Electrico Nacional, Sobre el SIC. < https://sic.coordinador.cl/sobre-sic/sic/ > .
[8] Generadoras de Chile, Generación Eléctrica en Chile. < http://generadoras.cl/generacion-electrica-en-chile > .
[9] Coordinador Electrico Nacional, Sistema Eléctrico Nacional. < https://www.coordinador.cl/sistema-electrico/ > .
[10] ELECTRICIDAD, La nueva realidad del Sistema Eléctrico Nacional. Electricidad – La revista energética de Chile. < http://www.revistaei.cl/reportajes/la-nueva-realidad-del-sistema-electrico-nacional/ > .
[11] Central Energía, Autoridad – coordinación. < http://www.centralenergia.cl/actores/autoridad-energia-chile/ > .
[12] Central Energía, Regulación. < http://www.centralenergia.cl/regulacion/ > .
[13] XM, Mercado de energía. < http://www.xm.com.co/Paginas/Mercado-de-energia/descripcion-del-sistema-electrico-colombiano.aspx > .
[14] MinMinas, Ministerio de Minas y Energía. < https://www.minenergia.gov.co/ > .
[15] UPME, La Unidad de Planeación Minero Energética. < http://www1.upme.gov.co/ > .
[16] CREG, Comisión de Regulación de Energía y Gas. < http://www.creg.gov.co/ > .
[17] Superservicios, La Superintendencia de Servicios Públicos Domiciliarios. < http://www.superservicios.gov.co/ > .

# 18

# Electricity markets and local electricity markets in Europe

*Zita Vale[1], Débora de São José[2] and Tiago Pinto[3]*

[1]School of Engineering, Polytechnic of Porto (ISEP/IPP), Porto, Portugal [2]Polytechnic Institute of Porto, Porto, Portugal [3]GECAD Research Group, Polytechnic of Porto (ISEP/IPP), Porto, Portugal

## 18.1 Introduction

The electricity sector is traditionally composed of large centralized power plants, which generate electricity that is transported through high-, medium-, and low-voltage transmission lines to the final users. Most of these are located in big urban centers, far away from the energy generation facility. This centralized organization leads to an economy of scale and made possible the rapid development of this sector as it was built, essentially, to be a monopoly industry, vertically integrated with governmental supervision aiming to promote efficiency with lower costs and losses [1].

However, many changes have taken place in the last decades, namely regarding the economy, technology, environment, and even end-users. Today electricity is considered a social necessity and consumers, in all developed and developing countries and regions of the globe, expect it to be available at all moments of the day [2]. This expectation is kept even considering increasing demand, environmental restrictions, increasing use of nondispatchable energy sources, technical issues, and many other factors that negatively impact the sector's efficiency [3].

Furthermore, electricity availably at reasonable costs is a crucial factor for todays' societies, enabling their economic development and adequate comfort levels for people. Scientific and social advancement brought more consciousness to society in energy-related matters. To keep up with the need of decreasing the electricity impact on the environment while coping with increasing demand levels, the industry has to adapt itself, this being one of the main goals of smart grids.

The technological change, climate crisis, and the need to replace nonrenewable fuels by renewable energy sources or low emission sources are imposing a major transformation

311

in the electricity system and in its organization. This transformation is decentralizing the system by including small generation units, closer to load centers, connecting them with medium- and low-voltage grids and reshaping the system's structure [4,5].

Environment-related targets and the subsequent technological advancements brought the possibility of widespread dissemination of small and very small local/residential renewable generation. This helps to address challenges that the system faces in terms of sustainability, regional self-sufficiency, and increase in demand. At the same time, it also overcomes difficulties for providing electricity in remote places.

Considering the current easiness of installing small power generation units (e.g., photovoltaic units in residences), the consumer is empowered and gains importance. At the same time, active participation and contribution to the system are expected from the demand. Consumers that produce energy, usually referred to as prosumers, can have active roles, not only because they can produce at least a part of the energy they need but also because they can provide energy and energy-related services to the system and even to other consumers and different players in the sector. Demand flexibility, that is, consumers' ability to adapt their consumption in the face of different factors, also put them at the center. While generation is becoming less flexible due to the increased use of nondispatchable energy sources, such as wind and solar, demand flexibility is a highly valuable asset to cope with energy resource management. Deploying flexibility through demand response (DR) programs has proved to be an efficient way to contribute to increased system efficiency while bringing benefits to consumers [6]. DR is defined by Albadi and El-Saadany [7] as changes performed by the electricity end-user that occur as a consequence of changes in electricity prices over time or decreases in electricity consumption as a result of incentive payments. Thus DR can be defined as end-users intentionally changing their consumption pattern (period of the day, instantaneous demand, or total consumption). Based on that, the traditional passive consumer can turn into an active and empowered one based on generation ability and demand flexibility if adequately supported by adequate information, communication, and technologies [8–11].

Distributed generation requires new methods for efficient energy resource management [12,13] and has opened the way for microgrids, bringing new visions and opportunities regarding local energy resource management [14,15].

Considering consumers' size and impact as small producers in a system with centralized large and medium producers, their access and participation in electricity markets is particularly difficult. To reduce this barrier, the concept of the local market started to be researched with different approaches [16,17,18,19].

A local electricity market (LEM) is defined in Ref. [20] as an electricity market in a community that allows residential households and other community members to trade electricity generated in the community using a local platform. This allows the community as a whole and its individual members to be more active in energy-related decisions and actions. They will strategically manage the individual and the community resources according to their goals, including the participation in external markets for energy and service provision and DR opportunities. Moreover, LEM is also defined by [20] as "marketplaces that enable prosumers and/or other local generating entities to trade energy volumes of their choosing within local communities."

Although LEMs seem like an ideal solution to promote stress reduction in the electricity sector and to give prosumer negotiation power, they also bring new challenges that must be taken into consideration.

The local market is a new concept that still needs more research to clarify all aspects and consequences of their implementation. Business models are as important as technical models and methods, both economic and technical aspects deserving more attention, significant evolution in state of the art, and practical experimentation before their rollout. Human behavioral aspects are also of high importance in the context of LEMs and energy communities [7], in particular having in mind that the involved players are, in most cases, of small or medium size, without significant knowledge and material resources regarding their decisions and action in energy-related matters, including the participation in energy markets. This means that the successful implementation of LEMs requires business and technical models' advancements and significant developments in terms of how medium and small size players, including domestic consumers, see their role in this new context. Wide and effective dissemination of the new possibilities must be ensured to be adequate to the target audiences. LEMs benefit the power and energy system and the small- and medium-sized players participating in those markets. From the point of view of these players, including different consumer types, LEMs should bring new possibilities to lower their energy costs and/or increase their energy-related profits. For power and energy systems, LEMs enable gathering of the huge potential of demand flexibility of those small- and medium-sized players that is currently locked. Unlocking this potential is crucial to bring the required flexibility to the system to enable the envisaged intensive use of renewable, nondispatchable primary energy resources. Explicit and implicit DR schemes, respectively, enabled by incentive-based DR programs [21] and by dynamic electricity prices [22], can then be put in place with benefits for the power and energy systems and for the consumers, leading to lower energy costs.

However, the actual active participation of those small- and medium-sized players requires different business and technical models and methods that should be studied regarding their adequacy and practical effectiveness. It also requires the seamless deployment of the required platforms and supporting technological equipment.

Although this may seem, at first sight, a huge and complex path to be done in the short term, a closer and informed analysis should be done. In the past years, significant advancements have been made that can be used in favor of the successful implementation of LEMs in a short time horizon. The building blocks are already a reality. It is now required to choose them carefully and build innovative, effective, and economically feasible solutions that enable LEMs widespread implementation.

Significant advancements have been made regarding energy resource management at the consumer and community levels [9], load and consumer profiling, demand flexibility and DR [6,21,23], aggregation [24,25], microgrids [14], peer-to-peer (P2P) transactions [26,27], transactive energy [28,18], electricity markets models and simulation [29–33], and strategic market negotiation [15,34–36] in the past years. There have also been advancements in several technological aspects such as smart metering, the Internet of Things, different devices enabling the supervision and control of energy loads and assets, such as smart plugs [10,11] and home assistants, and distributed processing using light single-board computers [27].

The effective implementation of LEMs also requires them to be as little intrusive as possible into the small- and medium-sized players that participate but for whom that participation is not a significant business of professional activity. This includes a large number of different residential, commercial, and small industrial consumers. Those decisions and actions related to their participation should be automated, both in terms of the market negotiation and on the subsequent required actions and control regarding their energy assets. For that, traditional methods are not adequate as the envisaged participation requires intelligent behavior and decision-making. Artificial intelligence (AI)-based models and methods have been advancing in this field, with consolidated advancements concerning the application in electricity markets.

The already existent and validated models, methods, and devices are the required ingredients to put LEMs in place. With the increasing number of pilots and their replication, equipment and operation costs will significantly decrease toward the rollout of LEMs-related solutions.

In the new context, the new roles must be well-defined, and responsibilities clarified so that it is ensured that LEMs come in favor of their players and the whole system.

After this introduction section, Section 18.2 presents an overview of electricity markets in Europe, with a special focus on the EU vision and policy, addressing the aspects and the legislation and regulation more closely related with LEMs. Section 18.3 discusses the challenges and benefits of LEMs, and Section 18.4 presents some relevant examples of LEM implementations in Europe. Finally, the main conclusions taken from this overview are discussed in Section 18.5.

## 18.2 Electricity markets in the European Union

Europe and, more particularly, the EU have been pursuing ambitious goals in terms of energy, with energy policy pushing for more clean and affordable energy and highly competitive electricity markets. The European Union (EU) "aims to build a more integrated, competitive European energy market (energy union)" [37]. The aim is "to ensure a reliable supply of energy and to keep prices affordable" while supporting "energy from renewable sources and the efficient use of energy, both of which help to cut greenhouse emissions" [37].

In the EU energy policy scope, markets are closely related to consumers, with rules protecting energy consumers' rights. The EU aims at establishing effective and competitive single markets for gas and electricity. This has its roots in the Single European Act of 1986 [38], which should also be applied to the energy sector, particular to electricity and gas markets.

### 18.2.1 European electricity markets: pioneering targets and legislation

The traditional power system, as we know it, relies on large power plants for electricity generation. These are located far away from the most important consumer centers to which the transmission power system connects them. This model presents huge benefits, and many inefficiencies, namely the losses resulting from transmission at long distances and the consequent need to use high voltage levels, lack of flexibility, small or no active participation from consumers, among others [16].

During the last decades, different factors, such as the increasing use of renewables driven by the need to decarbonize the power industry and increasing use of small and microgeneration, are pressuring the sector for changes, bringing new actors and changing how they interact [2]. Therefore, these challenges are leading the system to adapt and reorganize itself. According to Ref. [39], the EU's energy system, just like the rest of the globe, is in the middle of a transformation and the electricity market is at the center of this change. The move to renewable and decarbonized generation brings up challenges and opportunities for market participants.

In its decisions of October 23 and 24, 2014 on the *2030 Climate and Energy Policy Framework*, the European Council focused on the significance of a progressively interconnected energy market and the requirement for adequate help to coordinate consistently expanding a degrees of variable sustainable power source and along these lines permit the Union to satisfy its initiative desire for the energy change [40].

Furthermore, the Commission Communication of February 25, 2015, entitled *A Framework Strategy for a Resilient Energy Union with a Forward-Looking Climate Change Policy*, sets out a vision of an Energy Union with residents at its center, where residents take responsibility for energy progress, advantage from new advancements to decrease their bills and partake effectively in the market, and where rights of vulnerable consumers are ensured [39,41,42].

The creation of local energy markets in the European Union Energy System will take place according to the 2019/944 [39,41,42], the 2019/943 [39], and the Renewables Directive 2018/2001 [40], which set legal and market participation principles for local energy communities and renewable energy communities. These Directives and Regulations do not emerge without any notice. They result from a set of policies that started being implemented in 2009 and can be seen in Table 18.1, which summarizes the main European Directives and Regulations that are relevant for LEM.

Altogether, these legislations are paving the way toward competitive, fair, transparent, and nondiscriminatory electricity markets that can cope with the increasing share of renewable-based generation and distributed energy resources.

By June 2010, all State Members informed the European Commission about their National Renewable Energy Action Plan with a well-defined roadmap to guide expected transitions. Progress reports are also used to keep track of implementation [91].

Furthermore, the renewable energy progress report, a report from the European Commission, shows that European State Members were on track to achieve the aggregate 20% goal [92].

Directive 2009/72/EC came to close the group of three EU's directive (Directive 96/92/EC; Directive 2003/54/EC; and Directive 2009/72/EC) which started the liberalization and restructuration of the electricity markets and designed an efficient, competitive, and sustainable energy market across the EU [93].

Additionally, during the COP24 (United Nations Climate Change Conference) in December 2018, the Katowice package was adopted. It describes all the common and detailed rules, procedures, and guidelines to operationalize the Paris Agreement [94]. It also shows that all addressed EU directives, regulations, and legislation were adopted by the end of 2018 [94].

At last, even considering all the past efforts in the LEM direction, there are still steps to go, especially related to the EU framework on the internal electricity market. Regulation (EU) 2019/943 [39] and Directive (EU) 2019/944 [95] both from 2019 established rules for the internal electricity market and include constraints imposed by the development of renewable energy and environmental policies.

**TABLE 18.1** European Directives and Regulations relevant for LEM.

| Name | Date | Subject addressed | Implementation deadline | Related regulations/documents | Reference(s) |
|---|---|---|---|---|---|
| Directive 2009/28/EC | April 23, 2009 | Promotion of the use of energy from renewable sources Contracting Parties' binding national targets to be accomplished using renewable energy in the electricity, heating and cooling, and transportation areas by 2020 | January 1, 2014 | Directive 2009/28/EC Directive (EU) 2018/2001 General Policy Guidelines on the 2030 targets Decision 2018/02/MC-EnC amending Decision 2012/04/MC-EnC Decision 2012/04/MC-EnC Policy Guidelines 05/2015-ECS Policy Guidelines 01/2018-ECS | [43–49] |
| Regulation 714/2009 EC | July 13, 2009 | Conditions for access to the network for cross-border exchanges in electricity Set fair rules for cross-border exchanges in electricity to improve competition within the internal market | January 1, 2014 | Regulation (EC) 714/2009 Regulation (EU) 2019/943 Decision 2011/02/MC-EnC Procedural Act 2012/01/PHLG-EnC Decision 2015/01/PHLG-EnC Policy Guidelines 03/2015-ECS Policy Guidelines 02/2018-ECS Policy Guidelines 01/2019-ECS | [50–54] |
| Directive 2009/72/EC | July 13, 2009 | Common rules for the internal market in electricity, generation, transmission, distribution, and supply of electricity | January 1, 2015 | Directive 2009/72/EC Decision 2011/02/MC-EnC Policy Guidelines 02/2015-ECS Policy Guidelines 01/2015-ECS Policy Guidelines 01/2013-ECS | [51,55–58] |
| Regulation 838/2010 EC | September 23, 2010 | Guidelines relating to the intertransmission system operator compensation mechanism and common regulatory approach to transmission charging Transmission system operators will receive payment for cost caused as a result of hosting cross-border flows of electricity on their networks on the basis of the guidelines set out in this regulation | January 2014 | Regulation (EU) 838/2010 Decision 2013/01-PHLG-EnC | [59,60] |
| Regulation 1227/2011 | October 25, 2011 | Wholesale Energy Market Integrity and Transparency | May 29, 2020 | Regulation (EU) 1227/2011 Decision 2018/10/MC-EnC Corrigendum Decision 2018/06/PHLG-EnC Correlation table for Regulation (EU) 1227/2011 | [61–64] |

| Name | Date | Applicable | Description | Related documents | References |
|---|---|---|---|---|---|
| Regulation 543/2013 EC | June 14, 2013 | January 1, 2014 | Submission and publication of data in electricity markets Provides a much more complete and thorough set of definitions of the data to be published, prescribes roles and responsibilities, and establishes a central surface for data to be published | Regulation (EU) 543/2013 Decision 2015/01/PHLG-EnC | [53,65] |
| Regulation 347/2013 EU | October 16, 2015 | January 1, 2017 | Guidelines for trans-European energy infrastructure Accelerate the permitting procedure and help investments in the energy infrastructure to accomplish the Energy Community's energy and environmental policy goals | Regulation (EU) 347/2013 Decision 2018/11/MC-EnC Decision 2016/11/MC-EnC Decision 2015/09/MC-EnC Recommendation 2018/01/MC-EnC Recommendation 2016/01/MC-EnC Recommendation 2014/01/MC-EnC Secretariat's report on implementation of PECIs | [66–70] |
| Paris Agreement | December 12, 2015 | NA | First universal legally binding global climate change agreement adopted at the Paris Climate Conference (COP21) Corresponds to a bridge between today's policies and climate neutrality before the end of the century | Paris Agreement Paris Agreement ratifications | [71] |
| Regulation 2016/631 EU | April 13, 2016 | July 12, 2021 | Establish a network code on requirements for grid connection of generators | Regulation (EU) 2016/631 Decision 2018/03/PHLG-EnC Correlation table for Regulation (EU) 2016/631 Derogation criteria for Regulation (EU) 2016/631 – Bosnia and Herzegovina | [32,72–74] |
| Regulation 2016/1388 EU | August 17, 2016 | July 12, 2021 | Establish a network code on demand connection | Regulation (EU) 2016/1388 Decision 2018/05/PHLG-EnC Correlation table for Regulation (EU) 2016/1388 Derogation criteria for Regulation (EU) 2016/1388 – Bosnia and Herzegovina | [75–78] |
| Regulation 2016/1447 | August 26, 2016 | July 12, 2021 | Establish a network code on requirements for grid connection of high voltage direct current systems and direct current-connected power park modules | Regulation (EU) 2016/1447 Decision 2018/04/PHLG-EnC Correlation table for Regulation (EU) 2016/1447 Derogation criteria for Regulation (EU) 2016/1447—Bosnia and Herzegovina | [29,79–81] |

(Continued)

**TABLE 18.1** (Continued)

| Name | Date | Subject addressed | Implementation deadline | Related regulations/documents | Reference(s) |
|---|---|---|---|---|---|
| General Policy Guidelines on the 2030 targets | November 29, 2018 | Setting targets for 2030 on energy efficiency, renewable energy, and greenhouse gas emission reduction | NA | General Policy Guidelines on the 2030 Targets for the Contracting Parties of the Energy Community | [44] |
| Directive 2018/2001 (EU) | December 11, 2018 | Promotes the growth of the share of renewable energy sources in final energy consumption by 2030 | NA | Directive (EU) 2018/2001<br>Regulation (EU) 2018/1999<br>Regulation (EC) No 1099/2008<br>Regulation (EU) No 525/2013<br>Regulation (EU) No 182/2011 | [40,66,67,82–84] |
| Regulation 2019/943 EU | June 5, 2019 | Aimed at improving the EU regulatory framework governing the internal electricity market | December 31, 2030 | Regulation (EU) 2019/943<br>Regulation (EC) No 1228/2003<br>Regulation (EC) No 714/2009<br>Regulation (EU) 2015/1222<br>Regulation (EU) 2016/1719<br>Regulation (EU) 2016/631<br>Regulation (EU) 2017/2195<br>Regulation (EU) 2019/941<br>Regulation (EU) 2019/942 | [39,41,42,85–90] |
| Directive 2019/944 (EU) | June 5, 2019 | Aim to improve the EU regulatory framework governing the internal electricity market | July 5, 2027 | Directive (EU) 2019/944<br>Regulation (EU) 2019/943<br>Regulation (EU) 2018/1999<br>Regulation (EU) 2019/942 | [39,41,42,57,84] |

PECIs, Projects of energy community interest.

## 18.2.2 Electricity markets in the EU: from Directive 96/92/EC to Directive 2019/944

Directive 96/92/EC [96] of December 19, 1996, concerning common rules for the internal market in electricity, has made significant contributions to creating an internal electricity market. Pursuing the effective establishment of an internal market for electricity in the EU is coherent with the EU laying based on being "an area without internal frontiers in which the free movement of goods, persons, services, and capital is ensured" [96]. This Directive also aims at fostering competition as "a competitive electricity market is an important step towards completion of the internal energy market." Moreover, this Directive has been a further step in the electricity market liberalization. It has been well-founded on past experience and the respective identification of difficulties for electricity trading between the Member States.

In 2003, Directive 2003/54/EC (Directive 2003/54/EC), repealed Directive 96/92/EC and advanced the pathway for an effective and competitive single market for electricity. Directive 2003/54/EC is based on the implementation of Directive 96/92/EC that showed benefits from the internal market in electricity, particularly "efficiency gains, price reductions, higher standards of service and increased competitiveness" (Directive 2003/54/EC [97]). The new Directive aimed at overcoming the previously identified shortcomings and at opening new "possibilities for improving the functioning of the market remain, notably concrete provisions are needed to ensure a level playing field in the generation and to reduce the risks of market dominance and predatory behavior." For that, it advanced on "ensuring nondiscriminatory transmission and distribution tariffs, through access to the network on the basis of tariffs published prior to their entry into force and ensuring that the rights of small and vulnerable customers are protected and that information on energy sources for electricity generation is disclosed, as well as reference to sources, where available, giving information on their environmental impact." This Directive put in place rules to reinforce the competition in wholesale and retail electricity markets.

Later, in 2009, Directive 2009/72/EC (Directive 2009/72/EC), repealed in its turn Directive 2003/54/EC. The new Directive was based on the existence of "obstacles to the sale of electricity on equal terms and without discrimination or disadvantages in the Community. In particular, nondiscriminatory network access and an equally effective level of regulatory supervision in each Member State do not yet exist."

It is recognized that there is a lack of independence of energy regulators, namely from governments. In that sense, the 2009 Directive advances in the direction of more independent and empowered energy regulators. The role of the Agency for the Cooperation of Energy Regulators (ACER) [98] established by Regulation No. 713/2009 is reinforced assigning to this agency several functions. Among them, ACER is mandated to "cooperate with national regulatory authorities and transmission system operators to ensure the compatibility of regulatory frameworks between the regions with the aim of creating a competitive internal market in electricity."

In 2019, Directive 2019/944/EC [39] regarding common rules for the internal market for electricity amended Directive 2012/27/EU [99]. Directive 2019/944/EC recognizes the contribution made by Directives 2003/54/EC and 2009/72/EC to create the internal electricity market. The 2019 Directive continues in the previous 2003 and 2009 Directives' line addressing "the persisting obstacles to the completion of the internal market for electricity." It recognizes that the "regulatory framework needs to contribute to overcoming the current problems of fragmented national markets which are still often determined by a high degree of regulatory interventions" that "have led to obstacles to the supply of electricity on equal terms as well as higher costs in comparison to solutions based on cross-border cooperation and market-based principles."

However, it is the move of the 2019 Directive to address the gap between current market models and the possibilities open by advances at the technological and business levels that makes it a promising step toward seamless integration of the decarbonization and market-driven approaches in Europe. In fact, the intensive transaction of renewable-based electricity and seamless, fair, nondiscriminatory participation of consumers and small- and medium-sized distributed energy resources was not adequately supported by the previous Directives. The 2019 Directive addresses new challenges, recognizing that "the Union's energy system is in the middle of a profound change. The common goal of decarbonizing the energy system creates new opportunities and challenges for market participants. At the same time, technological developments allow for new forms of consumer participation and cross-border cooperation. There is a need to adapt the Union market rules to a new market reality."

Furthermore, and considering the different sizes of producers and the power to negotiate associated with that size, promoting fair competition and easy access for different suppliers is of the utmost importance for Member States to allow consumers to take full advantage of the opportunities of a liberalized internal market for electricity [39].

The new Directive is a natural consequence of the two Commission Communication of 15 July 2015, entitled *Delivering a New Deal for Energy Consumers* and *Launching the public consultation process on a new energy market design*. The first one brought to the forefront the need for a retail market that better serves energy consumers and that can ensure a better link between wholesale and retail markets. The second one "highlighted that the move away from generation in large central generating installations toward decentralized production of electricity from renewable sources and toward decarbonized markets requires adapting the current rules of electricity trading and changing the existing market roles." It also "underlined the need to organize electricity markets in a more flexible manner and to fully integrate all market players — including producers of renewable energy, new energy service providers, energy storage and flexible demand."

The Directive 2019/944 clearly recognizes the importance of consumers and of demand flexibility to meet the EU renewables and decarbonization targets. It states that "the Union would most effectively meet its renewable energy targets through the creation of a market framework that rewards flexibility and innovation. A well-functioning electricity market design is the key factor enabling the uptake of renewable energy."

The additional required flexibility can come from the demand side in the new context in which generation is losing flexibility due to the increasing trend toward higher shares of nondispatchable renewable-based generation technologies. Consumers can contribute to delivering flexibility through adequate DR schemes and providing a large set of different innovative services compatible with their possibilities.

The relevant role of consumers in the energy transition process is recognized in the new Directive, which lays the foundations for their intensive participation. The Directive states that "competition in retail markets is essential to ensuring the market-driven deployment of innovative new services that address consumers' changing needs and abilities, while increasing system flexibility." The Directive paves the path for increased and seamless consumers participation "By empowering consumers and providing them with the tools to participate more in the energy market, including participating in new ways, it is intended that citizens in the Union benefit from the internal market for electricity and that the Union's renewable energy targets are attained."

In no. 11 of its article 2, the Directive defines active customer as "a final customer, or a group of jointly acting final customers, who consumes or stores electricity generated within its premises located within confined boundaries or, where permitted by a Member State, within other premises, or who sells self-generated electricity or participates in flexibility or energy efficiency schemes, provided that those activities do not constitute its primary commercial or professional activity." The last aspect is very significant as consumers are entitled to sell energy and participate in flexibility or energy efficiency schemes as part of their customer position, not considering these activities as part of a formal business unless they constitute the customer's primary commercial or professional activity.

Article 15 is dedicated to active customers and states that "Member States shall ensure that final customers are entitled to act as active customers without being subject to disproportionate or discriminatory technical requirements, administrative requirements, procedures and charges, and to network charges that are not cost-reflective." It also states that active customers are "entitled to operate either directly or through aggregation," "to sell self-generated electricity, including through power purchase agreements," "to participate in flexibility schemes7 and energy efficiency schemes." It also determines that active customers are "financially responsible for the imbalances they cause in the electricity system" and that "they shall be balance responsible parties or shall delegate their balancing responsibility." The same articles states that Member States shall ensure that active customers that own an energy storage facility "are not subject to any double charges, including network charges, for stored electricity remaining within their premises or when providing flexibility services to system operators" and "are allowed to provide several services simultaneously, if technically feasible."

The 2019 Directive puts a new type of entity that should gain significant importance as a market player, the Citizen energy communities. The Directive aims to "provide them with an enabling framework, fair treatment, a level playing field and a well-defined catalogue of rights and obligations." Significantly, it states that "Household customers should be allowed to participate voluntarily in community energy initiatives as well as to leave them, without losing access to the network operated by the community energy initiative or losing their rights as consumers." The Directive makes practical implementation of the energy sharing concept by distinct consumers within communities possible stating that "Electricity sharing enables members or shareholders to be supplied with electricity from generating installations within the community without being in direct physical proximity to the generating installation and without being behind a single metering point." In no. 11 of its article 2, the Directive defines as follows:

"citizen energy community" means a legal entity that:
(a) is based on voluntary and open participation and is effectively controlled by members or shareholders that are natural persons, local authorities, including municipalities or small enterprises;

(b) has for its primary purpose to provide environmental, economic or social community benefits to its members or shareholders or to the local areas where it operates rather than to generate financial profits; and

(c) may engage in generation, including from renewable sources, distribution, supply, consumption, aggregation, energy storage, energy efficiency services or charging services for electric vehicles or provide other energy services to its members or shareholders;

In this way, these communities can adequately manage their energy resources and need, including the ones related to generation, storage, and electric vehicles.

Another important advancement is that the Directive imposes that Member States ensure that "the national regulatory framework enables suppliers to offer dynamic electricity price contracts. Member States shall ensure that final customers who have a smart meter installed can request to conclude a dynamic electricity price contract with at least one supplier and with every supplier that has more than 200 000 final customers." This fosters the actual use of the possibility of smart meters, enabling the implementation of dynamic electricity prices, which are still rarely used despite the huge investments made in the smart meters' rollout.

Another aspect in which the 2019 Directive makes significant advances is in aggregation contracts. In its article 13, it states that "Member States shall ensure that all customers are free to purchase and sell electricity services, including aggregation, other than supply, independently from their electricity supply contract and from an electricity undertaking of their choice." It also states that final customers are entitled to make aggregation contracts without the consent of their electricity undertakings and that "Member States shall ensure that customers are not subject to discriminatory technical and administrative requirements, procedures or charges by their supplier on the basis of whether they have a contract with a market participant engaged in aggregation." In article 17, the Directive also determines that "Member States shall allow and foster participation of demand response through aggregation."

Member States still to ensure the deployment of smart metering systems in their territories that may be subject to an economic assessment of all of the long-term costs and benefits to the market and the individual consumer or which form of smart metering is economically reasonable and cost-effective and which time frame is feasible for their distribution [39]. Number 1 of article 21 of this 2019 Directive, which regards the entitlement to a smart meter, determines that "where the deployment of smart metering systems has been negatively assessed as a result of the cost-benefit assessment (...) and where smart metering systems are not systematically deployed, Member States shall ensure that every final customer is entitled on request, while bearing the associated costs, to have installed or, where applicable, to have upgraded, under fair, reasonable and cost-effective conditions, a smart meter." In article 11, which regards the entitlement to a dynamic electricity price contract, the Directive states that "Member States shall ensure that the national regulatory framework enables suppliers to offer dynamic electricity price contracts. Member States shall ensure that final customers who have a smart meter installed can request to conclude a dynamic electricity price contract with at least one supplier and with every supplier that has more than 200 000 final customers."

Although being focused on the electricity market, Directive 2019/944 also addresses vulnerable customers' issues in article 28 and energy poverty in article 29. The same issues are addressed in Directive 2018/2001 that states that "empowering jointly acting renewables self-consumers also provides opportunities for renewable energy communities to advance energy efficiency at household level and helps fight energy poverty through reduced consumption and lower supply tariffs." It also states that "Member States should take appropriate advantage of that opportunity by, inter alia, assessing the possibility to enable participation by households that might otherwise not be able to participate, including vulnerable consumers and tenants."

However, Member States should make the most of that opportunity by surveying the likelihood to empower support by residences that may, in some way or another, not have the option to partake, including vulnerable consumers and residents [40].

Directive 2019/944 measures regarding distributed energy resources, including small-scale generation, storage, demand flexibility, and the active role of end customers and citizen energy communities, pave the way toward establishing of local markets.

The concept of LEM gained more relevance in Europe during 2016 when European Commission presented a new policy package titled "Clean energy for all Europeans" which has as a priority the empowerment of customers, who would get more active by taking advantage of local availability of energy resources [100,101].

Progressive electricity directives in 1996, 2003, and 2009 have altogether moved the power flexibly segment from predominance by national imposing business models toward an European market ruled by contending skilled European organizations whose plans of action have kept on changing drastically in the course of recent years [102].

Considering the European Commission's goal then established to generate half of its total electricity from renewable sources by 2030, to have a 100% carbon-free electricity generation by 2050 and the necessity to ensure renewable energy sources cost-effectiveness, it is crucial to guarantee market integration by establishing local energy markets [78].

## 18.3 Local electricity markets in the EU: challenges and benefits

Considering that LEM implies a structural transformation on the European Union Energy System, it brings some challenges in different areas, such as technology, regulation, legislation, economy, and actors/stakeholders involved in or affected by these changes [2].

Although pursuing its continuous support to renewables, it is the time for the EU to foster their integration using market-driven mechanisms. Directive 2018/2001 on the promotion of the use of energy from renewable sources [39] determines in its article 4, which regards the support schemes for energy from renewable sources, that "support schemes for electricity from renewable sources shall provide incentives for the integration of electricity from renewable sources in the electricity market in a market-based and market-responsive way, while avoiding unnecessary distortions of electricity markets as well as taking into account possible system integration costs and grid stability."

Pursuing the goal of intensive use of renewables, the EU recognizes the importance of small-scale projects and local transactions. Directive 2018/2001 clearly states that "support schemes for electricity from renewable sources or 'renewable electricity' have been demonstrated to be an effective way of fostering deployment of renewable electricity." It also determines that "when Member States decide to implement support schemes, such support should be provided in a form that is as nondistortive as possible for the functioning of electricity markets. To that end, an increasing number of Member States allocate support in a form by means of which support is granted in addition to market revenues and introduce market-based systems to determine the necessary level of support. Together with steps by which to make the market fit for increasing shares of renewable energy, such support is a key element of increasing the market integration of renewable electricity, while taking into account the different capabilities of small and large producers to respond to market signals." At the same time, this Directive states that "small-scale installations can be of great benefit to increase public acceptance and to ensure the rollout of renewable energy projects, in particular at local level" and that "In order to ensure participation of such small-scale installations, specific conditions, including feed-in tariffs, might therefore still be necessary to ensure a positive cost-benefit ratio, in accordance with Union law relating to the electricity market."

Directive 2018/2001 also states that "it is appropriate to allow the consumer market for renewable electricity to contribute to the development of energy from renewable sources." In this sense, it determines that "Member States should therefore require electricity suppliers who disclose their energy mix to final customers pursuant to Union law on the internal market for electricity, or who market energy to consumers with a reference to the consumption of energy from renewable sources, to use guarantees of origin from installations producing energy from renewable sources."

The Directive stresses the interest of local renewable energy communities to foster and increase renewables use. It states that these communities' characteristics may prevent them from competing with large-scale players fairly. Therefore the Directive states that the "participation in renewable energy projects should be open to all potential local members based on objective, transparent and non-discriminatory criteria" and that "measures to offset the disadvantages relating to the specific characteristics of local renewable energy communities in terms of size, ownership structure and the number of projects include enabling renewable energy communities to operate in the energy system and easing their market integration."

The Directive also states that "renewable energy communities should be able to share between themselves energy that is produced by their community-owned installations."

In its article 2, the Directive defines P2P trading of renewable energy as "the sale of renewable energy between market participants by means of a contract with pre-determined conditions governing the automated execution and settlement of the transaction, either directly between market participants or indirectly through a certified third-party market participant, such as an aggregator. The right to conduct peer-to-peer trading shall be without prejudice to the rights and obligations of the parties involved as final customers, producers, suppliers or aggregators."

Many potential benefits can be expected from the practical implementation of LEMs in the EU. One of the key aspects of the LEM scenario for prosumers and customers is that the collective use of the available energy resources, including the demand flexibility, improves sustainability and quality of energy at different levels. Also, putting in place the framework required for the new LEMs should be made with the cooperation of those influenced by the tasks, particularly local and regional populations [40]. That creates engagement from all the parts, which facilitates and increases the chance of succeeding.

Apart from the advancement of the market for sustainable energy sources, in particular in the long term, it is important to consider the positive effect at the regional and local levels, in economic and social terms, with new business opportunities and better conditions for energy reasonable prices, able to support business competition and people comfort.

In spite of the liberalization of the electricity retail market [103], prices in the retail market evidence difficulties in closely following the wholesale prices, even on a yearly or multiyear analysis. The responsiveness of the energy component of the retail price to wholesale energy price changes shows that only a small part of the wholesale market competition benefits is being passed to the consumers in the retail market. In the EU, the energy component of the electricity retail prices decreased, on average, only 9.4% over the 2008−19 period, while at the same time wholesale prices decreased by 19.5%, leading to a 49.1% increase in markups over this period [104].

Another main benefit that can be expected from LEM is related with network operation. Considering the restrictions that come from geographical imbalances between generation and consumption, Staudt et al. [3] point out that a local market is a good approach to mitigate this problem. In LEM the electricity cost is defined at local level and transmission occurs with less congestion [3]. Besides, local pricing also came with some other advantages as it is less complex, facilitates market operations, and is more transparent.

These significant changes in the energy sector have and will continue to have many impacts in existing actors, facilities, and stakeholders who will need to invest in new models and technologies to overcome the barriers and to adapt themselves to new models. Governments and local governments are key, altogether with universities, research centers, and different types of institutions that act in the sector, to facilitate and create the ways to establish the knowledge and the means necessary to overcome such challenges.

While many advances are still needed, new knowledge on LEMs is being produced at a high rate. The progress so far has proven that although the best models, methods, and technological implementations may be under discussion, the knowledge is mature enough to enable practical implementations and pilots as the ones in [8,16,25]. Some practical implementations and pilots are presented in Section 18.4.

The possible benefits that can come from those implementations for consumers and other small- and medium-sized players must be carefully analyzed, as well as the benefits for the power and energy system at different geographic scales. That analysis can only, for now, be done at a small scale, with the experience gained in the already implemented pilots. Rolling out these implementations requires robust models to ensure consumers' active participation and other small- and medium-sized players in LEMs. Looking at the current state of the art, the key to ensuring that participation in a seamless way seems to be a mix of adequate business models with technological solutions that strongly support the users' participation via automated, intelligent devices and platforms. Making those devices and platforms bring value to their users while not disturbing their activities and respecting their requirements must be ensured. Moreover, this must be attained with reasonable costs that should be largely compensated by the reduced costs and/or increased incomes, resulting from the users' active participation.

The targeted participation involves all types of consumers as well as a high number of other small and medium size players that should be actively involved in energy-related activities. The appearance of the new active players in a sector which is used to rely on highly specialized organizations and professionals brings new challenges. The electricity sector high-quality service and practices should not be compromised by the new actors. For that, the LEM supporting technologies and systems must deal with the behavior, translated in decisions and actions, of a significant number of heterogeneous players. To achieve a widespread participation of consumers, automated actions are required and must be according to the consumers' needs, goals, and interest. This aspect still needs significant advancement, requiring the use of advanced information and telecommunication technologies to tackle this particular problem. AI-based models are needed to support the players in their new roles according to their specific characteristics and needs. Machine learning techniques can be used to produce generation and load forecasts, and these may be tailored to cope with each player and installation particularities and to adapt the results to each specific context [105]. Moreover, data mining approaches can be used to support dynamic strategies leading to different and more efficient resource aggregation [24,106]. Different machine learning techniques are also useful to support the players decisions regarding their participation and adapting the behavior to specific contexts [29,35,107]. Decentralized actions should result from decentralized decision-making which requires distributed intelligence. Components and systems able to materialize the decisions into actions according to multiple goals and interests are needed. For that, the relevant components and systems should be able to understand each other in a seamless way and the effort to adapt the computer applications to changes in devices and communication protocols should be as low as possible. Semantic based approaches and ontologies have here a crucial application, given that the required real-time performance is ensured [8,32].

## 18.4 Some examples of local electricity market pilots in Europe

Significant research done in areas relevant for LEM can give support for experiments and pilot implementation. Implementation of LEMs projects started with the Brooklyn Microgrid [108], which is a notable LEM project that claimed, in 2016, to have executed the first blockchain-based power exchange between two households in Brooklyn, New York [108]. Since then, this and other projects worldwide have been bringing to reality some of the most recent concepts around LEM. LEM-related research and pilots are flourishing in Europe due to the EU's ambitious energy policy and the consequent funding of high-quality projects.

This section presents several innovative LEM projects and other implementations that address different aspects relevant for their successful in practice.

### 18.4.1 +CityxChange (November 1, 2018—October 31, 2023)

+CityxChange (Positive City ExChange) is a smart city project that has been granted funding from the EU's Horizon 2020 research and innovation program in the call for

"Smart cities and communities." The Norwegian University of Science and Technology leads the +CityxChange consortium together with the Lighthouse Cities Trondheim Kommune (Norway) and Limerick City and County Council (Ireland). Within the +CityxChange project, the cities of Trondheim (Norway), Limerick (Ireland), Alba Iulia (Romania), Písek (Czech Republic), Sestao (Spain), Smolyan (Bulgaria), and Võru (Estonia) will experiment how to become leading cities integrating smart positive energy solutions. Through the use of digital services, the quality of life for and together with the citizens shall be improved, more energy produced than consumed, and experiences with cities across Europe exchanged to learn faster together [109].

## 18.4.2 BioVill (March 1, 2016—February 28, 2019)

BioVill started in March 2016 with the objective to transfer and adapt experiences gained in countries where bioenergy villages already exist (Germany and Austria) to countries with less experience in this sector (Slovenia, Serbia, Croatia, Macedonia, and Romania, where seven potential bioenergy villages were selected with 13—202 connected consumers). The project fosters the development of the bioenergy sector in selected target countries by strengthening the role of locally produced biomass as the main contributor for energy supply at the local level, considering opportunities for market uptake or expansion for local farmers, wood producers, or small and medium-sized enterprise (SMEs). Core activities of BioVill include the technological and economic assessment of the target villages, the involvement and active participation of stakeholders and citizens, the development of local bioenergy value chains and technologies, as well as capacity building about financing schemes and business models. The outcome of BioVill is the initiation of five bioenergy villages in Slovenia, Serbia, Croatia, Macedonia, and Romania up to the investment stage. The project was conducted by Deutsche Gesellschaft für Internationale Zusammenarbeit, WIP Renewable Energies, KEA Klimaschutz- und Energieagentur Baden-Württemberg GmbH (KEA Climate Protection and Energy Agency of Baden-Württemberg GmbH), Austrian Energy Agency, North-West Croatia Regional Energy Agency, Macedonian section of the SDEWES Centre (SDEWES-Skopje), Green Energy Innovative Biomass Cluster, Slovenian Forestry Institute, and Standing Conference of Towns and Municipalities [110].

## 18.4.3 D3A (2016)

The Decentralized Autonomous Area Agent (D3A) is a blockchain-based, hierarchical energy market exchange engine under development since 2016 by Grid Singularity, in close collaboration with the Energy Web Foundation. It is an agent-based software optimizing the operation of every grid device through its representative agent. The D3A showcases and harnesses the potential of transactive smart grids, renewable energy sources, and P2P energy trading through spot and balancing market simulations using default 15-minute interval markets. The software is open source, and a user interface is being developed for utilities, Distributed System Operator, and Transmission System Operator [111].

### 18.4.4 DOMINOES (October 1, 2017–March 31, 2021)

DOMINOES is a European research project supported by Horizon 2020, conducted by Empower (coordinator, Finland), EDP (CNET and EDP Distribution, Portugal), ISEP-GECAD (Portugal), Lappeenranta University of Technology (LUT, Finland), VPS (United Kingdom), University of Leicester (United Kingdom), and University of Seville (Spain). It aims to enable the discovery and development of new DR, aggregation, grid management, and P2P trading services by designing, developing, and validating a transparent and scalable local energy market solution. The main goal is to understand how distribution system operators (DSOs) can dynamically and actively manage grid balance in a scenario where microgrids, ultradistributed generation and energy independent communities will be prevalent [112,113].

### 18.4.5 Energy Collective (2017)

The Energy Collective project is led by the Technical University of Denmark. It started in 2017 in Denmark. The experimental part of the project's main location is a community housing site called "Svalin." It has 20 families living in their own house, a community house, and some electric vehicle (EV) charging facilities. The community is energy positive, that is, it produces more electric energy (from rooftop solar) than it consumes. It aims at allowing direct trading and exchange of electric energy in the community. Also, aggregation toward the outside world, for example, for the provision of grid services to the DSO, is investigated. The project yielded alternative forms of consumer-centric electricity markets, focusing on their inherent properties (e.g., fairness, scalability) and ease of implementation [114].

### 18.4.6 E-REGIO (January 30, 2015–April 29, 2020)

E-REGIO project started in 2017 and aims to analyze, test, and validate a new way to implement Local Energy Markets around energy storage units and flexible assets supervised by the Local System Operator. The real Local Energy Markets, enabling actors to exchange local and renewable electricity in a neighborhood, will be implemented in Norway (Skagerak Energi) and Sweden (KIC-InnoEnergy) to demonstrate the feasibility and business potential of the concept. The project is being executed by Smart Innovation Norway, eSmart Systems, De Technische Universiteit Eindhoven, Skagerak Energi, and InnoEnergy [115].

### 18.4.7 LAMP (2017)

The Landau Microgrid Project (LAMP) is a German LEM implementation and research project implemented by the Karlsruhe Institute of Technology and the power utility Energie Südwest AG. It began in 2017 and targets investigating LEM behavior, advancing the utilization and extension of local renewable generation, setting up an energy network, and in the drawn-out structure of local energy adjusts to decrease grid growth. LAMP has been live since October 2018 and has 11 private buyers and two producers trading local energy on a 15-minute merit-order market [116].

## 18.4.8 P2PQ (2018)

Peer-to-Peer im Quartier (P2PQ) is a joint research project by Wien Energie, RIDDLE&CODE, and AIT, with funding from the Austrian Research Promotion Agency, which started in 2018. It aims to create a P2P energy market for prosumers to allocate locally produced energy and acts as the basis for a local energy community. It has a blockchain-based architecture to connect the local photovoltaic (PV) system, the EV charging station, and the household meters. The whole system for the presently 41 pilot households went live in late 2019 [117].

## 18.4.9 Pebbles (2018)

The Pebbles Project is a German demonstrator and research project by Allgäu Netz, Allgäu Überlandwerk, Siemens, Hochschule Kempten, and Fraunhofer FIT. It started in 2018 and aims to develop and demonstrate concepts for LEMs and grid services and new business models. Day-ahead and intraday trading shall enable the efficient allocation of decentralized flexibility, which primarily leads to an increased local self-consumption. Thus the LEM has the potential to strengthen the local community and to reduce grid congestion. Actual trading was expected to begin in 2020 by comprising 15 local and numerous virtual participants trading electricity on a 15-minute basis in a day-ahead market [118].

## 18.4.10 Quartierstrom (2017−20)

The project Quartierstrom is a LEM implemented in the Swiss town of Walenstadt. The 3-year research project is led by the Bits to Energy Lab at the Swiss Federal Institute of Technology (ETH Zürich). It is funded as a flagship project by the Swiss Federal Office of Energy. The project evaluates technical feasibility, market mechanisms, and user behavior in a blockchain-based LEM. In addition to the local utility company EW Walenstadt, the research consortium comprises several other industry partners (Sprachwerk, Bosch, SCS, Cleantech21, SWiBi, BKW, SBB, Planar) and academia (HSG, HSLU). The project started in 2017. Since December 2018, 37 households can trade solar energy [119].

## 18.4.11 RegHEE (March 1, 2019−February 28, 2022)

The project "Local Trade and Labeling of Electricity from Renewable Sources on a Blockchain Platform" (RegHEE) is a German research project led by the Technical University of Munich and Thüga AG. Since March 2019, it aims to research the design, investigation, and evaluation of a blockchain-based P2P market for distributed generation and storage units, including labeling. A reference system with a centralized architecture and exhibiting comparable functionality will be implemented. Both the blockchain-based platform and the reference system will be deployed in the field in cooperation with a municipal utility [120].

## 18.4.12 SoLAR (July 15, 2018—July 14, 2021)

The SoLAR, Smart Grid ohne Lastgangmessung Allensbach—Radolfzell, demonstrator project started in 2018 on a real under construction property in Allensbach. It researches how feasible and operational a complete decentralized real-time pricing system that maximizes the integration of fluctuating renewables and reduces emissions on a local basis with minimum effort for storage, grid, and ICT, could be. SoLAR has been successfully tested as a virtual demonstrator in 2018 and went into real operation after finishing construction (end of 2019). Twenty-two households are using flexibility potentials in their electricity use to increase self-consumption and benefit from lower tariffs. In the next stage, grid-friendly services, as well as virtual integration of remote fluctuating sources, will take place while enlarging the features of the current energy management market system. The project is financed by the state Baden Württemberg [121].

## 18.4.13 TradeRES (February 1, 2020—January 31, 2024)

The TradeRES project—Tools for the Design and modeling of new markets and negotiation mechanisms for a ~100% Renewable European Power Systems—is a project coordinated by the Laboratório Nacional de Energia e Geologia I.P. (Portugal). It has nine participants in its consortium, namely, Technische Universiteit DELFT (Netherlands), Teknologian Tutkimuskeskus VTT OY (Finland), Nederlandse Organisatie Voor Toegepast Natuurwetenschappelijk Onderzoek TNO (Netherlands), Instituto Superior de Engenharia do Porto (Portugal), Imperial College of Science Technology and Medicine (United Kingdom), Deutsches Zentrum Fur Luft—UND Raumfahrt EV (Germany), BITYOGA AS (Norway), SMARTWATT—Energy Services SA (Portugal), and ENBW Energie Baden-Wurttemberg AG (Germany). In this project, the consortium aim to develop and test an electricity market design that meets a near 100% renewable power system [122].

## 18.4.14 VPP (2017)

The Virtual Power Plant (VPP) project is a German research project realized by the University of Wuppertal, the local energy utility WSW, and the nonprofit organization "Aufbruch am Arrenberg." It started in 2017, and the research goal is the local energy supply through energy resources in urban structures and the flexibility of end consumers through DR. The areas are aggregated in urban districts in Wuppertal, for example, the real-world laboratory "Arrenberg." The project has 550 participants. The participating households receive incentives signals via a dashboard. These incentive signals represent energy scarcity and excess based on market prices and local generation data [123,124].

Besides the projects mentioned above and pilots, there are many research works with regard to different aspects relevant for LEMs. Table 18.2 summarizes some of those works which address European case studies.

TABLE 18.2   Research works relevant for LEM addressing European case studies.

| Description | Methodology | Year | Reference(s) |
|---|---|---|---|
| Analysis of different ways of increasing flexibility in the Danish energy system by the use of local regulation mechanisms | Compared LEM with the opposite extreme and presented solutions to all balancing problems via electricity trade on the international market | 2004 | [19] |
| Presents a thorough analysis of the motives and methods needed to achieve a single European energy market | Literature review | 2009 | [125] |
| Research the market power and information issues in the ongoing process of the European Electricity Market | Regulatory game played among a key group of utilities, customers, and regulatory authority | 2012 | [126] |
| Emphasizes the neglected European paradox of Germany's energy transition and presents working examples and possible solutions to uphold electricity supply in Europe's powerhouse | Literature review | 2014 | [127] |
| Analyses electricity prosumer model in the context of the UK market | Levelized cost of electricity for photovoltaic and wind energy were modeled, got self-consumption levels, and estimated technologies per site. Used HOMER for simulations | 2015 | [128] |
| Identifies and discusses prosumer markets related to prosumer grid integration, P2P models, and prosumer community groups | Literature review | 2016 | [129] |
| Analyses the need and costs of dispatch measures and proposes a LEM setup to create economic incentives for the expansion of generation capacity close to load centers | Analytical model to investigate small networks and a graph model simulation to deal with more sophisticated structures | 2017 | [3] |
| Presented best designs that satisfy the goals of short-run efficiency and long-run efficiency | Literature review | 2017 | [130] |
| LEM overview | Literature review | 2018 | [113] |
| Overview P2P markets | Literature review | 2019 | [26] |
| Tests the implications of recently proposed market designs under the current rules in the context of the German market and introduces a novel market design called Tech4all | Used a simplistic equilibrium model representing heterogeneous market participants in an energy community with their respective objectives | 2020 | [131] |

HOMER, Hybrid optimization of multiple energy resources.

From Table 18.2 and the projects presented above, it is possible to conclude that the LEM-related research has a wide set of challenges to tackle. The concepts, the methods, and the main technological requirements related to LEMs are still being discussed. However, the community working in the area starts to reach an agreement regarding many methodologic and technological aspects leading to consolidated knowledge and practical, real-world implementations.

## 18.5 Conclusion

In the past few decades, the electricity sector has been facing many challenges. The increasing use of fossil fuels brought widespread concerns regarding the environmental impact and reinforced the need for using renewables or carbon-free technologies for energy generation. The increase of renewable-based and small and microgeneration brought new actors to the sector and changed its dynamics alongside technical issues and challenges. With reduced flexibility on the generation side, demand flexibility and its delivery through adequate DR programs gained reinforced importance. Even the most recent electricity market models prove to constitute a barrier to the intensive use of renewables and distributed resources as well as to the active participation of the huge majority of consumers and other small and medium size players. The ongoing transformations are imposing new models and arrangements aiming at the adequate use of all the available resources and their contribution toward more efficient and clean electricity sector.

Given the distributed nature of the new and important resources, local approaches present themselves as enablers for the envisaged transformation, giving way to the concept of LEM.

A LEM is an electricity market in a community that allows the community members, including residential households, to trade electricity using a local platform. This enables the community and its members to assume an active role on the promotion, use, and management of the local resources. In such an arrangement, local energy generation and demand flexibility can be used in favor of the community members, keeping profit inside the community. P2P energy transactions can be done locally, supported by transactive energy schemes. Moreover, the community can participate in energy transactions and service provision to other local energy markets and deliver demand flexibility to local and system operators through DR programs.

Pursuing its ambitious and pioneer vision for clean and affordable energy and highly competitive electricity markets, the EU has been creating the conditions, through directives, policy guidelines, and regulations, for the new reality. This should ensure a reliable supply of energy, keep electricity prices affordable, foster the use of renewables, and increase energy efficiency. With consumers at the center of its energy policy, the EU markets rules not only aim at protecting energy consumers' rights but also at ensuring that they can adequately use their energy resources as efficiently as possible toward their own and the general benefit.

In spite of the significant benefits brought by LEMs, there are still significant barriers for their effective and efficient implementation. LEMs are still a relatively new research topic although some of the respective underlying concepts are already being used in practice and the current legislation brings high expectations for their full operation in the short term. The already obtained results bring hope that this is actually possible as new models and methods on the subject are being produced at higher pace. Pilots and practical implementations are proving that LEMs concepts can be brought to the daily electricity market operation and that the results cope with the European Commission's guidelines and targets. Looking at the results that have been delivered so far, it is not risky to say that the required technical and technological advances are either in place or very close to being real.

However, LEMs must deal with the behavior, translated in decisions and actions, of a significant number of players. All types of consumers as well as a high number of other small- and medium-sized players should be soon actively involved in energy-related activities. These new active players are seen for the first time in their new roles in a sector which is usually not used to rely in actors who do not master the power and energy aspects. The electricity sector praises its high-quality service and practices and has been mostly relying on a relatively low number of specialized organizations and people. Professional training, high-quality human and material resources are the usual tools of the sector to deal with its highly demanding requirements. How can the new actors, traditionally mere energy consumers or the new incomers attracted to this arena by the new business opportunities, contribute positively and not harm the well-established high-quality levels?

New approaches, able to model and cope with emerging issues, are required. Decentralized decision-making, distributed intelligence, and components and systems able to materialize the decisions into actions according to multiple goals and interests are needed. Consumers and the other new small and medium-sized players having active roles in the sector have not only to be able to participate and be fairly remunerated by that participation in legislative terms. They also have to be able to access the means for an efficient and rewarding participation. Either this is enabled in a seamless way or the whole concept of concretization is at risk. There are still significant advances to attain these goals. To achieve a widespread participation of consumers, automated actions are required and must be according to the consumers' needs, goals, and interest. This has been the less addressed aspect so far and the one that requires more attention, research, and advances in the coming times. With an increasing interest in the field, the ongoing and emerging research and development efforts are already giving good results. The use of advanced information and telecommunication technologies to tackle this particular problem is advancing, with AI-based models enabling the support of the new players according to their specific characteristics and needs.

# References

[1] D. Levi-Faur, On the "net impact" of Europeanization: the EU's telecoms and electricity regimes between the global and the national, Comp. Polit. Stud. 37 (1) (2004) 3−29. Available from: https://doi.org/10.1177/0010414003260121.

[2] E. Espe, V. Potdar, E. Chang, Prosumer communities and relationships in smart grids: a literature review, evolution and future directions, Energies 11 (10) (2018) 2528. Available from: https://doi.org/10.3390/en11102528.

[3] P. Staudt, F. Wegner, J. Garttner, C. Weinhardt, Analysis of redispatch and transmission capacity pricing on a local electricity market setup, in: 2017 14th International Conference on the European Energy Market (EEM), IEEE, 2017, pp. 1−6. https://doi.org/10.1109/EEM.2017.7981959.

[4] M.L. Di Silvestre, S. Favuzza, E.R. Sanseverino, G. Zizzo, How decarbonization, digitalization and decentralization are changing key power infrastructures, Renew. Sustain. Energy Rev. 93 (October) (2018) 483−498. Available from: https://doi.org/10.1016/j.rser.2018.05.068.

[5] B.P. Koirala, E. Koliou, J. Friege, R.A. Hakvoort, P.M. Herder, Energetic communities for community energy: a review of key issues and trends shaping integrated community energy systems, Renew. Sustain. Energy Rev. 56 (April) (2016) 722−744. Available from: https://doi.org/10.1016/j.rser.2015.11.080.

[6] P. Faria, Z. Vale, Demand response in electrical energy supply: an optimal real time pricing approach, Energy 36 (8) (2011) 5374−5384. Available from: https://doi.org/10.1016/j.energy.2011.06.049.

[7] M.H. Albadi, E.F. El-Saadany, A summary of demand response in electricity markets, Electr. Power Syst. Res. 78 (11) (2008) 1989−1996. Available from: https://doi.org/10.1016/j.epsr.2008.04.002.

[8] G. Santos, P. Faria, Z. Vale, T. Pinto, J.M. Corchado, Constrained generation bids in local electricity markets: a semantic approach, Energies 13 (15) (2020) 3990. Available from: https://doi.org/10.3390/en13153990.

[9] F. Fernandes, H. Morais, Z. Vale, C. Ramos, Dynamic load management in a smart home to participate in demand response events, Energy Build. 82 (2014) 592−606. Available from: https://doi.org/10.1016/j.enbuild.2014.07.067.

[10] L. Gomes, F. Sousa, Z. Vale, An intelligent smart plug with shared knowledge capabilities, Sensors 18 (11) (2018) 3961. Available from: https://doi.org/10.3390/s18113961.

[11] L. Gomes, F. Sousa, T. Pinto, Z. Vale, A residential house comparative case study using market available smart plugs and EnAPlugs with shared knowledge, Energies 12 (9) (2019) 1647. Available from: https://doi.org/10.3390/en12091647.

[12] J. Soares, B. Canizes, M.A.F. Ghazvini, Z. Vale, G.K. Venayagamoorthy, Two-stage stochastic model using benders' decomposition for large-scale energy resource management in smart grids, IEEE Trans. Ind. Appl. 53 (6) (2017) 5905−5914. Available from: https://doi.org/10.1109/TIA.2017.2723339.

[13] Z. Vale, H. Morais, P. Faria, C. Ramos, Distribution system operation supported by contextual energy resource management based on intelligent SCADA, Renew. Energy 52 (2013) 143−153. Available from: https://doi.org/10.1016/j.renene.2012.10.019.

[14] H. Morais, P. Kádár, P. Faria, Z.A. Vale, H.M. Khodr, Optimal scheduling of a renewable micro-grid in an isolated load area using mixed-integer linear programming, Renew. Energy 35 (1) (2010) 151−156. Available from: https://doi.org/10.1016/j.renene.2009.02.031.

[15] H. Morais, T. Pinto, Z. Vale, I. Praca, Multilevel negotiation in smart grids for VPP management of distributed resources, IEEE Intell. Syst. 27 (6) (2012) 8−16. Available from: https://doi.org/10.1109/MIS.2012.105.

[16] H. Le Cadre, On the efficiency of local electricity markets under decentralized and centralized designs: a multi-leader Stackelberg game analysis, Cent. Eur. J. Oper. Res. 27 (4) (2019) 953−984. Available from: https://doi.org/10.1007/s10100-018-0521-3.

[17] A. Sumper, Micro and Local Power Markets, Wiley, 2019. Available from: https://doi.org/10.1002/9781119434573.

[18] F. Lezama, J. Soares, P. Hernandez-Leal, M. Kaisers, T. Pinto, Z. Vale, Local energy markets: paving the path toward fully transactive energy systems, IEEE Trans. Power Syst. 34 (5) (2019) 4081−4088. Available from: https://doi.org/10.1109/TPWRS.2018.2833959.

[19] H. Lund, E. Münster, Integrated energy systems and local energy markets, Energy Policy 34 (10) (2006) 1152−1160. Available from: https://doi.org/10.1016/j.enpol.2004.10.004.

[20] B. Richter, E. Mengelkamp, C. Weinhardt, Maturity of blockchain technology in local electricity markets, in: 2018 15th International Conference on the European Energy Market (EEM), IEEE, 2018, pp. 1−6. https://doi.org/10.1109/EEM.2018.8469955.

[21] M.A.F. Ghazvini, J. Soare, N. Horta, R. Neves, R. Castro, Z. Vale, A multi-objective model for scheduling of short-term incentive-based demand response programs offered by electricity retailers, Appl. Energy 151 (2015) 102−118. Available from: https://doi.org/10.1016/j.apenergy.2015.04.067.

[22] M.A.F. Ghazvini, J. Soares, H. Morais, R. Castro, Z. Vale, Dynamic pricing for demand response considering market price uncertainty, Energies 10 (9) (2017) 1245. Available from: https://doi.org/10.3390/en10091245.

[23] M.A.F. Ghazvini, J. Soares, O. Abrishambaf, R. Castro, Z. Vale, Demand response implementation in smart households, Energy Build. 143 (2017) 129−148. Available from: https://doi.org/10.1016/j.enbuild.2017.03.020.

[24] T. Pinto, H. Morais, P. Oliveira, Z. Vale, I. Praça, C. Ramos, A new approach for multi-agent coalition formation and management in the scope of electricity markets, Energy 36 (8) (2011) 5004−5015. Available from: https://doi.org/10.1016/j.energy.2011.05.045.

[25] C.A. Correa-Florez, A. Michiorri, G. Kariniotakis, Optimal participation of residential aggregators in energy and local flexibility markets, IEEE Trans. Smart Grid 11 (2) (2020) 1644−1656. Available from: https://doi.org/10.1109/TSG.2019.2941687.

[26] T. Sousa, T. Soares, P. Pinson, F. Moret, T. Baroche, E. Sorin, Peer-to-peer and community-based markets: a comprehensive review, Renew. Sustain. Energy Rev. 104 (April) (2019) 367−378. Available from: https://doi.org/10.1016/j.rser.2019.01.036.

[27] L. Gomes, Z.A. Vale, J.M. Corchado, Multi-agent microgrid management system for single-board computers: a case study on peer-to-peer energy trading, IEEE Access 8 (2020) 64169−64183. Available from: https://doi.org/10.1109/ACCESS.2020.2985254.

[28] O. Abrishambaf, F. Lezama, P. Faria, Z. Vale, Toward transactive energy systems: an analysis on current trends, Energy Strategy Rev. 26 (2019) 100418.

[29] T. Pinto, Z. Vale, T.M. Sousa, I. Praça, G. Santos, H. Morais, Adaptive learning in agents behaviour: a framework for electricity markets simulation, Integr. Comput. Eng. 21 (4) (2014) 399−415. Available from: https://doi.org/10.3233/ICA-140477.

[30] I. Praca, C. Ramos, Z. Vale, M. Cordeiro, MASCEM: a multiagent system that simulates competitive electricity markets, IEEE Intell. Syst. 18 (6) (2003) 54−60. Available from: https://doi.org/10.1109/MIS.2003.1249170.

[31] G. Santos, T. Pinto, H. Morais, T.M. Sousa, I.F. Pereira, R. Fernandes, et al., Multi-agent simulation of competitive electricity markets: autonomous systems cooperation for European market modeling, Energy Convers. Manag. 99 (2015) 387−399. Available from: https://doi.org/10.1016/j.enconman.2015.04.042.

[32] G. Santos, T. Pinto, I. Praça, Z. Vale, An interoperable approach for energy systems simulation: electricity market participation ontologies, Energies 9 (11) (2016) 878. Available from: https://doi.org/10.3390/en9110878.

[33] Z.A. Vale, T. Pinto, A.T. Al-Awami, Electricity markets, in: K.Y. Lee, Z.A. Vale (Eds.), Applications of Modern Heuristic Optimization Methods in Power and Energy Systems, Wiley, 2020, pp. 775−818. Available from: https://doi.org/10.1002/9781119602286.ch7.

[34] T. Pinto, H. Morais, T.M. Sousa, T. Sousa, Z. Vale, I. Praca, et al., Adaptive portfolio optimization for multiple electricity markets participation, IEEE Trans. Neural Netw. Learn. Syst. 27 (8) (2016) 1720−1733. Available from: https://doi.org/10.1109/TNNLS.2015.2461491.

[35] T. Pinto, I. Praça, Z. Vale, H. Morais, T.M. Sousa, Strategic bidding in electricity markets: an agent-based simulator with game theory for scenario analysis, Integr. Comput. Eng. 20 (4) (2013) 335−346. Available from: https://doi.org/10.3233/ICA-130438.

[36] T. Pinto, R. Faia, M.A.F. Ghazvini, J. Soares, J.M. Corchado, Z. Vale, Decision support for small players negotiations under a transactive energy framework, IEEE Trans. Power Syst. 34 (5) (2016) 4015−4023. Available from: https://doi.org/10.1109/TPWRS.2018.2861325.

[37] European Commission, EC energy, Energy, Retrieved 23 July 2020, Available online at < https://ec.europa.eu/info/policies/energy_en >, 2019 (last accessed 16.11.20).

[38] European Parliament, Resolution on the Single European Act, Official Journal of the European Communities, Doc. A2-16(NoC 7), 1986, pp. 105−119.

[39] European Parliament, and Council of the European Union, Commission Regulation (EU) 2019/943 of the European Parliament and of the Council of 5 June 2019 on the internal market for electricity, Off. J. Eur. Union L 158 (2019) 54−124. < https://eur-lex.europa.eu/legal-content/EN/TXT/PDF/?uri = CELEX:32019R0943&from = EN >.

[40] European Parliament, and Council of the European Union, Directive (EU) 2018/2001 of the European Parliament and of the Council on the promotion of the use of energy from renewable sources, Off. J. Eur. Union L 328 (2018) 82−209. < https://eur-lex.europa.eu/legal-content/EN/TXT/PDF/?uri = CELEX:32018L2001&from = EN >.

[41] European Parliament, and Council of the European Union, Regulation (EU) 2019/941 of the European Parliament and of the Council of 5 June 2019 on risk-preparedness in the electricity sector and repealing Directive 2005/89/EC, Off. J. Eur. Union L 158 (2019) 1−21. < https://eur-lex.europa.eu/legal-content/EN/TXT/?uri = uriserv:OJ.L_.2019.158.01.0001.01.ENG >.

[42] European Parliament, and Council of the European Union, Regulation (EU) 2019/942 of the European Parliament and of the Council of 5 June 2019 establishing a European Union agency for the cooperation of energy regulators, Off. J. Eur. Union L 158 (2019) 22−53. < https://eur-lex.europa.eu/legal-content/EN/TXT/?uri = uriserv:OJ.L_.2019.158.01.0022.01.ENG >.

[43] European Parliament, and Council of the European Union, Directive 2009/28/EC of the European Parliament and of the Council of 23 April 2009 on the promotion of the use of energy from renewable sources and amending and subsequently repealing Directives 2001/77/EC and 2003/30/EC, Off. J. Eur. Union L 140 (2009) 16−62. < https://eur-lex.europa.eu/legal-content/EN/ALL/?uri = CELEX%3A32009L0028 >.

[44] Ministerial Council of the Energy Community, General policy guidelines on the 2030 targets for the contracting parties of the Energy Community, 2018.

[45] Ministerial Council of the Energy Community, Decision of the Ministerial Council of the Energy Community - D/2018/2/MC-EnC: amending Decision 20121041MG-EnC of 18 October 2012 on the implementation of Directive 20091281EG and amending Article 20 of the Energy Community Treaty, 2018.

[46] Ministerial Council of the Energy Community, Decision of the Ministerial Council of the Energy Community—D/2012/04/MC-EnC: decision on the implementation of Directive 2009/28/EC and amending Article 20 of the Energy Community Treaty, 2012.

[47] European Parliament, and Council of the European Union, Directive (EU) 2018/2001 of the European Parliament and of the Council of 11 December 2018 on the promotion of the use of energy from renewable sources (Recast), Off. J. Eur. Union L 328 (2018) 82–209. < https://eur-lex.europa.eu/legal-content/EN/TXT/?uri = CELEX%3A32018L2001 > .

[48] Energy Community Secretariat, Police guidelines by the Energy Community Secretariat on reform of the support schemes for promotion of energy from renewable sources, 2015. https://www.energy-community.org/.

[49] Energy Community Secretariat, Policy guidelines by the Energy Community Secretariat on the grid integration of prosumers, 2018. < https://www.energy-community.org/ > .

[50] Energy Community Secretariat, Regulation (EC) No 714/2009 on conditions for access to the network for cross-border exchanges in electricity and repealing Regulation (EC) No 1228/2003, Off. J. Eur. Union. 714 (2009) 113–136.

[51] Ministerial Council of the Energy Community, Decision of the Ministerial Council of the Energy Community D/2011/02/MC-EnC: decision on the implementation of Directive 2009/72/EC, Directive 2009/73/EC, Regulation (EC) No 714/2009 and Regulation (EC) No 715/2009 and amending Articles 11 and 59 of the Energy Community Treaty, 2011.

[52] Permanent High Level Group of the Energy Community. Procedural Act No. 01/2012 PHLG-EnC of the Permanent High Level Group of the Energy Community of 21 June 2012 laying down the rules governing the adoption of guidelines and network codes in the Energy Community, 2012.

[53] Permanent High Level Group of the Energy Community. Decision No 2015/01/PHLG-EnC of the Permanent High Level Group of the Energy Community of 24 June 2015 on the implementation of the Commission Regulation (EU) No 543/2013 of 14 June 2013 amending Annex I to Regulation (EC) No 714/2009 of the European Parliament, 2015.

[54] European Parliament and Council of the European Union, Regulation (EU) 2019/943 of the European Parliament and ff the Council of 05 June 2019 concerning the internal market for electricity, Off. J. Eur. Union L125/54 (2019).

[55] European Parliament, and Council of the European Union, Directive 2009/72/EC of 13 July 2009 concerning common rules for the internal market in natural gas and repealing Directive 2003/55/EC, Off. J. Eur. Union L 211 (2009) 94–136.

[56] Energy Community Secretariat, Policy guidelines by the Energy Community Secretariat for reforms of electricity market model, regulated electricity prices and electricity tariff reform in the contracting parties of the Energy Community, 2013. < https://www.energy-community.org/ > .

[57] Energy Community Secretariat. Policy guidelines by the Energy Community Secretariat for treatment of value added tax on transactions related to cross border trade of electricity in the contracting parties of the Energy Community, 2015. < https://www.energy-community.org/ > .

[58] Energy Community Secretariat, Policy guidelines by the Energy Community Secretariat on the independence of national regulatory authorities, 2015. < https://www.energy-community.org/ > .

[59] European Commission, Regulation (EU) 838/2010 of 23 September 2010 on laying down guidelines relating to the inter-transmission system operator compensation mechanism and a common regulatory approach to transmission charging, 2010.

[60] Energy Community Secretariat, Decision No 2013101/PHLG-EnG of 23 October 2013 on the incorporation of Commission Regulation (EU) No 838/2010 in the Energy Community, 2013.

[61] European Commission, Regulation (EU) No 1227/2011 of the European Parliament and of the Council of 25 October 2011 on Wholesale Energy Market Integrity and Transparency, 2011.

[62] Ministerial Council of the Energy Community, Decision of the Ministerial Council of the Energy Community - D/2018/10/MG-EnC: implementing Regulation (EU) No 122712011 of the European Parliament and of the Council on Wholesale Energy Market Integrity and Transparency, 2018.

[63] European Parliament, and Council on Wholesale Energy Market Integrity And Transparency, Decision 2018/10/MC-EnC of the Ministerial Council of the Energy Community of 29 November 2018 implementing Regulation (EU) No 1227/2011 of the European Parliament and the Council on Wholesale Energy Market Integrity and Transparency, 2018.

[64] European Parliament, and Council of the European Union, Commission implementing Regulation (EU) No 1348/2014 of 17 December 2014 on data reporting implementing Article 8(2) and Article 8(6) of Regulation (EU) No 1227/2011 of the European Parliament and of the Council on Wholesale Energy Market Integrity and Transparency, 2014.

[65] European Commission, Regulation (EU) 543/2013 of 14 June 2013 on submission and publication of data in electricity markets and amending Annex I to Regulation (EC) 714/2009, Off. J. Eur. Union L 163 (2013) 1–12.

[66] European Parliament and European Council, Regulation (EU) No 347/2013 of the European Parliament and of the Council of 17 April 2013 on guidelines for trans-European energy infrastructure and repealing Decision No 1364/2006/EC and amending Regulations (EC) No 713/2009, (EC) No 714/2009 and (EC) No 715/2009, 2013. < https://eur-lex.europa.eu/legal-content/en/TXT/?uri = celex%3A32013R0347 >.

[67] European Parliament and European Council, Regulation (EU) No 525/2013 of the European Parliament and of the Council of 21 May 2013 on a mechanism for monitoring and reporting greenhouse gas emissions and for reporting other information at national and union level relevant to climate change and repealing Decision No 280/2004, 2013. < https://eur-lex.europa.eu/legal-content/EN/TXT/?uri = CELEX%3A32013R0525 >.

[68] Ministerial Council of the Energy Community, Decision of the Ministerial Council of the Energy Community - D12018111/MG-EnC on the establishment of the list of projects of Energy Community interest ('Energy Community list'), 2018.

[69] Ministerial Council of the Energy Community, Decision of the Ministerial Council of the Energy Community - D/2016/11/MC-EnC on the establishment of the list of projects of Energy Community interest ('Energy Community list'), 2016.

[70] Ministerial Council of the Energy Community, Decision of the Ministerial Council of the Energy Community - D/2015/09/MC-EnC on the implementation of Regulation (EU) No 347/2013 of the European Parliament and of the Council on guidelines for trans-European energy infrastructure, 2015.

[71] United Nations, Paris Agreement, Available online at < https://unfccc.int/sites/default/files/english_paris_agreement.pdf >, 2015 (last accessed 15.11.20).

[72] European Commission, Regulation (EU) 2016/631 of 14 April 2016 establishing a network code on requirements for grid connections of generators, 2016. < https://eur-lex.europa.eu/legal-content/EN/TXT/?uri = OJ%3AJOL_2016_112_R_0001 >.

[73] Permanent High Level Group of the Energy Community, Decision No 2018/03/PHLG-EnC of the Permanent High Level Group of the Energy Community of 12 January 2018 on incorporating Commission Regulation (EU) 2016/631 Ol14 April 2016 establishing a Network Code on requirements for grid connection of generators in the Energy Community, 2018.

[74] State Electricity Regulatory Commission Bosnia and Herzegovina, Criteria for granting derogations from requirements for connection of generating modules, 2019.

[75] Permanent High Level Group of the Energy Community, Decision 2018/05/PHLG-EnC of the Permanent High Level Group of the Energy Community of 12 January 2018 on incorporating Commission Regulation (EU) 2016/1388 of 17 August 2016 establishing a Network Code on demand connection in the Energy Community, 2018.

[76] Permanent High Level Group Decision 2018/05/PHLG-EnC, Commission Regulation (EU) 2016/1388 of 17 August 2016 establishing a Network Code on demand connection - incorporated and adapted by Permanent High Level Group Decision 2018/05/PHLG-EnC of 12 January 2018, 2016.

[77] State Electricity Regulatory Commission Bosnia and Herzegovina, Criteria for granting derogations from requirements for connection of demand facilities, 2019.

[78] European Commission, Achieving global leadership in renewable energies, European Commission—Fact Sheet—MEMO/16/3987, November 2016, 2016. < http://europa.eu/rapid/press-release_MEMO-16-3987_en.htm >.

[79] European Commission, Commission Regulation (EU) 2016/1447 of 26 August 2016 establishing a code of conduct and trade requirements for grid connection requirements for high voltage dc systems and dc power plants, 2016.

[80] Permanent High Level Group of the Energy Community, Decision No 2018/04/PHLG-EnC of the Permanent High Level Group of the Energy Community of 12 January 2018 on incorporating Commission Regulation (EU)

201611447 of 26 August 2016 establishing a Network Code on requirements for grid connection of high voltage direct current systems and direct current connected power park modules in the Energy Community, 2018.

[81] State Electricity Regulatory Commission Bosnia and Herzegovina, Criteria for granting derogations from requirements for connection of existing and new high voltage direct current systems (HVDC systems and direct current-connected power park modules (PPMs), 2019.

[82] European Parliament and European Council, Regulation (EC) No 1099/2008 of the European Parliament and of the Council of 22 October 2008 on energy statistics, 2008. < https://eur-lex.europa.eu/legal-content/EN/ALL/?uri = CELEX%3A32008R1099 >.

[83] European Parliament and European Council, Regulation (EU) No 182/2011, 2011. < https://eur-lex.europa.eu/legal-content/EN/ALL/?uri = CELEX:32011R0182 >.

[84] European Parliament and European Council, Regulation (EU) 2018/1999 of the European Parliament and of Council of 11 December 2018 on the Governance of the Energy Union and Climate Action, amending Regulations (EC) No 663/2009 and (EC) No 715/2009 of the European Parliament and Council of the European Union, 2018. < https://eur-lex.europa.eu/legal-content/EN/TXT/?uri = uriserv:OJ.L_.2018.328.01.0001.01.ENG&toc = OJ:L:2018:328:FULL >.

[85] European Commission, Commission Regulation (EU) 2015/1222 of 24 July 2015 establishing a guideline on capacity allocation and congestion management, Off. J. Eur. Union. L 197 (2015) 24–72. < https://eur-lex.europa.eu/legal-content/EN/TXT/?uri = CELEX%3A32015R1222 >.

[86] European Commission, Commission Regulation (EU) 2016/1719 of 26 September 2016 establishing a guideline on forward capacity allocation, 2016. < https://eur-lex.europa.eu/legal-content/EN/TXT/?toc = OJ:L:2016:259:TOC&uri = uriserv:OJ.L_.2016.259.01.0042.01.ENG >.

[87] European Commission, Regulation (EU) 2016/631 of 14 April 2016 establishing a network code on requirements for grid connection of generators 2016/631, Off. J. Eur. Union L 112 (2016) 1–68. < https://eur-lex.europa.eu/legal-content/EN/TXT/?uri = OJ%3AJOL_2016_112_R_0001 >.

[88] European Commission, Commission Regulation (EU) 2017/2195 of 23 November 2017 establishing a guideline on electricity balancing, 2017. https://eur-lex.europa.eu/legal-content/EN/TXT/?uri = CELEX%3A32017R2195.

[89] European Parliament and European Council, Regulation (EC) No 1228/2003 of the European Parliament and of the Council of 26 June 2003 on conditions for access to the network for cross-border exchanges in electricity, 2003. < https://eur-lex.europa.eu/legal-content/EN/ALL/?uri = CELEX%3A32003R1228 >.

[90] European Parliament and European Council, Regulation (EC) No 714/2009 of the European Parliament and of the Council of 13 July 2009 on conditions for access to the network for cross-border exchanges in electricity and repealing Regulation (EC) No 1228/2003, 2009. < https://eur-lex.europa.eu/legal-content/EN/ALL/?uri = CELEX%3A32009R0714 >.

[91] European Parliament and Council of the European Union, Directive 2009/72/EC of the European Parliament and of the Council of 13 July 2009 concerning common rules for the internal market in electricity and repealing Directive 2003/54/EC, Off. J. Eur. Union L 211 (2009) 55–93.

[92] European Commission, Renewable energy progress report, Vol. COM (2019), 2019. < https://www.europeansources.info/record/renewable-energy-progress-report-2/ >.

[93] A. Meletiou, C. Cambini, M. Masera, Regulatory and ownership determinants of unbundling regime choice for European electricity transmission utilities, Util. Policy 50 (2018) 13–25. < https://doi.org/10.1016/j.jup.2018.01.006 >.

[94] United Nations, UNFCCC—United Nations Framework Convention on Climate Change, Paris Agreement, 2019. Available online at https://unfccc.int/ 2019 (last accessed 1511.20).

[95] European Parliament, and Council of the EU, Directive (EU) 2019/944 on common rules for the internal market for electricity and amending Directive 2012/27/EU, Off. J. Eur. Union L 158 (2019) 125–199. < http://eur-lex.europa.eu/pri/en/oj/dat/2003/l_285/l_28520031101en00330037.pdf >.

[96] European Parliament, & Council of the European Union, Directive 96/92/EC of the European Parliament and of the Council of 19 December 1996 concerning common rules for the internal market in electricity, Off. J. Eur. Union L 027 (1997) 20–29.

[97] European Parliament and European Council, Directive 2003/54/EC of the European Parliament and of the Council of 26 June 2003 concerning common rules for the internal market in electricity and repealing Directive 96/92/EC—statements made with regard to decommissioning and waste management activities, Off. J. Eur. Union L 176 (2003) 37–56.

[98] Agency for the Cooperation of Energy Regulators, ACER. Available online at < https://wp.acer.europa.eu >, 2020 (last accessed 16.11.20).

[99] European Parliament and Council of the European Union, Directive 2012/27/EU of the European Parliament and of the Council of 25 October 2012 on energy efficiency, amending Directives 2009/125/EC and 2010/30/EU and repealing Directives 2004/8/EC and 2006/32/EC, Off. J. Eur. Union L 315 (2012) 1−56.

[100] European Commission, Clean energy for all Europeans, Euroheat and Power Congress 2019, 2019. < https://doi.org/10.2833/9937 >.

[101] European Commission, Providing a fair deal for consumers, European Commission—Fact Sheet, Brussels, 2016.

[102] C.K. Chyong, M. Pollitt, Europe's electricity market design: 2030 and beyond, Centre on Regulation in Europe (CERRE). Available online at < https://cerre.eu/publications/europes-electricity-market-design-2030-and-beyond/ >, 2018 (last accessed on 15.11.20).

[103] F. Ghazvini, M. Ali, S. Ramos, J. Soares, R. Castro, Z. Vale, Liberalization and customer behavior in the Portuguese residential retail electricity market, Util. Policy 59 (2019) 100919. Available from: https://doi.org/10.1016/j.jup.2019.05.005.

[104] ACER/CEER, Annual Report on the results of monitoring the internal electricity and natural gas markets in 2019; Energy Retail and Consumer Protection Volume, European Union Agency for the Cooperation of Energy Regulators and the Council of European Energy Regulators. Available online at < https://www.ceer.eu/documents/104400/7065288/2019 + Retail + and + Consumer + Protection + - + Volume + 3/53f57f31-62b7-8d87-62f4-1d9df49d4acb >, 2020 (last accessed 13.11.20).

[105] D. Ramos, B. Teixeira, P. Faria, L. Gomes, O. Abrishambaf, Z. Vale, Use of sensors and analyzers data for load forecasting: a two stage approach, Sensors 20 (12) (2020) 3524. Available from: https://doi.org/10.3390/s20123524.

[106] P. Faria, J. Spínola, Z. Vale, Aggregation and remuneration of electricity consumers and producers for the definition of demand-response programs, IEEE Trans. Ind. Inform. 12 (3) (2018) 952−961. Available from: https://doi.org/10.1109/TII.2016.2541542.

[107] T. Pinto, T.M. Sousa, I. Praça, Z. Vale, H. Morais, Support vector machines for decision support in electricity markets' strategic bidding, Neurocomputing 172 (2016) 438−445. Available from: https://doi.org/10.1016/j.neucom.2015.03.102.

[108] C. Weinhardt, E. Mengelkamp, W. Cramer, S. Hambridge, A. Hobert, E. Kremers, et al., How far along are local energy markets in the DACH + region?, in: Proceedings of the Tenth ACM International Conference on Future Energy Systems, ACM, New York, NY, 2019, pp. 544−549. https://doi.org/10.1145/3307772.3335318.

[109] + CityxChange - Positive City ExChange, + CityxChange, + CityxChange Project. 2019. Available online at https://cityxchange.eu/, 2019 (last accessed 12.11.20).

[110] North-West Croatia Regional Energy Agency, BioVill, BioVill Project. 2017. Available online at < http://biovill.eu/ >, 2017 (last accessed 15.11.20).

[111] Bronski, P., J. Creyts, S. Gao, S. Hambridge, S. Hartnett, E. Hesse, et al., The Decentralized Autonomous Area Agent (D3A) market model. Available online at https://energyweb.org/D3A, 2018 (last accessed 12.11.20).

[112] DOMINOES Project Consortium, DOMINOES Project, DOMINOES smart distribution grid, 2018. Available online at http://dominoesproject.eu/ 2018 (last accessed 14.11.20).

[113] G. Mendes, J. Nylund, S. Annala, S. Honkapuro, O. Kilkki, J. Segerstam, Local energy markets: opportunities, benefits, and barriers, in: CIRED Workshop Proceedings, 2018, pp. 1−5. https://doi.org/10.34890/443.

[114] The Energy Collective Project, The Energy Collective Project, Toward direct sharing and trading of electric energy, 2018. Available online at http://the-energy-collective-project.com/, 2018 (last accessed 15.11.2020).

[115] Smart Innovation Norway, E-REGIO Project, E-REGIO Smart Community Markets, 2018. Available online at < https://www.eregioproject.com/ >, 2018 (last accessed 12.11.20).

[116] Landau Microgrid Project, "Landau Microgrid Project," LAMP Project, 2018. Available online at https://im.iism.kit.edu/english/1093_2058.php, 2018 (last accessed on 13.11.2020).

[117] Wien Energie GmbH, "P2PQ - Peer2Peer Im Quartier," P2PQ Project, 2018. Available online at < https://nachhaltigwirtschaften.at/de/sdz/projekte/peer2peer-im-quartier.php >, 2018 (last accessed 14.11.20).

[118] Pebbles Projekt Consortium, "Pebbles Projekt," Pebbles Project, 2018. Available online at < https://pebbles-projekt.de/en/ >, 2018 (last accessed 13.11.2020).

IV. Regulatory framework: Current trends and future perspectives

[119] L. Ableitner, I. Bättig, N. Beglinge, A. Brenzikofer, G. Carle, C. Dürr, et al., Community energy network with prosumer focus, Available online at <https://quartier-strom.ch/index.php/en/homepage/>, 2020 (last accessed 13.11.20).

[120] RegHEE Project, RegHEE—local trade and labeling of electricity from renewable sources on a blockchain platform, RegHEE Blockchain, 2019. Available online at https://www.ei.tum.de/en/ewk/forschung/projekte/reghee/, 2019 (last accessed on 15.11.20).

[121] International Solar Energy Research Center Konstanz, "SoLar - Smart Grid Ohne Lastgangmessung Allensbach − Radolfzell," SoLar Project, 2020. Available online at <http://isc-konstanz.de/en/isc/institute/public-projects/current-projects/other/solar.html>, 2020 (last accessed 13.11.20).

[122] TradeRES Project Consortium, TradeRES—tools for the design and modeling of new markets and negotiation mechanisms for a ~100% renewable European power system, TradeRES Project, 2020. Available online at https://traderes.eu/project/, 2020 (last accessed 19.11.20).

[123] WSW Wuppertaler Stadtwerke GmbH, "VPP - Virtual Power Plant - Hebung von Flexibilitäten in Großstädtischen Strukturen," VPP - Virtual Power Plant Project, 2020, Available online at <https://www.wsw-online.de/wuppertalspartwatt/>, 2020 (last accessed 14.11.20).

[124] H. Alexander, L. Seeger, M. Zdrallek, P. Biesenbach, Approach for multi criteria optimization and performance monitoring of a virtual power plant with urban structures, 25th International Conference on Electricity Distribution (2019). https://doi.org/http://dx.doi.org/10.34890/862.

[125] B. Chaudhuri, Electricity reform in Europe: toward a single energy market, Int. J. Regul. Gov. 10 (1) (2010) 59−62. Available from: https://doi.org/10.3233/IJR-120094.

[126] I. Soares, A. Faina, J. Lopez, L. Varela-Candamio, What's happening to the European Electricity Market? Eur. Res. Stud. J XV (2012) 145−156.

[127] T. Sattich, Germany's energy transition and the european electricity market: mutually beneficial? J. Energy Power Eng. 8 (2) (2014) 264−273. Available from: https://doi.org/10.17265/1934-8975/2014.02.008.

[128] P. Kästel, B. Gilroy-Scott, Economics of pooling small local electricity prosumers—LCOE & self-consumption, Renew. Sustain. Energy Rev. 51 (2015) 718−729. Available from: https://doi.org/10.1016/j.rser.2015.06.057.

[129] Y. Parag, B.K. Sovacool, Electricity market design for the prosumer era, Nat. Energy 1 (4) (2016) 16032. Available from: https://doi.org/10.1038/nenergy.2016.32.

[130] P. Cramton, Electricity market design, Oxf. Rev. Econ. Policy 33 (4) (2017) 589−612. Available from: https://doi.org/10.1093/oxrep/grx041.

[131] A. Lüth, J. Weibezahn, J.M. Zepter, On distributional effects in local electricity market designs—evidence from a German case study, Energies 13 (8) (2020) 1993. Available from: https://doi.org/10.3390/en13081993.

# Local electricity markets: regulation, opportunities, and challenges in the United Kingdom

*Karim L. Anaya*

Energy Policy Research Group (EPRG), Cambridge Judge Business School, University of Cambridge, Cambridge, United Kingdom

## 19.1 Introduction

The expansion of distributed energy resources (DER) enables the creation of local electricity markets. In the United Kingdom, the Feed-in-Tariff (FIT) scheme has been decisive in the proliferation of small-scale generation with around 30% of the energy capacity connected at distribution level. Solar photovoltaics (PV) is among those with the highest number of units installed, with 0.85 million installations in 2018—19.

Local electricity markets are having a more active role not only at local level but beyond, with the possibility to offer flexibility services[1] to other parties (e.g., distribution network operators—DNOs; National Grid Electricity System Operator—NGESO). However the offer of flexibility services to DNOs and NGESO is still a work in progress in the United Kingdom and in other jurisdictions. Depending on the regulatory framework and the type of flexibility to be offered, trading can be done with or without intermediaries (e.g., independent aggregators, suppliers[2]). Benefits from flexibility technologies (e.g., demand response, storage, interconnection) has been estimated to be worth between £17 and 40 billion cumulative to 2050 in consumers' bills [1], helping to meet the UK's carbon targets at lower costs.

---

[1] For further details on those services provided via demand-side response, energy storage, and distributed generation, see: https://www.ofgem.gov.uk/electricity/retail-market/market-review-and-reform/smarter-markets-programme/electricity-system-flexibility.

[2] Suppliers can also act as aggregators.

The aim of this chapter is to assess the current situation of local electricity markets in the United Kingdom. It explores the main drivers behind their deployment (e.g., regulatory, technical) and identifies the main issues that need to be addressed in order to capture the benefits that flexibility services can offer across all the parties. The structure of this chapter is as follows. Section 19.2 provides a brief background of local electricity markets. Section 19.3 discusses the main enablers of its deployment with a focus on subsidies, innovation programs, and technology. Section 19.4 explains the main challenges that local electricity markets are facing considering regulatory and technical issues. Section 19.5 provides a brief description of innovative local electricity markets in the United Kingdom. Section 19.6 concludes.

## 19.2  Background on local electricity markets in the United Kingdom

Local electricity markets help to manage more efficiently decentralized generation, storage, and flexible assets (i.e., from residential and business customers) and to allocate better the distribution of benefits across the different parties. The aggregation of flexibility services allows providers to participate not only in local markets but to provide support at a national level (e.g., congestion, ancillary services), reducing the need to reinforce the network or to defer an upgrade. Local electricity markets can be categorized into a set of archetypes, with the capability to trade flexibility services within the same local market or beyond. Ofgem [2] has identified five local energy archetypes, ordered in line with the current level of establishment: (1) local consumer services—which aim to improve energy outcomes for local people; (2) local generation—involves a local generation asset to benefit local consumers; (3) local supply—which supplies local communities with affordable/low-carbon energy; (4) microgrids—decentralized grids that operate off-grid or in parallel to the national grid; and (5) virtual private (VP) networks—seek to operate on the public distribution networks to balance supply and demand (local balancing) through commercial agreements. The deployment of local electricity markets, especially the ones with VP is still in development stage. In the United Kingdom, many of them are in the form of trials (i.e., Cornwall Local Energy Markets, discussed in Section 19.5).

Table 19.1 summarizes the different partners, services, and a combination of trading arrangements in the United Kingdom. There are different kinds of providers, from residential customers to those that own generation units, and suppliers/aggregators (which buy flexibility services from customers in order to sell them to DNOs or to the system operator). Buyers can be those that are part of the same grid (peer-to-peer) or represented by network and system operators. In some cases, trading is via a market platform such as Piclo and Cornwall Local Energy Market.[3] In terms of trading configurations, there are different options with and without intermediaries (e.g., independent aggregator/supplier/market platform[4]). Some trading configurations are in early stage of development [see Table 19.1 (6 and 7)].

---

[3] For further details about the two platforms see Section 19.5.

[4] A market platform can also act as intermediary (i.e., distribution utility procures flexibility via a marketplace instead of procuring this directly from flexibility providers).

TABLE 19.1 LEM specifications: participants, services, and trading configurations.

| Participants, services and configurations | Specifications |
| --- | --- |
| Providers | (1) Residential, industrial, commercial customers, (2) DER owners including storage, (3) aggregators, (4) suppliers |
| Buyers | Consumers within the same microgrid or local network (peer-to-peer), DNOs, NGESO |
| What is traded | Flexibility services, energy |
| Market platform operator | Does not own flexibility/generation assets, such as Piclo |
| Trading configurations | (1) Provider—DNO, (2) provider—aggregator—DNO, (3) provider—NGESO, (4) provider—aggregator/supplier—NGESO, (5) peer-to-peer, (6) provider—market platform operator—DNOs, (7) provider—market platform operator—DNOs and NGESO |

## 19.3 Enablers of local energy market

One of the main drivers behind the development of local electricity networks has been the subsidy schemes to renewable and low-carbon generation set by the Government, especially the ones at small scale such as FITs [3].[5] Other incentives schemes with a focus on innovation have also allowed to test new business models that encourage the participation of DER in different programs (from demand response to the provision of ancillary services). In addition to the subsidies, technological advances (i.e., in generation technologies and digitalization) have also made possible the acquisition of generating units especially by the residential sector and the possibility to trade services within the same local community (peer-to-peer) or with the network operator (via export tariff or via the trading of flexibility services). This section discusses each of these drivers.

### 19.3.1 Subsidies schemes for small-scale generators

The FIT program was introduced by the Government on April 1, 2010 in order to promote the installation of small-scale renewable and low-carbon electricity generation technologies. Depending on the technology and the size of installed capacity, up to 5 MW for solar PV, wind, hydro, and anaerobic digestion (and up to 2 kW for micro combined heat and power), installations get paid an amount (pence/kWh) which is adjusted annually (based on Retail Price Index—RPI) for a period of 20 years. Eligible installations are also subject to export payments (payment received for every kWh that is exported to the electricity grid). Tariff rates for generation have been also subject to deployment caps (available for each technology) since February 8, 2016, in order to limit the capacity that

---

[5] There are other subsidy schemes for large-scale generators such as Renewable Obligations (RO) and Contract for Difference (CfD). The last one is currently the main mechanism for supporting low-carbon electricity generation and generators need to compete against each other for a contract. For further details see: https://www.gov.uk/government/publications/contracts-for-difference/contract-for-difference.

TABLE 19.2  Development of solar PV (FIT).

| Year | Total # installations | New additions | % Solar PV installations | % Domestic installations | Total installed capacity (GW) | % Domestic capacity | Total scheme costs (million) |
|---|---|---|---|---|---|---|---|
| Year 1 (2010−11) | 30,201 | 30,201 | 94.40 | 96.90 | 0.108 | 76 | £12 |
| Year 9 (2018−19) | 849,026 | 28,998 | 98.87 | 95.63 | 6.21 | 46.50 | £1,410 |

*Data from Ofgem, Feed-in-Tariff Annual Report 2010-2011, Office of Gas and Electricity Markets, 2011 [4]; Ofgem, Feed-in-Tariff Annual Report 2018-2019, Office of Gas and Electricity Markets, 2019 [5].*

can receive a particular rate per tariff period. Solar PV installations are the only ones with three different categories of tariff rates (based on the energy efficiency requirement for the building and on the number of installations). The FIT scheme was closed to new applicants from April 1, 2019. Solar PV has the highest share in terms of number installations. The total number of installations has increased from 30,000 installations in 2010−11 to around 0.85 m in 2018−19. Table 19.2 shows the evolution of solar PV installations.

The elimination of FIT for new incomers and the continuous reduction for the incumbents may have some repercussion in the deployment and financial viability of local electricity markets especially for community energy organizations. In 2018 there were in England 231 community energy organizations, with 172 engaged in energy generation projects of which 132 own operational energy projects with a total capacity of 154.5 MW [6]. The Renewable Energy Association (REA) suggests that this measure can produce up to 40% of the UK's solar installation industry to leave the national market.[6] However, it looks like subsidy-free new investments in solar PV installations will deploy anyway such as the case of the first and largest UK's subsidy-free solar farm (40 MW) in Bedfordshire which came online in December 2019. Among other subsidy-free solar arrays installations to come online in 2020 are those from local governments (Warrington Borough Council and Swindon Council) and the 8.5 MWp (MW peak)[7] extension of the 5 MWp Kentishes solar plant.

### 19.3.2 Innovation programs

Innovation is instrumental for the transition to smarter local electricity markets. New developments in electricity networks (e.g., Active Network Management—ANM; Distributed Energy Resources Market Systems—DERMS) enable more efficient DER management to unlock their value streams. According to [7] the size of the benefits due to innovation projects by DNOs could deliver up to £2.3 billion.[8] Ofgem supports innovation

---

[6] https://www.edie.net/news/10/Green-Alliance--Feed-In-Tarrif-closure-a--major-blow--for-community-solar/

[7] MWp refers to the DC capacity of the solar array (all solar modules).

[8] Estimations made on the analysis of innovation projects funded by a prior scheme Low Carbon Network Fund (LCNF) funded by Ofgem over the period 2010-2015.

**TABLE 19.3** Summary of innovation programs.

| Mechanism | How is funded | What is funded | Fund available (2013−23) (million) | Amount awarded (million) |
|---|---|---|---|---|
| NIA | % of DNO's totex allowance | Innovation projects | £500 | £190 (March 2018) |
| NIC | Via auctions up to £70 million (elec.), £20 million (gas) | Large-scale low-carbon projects | £720 | £270 (End 2018) |
| IRM | Fund requested to Ofgem | Innovation trials for business as usual (BAU) activities | No limit | £32 |

*Data from Ofgem, State of the Energy Market 2019 Report, Office of Gas and Electricity Markets, 2019 [8].*

in electricity distribution networks via three mechanisms (as part of the RIIO-1 innovation stimulus): Network Innovation Allowance (NIA), Network Innovation Competition (NIC), and Innovation Rollout Mechanism (IRM).[9] Table 19.3 describes each program.

In agreement with the recommendations from the UK Smart systems and flexibility plan [2,9], the Department for Business, Energy and Industrial Strategy (BEIS) has committed up to £70 million to smart energy system innovation. The program funds different demonstrations and studies related to innovative demand-side response (DSR), energy storage, electric vehicle (EV) to grid, flexibility markets, nondomestic smart energy, and those with bilateral collaboration on smart energy innovation (e.g., UK−Canada, UK−South Korea).[10]

## 19.3.3 Technological advances

In the last 10 years the cost of renewables has gone down. Utility-scale unsubsidized solar PV generation costs (referred to levelized cost of energy comparison—LCOE) has decreased around 89% (CAGR) while unsubsidized wind generation (a more mature technology) has decreased around 70% (CAGR) from 2009 to 2019. The LCOE range has reduced especially for solar PV with values of $44/MWh (high end) and $36/MWh (low end) in 2019[11] (see Fig. 19.1). Lithium-ion battery packs prices fell 85% from 2010 to 2018, with a price of $176/kWh in 2018 (compared with $1160/kWh in 2010). Their demand has risen 100-fold between the same period [10]. Lower renewable generation costs are facilitating the integration of more decentralized energy resources which are instrumental for the deployment of local electricity markets.

Digitalization is also a key enabler for local electricity markets allowing IoT connected devices (e.g., DER, smart appliances, smart meters, EV batteries, machines, sensors, etc.) in combination with smart technologies (e.g., artificial intelligence, machine learning, blockchain[12]) to collect and process data that create value across the different parties

[9] https://www.ofgem.gov.uk/network-regulation-riio-model/current-network-price-controls-riio-1/network-innovation

[10] https://www.gov.uk/guidance/funding-for-innovative-smart-energy-systems

[11] https://www.lazard.com/perspective/lcoe2019

[12] For details about smart technologies see Küfeoğlu et al. [11].

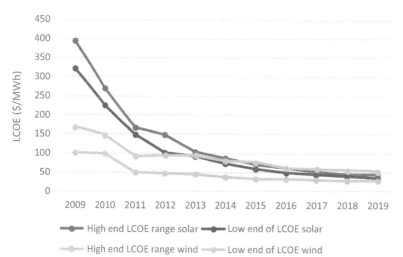

of the energy supply chain. It is expected that by 2025 there will be 41.6 billion IoT connected devices generating around 79.4ZB of data [12]. Digitalization opens new options of revenue streams to those able to provide flexibility services.


## 19.4 Challenges for the deployment of LEM and future steps

### 19.4.1 Access to the markets and the supplier hub model

The participation of local energy market (LEM) in the provision of flexibility services and electricity in local (peer-to-peer and distribution level) and national markets is still restricted. In terms of peer-to-peer, and with some exceptions (i.e., trials), LEM cannot trade directly with consumers without the intervention of suppliers. There is an ongoing revision of the current "supplier hub model." According to Ofgem, "the current supplier hub model may not be fit for purpose for energy consumers over the longer term" ([9], p. 4). The revision aims to modify the Balancing and Settlement Code (BSC) via modifications (1) P379—facilitating the trade of electricity from/to multiple suppliers through meter splitting; and (2) P375—using metering behind the site Boundary Point in order to separate balancing related services on site from imbalance-related activities more accurately [13]. At distribution level, it is observed some participation mainly via the different flexible programs that DNOs are promoting. The procurement is usually via Piclo Flex (an independent marketplace) with some exceptions. For instance, Western Power Distribution (WPD) procures using different ways: Piclo, Cornwall Local Energy Market, and its own flexibility program (Flexible Power). In terms of national markets, LEM can provide services to the ancillary services and capacity markets, usually via aggregators (greater capacities are needed to solve system issues in comparison with the local ones). However, in many cases depending on their size, the chances to participate are reduced due to

TABLE 19.4  Description of balancing services (selected services only).

| Type of service | Min. size (MW) | Notice period | Duration | Regularity (times called) | Tenders periodicity | Relative value |
|---|---|---|---|---|---|---|
| Static firm frequency response (FFR) | 1 | 30 s | Max 30 min (usually 5 min) | From 10 to 30 | Monthly tender | Medium |
| Dynamic FFR | 1 | 2 s | Max 30 min (usually 3-4 min) | Daily | Monthly tender | High |
| Short term operating reserve (STOR) | 3 | 20 min | 2−4 h (usually <20 min) | Required to deliver 3 × per week | 3 tenders per year | Low |
| Fast reserve | 25 | 2 min | 15 min | | Monthly tender | Low |

*Notes*: Relative value means the value to participants in providing the services, high means greater value.
*Data from NGESO, ESO Balancing Mechanism Access: A Guide to Entering the Balancing Mechanism, National Grid ESO, 2019 [14].*

conditions in capacity size (i.e., for fast reserve a minimum of 25 MW is required and for STOR 3 MW) (see Table 19.4).

In terms of capacity markets, DSR aggregators have been well represented, however there have been some concerns related to a differentiated treatment of DSR in comparison with generators [15]. Among these are the duration of the contract (1 year instead of 3−15 for refurbished and new-build generation, respectively), staking of contracts (which was clarified by Ofgem and BEIS in the Smart Systems and Flexibility Plan), and type of payments (independently aggregated DSR can get only availability payments while generating both availability and utilization payments, which produces an increase in the capacity price asked by the aggregated DSR provider). A major complaint by Tempus Energy to the General Court of the European Union, about the CM design regarding the treatment of DSR, produced the suspension of CM in November 2018 with no payment to energy providers. One year later the program was reinstated by the regulator with £1 billion in deferred payments to energy providers [16].

Participation of LEM in the other two markets is more challenging. Independent aggregators are deterred to participate in the wholesale market and small generators are subject to special rules (Central Volume Allocation, Supplier Volume Allocation) that may deter their participation due to the size of capacity required, administrative costs, and the need to partner with a supplier. In terms of balancing mechanism, there are some limitations too. Here larger generators are the ones with more chances to be called than the smaller ones (such as those from LEM). Then again, aggregation of services would be required. However, currently independent aggregators cannot participate directly in the balancing mechanism without an agreement with a supplier or without a supply license. There is an ongoing reform of the BSC (Modification P344, Project TERRE)[13] that will open the balancing market to independent aggregators as "Virtual Lead Parties." This will allow the creation of a new type of BMU (with individual or aggregated units) with a minimum size of 1 MW.

[13] https://www.elexon.co.uk/mod-proposal/p344/

## 19.4.2 Network charging reform

There is an ongoing revision of the current network charging network that aims to ensure a more efficient use and access to the electricity networks. The aim of the reform is to send the right signals to end users for both network costs recovery and use. The cost of the different components that make the grid and that are recovered via network charges amounts to around £9 billion per year reflected in the customers' energy bills. The revision has two components:

*Access and forward-looking charging SCR* evaluate the user's access to the networks and the way how rights to it are allocated, and "forward-looking charges" send signals to users regarding the effect of their actions on the networks. The evaluation covers different aspects such as (1) review of the distribution use of system charges in order to improve the locational accuracy of the charges, (2) access rights for transmission and distribution users, (3) review of distribution connection charging boundary, and (4) a focused review of transmission network use of system charges [17].

The review is still a work in progress with some parts of the reform potentially relevant to LEM. For instance depending on the type of access right, the export/import capacity of the service provided can be interrupted in specific times and locations (e.g., DER units placed in constraint zones can be exposed to higher level of restrictions), then revenues for the provision of services can be affected. However, this can be mitigated by the possibility of curtailment trade between users (contracted flexibility), which is also in evaluation [18]. This may represent an opportunity to LEM to trade network capacity rights. The other thing is in terms of charges. For instance, users of the distribution networks are subject to specific charges and the way they are estimated can affect the size of them; the better the granularity, the more cost-reflective the charges. Regarding connection charges, the move to a shallower connection boundary means that more cost will be treated as general reinforcement and upfront connection costs can be reduced. Another option in evaluation is to keep the current arrangement (shallow-ish boundary) but introduce connection charges to be paid over time.

*Targeted charging review (TCR)* evaluates the "residual charge," which is the difference between what is recovered from the forward-looking charges and the allowed revenue for the network owners under the price control settlement. It also examines the differentiated charges applied to small and larger generators. There are three points of discussion associated to this review: (1) who pays for the residual charges (e.g., demand, generation, both), (2) the mechanism to be used to collect charges (e.g., fixed, capacity charge), and (3) the how this should be implemented (e.g., at voltage level, user group, others).

Ofgem has published its final decision in December 2019 about TCR. It was decided that residual charges should be levied to users via a fixed rate (currently based on the individual user's consumption) and a single fixed band for domestic consumers and with refined bands only for nondomestic consumers. The fixed charges for transmission and for distribution will be implemented in stages, in 2021 and 2022, respectively. It was also decided to remove "nonlocational" embedded benefits (which eliminates any market distortion due to different treatment between small and large generators connected to the distribution network), to be implemented in 2021. According to Ofgem with these modifications the cost to maintain the electricity network will be recovered more fairly, with savings to consumers of £300 million per year from 2021 and with £4−5 billion consumer savings in total over the period 2040 [19].

From this, it is expected that DER owners will pay more for network charges. This is because the new fixed rate will be applied regardless of consumption, which is lower if the consumer has solar PV, storage units, etc., installed for their own consumption. In addition, consumers will not be able to shift demand with low incentives to Triad avoidance. According to Good Energy[14] this may imply up to 20% extra for some low-consuming domestic consumers. On the other hand, the decision to remove embedded benefits will affect revenue streams for all generators connected to the distribution network under 100 MW. Balancing Services Use of System embedded benefits will be removed in 2021 with an estimated revenue loss of around £3/MWh.[15]

### 19.4.3 Impact on electricity networks

The integration of renewables into the electricity system would require an accumulative investment of around $3.4 trillion (representing 30% of the total) by transmission and DNOs between 2018 and 2050 [10]. The deployment of local electricity markets means that more participants will be able to connect their devices, generating units (to export capacity), EVs, and trade energy and flexibility services between them (i.e., microgrids) or with network operators (e.g., DNOs, TSO). This may produce local constraints that need to be anticipated by networks operators sooner rather than later in order to evaluate potential investment scenarios (business as usual vs nonwires solutions). Then better planning, forecasting, and visibility (i.e., DER registration assets) are mandatory to mitigate any negative impact. Even though more DER connections can be a problem, they (in combination with smart technologies and actions) also bring opportunities to alleviate the stress of the electricity system. For instance, smart charging (the adaptation of the charging cycle) and the use of vehicle-to-grid (V2G) can alleviate the need to invest in additional network capacity, peaking plant requirements, and the need of network storage. Smart charging can provide both local (e.g., voltage control, local congestion, and capacity management) and system flexibility (e.g., peak shaving, ancillary services including nonfrequency ones) [20]. For instance, according to [21], smart charging in the United Kingdom could save around £180 million/year relative to a passive charging methods and an additional £90 million/year for V2G by 2030. Global savings can be between $100 and $280 billion due to smart charging (i.e., avoidance of investment in new electricity infrastructure) between 2016 and 2040 [22].

### 19.4.4 New roles and interactions: the transition to Distribution System Operators

The Open Networks Project, a major industry initiative launched in January 2017, is supporting the Government in the transition to a smart and flexible energy system. To make this possible, DNOs need to take a more active role in how networks are managed

---

[14] https://www.energylivenews.com/2019/11/22/ofgem-confirms-controversial-plans-for-power-network-charges/

[15] https://smartestenergy.com/info-hub/blog/ofgems-tcr-final-decisions/

in order to support their evolution to Distribution System Operators (DSO). Smart energy technologies can provide flexibility services that are needed to manage the network more efficiently. Then LEMs can play an important role in supporting the DNOs in this evolution. Different rules need to be defined, especially the ones that allow the trade of different services provided by LEMs for grid management. The ENA project involves a set of workstreams. "DSO Transition" (Workstream 3) proposes five potential industry structures ("Future Worlds") that assess different models to coordinate and trade flexibility services from DER and to support whole system optimization of investment and operation [23]. These are the following:

*World A (DSO Coordinates)*: DSO acts as a neutral market facilitator for all DER and provides services to NGESO.

*World B (Coordinated DSO-ESO)*: DSO and ESO work together to efficiency manage networks with coordinate procurement and dispatch of flexibility services.

*World C (Price driven flexibility)*: changes developed though Ofgem's reform of electricity network access and forward-looking charges have improved access arrangements and forward-looking signals to customers.

*World D (ESO Coordinate(s))*: ESO is the counterparty for DER and DSOs inform their requirements.

*World E (Flexibility Coordinator(s))*: a national third party acts as the neutral market facilitator for DER providing services to both ESO and DSOs.

Fig. 19.2 illustrates each World. The current situation is represented by the flag, with limited participation of LEM in the provision of flexibility services for system and local support, with few exceptions.[16] Participation of residential customers is much more restricted. The arrows indicate the different potential trends based on the Future Worlds proposed.[17] Results from the impact assessment suggest that there is not a preferred Future World and that there are specific trade-offs associated to each one. By 2030 World A and B perform relatively better than the others, however for longer periods (by 2040, 2050) the performance across different Future Worlds is similar [25]. The selection of the preferred option is still in evaluation.

The transition from DNO to DSO opens new opportunities to LEM. The transition will enable smarter ways to manage LEM and to maximize the value of flexibility services. Regardless of the World to be proposed, LEM should be designed to stack multiple revenue streams (to solve both local and system grid issues) and to be remunerated for this appropriately.

### 19.4.5 Improvements in LEM capabilities

LEM capabilities matter not only to harmonize the different solutions and products offered by them improving the integration of internal and cross-border markets, but also to mitigate the negative effect that their integration can have on the electricity system. The

---

[16] Power Responsive, ENWL in the provision of Frequency Fast Response (FFR), flexibility services from DER procured by DNOs via Piclo Flex or their own flexible platforms (e.g., Flexible Power from WPD).

[17] World C has been excluded because it is the one that can be incorporated in the rest of Worlds.

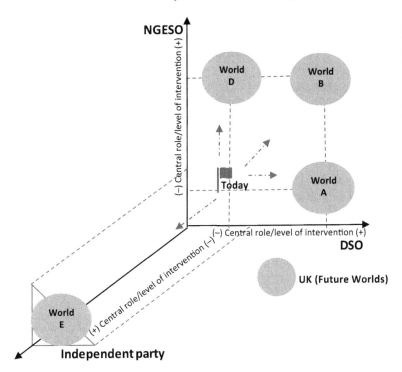

**FIGURE 19.2** Proposal of Future Worlds in the United Kingdom. *Source: Modified from CERRE, Smart Consumers in the Internet of Energy: Flexibility Markets and Services from Distributed Energy Resources, Centre on Regulation in Europe, 2019, [24].*

Requirements for Generators (RfG), which entered into force in April 2019 in GB for new generators [26], proposes a set of technical requirements arranged in four types A–D based on capacity and connection voltages (800 W to 50 MW with connection voltages lower than 110 kW and over 50 MW with connection voltage above 110 kV), for further details see Ref. [27]. It is important to LEM to align to the new RfG proposal in order to maximize the value of integrating more generators (with a focus on the renewable generation) and to minimize disruptions in the electricity network.

## 19.5 Example of innovative local electricity markets in the United Kingdom

There are different initiatives that are exploring new business models for local electricity markets in order to alleviate local and also system network congestions, voltage and thermal constraints, among others. This section describes key recent innovative examples of new opportunities for local electricity networks in the United Kingdom.

### 19.5.1 Cornwall Local Energy Market: peer-to-grid (DNO, NGESO)

The Cornwall LEM is operated by Centrica and was launched in late 2017. The project is part-funded by the European Regional Development Fund. The network consists of solar and batteries systems installed into 100 homes in Cornwall and low-carbon energy technologies installed in more than 125 Cornish business. The DNO WPD and NGESO are project partners.[18] The project involves two mechanisms for purchasing flexibility services. In Phase 1 "quote and tender" (that ran from May to August 2019), WPD procured flexibility services (i.e., demand response services) via tenders using pay as bid. WPD provided different types of payments depending on the type of service. Around 11 MVA were signed in Phase 1 and aggregation was allowed (i.e., Kiwi Power acted as aggregator for 12 sites and Centrica for 20 domestic customers with batteries installations). Phase 2 (that ran from August to December 2019) represented a spot market where WPD and NGESO could procure flexibility services. Phase 2 allowed simultaneously the procurement of flexibility services WPD and NGESO using the same pool of resources. The bids for flexibility services by both operators are matched with offers from flexibility providers (residential and business) via auctions (months ahead, intraday). The clearing solution was provided by N-SIDE, which evaluated the bids and found the optimal clearing solution considering potential constraints. LEM portal and the clearing solution facilitates the coordination of the DNO and NGESO in order to avoid any conflicts [28].

### 19.5.2 Piclo Flex: peer-to-grid and coordination with DNOs

Piclo operates a market platform created for the trading of local flexibility services (i.e., by exporting or importing power to the distribution network) between flexibility providers and DNOs. Flexibility providers usually submit two kinds of payments: availability and utilization. The potential providers register their assets online and submit key information about the kind of service they can provide. The Piclo platform registers over 200 flexibility providers offering more than 4.5 GW of flexibility volume. Flexibility services are only available to be procured by the DNOs within their respective Constraint Managed Zones. Among the DNOs with active procurement are UK Power Networks, WPD, and Scottish Power Energy Networks via Piclo. DNOs require minute-by-minute metering data and aggregation of resources is allowed (e.g., UK Power Networks calls the aggregated single unit a Flexible Unit with a minimum of 50 kW) [29]. Each DNO provides information on their respective websites about the flexibility services to be procured, however some variations are observed in terms of (1) names for the services to be procured, (2) duration of the service provided and response time, (3) minimum capacity (kW), and (4) maximum range of revenues depending of the service to be procured. Until now (Dec. 2019), UK Power Networks has procured 43 MW and WPD 123.1 MW (via Piclo and others) with future plans to contract flexibility services up to 170 MW[19] and 334 MW,[20] respectively, in 2020.

---

[18] https://www.centrica.com/innovation/cornwall-local-energy-market

[19] https://www.newpower.info/2019/11/ukpn-opens-doors-to-flexibility-options-on-high-and-low-voltage-networks/

[20] https://networks.online/gphsn/news/1001947/wpd-launches-largest-flex-auction-334MW

### 19.5.3 Local Energy Oxfordshire: the industry-first local energy system in the United Kingdom

Project Local Energy Oxfordshire, a £40 million initiative, is one of the most ambitious trials that aims to implement an integrated smart grid across Oxfordshire, redesigning the energy market at local level and contributing to the transition to a zero carbon energy future.[21] A new market platform will be created in order to enable flexibility services that help to maximize utilization of assets (e.g., generators, storage, network assets) and demand. This is a collaboration between one DNO (SSEN, project leader), market operators (including those that operate V2G platforms), county and city councils, academia, and community energy social enterprise. A set of plugin projects are expected to be implemented in the 3-year trial in order to test new market structures, enable network optimization and flexibility, and peer-to-peer trading.

## 19.6 Conclusions

This chapter evaluates the current situation of local electricity markets in the United Kingdom. It explores the main drivers behind their development and the challenges that they are facing now. From the regulatory point of view there are new proposals and ongoing evaluations that aim to allocate more fairly the cost to maintain the electricity network and to clarify roles and set new ones across all the parties involved. However, this also may alter the way that local electricity markets evolve over time. Some main issues are the elimination of the nonlocational embedded benefits and the determination of a fixed rate for residual charges (which means that revenues for shift demand may be reduced). There are also new opportunities to capture additional revenues by providing flexibility services to DNOs and system operators, individually or in aggregated mode. Then with an adequate regulatory framework and clear trading rules, local electricity markets will play an important role in the efficient and smarter integration of DER, facilitating the transition to a more flexible and low-carbon electricity system.

## References

[1] Carbon Trust and Imperial, An Analysis of Electricity System Flexibility for Great Britain, Carbon Trust and Imperial College London, 2016.
[2] Ofgem, Ofgem's Future Insights Series. Local Energy in a Transforming Energy System. Office of Gas and Electricity Markets, 2017.
[3] UKERC, The Evolution of Community Energy in the UK, UK Energy Research Centre, 2018.
[4] Ofgem, Feed-in-Tariff Annual Report 2010-2011, Office of Gas and Electricity Markets, 2011.
[5] Ofgem, Feed-in-Tariff Annual Report 2018-2019, Office of Gas and Electricity Markets, 2019.
[6] Community Energy, State of the Sector 2019. Insights, Opportunities and Challenges in a Changing Community Energy Landscape, Community Energy England, 2019.
[7] Pöyry, An Independent Evaluation of the LCNF. A Report to Ofgem, Pöyry Management Consulting, 2016.
[8] Ofgem, State of the Energy Market 2019 Report, Office of Gas and Electricity Markets, 2019.

[21] https://project-leo.co.uk/

[9] Ofgem, Upgrading Our Energy System: Smart Systems and Flexibility Plan, Office of Gas and Electricity Markets, 2017.

[10] BNEF, New Energy Outlook 2019, Bloomberg NEF, 2019.

[11] S. Küfeoğlu, L. Gaomin, K. Anaya, M.G. Pollitt, Digitalisation and New Business Models in Energy Sector, EPRG Working Paper 1829, 2019.

[12] IDC, Worldwide Global DataSphere IoT Device and Data Forecast, 2019-2023, International Data Corporation, 2019.

[13] Catapult, The Policy and Regulatory Context for New Local Energy Markets, ERIS Elective, Catapult Energy Systems, 2019.

[14] NGESO, ESO Balancing Mechanism Access: A Guide to Entering the Balancing Mechanism, National Grid ESO, 2019.

[15] R. Bray, B. Woodman, Barriers to Independent Aggregators in Europe, Energy Policy Group Working Paper EPG 1901, University of Exeter, 2019.

[16] BEIS, Capacity Market – Notification of Relevant Authority Decision and Confirmation of the Deferred Capacity Payment Trigger Event and T-1 Capacity Agreement Trigger Event, Department for Business, Energy & Industrial Strategy, 2019.

[17] Ofgem, Electricity Network Access and Forward-Looking Charging Review – Significant Code Review Launch Statement and Decision on the Wider Review, Office of Gas and Electricity Markets, 2018.

[18] Ofgem, Exec Summary – Summer 2019 Working Paper, Office of Gas and Electricity Markets, 2019.

[19] Ofgem, Targeted Charging Review: Decision and Impact Assessment, Office of Gas and Electricity Markets, 2019.

[20] IRENA, Innovation Outlook. Smart Charging for Electric Vehicles, International Renewable Energy Agency, Abu Dhabi, 2019.

[21] elementenergy, V2GB – Vehicle to Grid Britain. Requirements for Market Scale-Up (WP4). Public Report, Innovate UK, 2019.

[22] IEA, Digitalization & Energy, International Energy Agency, 2017.

[23] ENA, Open Networks Future Worlds. Developing Change Options to Facilitate Energy Decarbonisation, Digitisation and Decentralisation, 2018.

[24] CERRE, Smart Consumers in the Internet of Energy: Flexibility Markets and Services from Distributed Energy Resources, Centre on Regulation in Europe, 2019.

[25] Baringa, Future World Impact Assessment. Report prepared to Energy Networks Association, Baringa Partners, 2019.

[26] NGESO, Requirements for Generators (RfG) Fact Sheet, National Grid ESO, 2018.

[27] K.L. Anaya, M.G. Pollitt, Reactive power procurement: a review of current trends, Appl. Energy 270 (2020) 114939.

[28] WPD, Visibility Plugs and Socket. Phase 1 – Interim Learning Report, Next Generation, 2019.

[29] UKPN, Flexibility Services Invitation to Tender – 2019, UK Power Networks, 2019.

# Competition and restructuring of the South African electricity market

*Komla Agbenyo Folly*

Department of Electrical Engineering, University of Cape Town, Cape Town, South Africa

## 20.1 Introduction

In South Africa, coal and nuclear dominate the base-load supply power stations, whereas wind, hydro, solar (renewable energy-RE) and gas constitute the peaking contingent of the generation sources. Eskom is the only power utility in South Africa and is a state-owned company. Eskom is among the top 20 power utilities worldwide in terms of installed generation capacity [1]. It owns and operates a number of coal-fired, gas-fired, hydro, and pumped storage power stations, as well as one nuclear power station [1]. According to Eskom's Integrated Results 2017, the net maximum generating capacity as at March 2017 amounted to 44 GW, slightly higher than in 2013 by 0.9 GW [1,2]. It generates more than 90% of the electricity used in the country and the balance is supplied by municipalities and redistributors as well as private generators. It is estimated that the South Africa power utility generates between 30% and 40% of the electricity on the African continent [3]. It has until recently been one of the four cheapest electricity producers in the world [1]. Approximately 91.2% of the power generated by Eskom comes from coal-fired power stations, while about 8.8%, is generated from RE sources [2]. This is because coal in South Africa used to be plentiful and cheap. This has no longer been the case in the last few years. Eskom generates, transmits, and distributes electricity to industrial, mining, commercial, agricultural, and residential customers and to municipalities, who in turn redistribute electricity to businesses and households within their areas. It also purchases electricity from independent power producers (IPPs) in terms of various agreement schemes as well as electricity generating facilities beyond the country's borders [1]. The heavy reliance on fossil fuel energy generation by the growing South African economy has contributed immensely to increasing emissions of greenhouse gases [3]. South Africa is the 12th largest emitter of carbon dioxide ($CO_2$) in the world. It is responsible for nearly half of the $CO_2$ emission for the entire continent of Africa and for about 1.6% of global emissions [4].

355

In the foreseeable future, it is expected that the share of total capacity of conventional coal-fired power stations will decrease as more renewable generation comes online. There is no doubt that RE sources will contribute significantly to the production of electricity in the future South African grid. A transition from non-RE to RE will promote public awareness on energy saving, as well as on building a low-carbon society. The introduction of the Renewable Energy Independent Power Producer Procurement Program (REIPPPP) in 2011 by the South African government has facilitated the growing integration of RE into the South African grid. This program has been very successful, according to Refs. [4,5]. With abundant RE resources, South Africa can improve access to electricity services by adapting smart grid technologies to meet the electricity demand of the future [6].

South Africa's increasing economic growth, coupled with aggressive industrialization and a mass electrification program, has resulted in a steep increase in the demand for energy [1]. Existing production capacity cannot meet the growing demand for electricity. The main problem is that Eskom's generation, transmission, and distribution infrastructure is aging. These aging Eskom generation plants break down regularly. This is due to inadequate maintenance and operation of these plants over the years. The new build program which includes the commissioning of Medupi and Kusile power stations was supposed to bring an end to the energy crisis [7]. However, South Africans have been disappointed by the performance of these power stations. In fact, these new generators did not leave up to expectations as they are even more unreliable than the existing plants due to failed construction. Thirteen years or so after the first rolling blackout in South Africa, no adequate solution has yet been found to deal with the energy crisis in the country. To make the situation even worst, Eskom is said to have debt of over R 450 billion ($\approx$ US\$ 29 billion) and has to borrow money to service its debt. The government has suggested recently that Eskom be restructured and unbundled into three operating units, namely, generation, transmission, and distribution. The idea is to sell some of the generation plants to private companies. It is expected that by unbundling Eskom, this will reduce the operating costs and lead to increased efficiencies, especially if competition is allowed. However, not everyone believes that restructuring and privatizing Eskom will yield positive outcomes [8]. Some of the stakeholders such as the trade unions are opposed to the idea of privatizing Eskom due to the fear that this may lead to large-scale job losses. The question is whether restructuring and privatizing the power utility is the solution to eliminating the energy crisis. In this chapter, we will discuss the pros and cons of the restructuring and privatizing of Eskom and the role of the electricity market. Finally, we will look at whether Eskom can survive in its current form without some degree of restructuring.

## 20.2 Current energy situation in sub-Saharan African countries

Access to reliable electricity is the backbone of any development [7–10]. Sub-Saharan Africa's (SSA) socioeconomic development cannot be achieved without energy; and it cannot be sustained unless the energy is available, reliable, and secure. The continent's power-generation capacity is lower than that of any other world region and capacity growth has stagnated compared with other developing regions. In addition, the electricity transmission and distribution are ineffective with high technical and commercial losses.

The continent is faced with electricity shortages due to the lack of generation capacity and aging transmission and distribution infrastructure [11]. The rate of access to electricity is substantially lower than the rest of the world (43% compared to global access rate of 87%). Furthermore, the access rate is much lower in rural Africa (25%) than urban areas [10]. The total number of people without electricity has increased in recent decades as population growth has outpaced the growth in electrification. Nearly 600 million people are without electricity. In the last few years, there has been an increased pressure on sub-Saharan countries to electrify the deprived (rural) areas [7]. Efforts to promote electrification in SSA is gaining momentum, although the high population growth is still a challenge. One of the alternative solutions to the electric power crisis in SSA would be the use of a minigrid-based Distributed Generator coupled with the implementation of smart grid technologies. A Smart Grid is essentially an electricity network that uses digital and other advanced technologies, such as cyber-secure communication technologies, automated and computer control systems, in an integrated fashion to be able to monitor and intelligently and securely manage the transport of electricity from all generation sources to economically meet the varying electricity demands of end users. However, there are still significant challenges that must be overcome to deploy smart grids at the scale they are needed. Some of these challenges are technical, legal and regulatory, financial, and educational. Many regulatory policies in SSA are old and outdated and thus unable to deal with the consequences of smart grids; greater public engagement and participation is lacking; significant investments are required to purchase the new technologies envisioned for communicating information between the end users, electricity service providers, and to modernize the aging transmission and distribution infrastructure. Whether people can access electricity, and if so, how much they are able to consume are the two important metrics that can indicate the level of development. A holistic approach that ensures increased reliability, efficiency, transparency, and long-term sustainability needs to be adopted. This is particularly true for the South African and Nigerian electricity grids (the two largest economy on the African continent), which have been plagued in recent years with a lack of sufficient and reliable electricity generation. For example, Nigeria can generate 13,000 MW, but currently is only able to transmit about 4500 MW to the power grid with only 3000 MW of that getting to consumers. In the case of South Africa, the rolling blackouts, which were supposed to be last resort coping mechanism by the utility, have now been common place since 2007−08 [11]. In the next section, we explore the reasons for the rolling blackouts in South Africa.

## 20.3 Rolling blackouts in South Africa

The electricity sector in South Africa is dominated by the South African power utility, Eskom, which generates approximately 90% of the electricity used in the country. It generates between 30% and 40% of all electricity on the African continent making it one of the world's eleventh-largest power utilities in terms of generating capacity. The power-generation pool comprises the following sources: coal, hydro with pump storage, gas, wind, solar and nuclear (the only one in Africa). South Africa has the world's seventh largest coal reserves, so it is no surprise that about 69% of South Africa's primary energy comes from coal, followed by crude oil and solid biomass and waste. Eskom generates, transmits, and distributes electricity to industrial, mining, commercial, agricultural, and residential customers together with redistributors [1,7]. Although South Africa's 1998 energy white paper had predicted at the time that "for an assumed demand growth of 4.2%, Eskom's present generation capacity surplus will be fully utilized by about 2007" and stated that up to 30% of the country's generation could come from IPPs, the appropriate legislation was never enacted and no private generation was incorporated. In 2004 the power reserve margins dropped sharply as economic growth accelerated. The reserve margins continue to drop below 10%, and this led to rolling blackouts (i.e., load shedding) in 2007–08. Since 2007–08, the country has been experiencing rolling blackouts (i.e., load shedding).

After the 2007–08 rolling blackouts, the government called for expanded demand side management, where consumers were persuaded to use more efficient energy appliances and technology, or to switch to other forms of energy supply (i.e., gas, solar water heating, etc.) [7]. Demand market participation schemes were introduced where some industries were being paid for switching off machines at certain times of the day, and certain companies were compensated for self-generation. The government and Eskom realized the importance of generation mix, and diversification of energy sources was encouraged. It was decided to build new base load and peaking plants and return some old plants to service. Major projects that Eskom invested in include the construction of two new coal-fired power stations, Kusile and Medupi, and the construction of a pumped storage scheme in Drakensberg, Ingula. Furthermore, South Africa's renewable energy feed-in tariffs (REFITs) were introduced to attract independent renewable generation. The National Energy Regulator of South Africa (NERSA) approved the REFIT policy in 2009. However, in 2011 the REFIT was abandoned for a competitive bidding process for RE, known as the REIPPPP. REIPPPP projects are procured on a competitive tender basis with 70% of the scoring going to price and 30% to socioeconomic factors. Up to this point all these projects have been licensed by NERSA. The projects under the REIPPPP sell electricity to the national utility on a 20-year Power Purchase Agreement backed by the national government, with dispatch priority [12]. This program has been very successful in channeling substantial private sector expertise and investment into grid-connected RE in South Africa at competitive prices [5,12].

In 2010 the World Bank granted South Africa a US$3 billion loan for the construction of Medupi. However, Medupi took longer to build than expected. In July 2013 Eskom stated that construction of the coal-fired Medupi power station had fallen behind schedule, and the first of six 800 MW units came on line only in the second half of 2015. In the meantime, between 2014 and the first half of 2015, there were new waves of rolling blackouts. They go away to periodically return for longer periods and more severely each time. The rolling blackouts have resurfaced in force since November–December 2018.

The main problem is that the generation, transmission, and distribution infrastructure of Eskom is aging. These aging generation plants of Eskom are breaking down regularly. This is because of inadequate maintenance and operations over the years. With the announcement of the new build program, which includes the commissioning of Medupi and Kusile power stations, there were high expectations that we may see an end to the energy crisis. However, Eskom is scrambling to finish these plants, which have been hit by cost overruns and long delays. In fact, these new generators have not lived up to expectations as they are even more unreliable than the existing plants due to failed construction. Thirteen years or so after the first rolling blackout in South Africa, no adequate solution has yet been found to deal with the energy crisis in South Africa. It can be said that corruption and poor management have led to neglected maintenance, resulting in constant breakdowns. Electricity theft, a culture of nonpayment, and defaulting municipalities have also deepened the crisis.

Eskom is said to be in debt and has to borrow money to service its debt. Eskom has two major problems: its operating costs are too high, and it cannot pay its debt. It owes over R 450 billion ($\approx$ US$ 29 billion) and does not generate enough cash to pay even the interest on its debt [13]. Eskom's revenue remains flat as the power utility is unable to collect the debts from some of the municipalities as well as state-owned companies. It is estimated that the company debt surpasses revenue by about R 36 billion ($\approx$ US$ 2.25 billion) while debt continues to increase to a total of R 59 million ($\approx$ US$ 3.7 billion). Even though the electricity price has steadily increased since the first rolling blackout, this is not enough to recover every cost. At present Eskom is completely dependent on government bailouts and this is not sustainable in the long run.

## 20.4 Electricity Markets in Southern Africa

### 20.4.1 State of Electricity Market in Southern Africa

As stated earlier, without electricity, there is no development. The degree of development and the economy are highly correlated to a country's Gross Domestic Product which is also related to the state of the electricity transmission grid, as well as access to that grid. In addition, when the demand for electricity exceeds the supply, a country's economy suffers. This has been the case in South Africa for the past decade or so. This inadequacy in electricity supply can be extended to Southern Africa. The current state of the electricity market in Southern Africa is characterized by aging infrastructure, power cuts, increasing electricity prices, and high levels of $CO_2$ emissions.

The electricity supply industry (ESI) in Southern African countries consists of four main vertically integrated operations: generation, transmission, distribution, and supply (retail). Fig. 20.1 shows a simplified structure of the power utility in Eskom, South Africa. This structure is common to all power utilities in the Southern African Development Community's (SADC) region [3]. Because of the vertically integrated structure of these utility companies, any challenge experienced by one part of the business threatens the entire company and places the country's electricity supply at risk. Therefore, by allowing competitive markets, it is expected that more investment could be encouraged and efficient electricity pricing provided.

### 20.4.2 Southern African Power Pool

The Southern African Power Pool (SAPP) is a regional utility grouping that was created by SADC member states with the primary aim of providing reliable and economical electricity supply to the consumers of each of the SAPP member and also as a regional

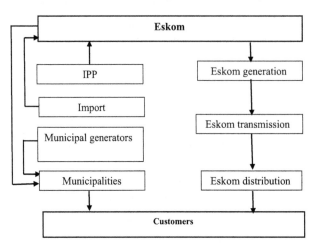

**FIGURE 20.1**   Structure of the South African power grid [24].

platform for trading electricity amongst themselves. The SAPP supplies the backbone of power that is traded across borders, supporting the economic development of communities and providing access to grid power. It was created with the primary aim to provide reliable and economical electricity supply to the consumers of each of the SAPP members. The expected future benefits are the development and the facilitation of the regional spot energy market that will attract investment in generation [14].

Eskom is a member of SAPP that provides the opportunity for the various utilities in the SADC region to ensure integrated planning and smooth and safe operation of the interconnected transmission system. Table 20.1 shows the list of SAPP members.

The members of the SAPP have undertaken to create a common market for electricity in the SADC region and to let their customers benefit from the advantages associated with this market. However, the SADC region is limited economically due to delayed or almost nonexistent investment in new electricity generating infrastructure, while much of the installed capacity is aging and in urgent need of refurbishment or replacement.

In terms of energy trading in SAPP, we can distinguish the following [14,15].

### 20.4.2.1 Bilateral and multilateral contracts

These were the pre-SAPP type markets and account for about 90%−95% of energy traded on the system. These contracts generally cover a period of 1−5 years, however, they could be longer. Recently, SAPP has started transitioning from cooperative pool to competitive pool which is bid-based instead of cost-based.

### 20.4.2.2 Short-term energy market

This market was developed and used over the period 2001−07 as a precursor to a full competition. It provided for hourly, daily, weekly, or monthly contracts and catered for a small proportion of the region's annual energy consumption (i.e., 300,000 GWh). The mechanism of this market was based on participants sending bids and offers to the SAPP

**TABLE 20.1** List of SAPP countries and related utilities.

| Countries | Full name of utility |
|---|---|
| Angola | Empresa Nacional de Electricidade |
| Botswana | Botswana Power Corporation (BPC) |
| Democratic Republic of Congo (DRC) | Societe Nationale d'Electricite (SNE) |
| Lesotho | Lesotho Electricity Corporation (LEC) |
| Malawi | Electricity Supply Corporation of Malawi (ESCOM) |
| Mozambique | Electricidade de Mocambique (EDM) |
| Namibia | NAMPOWER (Nam Power) |
| South Africa | ESKOM (Eskom) |
| Swaziland | Swaziland Electricity Board (SEC) |
| Tanzania | Tanzania Electricity Supply Company Ltd (TANESCO) |
| Zambia | Copperbelt Energy Corporation (CEC) |
| Zimbabwe | Zimbabwe Electricity Supply Authority (ZESA) |

Coordination Centre. The offers and bids were matched at the Coordination Centre and the results of successful bidders became firm short-term energy market (STEM) contracts.

### 20.4.2.3 Day-ahead market

This market scheme replaced STEM and has been in operation since 2008. It is a step to achieving a full energy trading on a SPOT market.

### 20.4.2.4 Development of new trading platform

This new trading platform includes day-ahead market, week-ahead, month-ahead, and intraday market which is an hour-ahead market. This is an internet-based trading system, where the respective power utilities can access the server to configure their participants. The wheeling and losses revenues calculations and settlements are incorporated in the system. This trading platform offers participants some flexibility in terms of maximizing profits and minimizing the exposure to balancing market costs.

## 20.5 Electricity pricing

South Africa had some of the least expensive electricity in the world in the early 2000s. Since the first rolling blackout in 2008, the electricity price has been rising sharply. It is estimated that from 2007 to 2015, electricity tariffs increased by more than 300%, while inflation over this period was 45% [16]. This has driven some customers off-grid and shut others down. It is estimated that since the first rolling blackout, Eskom's sales have been declining by about 1% per annum. The less it sells, the higher the tariff it wants, and thus in turn the less it sells. NERSA oversees electricity matters in the country, including issues related to pricing and the licensing of electricity generation, transmission, and distribution. The electricity pricing scheme employed by NERSA is based on the multiyear pricing determination (MYPD). The MYPD was implemented based on Eskom's cost recovery requirements, such that the utility remains functioning and sustains itself economically.

## 20.6 Local electricity markets

Africa is blessed with abundant RE. This includes hydro, geothermal, solar, wind, biogas, and biomass gasification. A number of pilot projects are underway throughout the continent to create micro-/minigrids [17]. Furthermore, RE will allow linkages between regional hubs of generation and consumption. With a high proportion of RE, there may be a limited need in the future to invest in the infrastructure of large centralized power pools, which take decades to build. In many parts of Africa, rural electrification is being achieved through a mixture of on-grid generation and dedicated microgrids, including solar systems to provide low-voltage DC to satisfy the basic demands for lighting and mobile phone recharging. With such excellent RE, one should look at creating localized power grids, or grid hubs, where energy is produced at or nearer the point of demand. Local entrepreneurs should be encouraged to create such localized power grids that rely as much as possible on local resources. This presents the opportunity to explore the application of smart grid technology even though there are challenges to overcome [6,18].

## 20.7 Restructuring of the South African power utility

For more than 100 years, the electric power industry worldwide has operated as government-sanctioned monopolies (i.e., regulated industry). However, over the last 25 years or so, the ESI has been undergoing continuous deregulation (reregulation) or restructuring. By electricity deregulation, we mean that the generation (wholesale) and distribution (retail) of the electric power is not under the restrictions and purview of the government [8]. In some countries restructuring the power industry is still a "work in progress" based on trial and error. The goal of competition is to reduce rates through the introduction of competition. The eventual goal is to allow consumers to choose their electricity suppliers [19–21]. "Deregulation" or restructuring creates a stronger integration between the two major goals of power system operation: economy and security. One of the drawbacks of deregulation is that it can lead to market power. Market power is the ability of an organization to control the price of a product by manipulating its supply, its demand, or both. However, market power can be eliminated by several means such as having independent supervisors of generation and transmission companies, ensuring that there are sufficient numbers of generation companies, eliminating transmission congestion, etc. [22,23]. Advocates of deregulation are adamant that competition in the industry will benefit consumers—that lowering electricity prices gives customers more choices and increases customer awareness [8]. However, this is not always the case.

Eskom was built as a vertically integrated monopoly. It owns almost all of South Africa's generating capacity, transmission, and part of the distribution. South Africa's cheap and abundant coal resources made coal-generated electricity an obvious choice for many years. This has served reasonably well in the past when coal was cheap and there was an excess of generation (in the late 1980s and early 2000s). It is now clear that this monopoly is not working anymore.

Due to the financial difficulties facing Eskom, the South African government has suggested recently that Eskom be restructured and unbundled into three operating units, namely, generation, transmission, and distribution. This is in line with the government policy since 1998. The idea is to sell some of generation plants to private companies. Transmission will remain a natural monopoly. According to Ref. [24]:

> Initially this transmission company would be a subsidiary of Eskom holdings and would be established as a separate state-owned transmission company before any new investments are made in generation capacity. Over time a multi-market model electricity market framework will ensure that transactions between electricity generators, traders, and power purchasers may take place on a variety of platforms, including bilateral contracts, a power exchange and a balancing mechanism.

For the market structure models and depending on the degree of restructuring, we can distinguish four phases and six basic models reflecting the degree of interaction among the phases, as shown in Fig. 20.2 [25].

It is expected that by unbundling Eskom, this will reduce the operating costs and lead to increased efficiencies, especially if competition is allowed. Some of the advantages in restructuring and unbundling Eskom include enhanced accountability, improved transparency, reduced operating costs, reduced opportunities for fraud and corruption, and open and nondiscriminatory access to the transmission system.

The question is whether restructuring the power utility is the solution to eliminating the energy crisis. According to Ref. [15]:

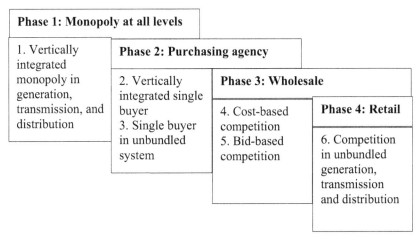

**FIGURE 20.2** Market structure models [25].

Unbundling deliver consistently superior results across the board of performance indicators when used as an entry point to implement broader reforms, particularly introducing a sound regulatory framework, reducing the degree of concentration of the generation and distribution segments of the market by attracting additional number of both public and private players and encouraging private sector participation.

However, not everyone believe that restructuring and privatizing Eskom will yield positive outcomes [8]. Some of the stakeholders, such as the trade unions, are vehemently opposed to the idea of privatizing Eskom due to fears that this will lead to large-scale job losses.

Complete restructuring may help deal with the energy crisis, but privatization alone without adequate restructuring will lead nowhere. While restructuring has been successful in many developing and developed countries, such as Brazil, Chile, the United Kingdom, the United States, and Australia [25], experience from other African countries such as Nigeria suggests that privatization alone will not be sufficient to solve Eskom's energy crisis [8,26]. The problem in Eskom is very complex and cannot be cast as technical issues only. For example, the issue of nonpayment by municipality and other state-owned entities will have to be resolved first. The restructuring should be thought through carefully and managed properly so that this does not lead to more problems.

By learning from the experiences of countries that have managed to successfully restructure their electricity sector, South Africa could also adopt a successful strategy for reform. According to Ref. [25], "A reformation model is the strategic plan consisting of multiple facets that is followed to achieve a specific market structure" (Table 20.2).

**TABLE 20.2**  Summary of possible reformation models.

| Phase | Market structure | Degree of restructuring | Degree of regulation | Private sector participation |
|---|---|---|---|---|
| Purchasing agency | Vertically integrated single buyer | Generation, distribution, and transmission vertically integrated | Regulation required to prohibit market power by single buyer | IPPs generate electricity and sell to single buyer |
| | Single buyer in an unbundled system | Generation unbundled from distribution and transmission | Regulation required in natural monopoly phases | Generation phase privatized |
| Wholesale competition | Cost- or bid-based wholesale competition | Generation and distribution unbundled from transmission | Regulatory institutions required to oversee Transmission network and competitive wholesale electricity market | Private participation in generation and distribution with open access to transmission network |
| Retail competition | Retail competition | Generation, distribution, and transmission unbundled | Normal sector specific regulation | Private participation in all stages of electricity services |

*Adapted from R. Minnie, Competition as a Means to Reform the South African Electricity Sector-A Policy Research Note, Centre for Competition Law and Economics, Stellenbosch University, BSc Project. Retrieved from: http://blogs.sun.ac.za/ccle/files/2018/02/CCLE-policy-research-document.pdf, 2018.*

Restructuring will require strong and committed government involvement, which is crucial in establishing a framework with the necessary incentives. In addition, competition will flourish only if multiple players compete in the market. Restructuring has its own set of problems that have to be considered and weighed against the benefits of the consumers. There is not one single winning market or "one size fits all" model. In each situation, important trade-offs must be made. The aim of the restructuring in South Africa should be to boost efficiency, create greater transparency over performance, give management focus, and minimize corruption as well as job creation.

## 20.8 Conclusions

Transparency is a prerequisite for developing competitive liberalized electricity markets. Restructuring (deregulation) and unbundling can only work when markets are competitive, adequate, and effective. Competitive and effective markets require the establishment of detailed market rules, design, and regulation. From the lessons learned from countries that have been successful in restructuring their electricity market, it can be inferred that restructuring yields substantial environmental benefits as cleaner gas generation replaces older inefficient coal-fired and thermal plants, leading to significant emission reductions. Future local markets cannot operate efficiently, if they are not liberalized and decentralized. In the deregulated electricity market, however, consumers can choose from a variety of options according to their needs and as per their choice and budget. In addition, consumers can also participate in demand response which is critical. Thus consumers become more aware of the process involved. In addition, consumers will benefit from the increased quality of service and possibly a reduction in electricity prices, though lower retail electricity prices are not guaranteed. In many developing countries, regulated prices are inefficiently low. There is no doubt that in some parts of SSA, the introduction of IPPs has led to a major increase in generating capacity and better technical performances as compared to the state-owned utilities. Africa offers opportunities to explore not only technologies for renewable generation, but also technologies and strategies for operating power system in smart ways that are tailored to the specific features and needs of Africa. Africa has the potential to become one of the most fertile regions for competitive local markets and smart grid innovations.

# References

[1] K. Ratshomo, R. Nembahe, South African Energy Sector Report, Department of Energy, Republic of South Africa, 2018. Retrieved from: http://www.energy.gov.za/files/media/explained/2018-South-African-Energy-Sector-Report.pdf.

[2] Power Africa in South Africa, 2018. Retrieved from: https://www.usaid.gov/sites/default/files/documents/1860/South_Africa_-_November_2018_Country_Fact_Sheet.pdf.

[3] A. Eberhard, Competition and Regulation in the Electricity Supply Industry in South Africa, 2001 Annual Forum, University of Cape Town, 2001. Retrieved from: http://www.tips.org.za/files/Competition_and_Regulation_in_the_Electricity_Supply_Industry_in_South_Africa.pdf.

[4] REN21, SADC Renewable Energy and Energy Efficiency Status Report, RENI Secretary, Paris, 2015. Retrieved from: https://www.ren21.net/2015-sadc-renewable-energy-and-energy-efficiency-status-report/.

[5] L. Baker, Governing Electricity in South Africa: Wind, Coal and Power Struggles. The Governance of Clean Development Working Paper Series, 2011. Retrieved from: http://www.tyndall.ac.uk/sites/default/files/gcd_workingpaper015.pdf.

[6] K.A. Folly, Challenge in Implementing Smart Grid Technology in Africa. Invited speaker at African utility week, Cape Town, 2013. Retrieved from: www.african-utility-week.com.

[7] K.A. Folly, Wind energy integration into the South African grid: prospect and challenges, in: D. Fleming (Ed.), Wind Energy: Development, Potential and Challenges, Nova Science Publisher, 2016. ISBN: 978-953-51-0214-4.

[8] K.A. Folly, Electricity Deregulation: Where Are We? ESI, Africa Magazine, 2017. Retrieved from: https://www.esi-africa.com/magazine-article/electricity-deregulationwhere-are-we/.

[9] A. Wyatt, Electric Power: Challenges and Choices, The Book Press Ltd, 1986.

[10] M.P. Blimpo, M. Cosgrove-Davies, Electricity Access in Sub-Saharan Africa: Uptake, Reliability and Complementary Factors for Economic Impact, International Bank for Reconstruction and Development/The World Bank, 2019. Retrieved from: http://documents.worldbank.org/curated/en/837061552325989473/pdf/135194-PUB-PUBLIC-9781464813610.pdf.

[11] K.A. Folly, Smart grid applications in Africa: opportunities and challenges, *Mathématique Appliquée à des Questions de Development* (MADEV), Plenary Talk Senegal, Dakar, 2019.

[12] A. Eberhard, J. Kolker, South Africa's Renewable Energy IPP Procurement Program: Success Factor and Lessons, Public-Private Infrastructure Advisory Facility, 2014. Retrieved from: http://www.gsb.uct.ac.za/files/PPIAFReport.pdf.

[13] B. Lindwa, Eskom debt nearing R450 billion, tariff increases 'not enough'. Retrieved from: https://www.the-southafrican.com/news/eskom-debt-nearing-r450-billion-tariff-increases-not-enough/, September 10, 2019.

[14] K.A. Folly, K. Awodele, L. Kapolo, N. Mbuli, M. Kopa, O. Obadina, Interconnected power grids, in: S. Santoso, H.W. Beaty (Eds.), Standard Handbook for Electrical Engineers, 17th ed., Mc Graw Hill Education, 2018, pp. 1329–1370.

[15] British High Commission Pretoria, Electricity Market Reform in South Africa, Promethium Carbon, Report-2016, 2016. Retrieved from: http://promethium.co.za/wp-content/uploads/2016/03/2016-03-21-Report-Electricity-Market-Reform-in-SADC-final.pdf.

[16] C. Fripp, South Africa Electricity Pricing Compared to the Rest of the Word. Retrieved from: http://www.htxt.co.za/2015/06/26/south-africas-electricity-pricing-compared-to-the-rest-of-the-world/, June 26, 2015.

[17] P.K. Ainah, K.A. Folly, Development of micro-grid in Sub-Saharan Africa: an overview, Int. Rev. Electr. Eng. 10 (5) (2015) 633–654.

[18] D. Manz, R. Walling, N. Miller, R. D'Aquila, B. Daryanian, The grid of the future, IEEE Power Energy Mag. 12 (3) (2014) 26–36.

[19] M. Ilic, F. Galiana, L. Fink, Power System Restructuring, Engineering and Economics, Kluwer Academic Publishers, 1998.

[20] L.L. Lai, Power System Restructuring and Deregulation: Trading, Performance and Information Technology, John Wiley & sons, Ltd, 2001.

[21] X.P. Zhang, Restructured Electric Power Systems: Analysis of Electricity Markets with Equilibrium Models, IEEE, John Wiley & Sons, Inc., Publication, 2010.

[22] K.A. Folly, J. Yan, On the impact of demand elasticity on electricity market, in: R. Douglas (Ed.), Electricity Markets: Impact Assessment, Developments and Emerging Trends, Nova Science Publisher, Inc, New York, 2016. ISBN: 978-1-63485-603-4.

[23] J. Yan, K. Folly, Investigation of the impact of demand elasticity on electricity market using extended Cournot approach, Int. J. Electr. Power Energy Syst, 2014, pp. 347–356.

[24] A. Eberhard, The Political Economy of Power Sector Reform in South Africa, 2004. Retrieved from: http://www.gsb.uct.ac.za/files/StanfordCUPBookChapterp215-253_6.pdf.

[25] R. Minnie, Competition as a Means to Reform the South African Electricity Sector-A Policy Research Note, Centre for Competition Law and Economics, Stellenbosch University, 2018. BSc Project. Retrieved from:. Available from: http://blogs.sun.ac.za/ccle/files/2018/02/CCLE-policy-research-document.pdf.

[26] Blomberg, Nigeria to review privatisation of power assets after supply fails to improve, Engineering News. Retrieved from: https://www.engineeringnews.co.za/article/nigeria-to-review-privatisation-of-power-assets-after-supply-fails-to-improve-2020-02-20, February 20, 2020.

# 21

# Asia electricity markets

*Panhong Cheng and Yan Gao*

School of Management, University of Shanghai for Science and Technology, Shanghai, P.R. China

## 21.1 China's electricity market

### 21.1.1 Market-oriented reform of China's electric power

#### 21.1.1.1 Reform process of China's electric power industry

The electric power industry is the basic industry related to the national economy and people's livelihood. The proportion of electric energy in the total energy and the speed of development of the electric power industry are taken to measure a country's comprehensive national strength and modernization degree. Therefore the development degree of the electric power industry and market has become an important link of national modernization and international cooperation [1].

China and emerging Asian countries such as the Philippines and India have witnessed great changes in electric power development in recent years. Compared with the mature electricity market in Western countries, the construction of the electricity market in China, India, the Philippines, and other Asian countries is steadily and rapidly advancing. A detailed introduction to the development process of electric power in these countries will be given.

For a long time, the development of China's electric power industry and market was restricted to a certain extent due to the problems in policy, system, and operation. Then the electric power industry increasingly became a "bottleneck" restricting the development of the national economy and the reform of the economic system. Therefore the market-oriented reform of the electric power industry is imperative.

The essence of the market-oriented reform of the electric power industry is to establish a competitive electricity market, promote the efficiency and reduce the cost of power generation (PG), transmission, distribution, and all links, so as to reduce the electricity price, improve the reliability of power supply, and improve the service to users. By using the regulation and incentive mechanism of the market, it can guide the investment to be more reasonable, realize the optimal allocation of power resources, and form the internal driving force of sustainable development.

The market-oriented process of China's electric power is a gradual process, which starts from deregulation based on the development needs, namely, deregulation of access. In the early 1980s the reform of power financing and investment system is carried out to mobilize social forces to raise funds to run electricity. The market-oriented sprout of China electric power industry is born in the process of the reform of the financing and investment system. The policy of "power plants run by all, power grids run by the state" and the price of "debt service" for new power plants were implemented from 1985 to 1991. Then the situation of exclusive power supply by the central government was broken. The enthusiasm of the central government, local governments, enterprises, and foreign investment in PG was greatly mobilized, and the development of PG was promoted. It not only raised a lot of funds for the development of the electric power industry, but also broke the situation of exclusive power operation from the system and promoted the deepening reform of the electric power industry. But the implementation of the policy also brought some new problems. One problem is that the construction plan of power plants is not scientific. To alleviate the power shortage, some areas have built a number of small thermal power plants with high energy consumption and serious environmental pollution. The other problem is backward transmission and distribution equipment. At that time, the power development policy tended toward power construction, ignoring the supporting construction of transmission and distribution in the power grid and restricting the rapid development of power to some extent. Another problem was the lack of effective restriction and supervision on new power projects, resulting in the phenomenon of "indiscriminate spending, chasing the government to increase the price of electricity."

Further deepening of the reform of power financing and investment promoted profound changes in the electric power industry, achieving diversification of investment subjects, financing channels, and investment methods after 1992. The main measures of reform were as follows: (1) Separation of government and enterprise. The State Power Company was established in 1997. In 1998 the Ministry of Electric Power was abolished after more than 1 year of operation and transition. The State Economic and Trade Commission performed the administrative functions of electric power. The State Power Corporation performed the management function of state-owned assets and became an electric power production operator. Then the electric power industry realized the separation of government and enterprise at the central level. The government functions of five regional electric power administrations (East China, Central China, North China, Northeast, and Northwest) and 27 provincial-level electric power bureaus have also been transferred. It marked the transition of China's electric power management system from a planned economy to a socialist market economy. It was a milestone in the development of China's electric power industry and laid down the foundation for the marketization of China's electric power. (2) Separating the power plant and the power grid, bidding on the Internet, breaking monopoly and introducing competitions. In 2002 the power system reform plan was issued, which clearly put forward the principles, overall objectives, and main tasks of power system reform. With the goal of "separating the power plant and the power grid, bidding on the Internet, breaking monopoly and introducing competition," the separation of power plants from the power grid was implemented. The PG enterprises and power grid enterprises were reorganized. At the same time, the reform of bidding on the Internet was carried out. With the establishment of the operation rules of the electricity market and the government supervision system, a competitive and open regional electricity market was initially built. The power distribution was opened to realize the overall competition in the electricity market. The pilot work of direct power supply to large users by PG enterprises was carried out to change the exclusive power purchase pattern of power grid enterprises. The reform of rural power management system was continued to be promoted. In 2002 the State Power Corporation was canceled, and a new power organization system consisting of State Grid Corporation of China and China Southern Power Grid, five PG groups, and four auxiliary industry groups was established [2]. (3) The establishment of the State Electricity Regulatory Commission. In 2005 a new power management system framework of China's electric power industry, moving from the past administrative management to market supervision, was gradually formed. (4) New stage of power reform. In 2015 on the basis of keeping the original payment mode, existing transmission line, power supply, and consumption mode, relevant enterprises in the market were allowed to participate in the power sale market, and the enterprises were subject to the deviation assessment of the executive power trading center. As a result, the power selling enterprises developed rapidly. All kinds of entities got into the electricity market, reducing the cost of power consumption. The highlights of this reform are as follows. Firstly, it was clearly put forward to "control the middle and open both ends," that is, control the transmission and distribution price, and open the market of PG and sale side. Secondly, the power grid corporation used the price model of cost plus reasonable profit which is different from the old price model (price difference between purchase and sale of electricity) by changing the profit model of power grid. Thirdly, the power sale side of power grid was opened to social capital, building a diversified power selling body and setting up a power selling company. In 2015 with the further development of the new round of electric power reform, the electric power industry ushered in a deep-seated change, promoting effective competition between the generation side and the sales side, promoting the coordinated and healthy development of the power grid, making the electricity market more dynamic, efficient, fair, and convenient, which released a huge dividend for China's economic development.

### 21.1.1.2 *Market-oriented reform program of China's electric power*

The electric power industry is related to the national economy and people's livelihood, and is an important pillar industry of the national economy. In the past 40 years of reform and opening up, the electric power industry has been developing continuously and rapidly. Great achievements have been made in terms of development speed, scale, and quality.

The electric power system reform plan in 2002 emphasized that it followed the development law of the electric power industry, played the basic role of allocating market resources, and established a power system suitable for the market economic system according to the actual situation. The overall goal of the reform was to break the monopoly, introduce competition, improve efficiency, reduce costs, optimize resource allocation, promote power development, promote grid interconnection, and build an electricity market. system with fair competition, open, orderly, and healthy development under the supervision of the government. At the same time, during the "Tenth Five-Year Plan" period, the task of electricity market reform was to implement the "separation of power plant and power grid," reorganize PG enterprises and power grid, implement competitive bidding, establish electricity market operation rules and government supervision system, preliminarily establish competitive and open regional electricity markets, and implement new power price mechanisms.

The main measures of electricity market reform are as follows:

(1) *Separation of power plant and power grid.* The generation assets and grid assets of the former state power company are separated by the separation of the power plant from the grid, as you can see in Fig. 21.1. According to the requirements of the modern enterprise system, the PG assets are reorganized into five national independent PG groups, the competitive bidding was gradually implemented and fair competition is carried out. Besides, 10% of the generation assets of state power company are reserved for peak load regulation and emergency standby of power grid corporation, and the other assets are allocated to the five generation groups. Power grid assets are organized into State Grid Corporation of China and China Southern Power Grid, and then State Grid sets up its subordinate regional power grid corporation, which is mainly responsible for electricity trading, dispatching, participating in power grid investment, and construction between regional power grids. Regional power grid corporations are responsible for operation and management of power grids, ensuring power supply safety, planning regional power grid development, fostering regional electricity markets, and managing the power dispatching trading center.

(2) *Bidding on the Internet.* The power dispatching and trading center was established, the generation bidding was carried out, and the past electricity price system of repayment

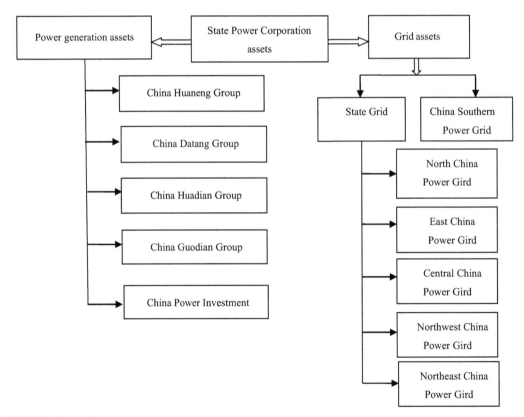

**FIGURE 21.1**  China's electric power industry structure after reform.

of principal and interest was changed. A new electricity price mechanism was formed, and the electricity price is divided into the on-grid electricity price, the transmission electricity price, the distribution electricity price, and the terminal sales electricity price. The on-grid and sale price is formed by market bidding, and the transmission and distribution price is determined by the government. The bidding on the Internet about local electric power companies is implemented according to the power grid structure, regional electricity price level, etc.

(3) *Electricity market planning*. China built a three-level electricity market system at the provincial, regional, and national levels with a reasonable price mechanism at the core. The provincial grid corporation is a subsidiary or branch of a regional grid corporation, and the regional grid corporation is a wholly owned or holding subsidiary of the State Grid Corporation of China, which is an independent legal person. Therefore the barriers for the provinces to maximize local interests are broken, and the coordination obstacles within the region are reduced.

(4) *Establishment of the State Electricity Regulatory Commission*. In China's original electric power regulatory system, the responsibility was scattered, inter department coordination was difficult, government intervention was too much, and enterprises lacked autonomy due to multidepartment decision-making. To effectively supervise electric power enterprises, the state electric power supervision and Administration Commission was established in 2005 and the functions and supervision mode of the Commission was clarified.

From the implementation of electric power system reform in 2002–14, the electric power industry has broken the system shackles of exclusive power operation, fundamentally changed the mandatory planning system and the problems of nonseparation between the government and enterprises, nonseparation between the plant and network, and preliminarily formed a multielement competition pattern of the main body of the electricity market. Firstly, it has promoted the rapid development of the electric power industry. In 2014 the installed PG capacity of the whole country reached 1.36 billion kW, the PG capacity reached 5.5 trillion kWh, the circuit length of power grid 220 kV and above reached 572,000 km, and the power transformation capacity of 220 kV and above reached 3.03 billion kVA. Secondly, the general service level of electric power has been improved. Through the transformation of rural power grid and the reform of rural power management system, the rural power supply capacity and management level have been significantly improved, the reliability of rural power supply has been significantly enhanced, the same price of urban and rural power grid has been basically achieved, and the problem of electricity consumption by the people without electricity has been basically solved. Thirdly, a diversified market system has been preliminarily formed. In terms of PG, multilevel, multiownership, and multiregion PG enterprises have been established. In terms of the power grid, in addition to the State Grid Corporation of China and China Southern Power Grid, local power grid enterprises such as Inner Mongolia power grid have been established. In terms of auxiliary industry, two design and construction integration enterprises, China Power Construction and China Energy Construction, have been established. Fourthly, the formation mechanism of electricity price has been gradually improved. At the PG stage, the benchmark price for PG is realized, and the transmission and distribution price of most provinces is gradually approved at the transmission and distribution stage. At the sales stage, the differential price, punitive price, residential step price, and other policies are successively issued. Fifthly, the market-oriented trading and supervision of electric power are actively explored. Experiments and explorations in the fields of bidding on the Internet, direct transactions between large users and PG enterprises, PG right transactions, and trans-provincial and transregional power transactions are successively carried out. Significant progress has been made in electricity market-oriented transactions, and important experience has been accumulated in power supervision.

At the same time, there are also some problems in the development of the electric power industry. The first problem is the lack of a transaction mechanism and the low efficiency of resource utilization. The effective competition mechanism of the electricity selling side has not been established, the market transaction between PG enterprises and users is limited, and the decisive role of market allocation resources is difficult to play. Energy saving, high efficiency, and environmental protection units could not be fully utilized. Water, wind, and light are often abandoned. In some areas, PG and power shortages coexisted. The second problem is that the price relationship is not straightened out and the market-oriented pricing mechanism has not been fully formed. The current electricity price management is still dominated by government pricing, and the adjustment of electricity prices often lag behind the cost changes, so it is difficult to reflect the electricity cost, market supply and demand, resource scarcity, and environmental protection expenditure in a timely and reasonable manner. The third problem is that the transformation of government functions is not in place, and various planning and coordination mechanisms are not perfect. There is too much deviation between various specific development plans, the actual implementation, and planning of power planning. The fourth problem is that the development mechanism is not sound, and faces difficulties regarding the development and utilization of new and renewable energy. The manufacturing capacity, construction, operation, and consumption demand of new energy industry equipment, such as photovoltaic PG, do not match, and a virtuous cycle of mutual promotion of research and development, production, and utilization has not been formed. The nondiscrimination of renewable energy and renewable energy PG and accessibility to the grid have not been effectively solved. The fifth problem is that the legislative work is relatively backward, which restricts the marketization and healthy development of electric power. Some existing power laws and regulations could not meet the actual needs of development, and some supporting reform policies could be issued, so it is urgent to revise relevant laws, regulations, policies, and standards to provide the basis for the development of the electric power industry.

Deepening the reform of electric power system is an urgent task, which is related to the overall situation of China's energy security and economic and social development. All sectors of the society are increasingly calling for accelerating the reform of the electric power system. The social demands and consensus to promote the reform are increasing. The reform has a loose external environment and a solid foundation for work. Therefore the Communist Party of China (CPC) Central Committee and the State Council issues several opinions on further deepening the reform of power system in 2015.

Deepening the reform of the electric power system is based on the further improvement of the separation of the government and the enterprise, the separation of the power plant and the power grid, and the separation of the main and the auxiliary. The core contents of this reform are as follows. The first is the electricity price reform. The transmission and distribution price shall be determined separately, and the sale price outside the public welfare shall be formed by the market step by step, and the formation mechanism of the price shall be straightened out. Secondly, the reform of the electric power trading mechanism improves the electricity market trading mechanism. The third is to establish a relatively independent power trading institution and form a fair and standardized market trading platform. The fourth is to promote the reform of PG and utilization plans and play the role of market mechanisms. The fifth is the reform of the electricity selling side. Opening the electricity distribution business to the social capital in an orderly way is to be carried out. The sixth is to ensure that renewable energy PG is purchased in accordance with the plan. The seventh is to open fair access to the power grid and establish a new mechanism for the development of distributed generation.

The years 2015–18 were a new regulatory cycle of electricity reform that made tremendous progress in all aspects. These mainly are as follows: (1) the establishment of trading institutions was basically completed, and a fair and standardized trading platform was set up for electric electricity market-oriented trading. At the regional level, Beijing Electric Power Trading Center, national electric power trading organization alliance, and Guangzhou electric power trading center were built. At the provincial level, all provinces established power trading institutions. (2) The reform of power transmission and distribution achieved the full coverage of 32 provincial (level) power grids, as well as the approval of power grid transmission and distribution prices in Shenzhen city. The total authorized revenue of power grid enterprises has been reduced by 48 billion RMB. All the price reduction space has been used to reduce the power price of industrial and commercial enterprises, effectively reducing the real economic burden. (3) The distribution and sale of electricity business were accelerated, the market construction of the power sale side achieved preliminary success and more than 3600 companies were registered. Two batches of pilot projects of incremental distribution business reform were successively organized and carried out, 195 pilot projects for social capital were launched, and a competition mechanism with multiagent participation was preliminarily established. At present, they are organizing and applying for the third batch of pilot projects of incremental distribution business reform. (4) The development and utilization of electricity plan were accelerated. And the scope of market participants and electricity scale of direct transactions were continuously expanded. Then electric electricity market-oriented transactions began to take shape. (5) The market rules system was preliminarily established, and market transactions are becoming increasingly active. In 2017, 1.6 trillion kWh of electricity was traded on the national market, reducing the cost of electricity consumption for the real economy by about 70 billion yuan. (6) The pilot construction of electric power spot market was carried out, and the construction of the spot market started steadily. In 2018 eight spot electricity markets were established, and in the first half of 2019 eight spot markets were put into trial operation. (7) The system framework of preferential consumption of clean energy was gradually established, and clean energy PG and trans-provincial clean energy power transmission are given priority to the protection. (8) The electricity market supervision plays a role and the legal consciousness of market subjects is improved. (9) The construction of the power industry credit system is promoted, and a new regulatory mechanism with credit as the core was built.

The reform of the electric power system is a new thing in China. The foreign model cannot be copied mechanically. China has no experience to follow, so it can only be explored in practice. The reform has entered the deep water area, and there are still some hard bones on which to chew. For example, to some extent and within a certain range, there are barriers between provinces, administrative interference in some provincial market-oriented transactions, the consumption of clean energy, the difficulty in promoting incremental distribution reform, and new challenges in market supervision. For these difficulties, we should not only have firm confidence and wisdom to actively promote reform, but also maintain the toughness and patience to deal with reform.

### 21.1.1.3 Market-oriented reform analysis of China electric power

In 2002 the competition mechanism was introduced into the generation side through the "separation of power plant and grid," which promoted the development of power in China. However, the limited "bidding on the Internet" led to the fact that the market cannot really play a fundamental role in the allocation of power resources. The contradiction between "planned coal" and "market power" intensified, the operation mechanism of electric power enterprises was not changed in place, and the profit and loss of the main body of the PG market was completely regulated by policies. Based on the principle of "control the middle and open the two ends," a new round of electric power reform in 2015 developed the generation and sale side, promoted the market-oriented construction of electric power trading, taking the mid and long-term electricity trading contracts in the province as the market prying point, and continued to deepen the transregional and trans-provincial markets, auxiliary service markets, spot markets, new energy alternative PG markets, distributed markets, and carbon trading markets. At the same time, the implementations of transmission and distribution price verification, a pilot of incremental distribution mix reform, and adjustment of new energy subsidies were aiming to make the market play a decisive role in resource allocation. At present, it mainly analyzes the supply and demand of electric power and the reform of electricity price in detail.

#### 21.1.1.3.1 The supply and demand of electric power

In 2018 the situation of national power supply and demand changed from general easing in the previous 2 years to general balance. Since 2015 with the continuous record of installed capacity of PG, the power supply capacity has been significantly enhanced and the PG capacity has been growing rapidly. In 2015—18, the cumulative increase of PG is about 1,254,100 million kWh [3—6]. In 2018 China's total generating capacity is 6,994,000 million kWh, a year-on-year increase (YOYI) of 8.4% (see Fig. 21.2). Among them, hydropower generation was 1,232,900 million kWh, YOYI of 3.2%; thermal PG was 4,923,100 million kWh, YOYI of 7.3%; nuclear PG was 294,400 million kWh, YOYI of 18.7%; wind PG was 366,000 million kWh, YOYI of 20.2%; solar PG was 177,500 million kWh, YOYI of 50.7%. The specific data for PG are shown in Table 21.1.

By the end of 2018 the installed capacity of full caliber PG in China was 1899.67 million kW, an increase of 5.9% over the previous year. This included 352.26 million kW of hydropower, an increase of 2.5% over the previous year; 1143.67 million kW of thermal power, an increase of 3.0% over the previous year; 44.66 million kW of nuclear power, an increase of 24.7% over the previous year; 184.26 million kW of wind power, an increase of 12.4% over the previous year; 174.63 million kW of solar power, an increase of 33.9% over the previous year. In addition, the PG installed structure was further optimized. According to Table 21.2, the proportion

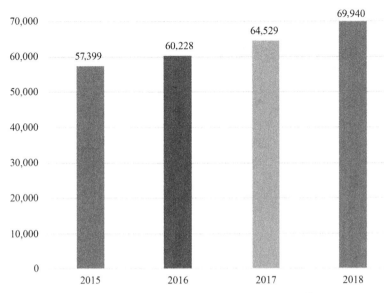

FIGURE 21.2 National power generation in 2015−18 (unit: 100 million kWh).

TABLE 21.1 National power generation in 2015−18 (unit: 100 million kWh, %).

| Year | Total | | Hydropower | | Thermal power | | Nuclear power | | Grid-connected wind power | | Grid-connected solar power | |
|---|---|---|---|---|---|---|---|---|---|---|---|---|
| | PG | YOYI | PG | YOYI | PG | YOYI | PG | YOYI | PG | YOYI | PG | YOYI |
| 2015 | 57,399 | 1.1 | 11,127 | 5 | 42,307 | −1.7 | 1714 | 28.7 | 1856 | 16.2 | 395 | 68 |
| 2016 | 60,228 | 4.9 | 11,748 | 5.6 | 43,273 | 2.3 | 2132 | 24.4 | 2409 | 29.8 | 665 | 69 |
| 2017 | 645,29 | 7.1 | 119,47 | 1.7 | 45,877 | 6 | 2481 | 16.4 | 3046 | 26.4 | 1178 | 77 |
| 2018 | 699,40 | 8.4 | 12,329 | 3.2 | 49,231 | 7.3 | 2944 | 18.7 | 3660 | 20.2 | 1775 | 50.7 |

of installed capacity of thermal PG is decreasing year by year, while the proportion of installed capacity of nuclear power, grid-connected wind power, and grid-connected solar PG is increasing (see Fig. 21.3). Specific data for installed capacity of PG in China are shown in Table 21.2.

Electricity consumption is a barometer of the operation of the national economy. On the one hand, the rapid growth of electricity consumption reflects the overall stability of China's economic operation. On the other hand, it reflects the positive progress made in the high-quality development of China's economy. Under the comprehensive influence of factors such as overall stable macroeconomic operation, rapid development of service industry, high technology and equipment manufacturing industry, cold wave in winter and high temperature in summer, rapid promotion of electric energy substitution, urban and rural power grid transformation, and upgrading to release electric power demand, the

**TABLE 21.2** National installed capacity of power generation in 2015–18 (unit: 10,000 kW, %).

| Year | Total Installed capacity | Hydropower Installed capacity | Proportion | Thermal power Installed capacity | Proportion | Nuclear power Installed capacity | Proportion | Grid-connected wind power Installed capacity | Proportion | Grid-connected solar power Installed capacity | Proportion |
|---|---|---|---|---|---|---|---|---|---|---|---|
| 2015 | 152,527 | 31,954 | 20.9 | 100,554 | 65.9 | 2717 | 1.8 | 13,075 | 8.6 | 4218 | 2.8 |
| 2016 | 165,051 | 33,207 | 20.1 | 106,094 | 64.3 | 3364 | 2 | 14,747 | 8.9 | 7631 | 4.6 |
| 2017 | 179,418 | 34,377 | 19.2 | 111,009 | 61.9 | 3582 | 2 | 16,400 | 9.1 | 13,042 | 7.3 |
| 2018 | 189,967 | 35226 | 18.5 | 114,367 | 60.2 | 4466 | 2.4 | 18,426 | 9.7 | 17,463 | 9.2 |

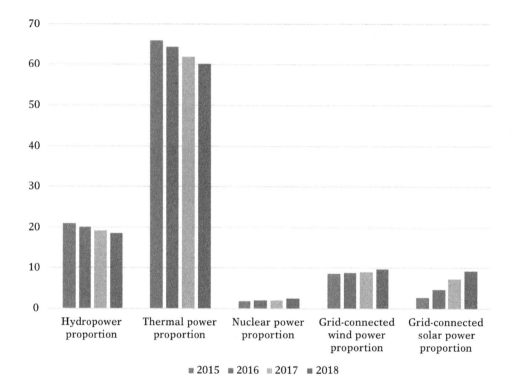

**FIGURE 21.3** National installed capacity of power generation in 2015–18 (unit: %).

power consumption of the whole society has achieved rapid growth. In 2018 the power consumption of the whole society was 6,844,900 million kWh, a YOYI of 8.5%, a new high in nearly 7 years. Among this, the power consumption of the second industry, the third industry, and urban and rural residents is increasing year by year (see Fig. 21.4). In terms of different industries, the power consumption of the primary industry is 72,800 million kWh, a YOYI of 9.8%. The power consumption of the secondary industry is 4,723,500 million kWh, a YOYI of 7.2%, 1.7 percentage points higher than the previous year, driving up

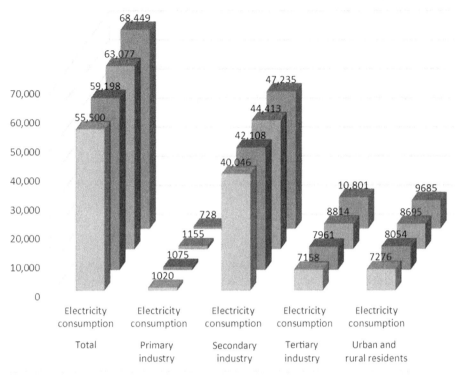

**FIGURE 21.4** National electricity consumption in 2015–18 (unit: 100 million kWh).

the power consumption of the whole society by 5.0 percentage points, 1.0 percentage points higher than the previous year, which is the main driving force for the growth of the power consumption of the whole society. This reflects the acceleration of the new momentum of economic development. The power consumption of the tertiary industry reached 10,801,000 million kWh, a YOYI of 12.7%. The growth rate is 2 percentage points higher than the previous year, and driving up the power consumption of the whole society by 1.9 percentage points, 0.5 percentage points higher than the previous year. The power consumption of the main industries in the tertiary industry has maintained a double-digit high-speed growth, supporting the rapid development of the modern service industry, and reflecting the accelerated formation of the modern service industry. The domestic power consumption of urban and rural residents is 968,500 million kWh, a YOYI of 10.4%, 2.6 percentage points higher than the previous year, and driving up the power consumption of the whole society by 1.5 percentage points. Both urban and rural residents have maintained double-digit growth in electricity consumption, reflecting the continuous improvement of people's quality of life [3]. The specific circumstances of national electricity consumption are shown in Table 21.3.

**TABLE 21.3**   National electricity consumption in 2015−18 (unit: 100 million kWh, %).

| Year | Total | | Primary industry | | Secondary industry | | Tertiary industry | | Urban and rural residents | |
|---|---|---|---|---|---|---|---|---|---|---|
| | Electricity consumption | YOYI | Electricity consumption | YOYI | Electricity consumption | YOYI | Electricity consumption | YOYI | Electricity consumption | YOYI |
| 2015 | 55,500 | 0.5 | 1020 | 2.5 | 40,046 | −1.4 | 7158 | 7.5 | 7276 | 5 |
| 2016 | 59,198 | 5 | 1075 | 5.3 | 42,108 | 2.9 | 7961 | 11.2 | 8054 | 10.8 |
| 2017 | 63,077 | 6.6 | 1155 | 7.3 | 44,413 | 5.5 | 8814 | 10.7 | 8695 | 7.8 |
| 2018 | 68,449 | 8.5 | 728 | 9.8 | 47,235 | 7.2 | 10,801 | 12.7 | 9685 | 10.4 |

① Since May 2018, the three industrial divisions have been adjusted in accordance with the notice of the National Bureau of Statistics on revising the provisions on three industrial divisions (2012). To ensure the comparability of data, the data in the same period have been reclassified according to the new standards. ② The power consumption index of the whole society is full caliber data.
*Data from National Energy Administration (NEA). [7−10].*

### 21.1.1.3.2 The development plan of electricity price

Electricity price is the leverage and core content of the electricity market [11]. Electricity price reform is the key link of power system reform. The goal of electricity price reform is to establish a scientific and reasonable electricity price formation mechanism and an electricity price system that are suitable for the electricity market, and they play an important role in guiding power supply and demand and allocating resources reasonably. At the same time, an efficient and powerful electricity price supervision system which is in harmony with the reform of the power system and the management of the electricity market should be established [2].

Since 2002 China has been deepening the reform of electricity price and the formation mechanism of electricity price is gradually optimized. According to the process of marketization and the timely adjustment and change of power supply and demand, the electricity price policy promotes the sustainable and healthy development of the electric power industry and improves the efficiency of power resource allocation. The main measures are as follows.

Firstly, the reform plan of electricity price and its supporting implementation measures are formulated. In 2003 the plan of the general office of the State Council on deepening the reform of the electric power system further emphasized that the price of PG and sale formed by market competition, and the price of power transmission and distribution to be determined by the government are the goals to be achieved step by step. As well as accelerating the construction of the electricity market, implementing the pilot project of separate transmission and distribution, and pushing forward the reform of electricity price in an all-round way are overall goals of power system reform. To solve the price problem of separation of power plants and power grid in power system reform, the National Development and Reform Commission has promulgated relevant supporting implementation measures, refined the measures of power price reform, and established the power price management measures to meet the new requirements of power system reform. In 2007 the implementation opinions on deepening the reform of electric power system during the "Eleventh Five-Year Plan" were issued, pointing out that the reform of the electric power system will adhere to the direction of market-oriented reform, promoting the reform of separation of main and auxiliary power grid enterprises. In 2009 the key tasks of electricity price reform included the establishment of a regional electricity market in Northeast China and the trial of bidding for grid access, which gained valuable experience; the measures for direct trading between power users and PG enterprises were formulated; and 20% of the electricity sales market was been opened up, allowing large users to directly trade power with PG enterprises and independently negotiate electricity prices. In 2010 "opinions on key work of deepening economic system reform in 2010" clearly proposed that the reform of the separation of main and auxiliary power grid enterprises should be completed, pilot work opinions on transmission and distribution system reform should be issued in 2011. These indicate that the approval of transmission and distribution electricity price is expected to become a breakthrough in electricity price reform.

The second measure is to promote the reform of transmission and distribution price. In 2015 the reform of the domestic electricity market started a new round, and the reform of transmission and distribution price continued. As of 2019, 32 provincial-level power grids and Shenzhen power grid transmission and distribution price approval have been fully covered, and the permitted revenue of power grid enterprises has been reduced by 48 billion yuan. A scientific and independent transmission and distribution price mechanism covering all levels of power grids has been established. In the future, with the continuation of power reform, the reform of transmission and distribution price is expected to continue to deepen and develop toward the pricing mode of permitted cost plus reasonable income.

The third measure is to regulate the price of transregional transactions. In 2005 the National Development and Reform Commission and the Electric Power Regulatory Commission issued the guiding opinions on promoting transregional electric energy trading, which clarified the transregional electric energy trading price, further standardized trans-provincial and transregional electric energy trading, played a regulatory role in the price leverage, broke down the barriers between provinces, developed the regional electricity market, and realized the optimal allocation of power resources.

The fourth measure is the implementation of time-of-use pricing system. To promote the balance between supply and demand of power system, cut peak and fill valley, and alleviate the shortage of electricity in a peak period, the time-of-use pricing system has been widely implemented in all power grids in China since 2002. The peak-valley time-of-use pricing method is implemented in the selling side of the power grid, which improves the load rate of electricity, promotes the rational allocation of power resources, and ensures the safety and reliability of the power system to some extent.

The fifth measure is to deepen the reform of the price formation mechanism of coal-fired PG bidding in the Internet. At present, the transmission and distribution price reform has achieved full coverage, and the price mechanism of "permitted cost + reasonable income" has been basically established; the scale of electricity market-oriented transactions in various regions has been expanded, about 50% of coal-fired PG bidding in the Internet has been formed through market transactions, and the spot market has begun to be established. The national power supply and demand are relatively loose, and the utilization hours of coal-fired PG are lower than the normal level. At the same time, with the deepening of the reform of the development and utilization plan, the common industrial and commercial power users or the power selling companies acting for their power purchase will adopt the same electricity price formation mechanism as the large industrial power users after entering the market, which will cause new cross subsidies and other problems. To solve these problems, it is urgently necessary to carry out market-oriented reform on the benchmark price of coal-fired PG and coal electricity linkage mechanism, so that a new path for industrial and commercial users to enter the market design will be open.

To steadily achieve the goal of fully liberalizing coal-fired PG bidding on the Internet, the current price mechanism is changed into a market-oriented price mechanism of "benchmark price + fluctuation." The benchmark price shall be determined according to the current benchmark price of local coal-fired PG, and the floating range shall be no more than 10% for upward floating and no more than 15% for downward floating in principle.

The implementation of "benchmark price + fluctuation" mechanism is different from the original "benchmark price of local coal-fired PG + coal price linkage mechanism." The essence is that the fluctuation range is formed by the market mechanism. It can not only effectively reflect the cost of power production, but also effectively reflect the elasticity of power demand. And the price mechanism has strong timeliness, which is helpful to promote the coordination and improvement of the whole industrial chain of electric power upstream and downstream. The price adjustment mechanism based on coal electricity linkage is replaced by market-based price formation mechanism. The new electricity price mechanism is more helpful to promote the coordinated development of the upstream and downstream industries and their markets, and push forward the construction, development, and improvement of China's modern energy market system. At the same time, considering the stable transition and the stability of the formation mechanism of the feed in price of other power sources, the benchmark price is determined according to the current benchmark price of local coal-fired PG during the transition period.

The mechanism of "base price + fluctuation" is mainly aimed at the medium and long-term electric power contracts formed by market entities such as PG enterprises, power selling companies, and power users through market-oriented ways such as over the counter (OTC) bilateral transactions or centralized bidding (including listing transactions) in the field. The specific scope and form of implementation should be adapted to local conditions. The price of coal-fired PG corresponding to the following two situations is still implemented. One situation is the coal-fired PG corresponding to the electricity consumption of industrial and commercial users who do not have the market trading conditions or participate in the market trading temporarily. The other situation is the coal-fired PG corresponding to the electricity consumption of residents and agricultural users.

The sixth measure is to standardize the management of renewable energy generation price and cost sharing. The trial measures for the management of the price and cost sharing of renewable energy PG stipulate that the price of renewable energy PG can be determined in the form of government pricing and government guided price, and power users are encouraged to purchase renewable energy power voluntarily, and the price of electricity shall be implemented according to the price of renewable energy PG plus the average transmission and distribution price of power grid. For the renewable energy bidding on the Internet, the current policy is that the grid shall pay the corresponding power purchase fees to the renewable energy PG enterprises according to the coal benchmark pricing, and the difference between the renewable energy price and the coal benchmark price shall be subsidized by the National Renewable Energy Development

Fund, which is funded by the renewable energy price surcharge. The regional pricing mechanism of onshore wind power and photovoltaic PG in China is determined by the distribution of wind and light resources in each region and the construction cost. With the development of renewable energy, the benchmark pricing adopts a declining mechanism, especially photovoltaic PG. The progress and upgrading of renewable energy technology represented by wind and solar energy is very fast, which brings efficiency improvement and cost reduction. According to the "Thirteenth Five-Year Plan" for renewable energy development, the future reform direction of the renewable energy generation pricing mechanism is to comprehensively implement the competitive allocation mode to determine the project and electricity price, promote the scale-up of nonsubsidized grid access projects, make renewable energy participate in the market-oriented competition, and gradually integrate into the electricity market.

## 21.1.2 Classification of China's electricity market

The concept of the electricity market was first proposed by the United Kingdom in 1989. Then countries around the world began to study the electricity market. A sound electricity market is composed of the electricity spot market and the electricity financial market. From the operation practice of the foreign mature electricity market, the electricity market is mainly divided into two modes: the centralized mode represented by the United States, and the decentralized mode represented by Europe. The decentralized mode is mainly applicable to countries and regions with little power grid congestion, a low proportion of new energy, a large number of flexible regulating units, and a high degree of marketization. In contrast, the centralized mode can take into account both the physical properties of power goods and the economic principles of market operation. The market efficiency and resource allocation effect under the centralized competition are better, the market price signal is more complete and accurate, and it is more suitable for the domestic unified and balanced scheduling production system.

### 21.1.2.1 Electricity spot market

The electricity spot market includes the day-ahead bidding market, real-time balance market, and frequency modulation (FM) auxiliary service market. The trading object is electric power. Electric power has not only commodity attributes, but also specific physical attributes. No matter the length of the bilateral contract for purchase and sale, and no matter what kind of contract the two parties have reached in the period of days or more, they must form a power curve that can be executed in the next day through the day-ahead market, or through negotiation between the buyer and the seller, and generate and use electricity according to the agreed curve, so as to ensure the real-time balance of PG and consumption The biggest change of marketization to the traditional generation plan is to determine the generation plan of each unit in different periods from the perspective of optimal allocation of resources, instead of the principle of average distribution. The basis of realizing the optimal allocation of resources is the unit's generation quotation in different periods. On the basis of the power curve agreed in the above contract, combined with the load forecasting information, the starting mode and real-time output of the generating unit are determined in the order of price priority, then the final generation plan is formed.

The key task of the new round of electric power system reform is the construction of the electricity spot market. By the end of June 2019, all eight pilot provinces of the national spot market had entered trial operation, marking a new stage of China's electricity market construction. With the steady development of the spot market pilot, it will accelerate the development of the electricity selling industry to become specialized and skilled, truly activate the electricity market, promote market competition, and put forward higher requirements for market risk prevention and control and market supervision. The construction of electricity spot market focuses on the following six aspects. The first is to promote the construction of medium and long-term trading and the spot market as a whole. In the process of the simulated trial operation of the provincial spot market, the connection between medium and long-term transactions and the spot market is still imperfect. At present, the medium and long-term transactions in various provinces are based on the direct transactions of large users before the establishment of the spot market. After the operation of the spot market, the medium and long-term transactions need to adapt to the spot market. The market subjects sign the medium and long-term contracts with transaction curves that play the role of stabilizing market supply and demand and prices and avoiding market risks. The second is to properly link up the plan with the market. In June 2019 the national development and Reform Commission and the National Energy Administration jointly issued the notice on comprehensively liberalizing the PG and utilization plan for operating power users, proposing to liberalize the PG and utilization plan in an all-round way except for some nonoperating ones. This greatly enriches the number and category of market subjects and provides a solid foundation for the construction and operation of the spot market. At the same time, the national conditions of planning and market dual track operation challenge the design of market rules. In the early stage of market construction, problems such as power grid operation safety, clean energy consumption, national economy, and people's livelihoods can be regulated and guaranteed through planned electricity. With the gradual expansion of the spot market, the capacity of planned electricity regulation gradually weakens. The functions carried out through planned electricity will need to be solved gradually through market-oriented mechanisms and means in the future. This requires that the connection between planning and market should be fully considered in the design of spot market rules, and the decisive role of the market in resource allocation should be played. Gradually, the traditional planning management should be transformed into the way that power resources are allocated through the electricity market. The third is to actively study the construction path of the national electricity market. From the experience of foreign countries, it is a general development trend to build a large-scale electricity market with optimized resource allocation. The western energy imbalance market in the west of the United States continues to attract new members, expand the market scope, and provide space for clean energy consumption in the west of North America. The European Union (EU) has continued to promote the construction of a unified electricity market. At present, a unified day-ahead and intraday cross-border market has been formed, fully realizing the complementary energy structure of all countries, and jointly helping to achieve the clean development goal of the EU. The reverse distribution of China's energy supply and demand, interregional energy resources endowment, significant differences in the level of economic development, and the rapid development of clean energy determine that the overall energy allocation must be carried out nationwide. It is an inevitable trend of China's future development to build a national electricity market with large-scale optimal allocation of resources. However, at present, the country has not yet formed a unified plan for the construction path of the national electricity market, and further prospective research is needed to provide the top-level design for the future development of the integration of provincial and inter provincial markets. The fourth is to explore the gradual integration of auxiliary service market and electric energy spot market. In the past few years, when the electricity spot market has not yet started, all regions have actively carried out market-oriented exploration and established regional or provincial peak shaving auxiliary service market to solve the peak shaving pressure brought by new energy grid connection, and achieved good results. In essence, peak load regulation is to regulate the output of generating units in the real-time operation of power grid. In the mature electricity market system, it is usually solved by balancing the market in real time. To realize the orderly connection of transaction varieties and ensure the efficient organization and operation of each market, the next step needs to explore the organic integration of provincial peak shaving auxiliary service market and electric energy spot market, carry out in-depth peak shaving transactions in a larger scope, realize the sharing of peak shaving resources, establish the FM auxiliary service market in the province, gradually realize the joint optimization and clearing with the electricity energy market, and improve the efficiency of market organization. The fifth is to reasonably guide renewable energy to participate in the electricity spot market. In May 2019 the notice on establishing and improving the energy consumption guarantee mechanism of renewable energy was issued. The notice set the target of the proportion of renewable energy in the power consumption of each provincial administrative region and further improved the consumption demand of renewable energy. To promote renewable energy to participate in the electricity spot market competition is an important way to promote the consumption of renewable energy, but it is necessary to consider the effective connection between the market transaction of renewable energy and the guiding price policy as a whole, and implement the relevant requirements of the state for the consumption of renewable energy. The sixth is to adapt to the characteristics of a diversified electricity market in the future. With the progress of artificial intelligence, blockchain, edge computing, and other technologies, as well as the development of distributed energy, such as distributed photovoltaic, energy storage, and electric vehicles, the main body of the emerging market, such as virtual power plants and load aggregators, has emerged. The main body of the emerging market has strong flexibility and expansibility, which can significantly improve the flexibility and economy of power system operation and have an increasingly significant impact on power system operation. The construction of the power market should gradually adapt to the characteristics of the main body of the emerging power market, guide the main body of the emerging market to actively participate in the power market, stimulate the vitality of the main body of the market, promote the good interaction between the load and storage of the source network, effectively guarantee the real-time balance between electricity supply and demand, and promote the realization of the goal of clean energy and low-carbon transformation by designing a reasonable market mechanism and business model and combining the advanced technologies such as blockchain and virtual power plant.

### 21.1.2.2 Electricity financial market

The electricity financial market is the financial derivative property of the electricity spot market. The electricity financial derivatives such as electricity futures and electricity options are traded by referring to the basic principles of futures and options trading. The trading object is the electricity financial derivatives, which does not belong to the regulatory scope of the electric power market rules. The electricity financial market is the perfection and supplement of the electricity spot market, which can attract a wide range of market participants, enhance the competitiveness of the electricity market, find the real electricity spot price in the electricity market, and provide reliable risk control for the electricity spot transaction. As the second largest power market in the world, China is also the fastest-growing power market. There is a huge space for development with diversified investors. With the deepening of the reform of electric power system, more and more operators of electric power enterprises have realized the importance of financial market. The traditional mode of production and operation has been transformed into the mode of both production and operation of financial capital. With the help of financial market, it can promote the development of electric power industry. Firstly, it can improve the financing efficiency, absorb idle funds to invest in power, realize the rapid expansion of enterprise scale, break the bottleneck and constraints restricting the development of electric power industry, and provide reliable energy security for economic development. Secondly, we can optimize the allocation of resources, integrate different types of power assets according to the principle of complementary advantages, and then promote the flow and reorganization of state-owned assets, so as to realize the adjustment of electric power industry structure and layout. In view of the current situation of "fighting" between electricity and coal, it is necessary to actively promote the coal power joint venture, reduce the transaction cost of electricity and coal, allow and encourage the mutual shareholding, equity participation and holding of coal power capital, and use the integration, merger, and reorganization of different capital to realize mixed operation. The third is to speed up the diversification of equity, improve the corporate governance structure and transform the management mechanism of enterprises, enhance the overall quality of enterprises, and consolidate the dominant position in the market competition. The fourth is to promote the formation of a scientific power regulatory system. The supervision and expectation of investors will make power enterprises face the pressure of the capital market. This pressure transmitted to the regulatory department is the electric power industry regulatory means and methods that can better meet the needs of the development of the electric power industry market. The fifth is that all participants in the electricity market can use different electricity derivative instruments to improve the operation status of enterprises and avoid electricity market risks. In a word, the use of electricity financial market can accelerate the realization of large company and group strategy, promote the reform of electric power industry, and promote the marketization process of electric power industry.

## 21.1.3 Development of China power grid

### 21.1.3.1 General situation of power grid

As an important energy, electric power is the foundation of national economy and people's livelihood. As a medium to deliver electric energy to all walks of life and thousands of households, the power grid has continuously played the role of pioneer and booster in the past 70 years, supporting the rapid development of the national economy, realizing the cross-regional configuration of energy, leading to the transformation of clean energy, ensuring the safe and reliable power supply of the whole country, and creating brilliant achievements that attract worldwide attention. China's economy has changed from a high-speed growth stage to a high-quality development stage. As an important infrastructure to serve the economic and social development and the people's good life, the power grid also shoulders the important mission of leading energy reform, promoting energy transformation, and promoting common development.

#### 21.1.3.1.1 The rapid growth of China's power grid has supported the rapid development of the national economy

After 1978 China insisted on making economic construction central, expanding the opening up, and maintaining a high economic development speed. By the end of 2018, the circuit length and transformation capacity of 220 kV and above transmission lines have reached 733,000 km and 4.02 billion kVA, respectively, with an average annual growth of 10.5% and 14.6%, respectively. The power grid has achieved leapfrog development, strongly supporting the high-speed growth of China's economy.

#### 21.1.3.1.2 The large scale of China's grid interconnection ensures the optimal allocation of energy across regions

China started with the development of urban isolated power grids, and then extended the construction of 110 kV and 220 kV power grids in various provinces and cities. As the interconnection of power grids can realize mutual support and exchange of power grids, reduce the reserve capacity, and improve the system's ability to withstand accidents, China has determined that the power industry should follow the road of interconnection. Since then, China's power grid has focused on the construction of 500 kV provincial and trans-provincial power grids. In 1989, the $\pm 500$ kV transmission line (transmission capacity of 1.2 million kW) from Gezhouba to Shanghai was put into operation, which was the prelude to cross-regional networking. In 1994 the Three Gorges power plant was built, gradually forming a national interconnected power grid pattern that took the North, Middle, and South power transmission channels as the main body and realized the West–East power transmission and North–South mutual supply of the multipoint interconnection between the north and south power grids. In 2011 with the commissioning of $\pm 400$ kV networking project from Qinghai to Tibet, the national interconnected power grid pattern was basically formed except for Taiwan. By the end of 2018, China has formed six regional or provincial alternating current (AC) synchronous power grids in North China, Central China, East China, Northeast, Northwest, South China, and Yunnan. The transregional transmission capacity of the whole country has reached 136.15 million kW, including 122.81 million kW of AC and direct current (DC) grid-connected transregional transmission capacity and 13.34 million kW of cross-regional point-to-network transmission capacity. The power grids support each other to provide a more solid platform for the optimal allocation of large-scale energy resources.

### 21.1.3.1.3 China's advanced grid technology promotes the transformation and development of clean energy

In 1972 China completed the first 330 kV ultrahigh-voltage (UHV) transmission and transformation project with a total length of 534 km. In 1981 the first 500 kV transmission line with a total length of 595 km was completed. In 2005 the first 750 kV transmission line with a total length of 146 km was built in Northwest China.

In the 21st century, to reduce environmental pollution and cope with climate change, China has vigorously developed clean energy and the State Grid Corporation of China has carried out research on the key technologies of the UHV power grid, equipment, and construction of demonstration projects. In 2009 the first 1000 kV high-voltage transmission line was built and put into operation, which is the AC transmission project with the highest voltage level at present. In 2010 two ± 800 kV high-voltage direct current (HVDC) lines were completed and put into operation. In 2018, the ± 1100 kV ultrahigh-voltage direct current (UHVDC) project with the highest voltage level and the longest power transmission distance from Zhundong to Wannan in the world was completed. The ± 800 kV high-voltage DC transmission project from Hainan to Zhumadian, which was started in 2018, will be the first UHVDC transmission channel specially built for clean energy export in China and even in the world. The UHV power grid has significant comprehensive benefits and long-term strategic significance for promoting energy transformation and green development, ensuring power supply, stimulating economic growth, etc.

### 21.1.3.1.4 The high security level of China's power grid promotes social stability and long-term stability

The installed PG capacity reached 1.9 billion kW by the end of 2018. At the same time, there has been no large-scale blackout in China's power grid in the past 20 years. This benefits from the power planning system at the national and provincial levels, the dispatching system at the five levels (national, regional power grid, provincial, prefecture, and county levels), and the implementation of multiple rounds of urban and rural power grid transformation and rural power grid upgrading. The work of the power grid in planning, dispatching, and people's livelihood management improves the overall reliability and power supply quality of the power grid, and provides a strong guarantee for economic and social development.

### 21.1.3.1.5 China's power grid has a large service population, which meets the needs of universal power services

Since the 1990s, through multiple rounds of rural power grid upgrading and transformation, household power supply and other projects, the "Twelfth Five-Year Plan" period has completely solved the problem of power consumption for people without electricity, "Thirteenth Five-Year Plan" period has solved the problem of "low voltage," and the quality of power supply has greatly improved. In 2018 the power supply reliability rate of urban users in China reached 99.946%, and that of some cities reached 99.999%. Although there is a gap in rural areas, it still reached 99.775%. Remarkable achievements have been made in the development of the power grid.

### 21.1.3.2 Transmission line and UHV

After the reorganization of state power company in 2002, the State Grid Corporation of China and China Southern Power Grid were established. The length of 110 kV and above transmission lines of the State Grid is 1,033,400 km, accounting for 77% of the total length of China's transmission lines. The 226,700 km of China's Southern Power Grid accounts for 17% of the total length of transmission lines. The remaining 6% of the transmission lines are mainly owned by local small power grids.

UHV is composed of 1000 kV and above AC and ± 800 kV and above DC transmission. It has the characteristics of strong transmission capacity and low transmission loss. It is suitable for large capacity and long-distance transmission of electric energy. It represents the highest level and development direction of transmission technology in the world. It provides a technical means to realize the optimal allocation of energy resources at a larger scope.

China's tracking of research on UHV technology began in the 1980s, and began to focus on large-scale research and demonstration, technical research, and engineering practice at the end of 2004. Through the joint efforts of all parties, China's UHV transmission technology has made continuous breakthroughs in development, and has successively completed and put the UHVAC test demonstration project and the UHVDC demonstration project into operation, and has kept safe and stable operation, indicating that China's UHV technology has matured. By 2019 China had put 11 DC UHV lines, 10 AC UHV lines, and 2 DC lines into operation and also has 4 AC UHV lines under construction. Among them, ± 1100 kV UHVDC project from Zhundong to Wannan and Sutong 1000 kV UHVAC GIL comprehensive pipe gallery project both occupy commanding heights in their respective fields of power transmission technology in the world today, promoting the new leap of China's electrical equipment manufacturing capacity, and greatly improving power transmission from West to East and large-scale optimization of energy allocation.

The Zhundong–Wannan project is the most advanced high-voltage DC project with the highest voltage level, the largest transmission capacity, the longest transmission distance, and the most advanced technical level in the world. The project passes through six provinces (regions), with a transmission capacity of 12 million kW and a total length of 3324 km, realizing the overall improvement of DC voltage, transmission capacity, and AC network side voltage. The project has an annual power transmission capacity of 60–85 billion kWh, which will effectively alleviate the contradiction between the medium- and long-term power supply and demand in East China, and reduce the annual coal consumption of about 38 million tons in East China. At the same time, the project will vigorously promote the development and delivery of energy bases in the West and North, expand the scope of clean energy consumption, promote the transformation of Xinjiang local resource advantages into economic advantages, and help Xinjiang's long-term stability and economic and social development.

The world's first UHV tunnel crossing, the Yangtze River comprehensive pipe gallery–Sutong GIL comprehensive pipe gallery project, is the first in the world to adopt UHV GIL technology in important transmission channels, and it is the world's longest distance GIL innovation project with the highest voltage level, the largest transmission capacity, and the highest technical level. The project is one of the components of the Huainan–Nanjing–Shanghai 1000 kV high-voltage AC project. It takes the lead in adopting the complete set of UHV GIL transmission technology in the world, reserving the technology for the advanced compact transmission across rivers, seas, and densely populated areas. After the project has been put into operation, the East China UHVAC ring network will realize the combined operation. The power receiving capacity of East China's power grid will be greatly improved, which can reduce 170 million tons of coal for PG and 310 million tons of carbon dioxide each year, and significantly improve the environmental quality of East China.

The production of two major projects is an important milestone in the process of the electric power industry. After the project is put into operation, clean power will be more reliable, efficient, and flexible into the Huadong UHVAC ring network, which has significant comprehensive benefits and long-term strategic significance for optimizing energy allocation, ensuring power supply, preventing air pollution, stimulating economic growth, and leading technological innovation.

In addition, the $\pm 800 \, \text{kV}$ HVDC project from Yazhong to Jiangxi under construction is the first UHVDC project in Southwest China. It is a major power transmission project that serves the national energy strategy of "power transmission from west to east," guarantees the water and electricity consumption in the west, and meets the green development needs of the central and eastern regions. The completion of the project will greatly alleviate the problem of wastewater from hydropower generation in Sichuan Province. It can achieve more than 40 billion kWh of external power supply and reduce 16 million tons of standard coal consumption and 40 million tons of carbon dioxide emissions each year. It is of great significance for the realization of clean hydropower out of Sichuan and the wider optimization of domestic energy allocation.

UHV not only blooms all over the country, but also goes abroad. In 2014 State Grid Corporation of China and State Power Corporation of Brazil jointly won the first phase project of Meilishan in Brazil, which is the first overseas UHVDC transmission project won by the State Grid Corporation of China, becoming a new important milestone in the field of power cooperation between China and Brazil. In 2015 the State Grid Corporation of China successfully won the bid for the second phase of the franchise project of Brazil's Meilishan hydropower $\pm 800 \, \text{kV}$ UHVDC transmission, which realizes the independent "investment, construction, and operation" of China's UHV overseas for the first time.

The second phase project of Meilishan in Brazil has a total investment of about 9.6 billion Real and a total transmission line length of about 2539 km. It is the $\pm 800 \, \text{kV}$ high-voltage transmission project with the longest transmission distance in the world. The State Grid Corporation of China has a 30-year franchise right for the project. The project was put into commercial operation in 2019 in Brazil. The second phase project of Meilishan is the delivery project of the Meilishan hydropower station, the second largest hydropower station in Brazil, and the main channel for interconnection between the north and the south of Brazil's power grid. The safe and high-quality operation of the project effectively solves the problem of clean hydropower transmission and consumption in the Amazon basin in northern Brazil, strongly supports and serves Brazil's economic and social development, and contributes to Brazil's energy security and stable supply program.

### 21.1.3.3 Smart grid and energy Internet

The problem of traditional energy shortage and environmental pollution is becoming more and more serious, which is the biggest challenge for the sustainable development of human society. To solve the energy crisis and environmental problems, the modern power grid has set up reliable, high-quality, efficient, green, and environmental protection goals. Therefore the concept of the smart grid comes into being. The smart grid can encourage users to participate in demand side management (DSM) through advanced two-way communication technology, and use incentive mechanisms and price response to play the role of the demand side power market. DSM was put forward in the 1980s. It aims to guide users to optimize the power consumption mode, improve the efficiency of terminal power consumption, avoid installing new PG and distribution infrastructure, and make better use of the existing PG capacity. The purpose is to cut the peak and fill the troughs, and promote the balance of power supply and demand. DSM can be roughly divided into energy efficiency measures (EEMs) and demand response (DR) [12]. EEM strategy is considered as permanent load reduction, while DR focuses on load flexibility. The increasing permeability of intermittent renewable energy generation, in particular, wind and solar energy, has led to the development of intermittent mitigation technologies, such as the application of DR strategies. As one of the solutions of DSM, DR refers to the market participation behavior of power users that changes their inherent power consumption mode according to the market price or the incentive measures of the power supply side. It can meet the PG and power demand in a more effective way, and has great potential in improving the reliability of the power system. The International Energy Agency emphasizes the important contribution of DR to the integration of multiple renewable energy sources, reaching nearly 185 GW of flexible demand in a cost-effective manner by 2040 [13]. In the power market, DR is mainly reflected in price response. Real-time pricing is an important pricing method in the existing pricing mechanism. In the smart grid environment, users and power suppliers exchange power price and demand information through the energy management center embedded in smart meters. Zethmayr and Kolata [14] conducted an empirical analysis on the effect of real-time electricity price, fixed electricity price, and time-of-use electricity price within the scope of large household users based on the smart electricity meter data in 1 year. It shows that the real-time electricity price is significantly better than the fixed electricity price and time-of-use electricity price in promoting the balance of supply and demand and cutting the peak and filling the troughs [14]. With the construction and development of the smart grid and energy Internet, in particular, smart meters and smart electrical equipment (mainly including smart sockets and smart appliances) will be gradually popularized in the future and the implementation of a real-time electricity price mechanism will be supported by effective hardware. Many Scholars (such as Samadi et al. [15]; Deng et al. [16]; Yang et al. [17]; Nge et al. [18]; Chaudhary and Rizwan [19]; Shahbazitabar and Abdi [20]; and Rahbari et al. [21]) have made in-depth research on the real-time electricity price in the case of the smart grid or grid connection of renewable energy and power storage equipment. The research results have a certain guiding role in the formulation of the electricity price and the implementation of the electricity consumption strategy.

In 2010 the State Grid put forward the idea of developing a strong smart grid, that is, to build a strong smart grid with UHV as the framework and coordinated development of all levels of the power grid, and to integrate the strong UHV grid with the intelligent power grid. The construction of the smart grid is divided into three stages: 2009−10 is the planning pilot stage, 2011−15 is the comprehensive construction stage, 2016−20 is the leading and upgrading stage. The specific investment situation is shown in Table 21.4.

TABLE 21.4   China's smart grid investment and construction stage analysis (unit: billion RMB).

| Stage | Time | Total grid investment | Intelligent investment | Annual intelligent investment |
|---|---|---|---|---|
| First stage | 2009–10 | 551 | 34.1 | 17 |
| Second stage | 2011–15 | 1500 | 175 | 35 |
| Third stage | 2016–20 | 1400 | 175 | 35 |

After nearly 10 years of continuous development, the reliability of urban power supply in the operation area of the State Grid has been basically stable at over 99.95%, and the reliability of the rural power supply has been basically stable at over 99.75%. The maximum power load in the operation area of the State Grid has increased from 649 million kW in 2015 to 810 million kW in 2018, with a compound annual growth rate of 5.7%. The above data fully show that the unified strong smart grid, with UHV power grid as the backbone grid, coordinated development of power grids at all levels and the characteristics of informationization, automation, and interaction, is basically formed. However, the strong smart grid only connects the power grid companies with the PG and sales enterprises, but does not connect to the end users, so ubiquitous power Internet of things construction to serve the end users is imminent by opening up the "last kilometer." The ubiquitous power Internet of things can connect power users and their equipment, power grid enterprises and their equipment, PG enterprises and their equipment, suppliers and their equipment, power customers and their equipment, share data, serve users, power grid companies, PG enterprises, suppliers and government societies, and constantly improve the energy resource allocation ability and intelligence level.

In the context of new power reform, with the continuous implementation of transmission and distribution price reform and deepening of power market reform in the future, the State Grid only collects the power grid fee in the process of transmission and distribution, and the profit space of the State Grid is compressed, so it is urgent to explore new achievement growth points. Therefore the essence of the development from smart grid to ubiquitous power Internet of things is the transition of State Grid from the construction of the main grid to the construction of the distribution grid, and the transformation from a single electricity seller to a comprehensive energy service provider.

Ubiquitous power Internet of things has played an important role in helping the modernization of national governance and meeting the energy needs of people's better lives. We can effectively support all kinds of clean energy access and improve the level of clean energy consumption by promoting holographic perception, intelligent analysis, and accurate prediction of clean energy PG. From January to August 2019, the operating area of the State Grid has absorbed 394.4 billion kwh of new energy, an increase of 15.9% year on year. In terms of serving the construction of a smart city, the company has built a city energy big data center, launched energy big data applications such as industrial energy use analysis, park activity analysis, etc., built a "City Smart Energy brain," and improved the ability of urban overall management and collaborative governance by promoting the deep integration of the energy and power system, government affairs, transportation, telecommunications, and other fields. The construction of the ubiquitous power Internet of things has also promoted a rapid response to services. At present, the "online State Grid" has been put into operation in five provincial companies, which can respond to various energy service needs of the people in an all-round and one-stop manner, and realize the full online processing of the energy utilization business for residents and enterprises. It is worth mentioning that the construction of the ubiquitous power Internet of things proposed by State Grid is a new strategy made by the power grid to conform to the development trend of the energy Internet.

In March 2019 the State Grid proposed to promote the construction of "three types and two networks" and build a world-class energy Internet enterprise with global competitiveness (the three types are "hub type, platform type, sharing type," the two networks are "strong smart grid" and "ubiquitous power Internet of things").

The energy Internet is a smart energy network with electricity at the center, the strong smart grid and ubiquitous power Internet of things as the basic platform, in-depth integration of advanced energy technology, modern information communication technology, and control technology to realize multienergy complementary, intelligent interaction, and ubiquitous interconnection. It is an important platform and a modern energy system for large-scale development, transmission and use of clean energy in the world. The energy Internet represents the development trend and direction of the power grid in the future. The construction of the energy Internet needs to strengthen the following aspects.

#### 21.1.3.3.1 To innovate and build the development pattern of energy Internet

A strong smart grid with UHV as the framework and coordinated development of power grids at all levels is persistently built and operated. At the same time, mobile Internet, artificial intelligence, and other modern information technology and advanced communication technology are fully used to realize the interconnection of everything and human—computer interaction in all aspects of the power system, to build a smart service system with comprehensive state perception, efficient information processing, and convenient and flexible application, and to achieve the integration of data flow, business flow, and energy flow.

#### 21.1.3.3.2 To promote the coordinated development of energy development and utilization

The coordinated development of the power grid and power supply is to realize the close connection and scientific layout of power grids at all levels, and to improve the overall utilization efficiency of the power grid. And the pivotal role of the power grid in multienergy conversion and utilization is fully played, so as to improve the comprehensive utilization efficiency of energy, and meet the various energy demands of users.

#### 21.1.3.3.3 To promote energy transformation and development

Clean and green transformation is the general trend of energy development. The role of the power grid as a platform is played to strengthen DR, promote clean energy transmission and consumption, promote the establishment of a transregional and trans-provincial market mechanism for the consumption of clean energy, and increase the enthusiasm of all regions for the consumption of clean energy.

#### 21.1.3.3.4 To increase the investment participants and attract the social fund help to build grid platform enterprises

Building an energy allocation platform, establishing a large power grid, and cultivating a large market promote large-scale optimal allocation of energy and power resources.

#### 21.1.3.3.5 To serve all the people and share the fruits of development

With the deepening of the reform of the electric power system, more social capital and various market entities will participate in the construction and value creation of the energy Internet, promote the common development of upstream and downstream enterprises, build a mutually beneficial and win−win energy Internet ecosystem, and share the development achievements with the whole society. In 2019 the first city-level energy Internet demonstration project was completed in Zhejiang, China. The core demonstration area realizes 100% access and consumption of renewable energy, wide open interconnection of clean energy, efficient power grid, low-carbon building, intelligent energy consumption and green transportation, and green sharing of the power grid side and consumption side.

### 21.1.3.4 Prospect of power grid development

After nearly 10 years of continuous development, the power grid has basically reached the goal of building a strong smart grid proposed in 2010. The growth space of the distribution grid side is very large. In recent years, according to the capacity data of transformers of various voltage levels in the State Grid, the compound growth rate of the distribution side (110 kV) in the recent 4 years is 20.94%, which is much higher than that of the transmission high-voltage side (220, 330, and 550 kV), but far lower than that of 750 kV (41.88%) and 1000 kV (122.48%) on the UHV side, indicating that the construction of the distribution side in recent years is not the primary goal of investment and development of the State Grid. Correspondingly, in recent years, with the rapid development of electric vehicles, and the significant increase of new energy installed capacity and grid-connected PG, the improvement of the distribution grid side, especially the end user, has become more and more urgent. In the past 4 years, the total number of electric piles connected to the State Grid has increased from 65,000 to 280,000 with an annual compound growth rate of 44%; the installed capacity of wind power and solar power of the State Grid has increased from 148 million kW to 300 million kW with an annual compound growth rate of 19.32%; the new energy PG has increased from 208.8 billion kWh to 439 billion kWh with an annual compound growth rate of 20.42%. In conclusion, the rapid growth of electric vehicles and new energy PG has increased the demand for the construction of the distribution grid side, so the current growth space of the distribution grid side is very large. China will continue to increase the investment in the distribution grid through the implementation of distribution grid construction and the transformation action plan. The investment in distribution gird construction and transformation will not be less than 2 trillion yuan in 2019−25. It is estimated that by 2020, the transformation capacity of the high-voltage distribution grid will reach 2.1 billion kVA, the line length of the high-voltage distribution grid will reach 1.01 million Li, the medium-voltage public distribution capacity will reach 1.16 billion kVA, and the line length of the medium-voltage public distribution grid will reach 4.04 million km. The construction of the distribution grid can effectively control the distributed power supply, realize the local consumption of high-density new energy, and promote the construction and development of the energy Internet.

In addition, due to the volatility and randomness of new energy generation output, it has become an urgent requirement to increase energy storage facilities on the generation side, grid side, and user side. As a new technology in China's electric power sector, energy storage can coordinate with shaving peaks and filling troughs in PG, and further reduce the wind and light waste of new energy through energy storage technology. The next development of energy storage will be closely related to energy reform and electric power system reform. In the future, power selling companies and energy service companies are expected to become purchasers and integrators of energy storage systems. Energy storage will be more closely combined with energy and electric power market.

## 21.2 Other Asian markets

### 21.2.1 India electricity market

In the 1960s the Indian electric power industry gradually changed from a private enterprise to a state monopoly. The Ministry of Industry, Central Electricity Authority, and other power management departments were established successively. State Electricity Boards with vertical integration of PG, transmission and distribution were established in the states. However, since the 1980s most state-level electricity boards have been losing money every year, and their development has been in trouble. To change the loss situation of power enterprises, the Indian government decided to implement the privatization reform of the power industry. The government began to carry out electricity market reform in 1991. The goal of the reform was to break the monopoly, introduce competition in the PG, transmission, and distribution, encourage private funds to enter the field of power production, transmission, and distribution, and promote the gradual reduction of power prices. The main measures are as follows:

1. To reform and restructure the State Electricity Board. The Andhra Punjab electricity reform law, which came into force in 1999, ended the monopoly position of State Electricity Board in PG, transmission and distribution. The State Electricity Board was reorganized into a Power Generation Co., Ltd. And a Power Transmission Co., Ltd., each of which is engaged in the PG, power transmission, and distribution business. The state government owns 100% of the shares of these two companies.
2. To encourage private investment and foreign investment. In 1991 India's electric power sector began to open to private investment. Private enterprises can invest in power plants and transmission fields. They can offer tariff preferences to imported equipment and provide support for project financing. They can also offer various preferences for the innovation, modernization, and renewal of old power plants invested in by private enterprises. To reduce the risk of private capital investing in the power sector, the Indian government also provides private investment involving eight fast lanes. The guarantee provided by the state government to the State Electricity Board to the PG company is counter guaranteed, enhancing the confidence of private investors and stimulating private investment to rush into the power sector.
3. To pay attention to hydropower development. Since 1997 India has announced its hydropower development policy, emphasizing the development of small hydropower and increasing private investment, simplifying the examination and approval procedures of the Central Electricity Authority for hydropower projects, increasing hydropower development funds through the central and state financial budgets and power finance companies, levying electricity consumption tax to establish power

development funds, two-thirds of which is used by the state government to promote power development, and one-third of which is used to promote hydropower development. The hydropower sector is allowed to adopt different price policies during peak power consumption to attract investment in hydropower projects.

4. To improve power laws and regulations. In 1991 the electricity amendment law was promulgated. In 1998 the electric power law was amended again. Based on the electricity law of 1998, the central government of India established the Central Electricity Regulatory Commission to establish the electricity charging benchmark. In 2001 the electric power Act 2001 was passed, which stipulated that the central government must consult with the state government to formulate national electric power policy; simplified the examination and approval procedure of Central Electricity Authority, and issued a license for private investment in the field of transmission. In addition to hydropower projects, the investment in power plants was exempt from license. The state electricity department was given more power, and the State Electricity Board could collect additional fees and gradually cancel subsidies. The electric power Act 2001 did not stipulate how the states should carry out the electric power reform mode, and gave the state government more flexible power to choose the reform mode suitable for their situation. To effectively use energy and reserve energy, the following programmatic bills were issued, such as the energy conservation law of 2001, the electricity law of 2003, and the national electricity policy of 2005.

5. To start the plan of accelerating the development and reform of electric power and attach importance to the construction of large-scale electric power projects. In February 2001 India launched the national "accelerated power development and reform plan", including the innovation, modernization, and renewal of old thermal power and hydropower plants, and the promotion of power reform and development in the state power sector. At the same time, to play the role of economies of scale and reduce costs, India formulated a large-scale power project policy, encouraging private and public sectors to invest in large-scale power station construction, especially large-scale hydropower.

At present, private capital has been introduced into every link of the distribution to improve the level of market competition. In terms of PG, central government enterprises, local government enterprises, and private capital enterprises all occupy a certain proportion of the PG market share. In terms of transmission, India's national, regional, and state grid three-level transmission systems are managed by different levels of power enterprises, transregional power transactions or trans-state power transactions in the region are managed by the Power Grid Corporation of India, and intrastate grid business is managed by the state transmission company. In terms of distribution, there are several distribution companies in each state, which can be owned by the government or invested in by private capital. In terms of dispatching, the national and regional dispatching centers are owned by the Power Grid Corporation of India, which schedules the national and regional transmission networks, while the state dispatching centers belong to the Power Grid Corporation of India or the Electricity Authority, which schedules the state grid. In terms of trading institutions, India's trading institutions are independent and have India Energy Exchange Limited and Power Exchange India Limited (PXIL), which are mainly responsible for trans-state trading.

As of fiscal year (FY) 2018, after years of silence, India's power industry has once again gained rapid growth momentum. It is the third largest power producer in the world after China and the United States.

In the past decade, India's total installed capacity and PG have grown rapidly, with cumulative annual growth rates of 9.3% and 6.6%, respectively. The cumulative annual growth rate of renewable energy is higher, reaching 20%, and this trend continues. India has steadily developed its PG, transmission, and distribution capacity to drive its national economic growth. In FY2018 the total installed capacity is about 350 GW, and the total generating capacity reaches 1376 billion kWh. Thermal energy, renewable energy, hydropower, and nuclear energy account for 65.3%, 19.7%, 13.0%, and 2.0% of the total installed capacity, respectively. The growth in availability of electricity during 2018−19 is 5.2% as compared to the prior year. During the year 2018−19, the peak shortage is 0.8% and the energy shortage is 0.6% as compared to 2.0% and 0.7%, respectively, in the prior year.

As on September 30, 2019, the peak demand for installed capacity in India is about 184 GW, and the total installed capacity of the power industry reaches 363 GW, among which the central government, local state, and private power companies account for 25%, 28%, and 47%, respectively. Thermal energy, renewable energy, hydropower, and nuclear energy account for 62.9%, 22.7%, 12.5%, and 1.9% of the total installed capacity, respectively. Compared with FY2018, India's power industry has changed its focus from fossil fuel to more environment-friendly, low-carbon, and sustainable energy, especially renewable energy [22]. The specific installed capacities sector-wise and region-wise are shown in Tables 21.5 and 21.6, and Figs. 21.5 and 21.6.

The overall generation (including generation from grid-connected renewable sources) in the country has been increased from 1110.458 billion kWh during 2014−15 to 1173.603 billion kWh during the year 2015−16, 1241.689 billion kWh during 2016−17, 1308.146 billion kWh during 2017−18, and 1376.095 billion kWh during 2018−19. The conventional generation during 2018−19 is 1249.337 billion kWh as compared to 1206.306 billion kWh generated during 2017−18, representing a growth of about 3.57%. The annual growth (2014−18) in PG is shown in Table 21.7.

**TABLE 21.5** All India's installed capacity sector-wise as on September 30, 2019 (unit: MW).

| Sector | Thermal | | | | | Nuclear | Hydro | RES | Grand total |
|---|---|---|---|---|---|---|---|---|---|
| | Coal | Lignite | Gas | Diesel | Total | | | | |
| State | 65,061.5 | 1290 | 7118.71 | 236.01 | 73,706.22 | 0 | 26,958.5 | 2349.98 | 103,014.7 |
| Private | 74,173 | 1830 | 10,580.6 | 273.7 | 86,857.3 | 0 | 3394 | 78,606.67 | 168,857.97 |
| Central | 57,660 | 3140 | 7237.91 | 0 | 68,037.91 | 6780 | 15,046.72 | 1632.3 | 91,496.93 |
| All India | 19,6894.5 | 6260 | 24,937.22 | 509.71 | 228,601.4 | 6780 | 45,399.22 | 82,588.95 | 363,369.6 |

① Renewable energy sources (RES) include Small Hydro Project, Biomass Gasifier, Biomass Power, Urban and Industrial Waste Power, Solar and Wind Energy. ② Data Source: Central Electricity Authority (CEA).
Unless otherwise specified, RES and data sources are given by ① and ②, respectively.

**TABLE 21.6**   All India's installed capacity region-wise as on September 30, 2019 (unit: MW).

| Region | Thermal | | | | | Nuclear | Hydro | Renewable energy sources | Grand total |
|---|---|---|---|---|---|---|---|---|---|
| | Coal | Lignite | Gas | Diesel | Total | | | | |
| Northern | 50,811.97 | 1580 | 5781.26 | 0 | 58,173.23 | 1620 | 19,707.77 | 15,829.74 | 95,330.74 |
| Western | 72,753.62 | 1540 | 10,806.49 | 0 | 85,100.11 | 1840 | 7547.5 | 24,648.49 | 119,136.1 |
| Southern | 43,042.02 | 3140 | 6473.66 | 433.66 | 53,089.34 | 3320 | 11,774.83 | 40,252.77 | 10,8436.94 |
| Eastern | 29,516.87 | 0 | 100 | 0 | 29,616.87 | 0 | 4942.12 | 1476.72 | 36,035.71 |
| Northeast | 770.02 | 0 | 1775.81 | 36 | 2581.83 | 0 | 1427 | 363.05 | 4371.88 |
| Islands | 0 | 0 | 0 | 40.05 | 40.05 | 0 | 0 | 18.19 | 58.24 |
| All India | 196,894.5 | 6260 | 24,937.22 | 509.71 | 228,601.43 | 6780 | 45,399.22 | 82,588.96 | 363,369.61 |

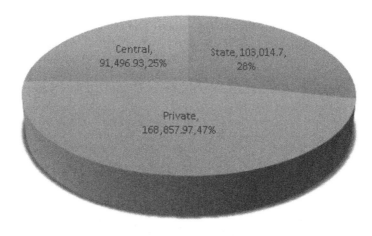

**FIGURE 21.5**   All India's installed capacity sector-wise as on September 30, 2019 (unit: MW).

We can see that the generation from conventional sources and from renewable sources during 2014−15 to 2018−19 are both increasing year by year, but the ratio of year-on-year growth is decreasing year by year, as shown in Figs. 21.7−21.9.

These are the results of India's efforts to adjust its energy structure, improve low-carbon new energy such as wind energy, solar energy, and biomass energy, and establish a green energy security system to support sustainable economic and social development. India is rich in renewable energy resources, especially solar and wind energy resources. Therefore India regards the development of renewable energy as an important national energy strategy, has set up a specific government agency responsible for renewable energy management, and has set the goal of large-scale development of renewable energy. In recent years, India

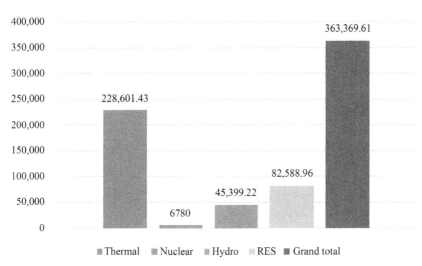

FIGURE 21.6  All India's installed capacity region-wise as on September 30, 2019 (unit: MW).

TABLE 21.7  2014–18 Power generation, India (unit: billion kWh).

| Year | Energy generation from conventional sources | Growth in conventional generation (%) | Energy generation from renewable sources | Growth in renewable generation (%) | Total energy generation |
|------|------|------|------|------|------|
| 2014–15 | 1048.673 | 8.43 | 61.785 | – | 1110.458 |
| 2015–16 | 1107.822 | 5.64 | 65.781 | 6.47 | 1173.603 |
| 2016–17 | 1160.141 | 4.72 | 81.548 | 23.97 | 1241.689 |
| 2017–18 | 1206.306 | 3.98 | 101.84 | 24.88 | 1308.146 |
| 2018–19 | 1249.337 | 3.57 | 126.758 | 24.47 | 1376.095 |

has continued to improve its renewable energy development goal, increasing the installed capacity of renewable energy in FY2022 to 175 GW, including 100 GW of photovoltaic power, 60 GW of wind power, 10 GW of biomass, and 5 GW of small hydropower. In the future, if India achieves the new target of 100 GW solar energy development, it will not only promote its own long-term energy security, but also reduce its dependence on fossil fuels, and reduce carbon dioxide emissions of more than $1.7 \times 10^8$ tons [23].

Most of India is located in the tropics and subtropics, and it is one of the countries with the most abundant solar energy resources in the world. According to the solar energy project, most parts of India have about 250–300 days of sunshine every year, especially in the central and northwest parts of India, which can reach 300–330 days of sunshine every year [24]. In terms of the average solar radiation intensity, India has reached 20 MW/km² and the daily sunshine intensity is as high as 4–7 kWh/m² [25]. According to the statistics of the global energy network institute, India's solar energy contains up to $5000 \times 10^{12}$ kWh

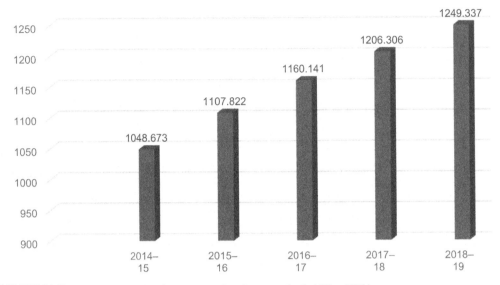

FIGURE 21.7    Energy generation from conventional sources (unit: billion kWh).

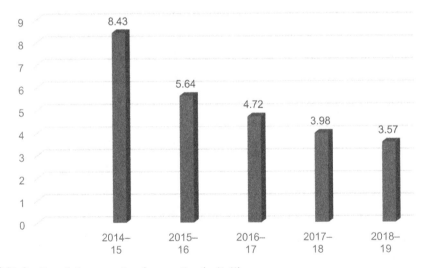

FIGURE 21.8    Growth in conventional generation (unit: %).

of power resources per year, which is higher than the total energy consumption of India in 1 year [24]. In addition, the PG potential of India's solar energy resources (750 GW) is far greater than that of the renewable energy sources such as wind energy (102 GW, 80 m height, excluding offshore wind resources), small hydropower (20 GW), and biomass energy (25 GW) [25]. Therefore India will focus on the development of new energy in the field of solar energy.

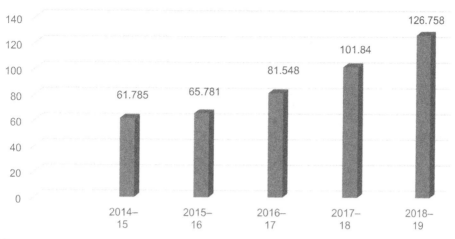

FIGURE 21.9   Energy generation from renewable sources (unit: billion kWh).

In view of the Indian government's ambitious solar energy development goals, the government has introduced a series of supporting subsidies and incentive policies. For example, in terms of investment, the government is expected to invest about 6 trillion rupees in the development and utilization of solar energy. In terms of government subsidies, the Indian solar energy company has decided to provide subsidies for 4835 MW solar projects with adaptive compensation fund. The fund aims to provide financial support for infrastructure projects that are difficult to obtain financial support in the short term. In terms of preferential tax policies, the government stipulates that fixed assets engaged in solar energy industry development investment can be calculated at an accelerated depreciation rate of up to 80%. In addition, tax relief policies shall be implemented for domestic solar energy enterprises and projects, central sales tax shall be exempted, and soft loan tariff relief shall be implemented for imported materials, components, and equipment. These preferential policies encourage investment, financing enterprises, and equipment suppliers to vigorously develop solar photovoltaic PG.

Although India has made some achievements in the development and utilization of solar energy and has a bright future, there are also many important factors restricting the further development of the solar energy industry. For example, the first is that the cost of land acquisition is high. As one of the most difficult land acquisition countries in the world, the high land acquisition cost in India restricts the development of its solar energy industry. The second is the lack of stable purchasing power. The limited purchasing power of the Indian people has limited the popularity of solar energy equipment. In addition, most of the government's subsidy funds for solar energy are used by solar panel manufacturers, rather than people's ability to purchase solar equipment. The third is that the local solar energy manufacturing industry is weak. Polysilicon materials used to make solar panels have been dependent on imports, and there is a lack of necessary raw materials at home. The fourth is lack of relevant basic data. The government's lack of data on the location and feasibility of solar energy use has led to the waste of resources and discourages investors.

## 21.2.2 Philippines electricity market

Due to the impact of the financial crisis in 1998, the PG capacity of power plants in the Philippines declined sharply. However, the state power company still had to purchase power according to the "pay as you go" clause, which eventually led to the insolvency of the state power company. Therefore the Philippines government was determined to implement power reform [26]. In 2001 the Philippines launched the reform of the power industry. The electric power market company was established in 2003. From 2004 to 2005, the market management system supporting the wholesale electricity spot market was introduced. In 2006 and 2010 the spot markets of Luzon and Visayas were put into commercial operation, respectively. In 2007 the transmission and transformation links of the transmission grid were franchised by the private sector. In 2013 the introduction of retail competition and grid opening policy made the Philippines power market shift from wholesale competition mode to retail competition mode. Since 2016 the number of competitive consumers (i.e., consumers who have the right to choose their own power suppliers) has dropped from the minimum threshold of 1 MW to 75 kW. At present, the Philippines has become the Asian country with the fastest pace of power market reform after Singapore. The Philippines basically copied the New Zealand power market reform mode, that is, the dispatching organization is subordinate to the transmission company, and the market operator is independent. At the same time, the vertical integrated management mode was broken, and the whole power industry was divided into four links: generation, transmission, distribution, and sales, and a unified wholesale electricity spot market was established in Luzon and Visayas.

The wholesale electricity spot market consists of three main bodies: dispatching organization, market operator, and trading participant [27]. The dispatching organization is responsible for providing market operators with data related to the system status, such as grid parameters, unit operation parameters, power outage plan, providing tests for all generators and users who can provide auxiliary services, issuing auxiliary service certificates for qualified generators and users, and signing bilateral purchase and sale of auxiliary services with auxiliary service suppliers according to the auxiliary service procurement agreement contract. The market operator is responsible for defining and maintaining the market network model, forecasting the system compliance, developing and sending real-time generation scheduling plan and generation priority list to the dispatching authority, carrying out the billing and settlement of transactions, and publishing market information in accordance with the rules of the wholesale electricity spot market. Market participants are responsible for complying with spot market registration requirements, participating in generation bidding, submitting bilateral contract data, and complying with dispatching instructions.

The Philippines initially designed the power market to launch the energy market and auxiliary service market simultaneously, but to reduce the risk, the Ministry of Energy finally decided to take the lead in launching the energy market instead of the auxiliary service market. However, the preliminary construction idea of auxiliary service market has been formed. At present, the energy market in the wholesale electricity spot market adopts the mode of "unilateral bidding on the generation side" and the mechanism of "full power bidding, net settlement, node marginal price," and has a ceiling price (initially, the ceiling price in the energy market was 64 pesos/kWh), which was lowered to 32 pesos/kWh in 2013. In 2015 the Philippines electricity market corporation further recommended that the minimum price be set at 10 pesos/kWh, with effect from 2016 [26]. At present, the trading period of the wholesale electricity spot market is 1 hour, and each independent and controllable generating unit is quoted separately in 10 sections according to its available generating capacity in each trading period. And the quotation of the last unit meeting the network constraints of power system safe operation and load demand at the corresponding network node is used as the market clearing price. For the planned auxiliary service market, Philippines electricity market corporation proposes to adopt the "generation side and interruptible load bidding" mode, and the mechanism of "total warehouse market, net price settlement in advance, regional reserve price" [28]. Philippines electricity market corporation proposes to carry out auxiliary service transactions of frequency modulation reserve, accident reserve, and adjustable reserve in the auxiliary service market of the wholesale electricity spot market. In addition, reactive power support and black start auxiliary service transactions shall be approved by the transmission company according to the Energy Regulatory Commission. The auxiliary service procurement agreement continues to be purchased by signing bilateral contracts. The transaction interval of auxiliary service is also 1 hour, and the auxiliary service market and energy market are jointly optimized and integrated.

The Philippines is a major emerging market for energy investment in Southeast Asia. Power demand in the Philippines has been rising steadily over the past decade. The total electricity sales and consumption all over the country increased from 94,370 GWh in 2017 to 99,765 GWh in 2018, equivalent to 5.7% growth. Out of these total sales and consumption, 56,036 GWh is contributed by private investor owned utilities, while 21,486 GWh is from the electric cooperatives' (ECs') contributions. Nonutilities are 4318 GWh. "Others" which refer to public buildings, streetlights, irrigation, agriculture, and "others not elsewhere classified" are 2203 GWh. Total system loss of the Distribution Utilities (DUs) accounts for 9%, while the utilities' own-use for office and station use of the power plants accounts for 7%. The total sales are 84,043 GWh, accounting for 84.2% of the total consumption [29]. The ratio of 2018 electricity sales and consumption by sector are as shown in Fig. 21.10.

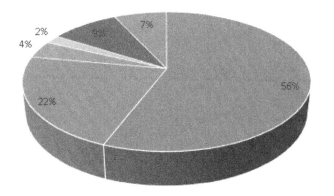

FIGURE 21.10 2018 Electricity sales and consumption by sector, Philippines (unit: %).

The total power supply, in terms of installed capacity, grew by 4.8% from 22,730 MW in 2017 to 23,816 MW in 2018. The country's total installed capacity grew by 6.1% from 21,425 MW in 2016 to 22,730 MW in 2017 [30]. As shown in Table 21.8.

A total of 933.6 MW new capacities are added to the country's supply in 2018 which include coal-fired (720 MW), oil-based (87 MW), geothermal (12 MW), hydropower (80 MW), and biomass (34 MW) power plants. A total of 835 MW new capacity was added to the country's supply base in 2017 which included coal-fired (630 MW), solar (127 MW), oil-based (77 MW), and hydropower (1 MW). In terms of share by grid, Luzon contributed 392 MW or 47%, Mindanao at 337 MW or 40%, and Visayas at 106 MW or 3%. The specific data are as shown in Table 21.9.

The country's total peak demand in 2018 was recorded at 14,782 MW, which is 993 MW or 7.2% higher than the 13,789 MW in 2017. The Luzon grid contributed 74% (10,876 MW) of the total demand, while Visayas and Mindanao had shares of 14% (2053 MW) and 13% (1,853 MW), respectively. Among the three grids, Luzon grid showed the highest increase of 822 MW or 8.2% from the 2017 peak demand of 10,054 MW, while Visayas and Mindanao grew by 3.9% and 5.3%, respectively. The country's total peak demand in 2017 grew steadily by 3.9% from 13, 272 MW in 2016 to 13,789 MW in 2017. Mindanao's peak demand growth rate at 6.5% is the highest among the three grids. Luzon's peak demand growth rate at 3.4% is the lowest among the three grids. The specific data are as shown in Fig. 21.11.

The Department of Energy (DOE) has published power-related development plans, such as Power Development Plan (PDP) 2016–2040, Distribution Development Plan 2016–2025, and Missionary Electrification Development Plan 2016–2020. They lay down the basic data and information of the Philippines Power System; power supply and

**TABLE 21.8**    2016–18 Installed and dependable capacity, Philippines (unit: MW).

| Fuel type | Installed | | | Dependable | | |
|---|---|---|---|---|---|---|
| | 2016 | 2017 | 2018 | 2016 | 2017 | 2018 |
| Coal | 7419 | 8049 | 8844 | 6979 | 7674 | 8368 |
| Oil based | 3616 | 4154 | 4292 | 2821 | 3287 | 2995 |
| Natural gas | 3431 | 3447 | 3453 | 3291 | 3291 | 3286 |
| Renewable energy | 6959 | 7080 | 7227 | 6004 | 6263 | 6592 |
| Geothermal | 1916 | 1916 | 1944 | 1689 | 1752 | 1770 |
| Hydro | 3618 | 3627 | 3701 | 3181 | 3268 | 3473 |
| Biomass | 233 | 224 | 258 | 157 | 160 | 182 |
| Solar | 765 | 886 | 896 | 594 | 700 | 740 |
| Wind | 427 | 427 | 427 | 383 | 383 | 427 |
| Total | 21,425 | 22,730 | 23,816 | 19,095 | 20,515 | 21,241 |

**TABLE 21.9**  2016–18 Newly operational capacities, Philippines (unit: MW).

| Power plant facility name | Installed | | | Dependable | | |
|---|---|---|---|---|---|---|
| | 2016 | 2017 | 2018 | 2016 | 2017 | 2018 |
| Coal | 1492 | 630 | 720 | 1369 | 594 | 690 |
| Oil based | 36 | 77 | 87 | 34 | 67 | 83 |
| Natural gas | 550 | 0 | 0 | 511 | 0 | 0 |
| Renewable energy sources | 613 | 128 | 126 | 475 | 104 | 122 |
| Geothermal | 0 | 0 | 12 | 0 | 0 | 12 |
| Hydro | 1 | 1 | 80 | 1 | 1 | 80 |
| Biomass | 12 | 0 | 34 | 11 | 0 | 30 |
| Solar | 600 | 127 | 0 | 463 | 103 | 0 |
| Wind | 0 | 0 | 0 | 0 | 0 | 0 |
| *Total* | *2691* | *835* | *933* | *2389* | *765* | *895* |

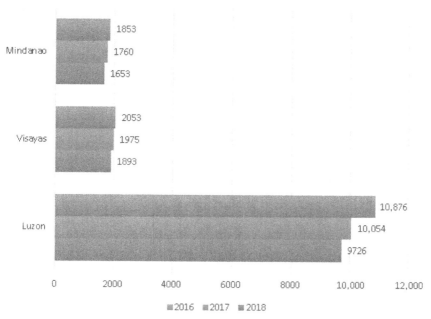

**FIGURE 21.11**  2016–18 Peak demand per grid (unit: MW).

demand outlook in grid and off-grid areas; and power sector roadmaps, policies, and programs for the short-, medium-, and long-term horizon. These serve as investors' references in business development in the country.

The DOE envisions to implement its policy thrusts and strategic directions which are geared toward full restructuring and reform of the electric power industry by 2040. For the four major subsectors of the power industry: generation, transmission, distribution, and supply, the DOE identifies its goals and strategies for implementation in the short-term (2017–18), medium-term (2019–22), and long-term (2023–40) planning horizons. Separate roadmaps for off-grid and missionary areas down to household electrification are also included in this plan toward the holistic development of the electric power industry. These roadmaps embody continuing policies and programs and future action plans focusing on ensuring quality, reliable, affordable, and secure supply; expanding access to electricity; and ensuring a transparent and level playing field in the power industry [30].

Compared with the PDP 2016–40, it is found that the total peak power demand, power sales, installed capacity, and stable use capacity of the Philippines in 2016–18 basically reached or slightly exceeded the target [31].

For the generation subsector, in the medium to long term, the DOE will encourage and facilitate the entry of new and emerging PG options such as nuclear technology, energy storage technologies, fuel cells, and ocean thermal energy conversion. To make this happen, the DOE will establish strong cooperation among government line agencies, such as the Department of Science and Technology and Philippines Nuclear Research Institute, and reinforce partnerships with private research institutions that undertake research, development, and demonstration of energy technologies. From 2019 to 2040, the DOE will lead in performance assessment and benchmarking of PG facilities to review and develop policies to improve the operation of the power plants. In line with this, the DOE will also encourage compliance to international standards for constructing power plants and accreditation of contractors. It will also develop resiliency policies for all generating assets. As part of the continuing activities, the DOE will conduct daily and periodic monitoring of the power demand–supply situation and provide technical support.

For the transmission subsector, the National Grid Corporation of the Philippines (NGCP) must have adequate reserve capacities to ensure uninterrupted power supply. Similar to the generation sector, DOE will likewise lead the conduct of regular performance assessment and benchmarking of transmission facilities and operations to address inefficiencies and system congestions. Apart from the continuing activities underlining transmission system improvement, transmission corporation and NGCP will jointly undertake the identification and development of new transmission backbones and alternative transmission corridors to lessen if not eliminate line congestions in view of the increasing demand. The DOE will closely monitor the implementation of the Transmission Development Plan as well as the Visayas–Mindanao interconnection which is targeted to be implemented by 2020. Similarly, the interconnection of the main grids and emergent island grids such as Mindoro is also envisioned to be implemented.

For the distribution and supply subsector, the DOE will continually focus on increasing the capacities of DUs, particularly ECs, in formulating their respective Distribution Development Plans. Over the planning horizon, DUs are mandated to undertake continuous upgrades and expansion of distribution infrastructure in their respective franchise areas. The sector shall utilize and invest in more efficient technologies such as the smart grid (e.g., prepaid electricity, etc.). Likewise, the distribution sector shall move toward the transformation of its current system and infrastructure to an energy-resilient one by implementing enabling resiliency policies. For the medium term, the integration of the supply development plan in the PDP is envisioned considering the transition of the electric power industry into a market-driven and competitive sector, wherein customers (contestable) are given the power of choice. The DOE, together with the Energy Regulatory Commission, will develop policies for lowering the threshold to 500 kW including aggregation for the same threshold.

## 21.3 Conclusion

Electric energy is the important material basis of social production and improvement, which relates to a country's security and national economy. For a long time, the problems in policy, system, and management have restricted the development of the electric power industry in various countries to a certain extent, thus affecting the economic and social development. Therefore the market-oriented reform of electric power industry is the only way.

The chapter mainly introduces China's electricity market, including market-oriented reform of China's electric power, the classification of China's electricity market, and the development of China's power grid. According to the aforementioned analysis about market-oriented reform of China's electric power, we know it plays an important role to explore the development of electricity futures trading and the establishment of the electricity derivatives market. Meanwhile, with the development from smart grid to ubiquitous power Internet of things, the State Grid proposes to build a world-class energy Internet enterprise with global competitiveness. The energy Internet is an important platform and modern energy system for large-scale development, transmission, and use of clean energy in the world. The energy Internet represents the development trend and direction of power grids in the future. The chapter also makes a brief analysis of the market-oriented reform of electric power, the current situation of electricity market, and the prospects for power grid development in India and the Philippines.

## References

[1] J. Shen, Research on Market Reform of China Power Industry [M], Xinhua Press, Beijing, 2004. in Chinese.
[2] Q. Shi, J. Li, Electricity Marketization and Financial Market [M], Shanghai University of Finance and Economics Press, Shanghai, 2009. in Chinese.
[3] CEC, List of basic data of 2018 electric power statistics annual Express [EB/OL], January 19, 2019. <http://www.cec.org.cn/guihuayutongji/tongjxinxi/niandushuju/2019-01-22/188396.html>.
[4] CEC, List of basic data of electric power statistics in 2017 [EB/OL], October 9, 2018. <http://www.cec.org.cn/guihuayutongji/tongjxinxi/niandushuju/2018-12-19/187486.html>.
[5] CEC, List of basic data of electric power statistics in 2016 [EB/OL], March 21, 2018. <http://www.cec.org.cn/guihuayutongji/tongjxinxi/niandushuju/2018-03-21/178791.html>.
[6] CEC, List of basic data of electric power statistics in 2015 [EB/OL], September 22, 2016. <http://www.cec.org.cn/guihuayutongji/tongjxinxi/niandushuju/2016-09-22/158761.html>.
[7] NEA, Electricity consumption of the whole society increased by 8.5% year on year in 2018 [EB/OL], January 18, 2019. <http://www.nea.gov.cn/2019-01/18/c_137754978.htm>.
[8] NEA, Electricity consumption of the whole society increased by 6.6% year on year in 2017 [EB/OL], January 2, 2018. <http://www.nea.gov.cn/2018-01/22/c_136914159.htm>.
[9] NEA, Electricity consumption of the whole society increased by 5.0% year on year in 2016, January 16, 2017. <http://www.nea.gov.cn/2017-01/16/c_135986964.htm>.

[10] NEA, The National Energy Administration issued the electricity consumption of the whole society in 2015, January 15, 2016. < http://www.nea.gov.cn/2016-01/15/c_135013789.htm >.

[11] M. Zeng, Theory and Application of Electric Power Market, China Electric Power Press, Beijing, 2000. in Chinese.

[12] A.F. Meyabadi, M.H. Deihimi, A review of demand-side management: reconsidering theoretical framework, Renew. Sustain. Energy Rev. 80 (2017) 367–379. Available from: https://doi.org/10.1016/j.rser.2017.05.207.

[13] G.D. Geremi, F. Paula, Review and assessment of the different categories of demand response potentials, Energy 179 (2019) 280–294.

[14] J. Zethmayr, D. Kolata, The costs and benefits of real-time pricing: an empirical investigation into consumer bills using hourly energy data and prices, Electr. J. 31 (2) (2018) 50–57. Available from: https://doi.org/10.1016/j.tej.2018.02.006.

[15] P. Samadi, A.H. Mohsenian-Rad, R. Schober, V.W.S. Wong, J. Jatskevich, Optimal real-time pricing algorithm based on utility maximization for smart grid, in: Proceedings of the IEEE International Conference on Smart Grid Communications, IEEE, 2010, Available from: https://doi.org/10.1109/SMARTGRID.2010.5622077.

[16] R. Deng, Z. Yang, M.Y. Chow, J. Chen, A survey on demand response in smart grids: mathematical models and approaches, IEEE Trans. Ind. Inform. 11 (3) (2015) 570–582. Available from: https://doi.org/10.1109/tii.2015.2414719.

[17] C. Yang, C. Meng, K. Zhou, Residential electricity pricing in China: the context of price-based demand response, Renew. Sustain. Energy Rev. 81 (Pt 2) (2018) 2870–2878. Available from: https://doi.org/10.1016/j.rser.2017.06.093.

[18] C.L. Nge, I.U. Ranaweera, O.M. Midtgard, L. Norum, A real-time energy management system for smart grid integrated photovoltaic generation with battery storage, Renew. Energy 130 (2019) 774–785. Available from: https://doi.org/10.1016/j.renene.2018.06.073.

[19] P. Chaudhary, M. Rizwan, Energy management supporting high penetration of solar photovoltaic generation for smart grid using solar forecasts and pumped hydro storage system, Renew. Energy 118 (2018) 928–946. Available from: https://doi.org/10.1016/j.renene.2017.10.113.

[20] M. Shahbazitabar, H. Abdi, A novel priority-based stochastic unit commitment considering renewable energy sources and parking lot cooperation, Energy 161 (2018) 308–324. Available from: https://doi.org/10.1016/j.energy.2018.07.025.

[21] O. Rahbari, N. Omar, Y. Firouz, M.A. Rosen, S. Goutam, V.D.B. Peter, et al., A novel state of charge and capacity estimation technique for electric vehicles connected to a smart grid based on inverse theory and a metaheuristic algorithm, Energy 155 (2018) 1047–1058. Available from: https://doi.org/10.1016/j.energy.2018.05.079.

[22] CEA, Annual reports [EB/OL], October 8, 2019. < http://www.cea.nic.in/annualreports.html >.

[23] Ministry of Science and Technology of the People's Republic of China, India revises national solar energy utilization plan objectives [EB/OL], December 7, 2015. < http://www.most.gov.cn/gnwkjdt/201512/t20151204_122632.htm >.

[24] GENI, Integrating Wind and Solar Energy in India for a Smart Grid Platform [R], 42, Global Energy Network Institute, San Diego, CA, 2013.

[25] The Ministry of New and Renewable Energy, Annual report 2015-16 [EB/OL], 2016. < http://mnre.gov.in/file-manager/annual-report/2015-2016/EN/Chapter%201/chapter_1.htm >.

[26] WESM, 2006-2016 Igniting transformation: special 10th anniversary edition [EB/OL], March 12, 2017. < http://www.wesm.ph/ >.

[27] WESM, Revised PDM [EB/OL], July 12, 2012. < http://www.wesm.ph/ >.

[28] WESM, Pricing and cost recovery mechanism for reserves in the Philippines [EB/OL], January 1, 2017. < http://www.wesm.ph/ >.

[29] DOE, 2018 Power supply and demand highlights [EB/OL], June 6, 2019. < https://www.doe.gov.ph/electric-power/2018-power-supply-and-demand-highlights >.

[30] DOE, 2017 Power supply and demand highlights [EB/OL], July 17, 2018. < https://www.doe.gov.ph/electric-power/2017-power-supply-and-demand-highlights-january-december-2017 >.

[31] DOE, Power Development Plan 2016-2040 [EB/OL], July 17, 2018. < https://www.doe.gov.ph/electric-power/power-sector-situation?q = power-development-plan >.

# Current trends and perspectives in Australia

## Alan Moran
### Regulation Economics, Melbourne, Australia

## 22.1 Introduction

### 22.1.1 Policy scope

There are several different dimensions of energy policy:

- exploration and development of coal, gas, petroleum, and uranium resources;
- export policies; and
- policies concerning the transformation delivery and sale of these and other energy sources (principally hydro, wind, and solar).

It is this third facet that is the prime concern of this chapter. Although exploration, development, and exports are vital to the national wealth (energy comprises over 35% of exports) they are addressed only in so far as they have a bearing on the supply of electricity and gas to domestic consumers.

Electricity supply has four components: generation, comprising 30%−50% of costs; local distribution, with 40%−50%; long distance transmission and retail (billing etc.) each with about 10% of the costs. Gas has a similar structure.

In the space of 30 years, policy interventions in electricity generation and gas exploration have shifted Australia from world leader to world laggard in the efficiency of its reticulated energy supply industries. The cause of this stands squarely with government. As well as being of considerable direct importance to households, energy prices and reliability are vital to the costs of all activities, hence policies of federal and state governments covering the domestic gas and electricity industries have created an economic tragedy for the nation.

## 22.1.2 The energy policy journey

With regard to the electricity industry, 25 years ago, the nation was a pioneer in global moves to replace central control of electricity supply with a market-based system based on competitive provision in generation and retailing.

Observing the benefits of electricity provision from a competitive market in several US jurisdictions and in the newly liberalized and privatized UK market, from the early 1990s Australia embarked upon the series of reforms. One catalyst of these reforms was, ironically, the result of the near bankruptcy of Victoria's state government, leading to the election of the Kennett government, which reformed and privatized the state's gas and electricity assets.

Learning from some unwanted effects of inadequate competition followed Britain's electricity privatization, the Victorian government tempered its goal of maximizing the return from asset sales by a strategy that created sufficient entities to ensure competitive tensions. This was important to prevent market power and excessive prices and to drive cost cutting. The disaggregation of the assets into separate entities was accompanied by the adoption of a bidding process for generator scheduling, modeled on those in place in Pennsylvania, New Jersey, and Maryland (PJM) and the England and Wales markets.

Partly as a result of transmission links between the different state systems and partly because of government policy, other states adopted similar reforms to that of Victoria. The previous monopolistic supply by five government entities in the four interconnected states (all except Western Australia, the Northern Territory, and Tasmania) was replaced by two dozen independent rival generator businesses. Though several of these were government owned (and this remains the case in Queensland) competition and less direct political control due to the corporatization and disaggregation of the government generators brought about vast improvements in availability of supply and considerable cost reductions.

The outcome was an upgraded electricity supply that was already relatively cheap because of the nation's abundant, low-cost coal reserves that are conveniently located close to major markets. The introduction of competition and profit-oriented private owners brought great changes in generator costs, including a more than fivefold saving in the heavily unionized labor force employed within the industry.

Competitive provision was also introduced in the retail sector with the individual state monopolies being replaced by some 30 rival suppliers. Distribution and transmission, as natural monopolies, were subjected to independent regulation over their prices and connection conditions.

Hence at the turn of the present century Australian electricity supply was cheaper than that of any other major nation and highly reliable. Generation was around 85% coal with the rest split between gas and hydro.

### 22.1.3 The resurgence of the regulatory state

#### 22.1.3.1 Electricity power costs

Uniquely throughout the world, and reflecting a governmental response to concerns, however unfounded, that links nuclear energy to nuclear war, nuclear power is illegal in Australia. But the high costs of nuclear power, in the context of Australia's low-cost coal, has probably meant Australia's ban amounted to virtue signaling. This may no longer be the case now that smaller modular reactors [1] are becoming available.

However, the Australian regulatory malady soon spread beyond nuclear power anxieties. No sooner had Australia achieved its peerless position in electricity supply, when government regulatory initiatives started to undermine it. By far the most important of these initiatives was the reinvigoration of measures, which were originally put in place in response to fears about resource depletion, to require retailers to assist in reducing demand for fossil fuels. Such fears about a coming resource depletion have a long history dating back from Jevons [2], an economist of the late 19th century, and were seen in Meadows et al. with the influential Club of Rome [3] notions of 1973. The famous wager [4] between Julian Simons and Paul Ehrlich put to bed the issue of impending resource depletion.

Commencing in 2002, in a measure which then Prime Minister John Howard has described as his worst political blunder, requirements were introduced for an increasing share within generation of renewables (excluding large-scale hydro, new supplies of which were banned on environmental grounds and remain so). The level of renewables required—the Renewable Energy Target (RET)—was set at 9500 GWh by 2012, ostensibly 2% of "additional" energy [5].

Once in place, there was unrelenting pressure to increase the renewables share and its associated subsidies. The Howard government itself commissioned former Senator Grant Tambling to review the program. His 2006 report, which the government did not accept, recommended a 50% lift in the mandatory renewables level.

The Rudd ALP government was elected in 2007. Rudd's first major act was the ratification of the Kyoto agreement to limit greenhouse gas emission growth. The Howard government had signed this and, though not having ratified it, was abiding by it with the renewable energy program and measures to prevent land clearing. Rudd expanded the renewables requirements to gradually grow to 41,000 GWh annually together with acceding to an unlimited number of small-scale units (rooftop panels), subsidized at a lesser rate; these are running at over 10,000 GWh per year. State governments introduced additional subsidy-dependent renewable energy requirements.

The subsidies that these measures entailed came, via retailer obligations, from consumers, impacting on bills and on the competitiveness of (subsidy-free) fossil fuel generators. For several years, renewable energy subsidies may have contributed to keeping prices low by adding capacity with low marginal costs to an inflexible existing coal-based supply. The price outcome of renewables' guarantee subsidy varied between $35 and $90 per MWh, far in excess of the

$30–$50 per MWh electricity wholesale price.[1] This meant renewable generators would automatically run, if available, irrespective of prices offered by other generators.

The most reliable estimate of different power source costs was undertaken for the Minerals Council by Solstice/GDH [6]. Table 22.1 summarizes the costs of alternative power supplies.

Subsidies have led to an extraordinary growth of new capacity in wind and solar as illustrated in Fig. 22.1.

The initial effect of increased renewable supply was reduced incentives for additional commercial supply (the latest major coal-fueled electricity supply source was commissioned in 2006). This injection of subsidized renewable supply impacted upon the profitability of coal plant, both by reducing the hours it could operate and by imposing stop–start operating costs on plant designed for continuous operation. These factors disincentivized major maintenance expenditure.

In 2016 the fermenting effect of subsidies to wind and solar on coal plant profitability brought closures of South Australia's Northern and Victoria's Hazelwood power stations.

These closures took out about 4% of national electricity capacity from stations that had been supplying around 7% of demand. The upshot (see Fig. 22.2) was a two-and-a-half-fold increase in the wholesale price compared with 2008 in a period during which overall inflation was 30%.

---

[1] The renewables' price advantage is tempered by wind's reduced availability during (hot, still) high price events, which brings about a discounted average price on the spot market. This is illustrated below for South Australia.

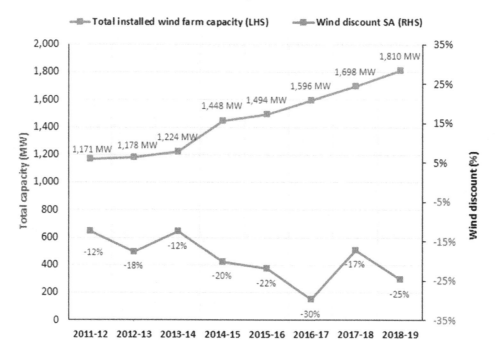

**TABLE 22.1** Cost estimates of different sources of generation.

| | Capacity factor (%) | Capex ($/MWh) | Fuel ($/MWh) | O&M ($/MWh) | Tax 30% company ($/MWh) | Total ($/MWh) |
|---|---|---|---|---|---|---|
| 650 MW black coal low | 87 | 17 | 11 | 8 | 4 | 40 |
| 650 MW black coal high | 87 | 22 | 35 | 15 | 6 | 78 |
| 650 MW gas ccgt low | 82 | 9 | 55 | 3 | 2 | 69 |
| 650 MW gas ccgt high | 82 | 14 | 86 | 12 | 3 | 115 |
| 650 MW solar low | 20 | 62 | | 12 | 16 | 90 |
| 650 MW solar high | 20 | 127 | | 19 | 26 | 171 |
| 650 MW wind low | 37 | 42 | | 12 | 14 | 64 |
| 650 MW wind high | 37 | 68 | | 33 | 12 | 115 |
| 650 MW solar + battery low | 96 | 263 | | 22 | 44 | 328 |
| 650 MW solar + battery high | 96 | 782 | | 29 | 102 | 913 |
| 650 MW wind + battery low | 96 | 156 | | 20 | 36 | 211 |
| 650 MW wind + battery high | 96 | 577 | | 43 | 73 | 693 |

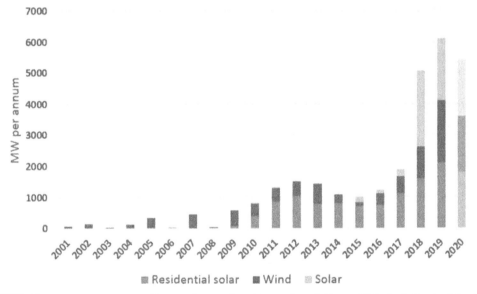

**FIGURE 22.1** New capacity installations. *Source: Clean Energy Council, Report 2020. Available at: <https://assets. cleanenergycouncil.org.au/documents/resources/reports/clean-energy-australia/clean-energy-australia-report-2020.pdf> [7].*

The increased prices (which collapsed in a COVID-dominated 2020) have been accompanied by much greater price volatility as a result of the intermittent nature of wind.

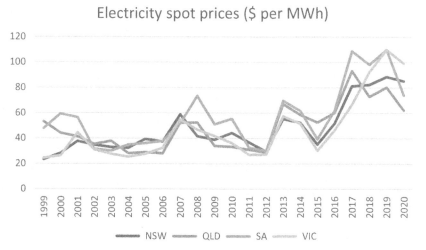

FIGURE 22.2   Australia's electricity spot prices from 1999 to 2020. NSW, New South Wales; QLD, Queensland; SA, South Australia; VIC, Victoria. *Source: Australian Energy Market Operator (AEMO). Available at: <https://www.aemo.com.au/> [8].*

The large share of subsidized wind capacity in South Australia, where wind supplied 44% of the state's energy in 2018, has brought increasingly common negative prices which prevailed for 9.9% of the time in August 2019 in that state. Such events are also seen in other states—on September 4, 2019, the electricity spot price in Queensland was stuck at the −$1000 per MWh regulated floor price for several hours.

In addition, the displacement of coal generation by wind and solar has reduced stability and brought a deterioration of reliability. One result of this was the complete loss of power in South Australia during September 2016. Although it is claimed that new requirements (or the proper application of extant requirements) will prevent a recurrence of that blackout, at a minimum the new supplies have necessitated a considerable intervention (and associated costs—amounting to $44 million in 2019/20) by the market manager [9], as illustrated in Fig. 22.3, to shore up system security.

Reliability concerns remain and have spread to Victoria, which on one isolated hot day, Friday December 20, 2019, came close to failure when output from wind farms gradually fell to one third of the earlier levels. This was in spite of the fact that 9 out of the 10 major fossil units were on line, and demand was 15% below previous peaks.

Government commissioned studies have usually shown a high degree of optimism over future renewable costs. This, together with sanguine expectations that the existing generators would continue to run, even with mounting financial losses, brought consultants to consistently forecast imminent falls in prices. As indicated in Fig. 22.4, forecasts commissioned by governments failed to pick the doubling of wholesale prices 2015—18 and estimated that such levels would not even have been reached with an intensification of the regulatory measures then in place.

Future prices (see Fig. 22.5), though likely to moderate in the near term as a result of a bulge in new renewable installations, indicate no long-term decline.

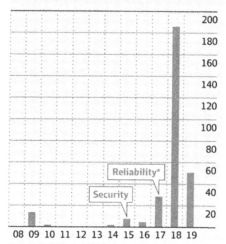

Energy interventions by type (number)

SOURCE: ENERGY SECURITY BOARD

**FIGURE 22.3**    Energy interventions by type. *Source: Energy Security Board.*

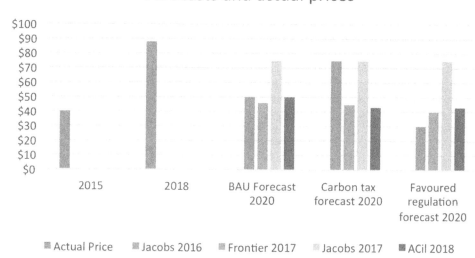

**FIGURE 22.4**    Comparison between the forecasted and actual energy prices.

The higher wholesale prices are reflected in cost to the final customer, though the increased wholesale and environmental cost are muted by the relatively large share of total costs accounted for by networks, costs which have not shown the price escalation seen with generation. Even so, as evidenced below, Australian electricity prices have increased far more than those of other countries (Fig. 22.6).

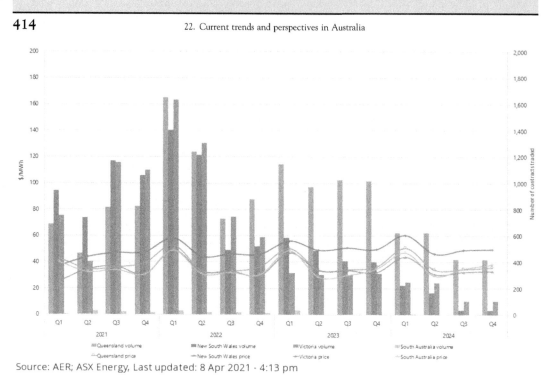

Source: AER; ASX Energy, Last updated: 8 Apr 2021 - 4:13 pm

**FIGURE 22.5**    Future energy prices. *Source: Marsden, Jacobs.*

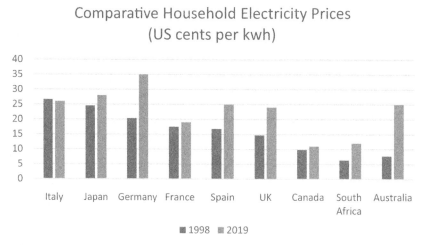

**FIGURE 22.6**    Household electricity prices during 1998 and 2015. *Source: 1998 ESAA, 2019. <https://www.global-petrolprices.com/electricity_prices/>.*

Comparative data was not available for the United States, China, and India in 1998 but average household prices for China and India in 2019 were one third of those in Australia; US average prices were boosted by high prices in some states but averaged a little under one half of those in Australia.

While subsidies to renewables have been a feature of all OECD nations' electricity policy, as well as that of some developing economies, Australia is being far more indulgent in this regard. Per capita investment in Australia was almost twice that of the next two highest countries, the United States and Japan, fivefold that in China, and almost fortyfold that of India. Fig. 22.7 illustrates this.

Australia also leads the world in rooftop solar [11], as shown in Fig. 22.8, for which subsidies provide 30%−40% of the capital cost. AEMO's draft 2019 Integrated System Plan [12] noted that, "Some 3700 megawatts (MW) of new capacity has entered the NEM since summer 2018−19. The bulk of this new capacity (some 90%) is rooftop PV and grid-scale solar generation." AEMO's modeling projects that these facilities could provide 13%−22% of total underlying annual NEM energy consumption by 2040.

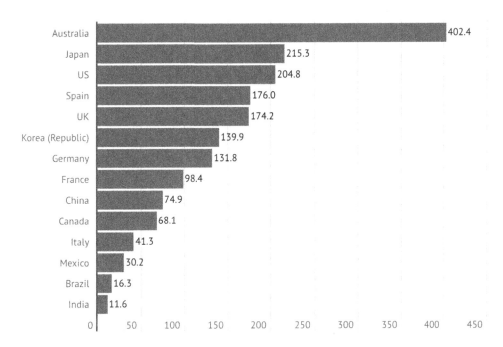

## Investment in clean energy ($US/per capita)

SOURCE: BLOOMBERG

**FIGURE 22.7** Per capita investment in clean energy. *Source: Supplied by Minister Taylor AFR. Chanticleer, Australian financial review, October 9, 2019. Available at: <https://www.afr.com/chanticleer/solving-the-energy-problem-20191009-p52z6q> [10].*

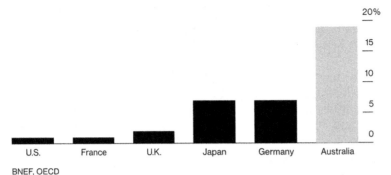

**FIGURE 22.8** Rooftop solar generation capacity.

BNEF, OECD
Note: 2018 data

Australian high take-up of (heavily subsidized) wind and solar energy has not prevented Australian policy being criticized for its inadequacy by The Greens and the Labor Party Opposition and being rightly lampooned with "fossil of the day" awards by activists attending UN Climate Conferences like that in Madrid in December 2019. The Australian Government has also been criticized for its reluctance to formally commit to "zero net emissions by 2050."

## 22.1.4 Electricity network and other costs

The AEMC [13] put the 2019–20 national weighted average electricity bill for the representative household consumer at approximately $1375 exclusive of GST. This was made up of:

- 44% regulated network component
- 40% wholesale market component
- 6% environmental policy component
- 10% residual component (comprising retailer's operating costs, customer acquisition and retention costs, return for investing in the business, and estimation errors)

(Note: the environmental component includes only the direct expenditures and not the effect of these in boosting wholesale prices two-and-a-half-fold.)

Network charges have also increased over recent years but at less than the general rate of inflation in those states where there is private ownership of networks (Victoria, South Australia). Higher increases have taken place in Queensland, Western Australia, and Tasmania, where continued government ownership of the networks remains and in NSW where partial privatization took place only in December 2015.

This is not coincidental. There are greater disciplines on costs in privately owned networks. All networks will approach the regulator (the AER) with ambit claims for future expenditures and therefore the prices they are permitted to charge. For private investor-owned networks, the shareholders' representatives (the Board of Directors) will force economies in management's actual expenditures in order to boost profits. This provides a base on which the regulator can determine future permissible expenditures for rate-setting purposes. With government ownership, these disciplines are much less forceful: having set a budget based on the allowable expenditures sanctioned by the regulator, government-owned businesses face far fewer pressures to economize on these expenditures.

Privatization, however, remains unpopular based on notions that it involves loss of public assets and a replacement of public service by commercial motives.

Depending upon the size of load, the cost of generation for business customers tends to comprise a larger component—up to 70% for smelters—than is the case for households. The pattern of increased customer costs for electricity is, however, evident with prices to businesses as well as households. Indeed, the uplift in electricity prices has impacted severely upon energy-intensive industries, especially aluminum smelters, which formerly spearheaded the nation's industrial competitiveness. Although the smelters' electricity is largely exempt from the renewable requirements, and is on long-term contract, these contracts are facing renewal. Their replacement at threefold former prices is leaving the smelters dependent on government support for their ongoing operations.

In this respect, Australia's strong relationship with the Trump Administration may have averted further pressure. Notwithstanding the overt government life-support, Australian aluminum exports to the United States have avoided the countervailing tariffs imposed on subsidized aluminum from other nations in spite of having made sales gains.

One aspect of the industry that has boomed is the bureaucracy governing it, both at the formal level of control and in the political oversight. The initially relatively small agencies responsible for operational management (AEMO), the legal features of trading (AEMC), and price fixing on the monopoly poles and wires (AER) have all sought and been given expanded resources and responsibilities for policy advice. New agencies have also been created including the Energy Security Board (ESB) and expanded roles have been given to the Australian Competition and Consumer Commission (ACCC) and various technical agencies like the CSIRO. Ministerial councils and individual state governments have also assumed considerable controls. Quantifying the costs of increased oversite is difficult because the regulatory agencies, perhaps understandably, do not assemble the material in a way that enables easily comparisons.

Unsurprisingly, the expanded oversight over the industry by bureaucrats and politicians has been inversely correlated to its efficiency.

### 22.1.4.1 Gas

For gas, a similar pattern of price increases to that of electricity is evident. Gas is now responsible for close to 20% of electricity generation (coal is 60%, hydro 7%, and wind/solar 13%). Its availability has been progressively squeezed by state government policies (in this respect there is little difference between the ALP and the LNP).

Responding to unsubstantiated scares about safety of fracking as a means of tapping gas reserves, exploration (let alone new production) for gas that would require this process has been virtually banned by all governments except that of Queensland (where most new gas is contracted to overseas markets). Fracking itself has been around for 65 years and two million wells have been sunk worldwide—mainly in North America. Although green groups raise scare campaigns, the US EPA [14] has found the practice, with appropriate care, is safe with only "isolated cases of water contamination." The Chief Scientist of New South Wales [15] came to a similar view. However, Deb Haarland [16], President Biden's nominee for Secretary of the Interior, in 2017 declared, "Fracking is a danger to the air we breathe and water we drink."

The Victoria government banned all gas exploration but is relaxing this for conventional gas.

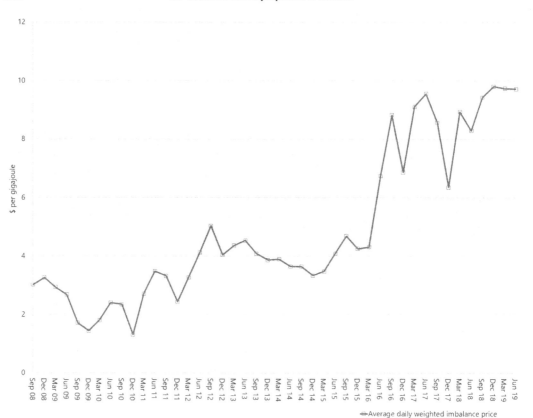

Source: AER; AEMO, Last updated: 19 Jul 2019 - 9:33 am

**FIGURE 22.9** Victorian gas market average daily weighted prices by quarter. *Source: AER; AEMO, Last updated: July 19, 2019—9.33 a.m.*

The upshot saw supplies progressively becoming scarcer and the price rising from under $3 per Gj to over $8 per Gj. This was more than double the US price where impediments to the exploration and production of "unconventional gas" have not been effective. Due to the collapse of international demand, prices fell in 2020 but are not expected to remain low (Fig. 22.9).

## 22.1.5 Measures introduced by the LNP government since 2013

The LNP has followed different paths under its three Prime Ministers: Tony Abbott, September 2013–September 2015; Malcolm Turnbull, September 2015–August 2018; and Scott Morrison since August 2018.

The Abbott Government sought to reduce the renewable subsidies requirements (and abolished a carbon tax introduced in 2012). Businessman Dick Warburton was appointed to recommend future policies; his report sought a de facto halving of the 41,000 GWh RET.

Without control over the Senate, the government was forced to compromise, with the RET being reduced to 33,000 GWh.

The LNP government led by Malcolm Turnbull, supported by his Energy Minister Josh Frydenberg, sought to introduce a version of a carbon tax on electricity, which it disarmingly and inaccurately called the National Energy Guarantee. State governments, the Opposition, the energy bureaucracy, and many in industry largely supported this goal. It was however the issue which caused the Liberal Party to replace Turnbull as Prime Minister in 2018. This same issue resulted in the replacement of Turnbull as Leader of the Opposition in 2009.

Many continue to press for new subsidies and/or carbon taxes for renewables—indeed, Josh Frydenberg as Treasurer remains influential in energy policy and has appointed Steven Kennedy, the author of the 2008 Garnaut Report [17], as Treasury Secretary.

The appointment of Angus Taylor as minister for energy under Scott Morrison brought a somewhat greater resolve to remove the nation from the almost two decades of subsidies and taxes in support of environment-friendly technologies.

### 22.1.6 Current policy approaches

Subsidies and other regulatory instruments remain the dominant factors in the energy industries' structure and cost. Total annual subsidies in 2019 were estimated at $6.9 billion.

| Subsidies to wind and solar for the year ending June 2019 ($M) | |
| --- | --- |
| Federal regulatory support | 3087 |
| State regulatory support | 951 |
| Federal fiscal support | 2418 |
| State fiscal support | 457 |
| Total | 6913 |

*AEMC, Commonwealth and state budget papers. For details see A. Moran, The hidden costs of climate policies and renewables. Available at: <https://35b1ca50-ea91-45c2-825d-3e16b7926e46.filesusr.com/ugd/b6987c_afd260bfd8284f9db5d97d73fe52cedb.pdf> [18].*

Particularly rapid growth has taken place with the small-scale renewable energy scheme (SRES) (rooftop solar), which has increased to an estimated cost of $1.5 billion [19].

As previously discussed, more important than the direct cost of these subsidies is their effect in lifting prices. With static demand, over the 4 years since the withdrawal of coal capacity due to the subsidies to renewables, the annual wholesale cost of electricity increased from $7.5 billion to approaching $20 billion in 2019. In the context of the COVID instigated price collapse in 2020, generation assets values have been undermined, with the major generator—retailers being forced into write-downs, in AGL's case by $3.5 billion, almost half of its asset value, mainly due to its renewable energy contracts.

Minister Taylor's approach is to avoid any further expansion of the requirement on retailers to increase the amount of (subsidized) large-scale wind and solar. Incongruously, however, he is maintaining the SRES subsidy to rooftop solar, that has resulted in Australia leading the world in these installations and that even interventionist-minded bodies like the ACCC have recommended closing. Additional policies are

- Trying to prevent further closures of coal generators, including by requiring a 3-year notice of closure, and to foster new ones.
- Jawboning retailer—generators into lowering prices, partly through ensuring customers are made better aware of lower cost options, requiring retailers to have adequate contract coverage to supply their customers, and setting in train "big stick" laws that can bring asset divestiture.
- Promoting Snowy 2, a major expansion of pumped hydro that aims to improve the Snowy scheme's ability to counterbalance the increase in intermittent power.
- Encouraging transmission designs that will avoid further subsidies to remotely located renewables.
- Requiring a form of domestic gas reservation to keep prices lower/availability higher than might otherwise occur.

All this is far less harmful than the redoubling of the subsidies to renewables that was the ALP platform taken to the nation in the May 2019 election. It will not however bring lower prices nor markedly greater reliability. Still less will it return Australia to its former world leadership in low price/high reliability. Such an outcome is possible only if action is taken to remove the existing subsidies that are to remain in place for a further 10 years as well as to cease issuing new ones under the SRES scheme for rooftop solar installations.

The government's unwillingness to repeal the SRES scheme is one indicator of a lack of resolve (or political capital) to take even modestly unpopular decisions necessary to repair the broken supply system.

Ironically, virtually all parties now consider there is little alternative but to have further government intervention in the market that has been undermined by such intervention. Many within the energy bureaucracies and the renewables industry call for further support for what they see—or claim to see—as a renewables-dominated energy future. Characteristically, those seeking intervention in that direction often express outrage at interventions that do not support their favored industries.

The costs imposed on conventional reliable plant by renewables include the "hollowing out" of demand during the daytime, which wrecks the economics of baseload plant that is designed to amortize capital costs by continuous operations. Even though the government has won a grudging deferral of the next scheduled major plant closure, Liddell, owned by AGL, the largest energy company, the market operator's forecast future pattern of closures (Fig. 22.10), shows the difficulties in reversing course.

The difficulties are further exemplified in the outcome of Minister Taylor's plan for a form of government support for "dispatchable" plant, which many saw as code for coal, especially in view of gas supply constraints resulting from state government exploration bans. Part of this has been a plan to require retailers to hold "firming" contracts to provide assurances that power will be available to their customers; wind and solar cannot supply such contracts. But these requirements are unnecessary, as retailers' internal risk management procedures already insist on such contracts to avoid exposure to spot electricity prices of up to $14,000 for a product that is retailed at perhaps $150.

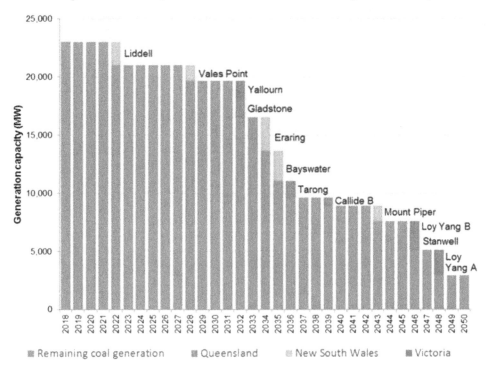

**FIGURE 22.10** Scheduled coal power station decommissionings. *Source: Energy Council, Coal power station decommissionings. Accessible at: <https://www.energycouncil.com.au/analysis/power-plant-shutdowns-how-do-you-prepare-for-the-transition/> [20].*

Moreover, the fact is that with existing policy settings favoring wind/solar, any new coal power station would simply expedite the closure of an existing coal station, adding little to the increased security and reliability that is the initiative's immediate goal.

Furthermore, the committee appointed to advise on the best prospects, recommended only one small coal fueled generator upgrade, which perhaps underlines the difficulties governments now have in obtaining advice that is untarnished by the dominant green paradigm. Indeed, the Commonwealth Government supports the Carbon Marketplace Initiative [21] under which firms are invited to donate up to $12,500 per year in what amounts to an assisted suicide pact involving a "journey toward net zero emissions."

The Department of Environment and Energy [22] sees an ongoing increase in renewable supplies at the expense of coal and gas. It envisages (Fig. 22.11) the following fuel supply shares, where renewables (with no growth in hydro) lift their share of generation from 27% in 2020 to 48% in 2030.

The deep hostility to coal throughout official circles is also illustrated by a decision in February 2019 of the senior judge in the NSW Land and Environment Court, Mr. Justice Brian Preston, who rejected the application to operate for a new coal mine because, among other reasons, he said it would be unable "to meet generally agreed climate targets" for

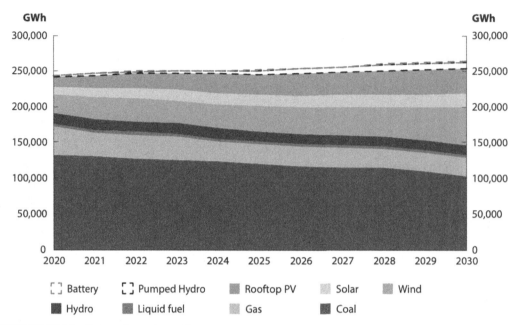

FIGURE 22.11    Projected supply to 2030.

a "rapid and deep decrease" in emissions. The case against the mine [23] was run by the Environmental Defenders Office NSW, which is funded in part by the state government and at which Preston once served as the founding principal solicitor.

In NSW there are gas and coal projects including Rocky Hill and Wollongong that have been derailed through the planning system. where local interests and activists are allowed to dominate. With such views prevalent throughout the judiciary as well as in political circles, it would require some form of government guarantee as indemnification against any new measure that might prejudice fossil fuels in the market.

It is notable that there is a high public approval of green energy, with a September 2019 Omnipoll finding [24] (Fig. 22.12) that 81% of Australians support a role for it. Perhaps more significant is the importance younger people attached to reducing carbon emissions compared to reliability or price.

Many see these considerations and claims by the big four domestic banks that they will not invest in new coal as evidence that global warming anxiety would prevent new mines from obtaining finance. This is untrue—the domestic banks rarely lend for major projects and the hundreds of coal facilities planned or under construction around the world are testimony to the availability of finance. Australia, as illustrated by the 9-year-long proposal by Adani for a new coal mine (and the aforementioned views of a senior NSW judge) is vulnerable to political and judicial activity.

Australia also is somewhat hostage to labor unions with regard to major programs. For this reason, in assessing the costs entailed in new greenfield coal generators, Solstice (see Ref. [6]) in a meticulous report commissioned by the Minerals Council placed a loading of 25% on labor costs compared to those prevailing in the United States. Even so, the report

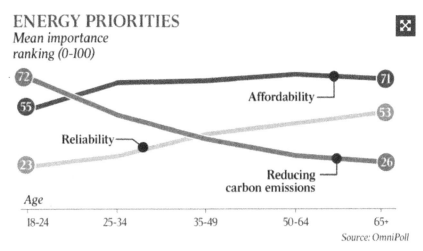

FIGURE 22.12   Energy-related priorities per age group. *Source: OmniPoll.*

still found a new coal generator could profitably operate at an electricity price of $50 per MWh if operated continuously as a baseload generator.

The government, with support from the Opposition, is seeking to curb what it sees as market power from generator–retailers. This is misplaced in the case of retailing. Though not all are active in every state market, there are 33 different brands among electricity retailers (16 for gas). Even though the big three (AGL, EnergyAustralia, Origin) have a 75% market share, this still means there is a very high degree of competition. It was, however, a mistake by the NSW government to prefer an asset sale program in 2014 that left the state market with too few competitors by selling the Bayswater and Liddell stations to the same buyer, AGL. This has left that firm with profitable opportunities in closing Liddell and gaining from the resultant price increase. The government is pressuring AGL to on-sell Liddell.

The ACCC has expressed concerns about vertical integration. These are unwarranted. Vertical integration is just a variation of long-term contracting and, as with other industries involved in make-or-buy strategies, offers firms different degrees of certainty.

Politicians and regulators alike have also expressed concerns about price insensitive customers, who have remained loyal to their retailers and have accepted uncompetitive price packages. They have done so in spite of an abundance of comparator sites, requirements to facilitate easy switching of retailers and a crescendo of information about how to get a better deal. In spite of sticky customers, the churn rate is very high, typically 30% a year [25], though policies of the ACCC to restrict aggressive sales campaigns and cold-calling by retailers has tended to mute this.

Other issues being debated largely surround measures that might prevent a further deterioration in reliability. The renewable sector makes considerable noise about breakdowns of "archaic" fossil fuel plant. This is unfounded. Although, as Fig. 22.13 shows, there has been an increase in fossil fuel plants' forced outage levels in NSW, this has not been substantial and is not seen in other states.

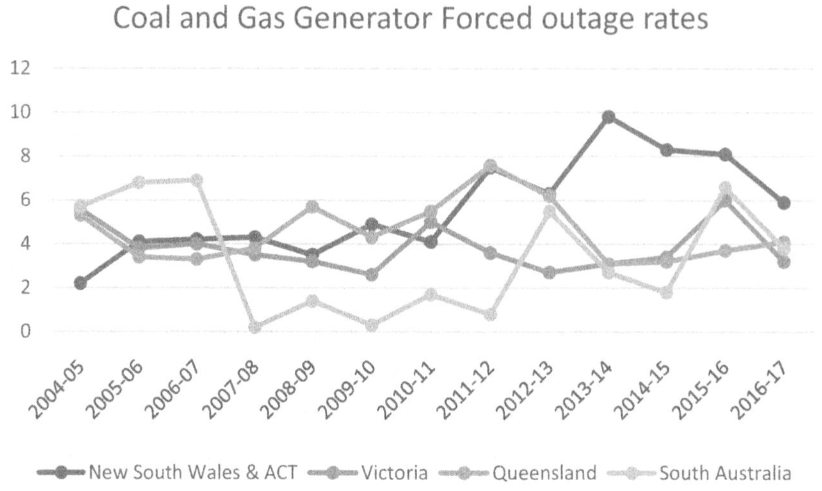

## Coal and Gas Generator Forced outage rates

FIGURE 22.13  Coal and gas generator forced outage rates from 2004 to 2017. *Source: EGA, Derived from Australian Energy Council, ESOO and reliability: what does it tell us?, August 29, 2019. Available at: <https://www.ener-gycouncil.com.au/analysis/esoo-and-reliability-what-does-it-tell-us/> [26].*

The market operator, AEMO, is calling for new expenditures to offset the deleterious effects on grid reliability caused by household and other small-scale weather-dependent generating facilities as well as by dispersed and remotely located wind. It, like the remotely located renewable facilities, favors the costs to this being incurred directly by customers rather than by the generators themselves. This would be a new form of subsidy for renewables and large-scale solar and provide little incentive for new facilities to choose locations that take the costs of transmission into account.

The AEMC, the rule-making body, had taken a different view and was calling for a policy change which would see new generators which choose to locate away from established transmission (or on transmission lines where their presence might cause congestion) to incur the costs. This latter view is preferable but faced with opposition from an industry and other regulators which prefer to have central direction and cost, this position has been largely abandoned.

In addition, there is the plan to change the nature of the Snowy and the Tasmanian hydroelectricity systems. These have assumed greater importance in a network with far greater price volatility. Snowy is to be redesigned as a pumped storage facility to allow better balancing of the intermittent renewable supplies, even though this will reduce gross output by 40%. Initiated by Prime Minister Turnbull (under whom the Commonwealth bought out the NSW and Victorian governments' shares of Snowy Hydro) the government is sees the Snowy scheme becoming "the battery of the nation."

Like so many government schemes, the cost of the pumped storage initiative has blown out and its claimed effectiveness has been questioned, especially by competitive solutions, including those involving actual batteries. Originally foreshadowed at $2 billion, costs including for beefed up transmission are now credibly estimated [27] to be some $10 billion.

The Marinus Link with Tasmania is another transmission plan, which is to add an additional 1200 MW of capacity to the existing 400 MW of constrained generation to the mainland.

The Prime Minister [28] has announced a new $1 billion fund for the Clean Energy Finance Corporation to "future proof the electricity grid." While such power-sharing vehicles may help prevent regional blackouts, they also deter new private investment and therefore further raise prices.

It is doubtful that such expenditures would pass a test of commerciality under any circumstances and they can only approach such a standard because of the price volatility subsidies to renewables have created.

## 22.2 Concluding comments

The bottom line of policies under consideration is that none of them are going to restore the former low-cost electricity and gas supply. Unless the Commonwealth government finds a way to renege on the subsidies to renewables, stops all new ones immediately, and can pressure state governments to cease impeding gas exploration and production, the present tragedy of high energy prices amidst an abundance of supply potential will continue.

Australia will certainly be the poorer for destroying its energy advantages. It is only the nation's vast natural wealth that has enabled a steady increase in living standards, an increase achieved despite government energy policies.

The bushfires in late 2019 and early 2020, though due to inadequate cold season burning [29], have been blamed by activists and vested interests on Australia not doing enough to replace fossil fuels by renewables. The resultant media pile-on has made for further difficulties in the government's ability to pursue sensible polices.

While consumer prices are the most visible and publicized aspect of the deterioration in electricity supply costs and reliability, it is the effect on commercial customers that is most serious. Mention has already been made of the aluminum plants, standing for many years at the apex of the nation's world-class manufacturing facilities. The three major facilities in Victoria (Portland), NSW (Tomago), and Queensland (Boyne Island) account for 10% of electricity demand. All face difficult contract negotiations for this supply, which comprises some 30% of their costs. These plants cannot rely on government subsidies over the long-term and, in any event, this leaves them vulnerable to countervailing trade measures of the sort that the Trump Administration had imposed on some exporters.

Moreover, electricity and to a lesser degree gas is ubiquitous in all commercial activities. In addition to its vital importance to aluminum smelting, it comprises over 5% of costs in other smelting, iron and steel, wood and paper, glass and ceramics. And while these industries might see a silver lining in the power price collapse that would accompany the closure of an aluminum smelter, that closure would undermine the economics of baseload coal plants. What would follow is the closure of one of them and the return of high prices, accompanied by a further diminution of reliability across the system.

Similarly, government, judicial, and activists' opposition to new gas saw its prices increase two-and-a-half-fold compared to those in the United States. Five years ago, they were similar. Industries highly dependent on gas include pulp and paper, metals, chemicals, stone, clay and glass, plastic, and food processing.

Present energy policies will therefore, at best, mean a serious underperformance of the economy. The core determinant of this, the attack on fossil fuels and uses of other resources including water and land itself, has been boosted by activists' populist slogans like "climate emergency" and "extinction rebellion." The wider tacit support for such refrains, even within the LNP, make it difficult to assemble the political will to reverse course. And government planning documents, like *Australia's emissions projections 2019*, are predicated on a continued increase in wind/solar displacing coal and increasing the overall renewables share from 27% in 2020 to over 48% of supply in 2030.

# References

[1] A. Cho, Smaller, safer, cheaper: one company aims to reinvent the nuclear reactor and save a warming planet, Science (2019). Available at: <https://www.sciencemag.org/news/2019/02/smaller-safer-cheaper-one-company-aims-reinvent-nuclear-reactor-and-save-warming-planet>.

[2] A. Messimer, William Stanley Jevons The Coal Question (1865), beyond the rebound effect, Ecol. Econ. 82 (2012) 97–103. Available at: <https://www.sciencedirect.com/science/article/abs/pii/S0921800912002741#!>.

[3] D.H. Meadows, D.L. Meadows, J. Randers, W.W. Behrens III, The Limits to Growth; A Report for the Club of Rome's Project on the Predicament of Mankind, Universe Books, New York, 1972. ISBN 0876631650. Retrieved 26 November 2017.

[4] Simon–Ehrlich wager, see <https://en.wikipedia.org/wiki/Simon%E2%80%93Ehrlich_wager>.

[5] See Australian Energy Council, Australia's renewable energy target. Available at: <https://www.energy-council.com.au/media/11565/ret-factsheet-final.pdf>.

[6] Prospects for a HELE USC Coal-fired Power Station, Solstice Development Services (SDS) Pty Ltd, 2017. Available at: <https://apo.org.au/node/96821>.

[7] Clean Energy Council, Report 2020. Available at: <https://assets.cleanenergycouncil.org.au/documents/resources/reports/clean-energy-australia/clean-energy-australia-report-2020.pdf>.

[8] Australian Energy Market Operator (AEMO). Available at: <https://www.aemo.com.au/>.

[9] Energy Security Board, Health of the NEM, 2021, p. 7. Accessible at: <http://www.coagenergycouncil.gov.au/sites/prod.energycouncil/files/publications/documents/The%20Health%20of%20the%20National%20Electricity%20Market_V01_Final%20.pdf>.

[10] Chanticleer, Australian financial review, October 9, 2019. Available at: <https://www.afr.com/chanticleer/solving-the-energy-problem-20191009-p52z6q>.

[11] See BloombergNEF. Available at: <https://about.bnef.com/blog/solar-and-wind-reach-67-of-new-power-capacity-added-globally-in-2019-while-fossil-fuels-slide-to-25/>.

[12] Australian Energy Market Operator, Draft Integrated System Plan December 2019. Accessible at: <https://www.aemo.com.au/-/media/Files/Electricity/NEM/Planning_and_Forecasting/ISP/2019/Draft-2020-Integrated-System-Plan.pdf>.

[13] Australian Energy Market Commission, Residential electricity price trends, 2020. Available at: <https://www.aemc.gov.au/sites/default/files/2020-12/2020%20Residential%20Electricity%20Price%20Trends%20report%20-%2015122020.pdf>.

[14] V. Volcovici, T. Gardner, Fracking not a 'widespread risk' to drinking water: U.S. EPA, Reuters, June 2015. Available at: <https://www.reuters.com/article/us-usa-water-fracking-idUSKBN0OK1X220150604>.

[15] NSW Chief Scientist, Independent review of coal seam gas activities in NSW information paper: fracture stimulation activities, September 2014. Available at: <http://www.chiefscientist.nsw.gov.au/__data/assets/pdf_file/0008/56924/140930-Final-Fracture-Stimulation.pdf>.

[16] T. Puko, Interior secretary nominee on collision course with oil industry, Wall Street Journal, February 14, 2021. Available at: <https://www.wsj.com/articles/interior-secretary-nominee-on-collision-course-with-oil-industry-11613318400?mod = hp_lead_pos3>.

[17] The Garnaut Review, Available at: <https://en.wikipedia.org/wiki/Garnaut_Climate_Change_Review>.

[18] For details see A. Moran, The hidden costs of climate policies and renewables. Available at: <https://35b1ca50-ea91-45c2-825d-3e16b7926e46.filesusr.com/ugd/b6987c_afd260bfd8284f9db5d97d73fe52cedb.pdf>.

[19] see Demand Manager, Australian solar rooftop subsidy 2019. Available at: <http://www.demandmanager.com.au/wp-content/uploads/2019/02/Australian-Rooftop-Solar-Subsidy-2019-Outlook.pdf>.

[20] Energy Council, Coal power station decommissionings. Accessible at: <https://www.energycouncil.com.au/analysis/power-plant-shutdowns-how-do-you-prepare-for-the-transition/>.

[21] Clean Energy Regulator, Buying ACCUs, 2020. Available at: <http://www.cleanenergyregulator.gov.au/Infohub/Markets/Pages/Buying-ACCUs.aspx>.

[22] Department of Industry, Energy, Science, Environment and Resources, Australia's emissions projections 2019 report. Available at: <https://www.industry.gov.au/data-and-publications/australias-emissions-projections-2019>.

[23] For details see <https://www.edo.org.au/annual-reports/>; and <https://d3n8a8pro7vhmx.cloudfront.net/edonsw/pages/5990/attachments/original/1541379654/EDO_NSW_2017-18_Annual_Financial_Report.pdf?1541379654>.

[24] G. Chambers, Younger Australians prioritise reducing carbon emissions over energy affordability, reliability, October 12, 2019. Available at: <https://www.theaustralian.com.au/nation/politics/younger-australians-prioritise-reducing-carbon-emissions-over-energy-affordability-reliability/news-story/bd40d16f9dd636500ab62bb41c57608f>.

[25] see Australian Energy Market Commission, Retail energy competition review (see Table 6.1), 2018. Available at: <https://www.aemc.gov.au/sites/default/files/2018-06/Final%20Report.pdf>.

[26] Derived from Australian Energy Council, ESOO and reliability: what does it tell us?, August 29, 2019. Available at: <https://www.energycouncil.com.au/analysis/esoo-and-reliability-what-does-it-tell-us/>.

[27] Bruce Mountain, Snowy 2.0 will not produce nearly as much electricity as claimed. The Conversation, August 15, 2019. Available at: <https://theconversation.com/snowy-2-0-will-not-produce-nearly-as-much-electricity-as-claimed-we-must-hit-the-pause-button-125017>.

[28] The Prime Minister, $1 Billion boost for power reliability, October 30, 2019. Available at: <https://www.pm.gov.au/media/1-billion-boost-power-reliability>.

[29] G.L. Morgan, et al, Prescribed burning in south-eastern Australia: history and future directions, Australian Forestry 83 (1) (2020). Available at: <https://www.tandfonline.com/doi/abs/10.1080/00049158.2020.1739883?journalCode=tfor20&>.

# Conclusions and paths for future research and development

*Tiago Pinto[1], Zita Vale[2] and Steve Widergren[3]*

[1]GECAD Research Group, Polytechnic of Porto (ISEP/IPP), Porto, Portugal [2]School of Engineering, Polytechnic of Porto (ISEP/IPP), Porto, Portugal [3]Pacific Northwest National Laboratory (PNNL), Richland, WA, United States

Local electricity markets (LEM) are emerging as one of the most promising solutions to boost the widespread deployment of distributed renewable energy generation. LEM enable the consumption of locally produced energy in nearby geographical areas, while fostering the active participation of consumers [1]. Although there is no global consensus on the definition of a LEM, it is widely accepted that LEMs should provide mechanisms for accommodating the negotiations between customers, prosumers, and small-scale producers [2]. Such mechanisms envisage the trading of locally produced energy, typically provided by distributed energy resources, while considering an auxiliary system to provide security in the energy supply. Therefore the balance between the demand and supply can be maintained at the local level while benefiting from the increasing penetration of distributed energy sources. LEMs envisage not only the direct trading of energy, but also of flexibility among community members. By trading energy produced locally, LEMs limit the dependency on incumbent suppliers and enhance their competitiveness, while avoiding or delaying investments in distribution network reinforcements by local demand—supply matching. Moreover, the engagement of end users is enhanced, creating a local identity and promoting social cooperation [3].

By engaging directly with each other, consumers, prosumers, and small-scale generators contribute to a sustainable power network by incentivizing the energy effectiveness at residential and building levels, while contributing to reduce the impact of market power by decreasing the need for large-scale generation sources and intermediary market actors [4]. In fact, local market design can mitigate local market power in several ways. Automated demand response is already implemented at a wide scale; it introduces elasticity into the market demand curve, which may lead to a mitigation of market power [5]. In specific grid services markets, such as voltage regulation, auction designs like the uniform price double auction decrease bid shading. For other grid services that are more like single-buyer procurement markets, such as reserves, a second price auction design can decrease the winner's curse while still fostering low-cost participation. Another market design that may be useful for LEMs is a call option market, in which participants are paid some amount to be available (dispatchable) and then paid for actual services that are rendered.

Call option markets harness the benefits of contestability to mitigate market power. Combining digitization and energy resource innovations creates the potential for coordinating the operation of these intelligent energy resources in a distributed decision-making framework that uses market-based techniques for negotiating price-for-service agreements and automation for energy resource response. These combined innovations drive, distributed generation and storage resources to become more energy efficent with lower costs of production and installation.

Another main benefit that can be expected from LEM relates to network operations. Considering the energy transport constraints arising from geographical imbalances between generation and consumption, Staudt et al. [6] point out that a local market is the most prominent solution. In nodal LEM approaches, the electricity cost incentives are developed at the local level to mitigate the congestion at the transmission level [6]. Additionally, local pricing is also less complex and more transparent than traditional market models, thus facilitating market operations and global acceptance.

Acceptance and widespread implementation of LEM are, however, deeply dependent on regulatory and governmental policies and incentives. The key to enable new business models, processes for efficient interaction between the many involved stakeholders, and taking advantage of progress toward different local goals will be the collaboration between energy service providers and local governments. Through policy actions, governments can minimize or eliminate blockages to boost local renewable integration by creating opportunities to bring together local government, energy utility companies, universities, research centers, and other entities to share knowledge and perspectives that can bring significant improvements to the sector and contribute to mitigating the upcoming challenges [7].

Overcoming such challenges and finding efficient LEM operation models require extensive investment in research and experimentation. Many initiatives have been undertaken during the last few years, and most of them are still under way. They are investigating some of the most prominent issues in the domain while implementing actual LEM experiments in the field. Some of the most relevant topics addressed by these works are the study of network constraints in LEM simulations, the study of regulation, policy, and social issues regarding LEM implementations, the design of bidding strategies to participate in LEM, the study of interactions between different types of players involved in LEM, the integration of LEM with other markets, the incorporation of uncertainties and risk in LEM participation, the development of distributed approaches in LEM simulations, the extension of LEM to distribution network level, the redesign of tariffs in LEM, the exploration of communications issues and the study of the influence of forecast methods on LEM results.

However, and despite the many advances that have already been accomplished, the current regulatory policy in many jurisdictions are a barrier to enabling the implementation of LEM. This means that all the components that are considered essential for effective LEM operation cannot be satisfied yet. In any case, such initiatives are fundamental for the practical applications of LEM. The knowledge gained and relationships made through real-world demonstrations will pave the way toward practical deployments of transactive energy systems and LEM, identifying barriers along the way that can only be studied and mitigated once such mechanisms are put in practice.

Learning from the insights brought by experiences in different areas of the globe, it is evident that the widespread deployment of LEM is profoundly reliant on the characteristics, needs, and priorities of the different countries. The incumbent market models for integrating distributed energy resources, typically characterized by the centralized models, for example, in the United States, and decentralized models, for example, in Europe, are starting points that require different pathways of evolution to enable a transition toward LEM.

In Europe, the concept of LEM gained the spotlight during 2016, when the European Commission presented a new policy package titled "Clean energy for all Europeans." A priority of this package is the empowerment of customers, who become active players in the European Union Energy System by taking advantage of the local availability of energy resources [8,9]. The creation of LEM in Europe is envisaged according to the Internal Electricity Market Directive [10,11] from 2019, and the Renewables Directive [12] from 2018, which set legal and market participation principles for local energy communities and renewable energy communities. Hence, the European Union is already creating the conditions, through directives, policy guidelines, and regulations, toward a progressively interconnected energy market, endowing consumers with a central, active role in the system.

The active participation of consumers is, however, still limited by regulatory guidelines, which prevent the implementation of most initiatives that foster consumption flexibility. In the United Kingdom for instance, the participation of flexibility services and transaction of electricity in local (peer-to-peer and distribution level) and national markets is still restricted. In terms of peer-to-peer, and with some exceptions (i.e., trials), LEM cannot trade directly with consumers without the intervention of suppliers [13]. There are, nevertheless, new opportunities to capture additional revenues by providing flexibility services to distribution network operators and system operators either individually or in aggregation. Thereby, it with an adequate regulatory framework and clear trading rules, LEM will likely play an important role in the UK to enable efficient and smarter integration of distributed energy resources, facilitating the transition to a more flexible and low carbon electricity system.

The United States regulatory context and the utility industry structure differ from the distribution system operator-based structure that is evolving in Europe. Those differences may enable more diverse experimentation that could contribute meaningfully to the evolution of LEMs in different models and different contexts. In the United States regulatory context, market design is part of a jurisdiction's institutional framework, and institutions will affect incentives for the development and adoption of LEMs. Federalism along with its heterogeneity means that multiple parallel institutional experiments, rather than a uniform adoption rule, will inform the growth of LEMs and the specifics of their market design. In the vertically-integrated, investor-owned utility model of regulation in many U.S. states, LEMs would have to be established within the conventional structure of the regulated utility and with regulatory approval. Doing so entails bringing LEM into the utility's regulated tariff structure and determining a specific tariff for LEM transactions. Reaching a regulated rate tariff for LEM transactions in this context will be challenging, and include the political challenges posed by powerful regulated utility companies wary to upset their incumbent positions and embrace LEM [14]. The market evolution towards the effective accomodation of distributed energy resources is, however, already taking shape, as most electric energy is traded in the wholesale markets of Independent System

Operators (ISO)/Regional Transmission Organizations (RTO). ISO/RTOs are regulated by the Federal Energy Regulatory Commission (FERC), which has opened wholesale markets for distributed energy resources aggregators and is doing more work to bolster that with FERC Order 2222 [15].

The benefits of electricity provision from a competitive market observed in several US jurisdictions and in the UK market, has led Australia to embark upon the series of reforms in the energy sector. Twenty-five years ago, Australia was a pioneer in global moves to replace central control of electricity supply into a market-based system based on competitive provision in generation and retailing. Currently, Australia's emissions projections are foreseeing a continued increase in wind and solar generation, displacing coal and increasing the overall renewables share from 27% in 2020 to over 48% of supply in 2030 [16]. This scenario calls out for urgent measures to enable a large-scale integration of variable renewable energy sources, which will undoubtedly require new models for local dispatch and distribution grid level generation-consumption balance, possibly forcing Australia to embark on the LEM initiative. However, the transition toward LEM is not envisaged to run smoothly, as Australian politicians and regulators have expressed concerns about price insensitive customers. Some customers have remained loyal to their retailer in spite of an abundance of comparator sites, requirements to facilitate easy switching of retailers and a crescendo of information about how to get a better deal. The failure in capturing consumers' flexibility is a major setback for the success of LEMs, which so largely depend on end users' flexibility to balance the variations from the generation side.

Measures to address the deficient response of consumers' flexibility learn from successful past experiences. Brazil has dealt with a large number of dramatic incidents over the years, such as blackouts, due to severe droughts and their impact on the Brazilian electricity sector, which is mainly hydroelectric energy [17]. During the 2001 blackout crisis caused by a water shortage in the hydro-electric system, Brazil imposed electricity rationing in which Brazilians were obliged to restrict electricity usage by 20%, on average [17]. Although this consumption reduction was drastic, it paved the way to the implementation of demand response initiatives that have enabled Brazilians to endow the system with a startling level of consumption flexibility, for example, the implementation of the Tariff Flag System (2013), the White Hourly Tariff (2016), and the consumer's opportunity to opt for electricity prepayment [18]. Consumption flexibility is, however, not the only requirement to support a potential successful implementation of LEM. In the South American context, the significant role of hydropower to supply large areas reduces the need for locally managed electrical energy, even when considering the introduction of mechanisms through which hydropower and other renewables can complement each other and can improve the cost-efficiency and reliability of power systems. Moreover, the very different scenarios faced by several South American countries make it very hard to foresee the paths for electric sector evolution toward LEM. As such, LEM is not envisaged to be emerging there as soon as in other places of the world.

The other extreme regarding the need for LEMs is Africa. The continent is faced with an electricity shortage due to the lack of generation capacity and aging transmission and distribution infrastructure [19]. The rate of access to electricity is substantially lower than the rest of the world (43% compared to global access rate of 87%). This is mostly due to the dispersed nature of people in rural Africa, who live mostly in small villages in geographically remote areas. Local generation and energy management are thereby envisaged as the most promising way to mitigate the problem, avoiding prohibiting investments in transmission infrastructure, both in terms of costs and building time. In many parts of Africa, rural electrification is being achieved through a mixture of on-grid generation and dedicated microgrids, including solar systems to provide low-voltage DC to satisfy the basic demands for lighting and mobile phone recharging [20]. Additionally, Africa is blessed with abundant renewable energy. This includes hydro, geothermal, solar, wind, biogas, and biomass gasification. With such excellent renewable energy, the path forward is in creating localized power grids, or grid hubs, where energy is produced at or nearer the point of demand [21]. Africa has, the potential to become one of the most fertile regions for competitive LEM smart grid innovations. However, there are still significant challenges that must be overcome to enable deploying LEM solutions at the scale needed. Some of these challenges are technical, legal and regulatory, financial, and educational. Many regulatory policies are outdated and not amenable to deal with the consequences of such a paradigm change. Public engagement and participation is lacking. Significant investments are required to purchase the new technologies envisioned for communicating information between the end users, electricity service providers, and to modernize and significantly expand aging transmission and distribution infrastructure.

Challenges related to policy, system, and management are also identified in the Asian context. These have restricted the development of the electric power industry in various countries to a certain extent, thus affecting the economic and social development. Therefore, the market-oriented reform of the electric power industry is identified as the path to take. China and emerging Asian countries such as the Philippines and India have witnessed great changes in electric power development in recent years. Compared with the mature electricity market in western countries, the construction of the electricity market in China, India, the Philippines, and other Asian countries is steadily and rapidly advancing. The market-oriented reform of electric power in China plays an important role to explore the development of electricity futures trading and the establishment of the electricity derivatives market. From the local energy management and trading perspective, efforts are being put into the development of a strategy to move from the smart grid to a ubiquitous power internet of things. The state grid proposes to build a world-class energy internet enterprise with global competitiveness. The energy internet is seen as an important platform for large-scale development of distributed energy management and trading approaches.

It is evident that the global transition toward wide scale implementation of LEMs is happening all around the world; however, multiple barriers and challenges are pointed out. Most of them are of a regulatory and policy nature, which can only be overcome by demonstrating to governments, regulators, and policy makers that LEMs are feasible, efficient, and advantageous.

These significant changes in the energy sector brought by the increasing penetration of renewable energy sources of variable nature are causing and will continue to cause relevant impacts in existing actors, facilities, and stakeholders who will need to invest, not only in new technologies but also in knowledge to overcome and adapt during the transitional period. Governments and local governments are key contributors to facilitate and pave the way, along with universities, research centers, and institutions that can access and create the means and knowledge necessary to overcome such challenges.

LEMs are still an emerging research topic, and despite the increase interest during the last few years, there are few research initiatives that address the LEM subject. Thus the current focus is to more thoroughly understand the concept implications and the possible benefits that can come from wide scale LEMs implementation. The research community should also focus on analyzing effective solutions to avoid the new issues related to the introduction of the new organizational and operational processes brought by LEMs.

The general economic foundation for LEM is essential to solidify. This requires attention to market design, which establishes market rules and shapes the incentives to participate, the trading incentives, and engagement strategies of participants. To align participant and system objectives and to enable participants to maximize value (i.e., to enable economic efficiency), LEM market design should focus on maximizing economic value subject to the physical constraints of the network [22]. LEMs will also be experiential platforms – parties need learn how to value the platform and the grid services that participant's resources can provide at different times and in different contexts and conditions.

Local market power in LEMs will be an issue due to the spatially distributed nature of many grid services and the potential for low liquidity and a low volume of transactions. This concern about market power is particularly relevant in a market design in which aggregators and retail energy providers participate in LEMs. They are more likely to operate several resources and have control over a larger portion of the overall portfolio in an LEM. Furthermore, considering the different sizes of producers, promoting fair competition and easy access to different suppliers is important to enable consumers to take full advantage from their active participation and flexibility [11]. Another particularly sensitive point is understating how potentially conflicting parties will be contending in a LEM scenario, for example, local energy communities, energy utilities, and retailers. This competition is, however, happening in a developing sector, which means that there is sufficient space for the different models to coexist without causing distress [23]. In fact, the positive effect brought by LEMs on regional and local improvement opportunities, export possibilities, social union, and business openings will force the existing players to rethink their position in the market and explore novel business models. These novel opportunities also have potential benefits, and hence should be explored by other involved players, such as independent energy producers, including renewables self-serve customers [12].

Another key aspect that requires further development to enable the widespread implementation of LEMs is related to the development of infrastructure that allows local transactions in a reliable and transparent manner. Currently, some open platforms for local electricity trading are available and can serve as a basis for the study of new business models related to the local use of flexibility, energy, and electricity. The deployment of smart metering systems is essential, requiring an economic assessment of all the long-term costs and benefits to the market and to the individual consumer, including which models should be used to ensure that smart metering is cost-effective and available in feasible time frames [11]. Digital communication and automation that guarantees the necessary security and reliability requirements are satisfied are further related topics that require deeper development. Standardization in this area and all LEM related processes will be needed in any transition scenario.

These challenges and related paths for future research and development will contribute to confronting the current identified barriers and to creating the conditions for a widespread implementation of LEMs.

# References

[1] E. Mengelkamp, J. Diesing, C. Weinhardt, Tracing local energy markets: a literature review, (2019). Available from: http://doi.org/10.13140/RG.2.2.17644.21128.

[2] F. Lezama, et al., Local energy markets: paving the path towards fully transactive energy systems, IEEE Trans. Power Syst. (2019) 1. Available from: https://doi.org/10.1109/TPWRS.2018.2833959.

[3] G. Mendes, et al., Local energy markets: opportunities, benefits, and barriers, (2018).

[4] E. Mengelkamp, et al., Trading on local energy markets: a comparison of market designs and bidding strategies, International Conference on the European Energy Market, EEM, IEEE, 2017. Available from: http://doi.org/10.1109/EEM.2017.7981938.

[5] F. Teotia, R. Bhakar, Local energy markets: Concept, design and operation, 2016 National Power Systems Conference (NPSC), IEEE, 2016, pp. 1–6. Available from: http://doi.org/10.1109/NPSC.2016.7858975.

[6] P. Staudt, et al., Analysis of redispatch and transmission capacity pricing on a local electricity market setup, 2017 14th International Conference on the European Energy Market (EEM), IEEE, 2017, pp. 1–6. Available from: http://doi.org/10.1109/EEM.2017.7981959.

[7] G.W. Braun, State policies for collaborative local renewable integration, Electr. J. 33 (1) (2020) 106691. Elsevier. Available from: https://doi.org/10.1016/j.tej.2019.106691.

[8] European Commission, Providing a fair deal for consumers, (MEMO/16/3961), 2016.

[9] European Commission, Clean energy for all Europeans, Euroheat Power 14 (2) (2019) 3. Available from: https://doi.org/10.2833/9937.

[10] The European Commission, Comission regulation (EU) 2019/943 of the European parliament and of the council of 5 June 2019 on the internal market for electricity, Off. J. Eur. Union. 2019 (714) (2018) 54–124.

[11] European Parliament and Council of the EU, Directive (EU) 2019/944 on common rules for the internal market for electricity and amending directive 2012/27/EU, Off. J. Eur. Union. (L 158)(2019) 18. < http://eur-lex.europa.eu/pri/en/oj/dat/2003/l_285/l_28520031101en00330037.pdf > .

[12] EU, Directive (EU) 2018/2001 of the European parliament and of the council on the promotion of the use of energy from renewable sources, Off. J. Eur. Union. 2018 (L 328) (2018) 82–209.

[13] UK Office of Gas and Electricity Markets (Ofgem) report, "State of the Energy Market 2019", 2019, 1–181. Available from: https://www.ofgem.gov.uk/system/files/docs/2019/11/20191030_state_of_energy_market_revised.pdf

[14] R. Ghorani, et al., Market design for integration of renewables into transactive energy systems, IET Renew. Power Gener. 13 (14) (2019) 2502–2511. Available from: https://doi.org/10.1049/iet-rpg.2019.0551.

[15] Federal Energy Regulatory Commission (FERC), Order 2222, FERC Opens Wholesale Markets to Distributed Resources: Landmark Action Breaks Down Barriers to Emerging Technologies, Boosts Competition. 2020. Available: https://www.ferc.gov/news-events/news/ferc-opens-wholesale-markets-distributed-resources-landmark-action-breaks-down.

[16] Australian Government, Department of Industry, Science, E. and R., Australia's emissions projections 2019, 2019. Available at: <https://www.industry.gov.au/data-and-publications/australias-emissions-projections-2019>.

[17] J.D. Hunt, D. Stilpen, M.A.V. de Freitas, A review of the causes, impacts and solutions for electricity supply crises in Brazil, Renew. Sustain. Energy Rev. 88 (2018) 208–222. Available from: https://doi.org/10.1016/j.rser.2018.02.030.

[18] G.G. Dranka, P. Ferreira, Load flexibility potential across residential, commercial and industrial sectors in Brazil, Energy 201 (2020) 117483. Available from: https://doi.org/10.1016/j.energy.2020.117483.

[19] K. Folly, Smart grid applications in Africa: opportunities and challenges, Mathématique Appliquée à des Questions de Development (MADEV), Dakar, 2019.

[20] P.K. Ainah, K.A. Folly, Development of micro-grid in sub-Saharan Africa: an overview, Int. Rev. Electr. Eng. (IREE) 10 (5) (2015). Available at: <https://www.praiseworthyprize.org/jsm/index.php?journal = iree&amp>.

[21] D. Manz, et al., The grid of the future: ten trends that will shape the grid over the next decade, IEEE Power Energy Mag. 12 (3) (2014) 26–36. Available from: https://doi.org/10.1109/MPE.2014.2301516.

[22] K.K. Küster, A.R. Aoki, G. Lambert-Torres, Transaction-based operation of electric distribution systems: a review, Int. Trans. Electr. Energy Syst. (2019) e12194John Wiley & Sons, Ltd, n/a(n/a). Available from: https://doi.org/10.1002/2050-7038.12194.

[23] F. Simon, EU's electricity market revamp hailed as "game-changer" for clean energy, Euractiv 2018 (2018). Available at: <https://www.euractiv.com/section/energy/news/eus-electricity-market-revamp-hailed-as-game-changer-for-clean-energy/>.

# Index

Printed in the United States
by Baker & Taylor Publisher Services